Two Manuals Combined into One Book

Physical Security

And

Military Role in Civil Disturbance Operations

Department of Defense

Pentagon
Publishing

Field Manual
No. 3-19.30

Headquarters
Department of the Army
Washington, DC, 8 January 2001

PHYSICAL SECURITY

Contents

Distribution Restriction: Approved for public release; distribution is unlimited.

---

*This publication supersedes FM 19-30, 1 March 1979.

Page

Preface

This field manual (FM) sets forth guidance for all personnel responsible for physical security. It is the basic reference for training security personnel. It is intended to be a "one-stop" physical-security source for the Department of Defense (DOD), the Department of the Army (DA), and other proponents and agencies of physical security.

Prevention and protection are the two primary concerns of physical security. Both serve the security interests of people, equipment, and property. These interests must be supported at all staff and command levels; and this support must be unified in joint, multinational, and interagency operations.

Support to joint, multinational, and interagency operations relies on the fact that the Army will not conduct operations alone. Additionally, force-projection operations conducted by the military will involve the integration of war-fighting capabilities with stability and support operations. This manual's primary focus is the articulation of a balanced understanding of physical security for joint, multinational, and interagency operations throughout the environments of peacetime, conflict, and war (whether in the continental United States [CONUS] or outside the continental United States [OCONUS]).

Physical security must integrate the various capabilities of joint, multinational, and interagency operations in pursuit of a seamless connection between the strategic, operational, and tactical levels of war. Physical security must also address an expanded range of threats that embraces not only traditional threat components of war, but also nontraditional threats generated by guerrillas, terrorists, criminals, and natural or man-made disasters. In addition, physical security must address the concept of Homeland Defense due to the aforementioned threats.

Homeland Defense is the military's role in the United States (US) government's principal task of protecting its territory and citizens. This is accomplished by joint, interagency, and multijurisdictional organizations. Homeland Defense includes—

- Supporting domestic authorities for crisis and consequence management with regard to weapons of mass destruction (WMD).
- Protecting national-security assets (such as installations) and deploying forces and ensuring the availability, integrity, and adequacy of other critical assets.
- Deterring and defending against strategic attacks while maintaining freedom of action through antiterrorism and force-protection operations.

With this in mind, it is essential to address the five pillars of force protection—combating terrorism, physical security, personal security, law enforcement, and operations security (OPSEC). Physical security is a central component of force protection and provides an integrated venue to express support for operations. Physical security is a primary-leader task and an inherent part of all operations to protect soldiers, family members, civilians, and resources. This function directly supports the Army's universal task list.

While the effects of these changes (when viewed individually) appear revolutionary, the basic activities remain relatively unchanged, though executed under different conditions and standards. Another component that remains unchanged is our reliance upon quality soldiers and leaders well versed in physical-security fundamentals. Leaders will be challenged to ensure that they are functionally proficient; possess an understanding of physical-security operations; are educated in joint, multinational, and interagency operations; and have the ability to perform physical-security functions in support of full-dimension operations.

Appendix A contains an English-to-metric measurement conversion chart. Appendix B is a sample installation crime-prevention handbook. This handbook is designed to assist commanders in developing crime-prevention programs for their installation and units.

The proponent of this publication is HQ TRADOC. Send comments and recommendations on DA Form 2028 directly to Commandant, US Army Military Police School (USAMPS), ATTN: ATSJ-MP-TD, Directorate of Training, 401 Engineer Loop, Suite 2060, Fort Leonard Wood, Missouri 65473-8926.

Unless this publication states otherwise, masculine nouns and pronouns do not refer exclusively to men.

Chapter 1

Physical-Security Challenges

Physical security is defined as that part of security concerned with physical measures designed to safeguard personnel; to prevent unauthorized access to equipment, installations, material, and documents; and to safeguard against espionage, sabotage, damage, and theft. As such, all military operations face new and complex physical-security challenges across the full spectrum of operations. Challenges relative to physical security include the control of populations, information dominance, multinational and interagency connectivity, antiterrorism, and the use of physical-security assets as a versatile force multiplier.

OVERVIEW

1-1. Reductions in manpower and funding are critical challenges to physical security. Manpower for supporting physical-security activities is reduced through deployments and cutbacks. The rapid evolution of physical-security-equipment technology also lends to physical-security challenges, which are exponentially multiplied by the introduction of the information age.

1-2. Physical-security challenges must be understood, and measures must be taken to minimize them to enhance force protection. Leaders must create order when coming upon a situation; and when they depart, some semblance of that order must remain. They must be aware of the human-dimension factors and ensure that their soldiers do not become complacent. It was human error rather than modern technology that took lives in the bombings of the African embassy. Warning was given, but not heeded. Complacency became a physical-security challenge.

AUTOMATED INFORMATION SYSTEMS

1-3. Success on past battlefields has resulted not so much from technological advances, but from innovative ways of considering and combining available and new technologies as they apply to war fighting. Some of these technologies dealt with disseminating and processing information. For example, the telegraph, the telephone, the radio, and now the computer have redefined the fire-support paradigm.

1-4. As the armed forces move into the technological age, a greater need for physical-security measures is required. The risks associated with automated information systems (AISs) are widespread because computers are used for everything. Army Regulation (AR) 380-19 outlines the requirements that commanders and managers need for processing unclassified and classified information and for securing media, software, hardware, and different systems.

1-5. The threat to AISs and information systems security (ISS) involves deliberate, overt, and covert acts. This includes the physical threat to tangible property, such as the theft or destruction of computer hardware. Also included is the threat of electronic, electromagnetic-pulse, radio-frequency (RF), or computer-based attacks on the information or communications components that control or make up critical Army command and control (C^2) infrastructures. In most cases, the threat's target is the information itself rather than the system that transmits it. The threat comes from a range of sources, including the following:

- Unauthorized users (such as hackers) are the main source of today's attacks, primarily against computer-based systems. The threat they pose to AIS networks and mainframe computers is growing.
- Insiders are those individuals with legitimate access to an AIS. They pose the most difficult threat to defend against. Whether recruited or self-motivated, the AIS insider has access to systems normally protected by ISS against an attack.
- Terrorists once had to operate in the immediate vicinity of a target to gain access to or collect intelligence on that target. The proximity to the target risked exposure and detection. Today, a terrorist can accomplish most target selection, intelligence collection, and preoperational planning by gaining access through a computer network. He can increase his probability of success by using computer systems to reduce his "time on target." Terrorist access to an AIS also increases the threat of critical-data destruction or manipulation. Although his presence would be virtual, the potential for damage to Army C^2 systems could be equal to or greater than that achieved by physical intrusion, especially when used as a force multiplier in conjunction with a traditional terrorist attack. Therefore, while traditional preventive measures are still needed to protect unwanted access to information, the information age has added additional concerns for the commander and new opportunities for those with hostile intent.
- Non-state- and state-sponsored groups provide additional challenges. In many cases, it is difficult to confirm state sponsorship of threat activity against an AIS, no matter how apparent the affiliation might seem. Activists of all persuasions are increasingly taking advantage of information-age technology. Neither AISs nor ISS are immune from an adversary's interest in exploiting US military information systems or disrupting communication infrastructures. The availability of low-cost technology and the proliferation of an AIS increase the risk to the Army by potential adversaries.
- Foreign-intelligence services (FIS), both civil and military, are continually active and are another source of contention concerning information systems. In peacetime, they are increasingly targeted against US commercial and scientific interests, rather than military information. With little effort, this peacetime intrusiveness could easily be refocused on AISs and ISS using a wide range of information operations tactics.

- Political and religious groups are other potential adversaries to AISs and ISS. The world's political climate is diverse and complicated. It embraces traditional mainstream political values, as well as radical religious fundamentalism and political extremism. When political or religious viewpoints also incorporate anti-US sentiment, US information infrastructures (including AISs) are increasingly at risk of penetration or exploitation by these potential adversaries.

1-6. When considering an AIS, physical security is more than just safeguarding the equipment. It includes the following elements:

- Software is marked for each system and secured when not in use.
- Initial logon is password-protected (at a minimum).
- Passwords are a minimum of eight characters, using a mixture of letters and numerals.
- Access to an AIS is allowed only to authorized and cleared personnel (per AR 380-19).

1-7. Classified material is entered and transmitted only on approved devices with the following considerations:

- Approved classified devices are operated in a secured environment.
- Classified devices are secured in appropriate containers when not in use.
- Secure telephone unit–III (STU-III) keys are secured in an appropriate safe when not in use (as outlined in AR 380-19).

1-8. Additional information regarding AISs can be found in ARs 380-5 and 380-19. Required training of personnel working with an AIS is located in AR 380-19.

OPSEC AND THE THREAT

1-9. OPSEC is a process of identifying critical information and subsequently analyzing friendly actions attendant to military operations and other activities. The threat is identified using the factors of mission, enemy, terrain, troops, time available, and civilian considerations (METT-TC). The threat defines the physical-security challenges. Implementing physical-security measures supports OPSEC. Providing soundproof rooms for conducting briefings is a simple but invaluable measure.

1-10. Another issue to consider when evaluating physical-security challenges is what actions to take in case of political implications interfering with physical-security measures. In the devastating event at Khobar Towers, a warning was given but not everyone received it. It took too long to evacuate the building after the warning was issued because a cohesive plan was not in place.

1-11. Commanders can minimize the challenges to physical security through proactive measures. They should periodically change the physical-security posture of their area of responsibility to throw off perpetrators.

Chapter 2

The Systems Approach

Commanders must ensure that appropriate physical-security measures are taken to minimize the loss of personnel, supplies, equipment, and material through both human and natural threats. Commanders commonly exercise those protective responsibilities through the provost marshal (PM) and/or physical-security officer and the force-protection officer. The force-protection officer must coordinate with several different agencies to complete his mission. For example, the Army's Intelligence and Counterintelligence Program (see Appendix C) provides information that will be used to complete the unit's crisis-management plan (see Appendix D).

PROTECTIVE SYSTEMS

2-1. The approach to developing protective measures for assets should be based on a systematic process resulting in an integrated protective system. The protective system focuses on protecting specific assets against well-defined threats to acceptable levels of protection. The system is organized in-depth and contains mutually supporting elements coordinated to prevent gaps or overlaps in responsibilities and performance.

2-2. Effective protective systems integrate the following mutually supporting elements:

- Physical protective measures, including barriers, lighting, and electronic security systems (ESSs).
- Procedural security measures, including procedures in place before an incident and those employed in response to an incident. (These include procedures employed by asset owners and those applied by and governing the actions of guards.)
- Terrorism counteraction measures that protect assets against terrorist attacks.

2-3. The following determinations are made when considering system-development procedures:

- The resources available.
- The assets to be protected.
- The threat to those assets.
- The risk levels applicable to those assets.
- The applicable regulatory requirements for protecting the assets.
- The applicable level of protection for those assets against the threat.
- Additional vulnerabilities to the assets (based on the threat).

SYSTEMS DEVELOPMENT

2-4. AR 190-51, DA Pamphlet (Pam) 190-51, and Technical Manual (TM) 5-853-1 are useful tools for developing protective systems using the systems approach. The key to applying these tools successfully is to use a team approach. A team may include physical-security, intelligence, and operations personnel; the installation engineers; and the user of the assets. It may also include representatives from the multinational, host-nation (HN), and local police as well as the regional security office from the embassy.

ASSETS

2-5. Protective systems should always be developed for specific assets. The goal of security is to protect facilities and buildings and the assets contained inside. The risk-analysis procedure in DA Pam 190-51 is used to identify assets. This procedure is applied to all mission-essential or vulnerable areas (MEVAs) according to AR 190-13. It represents the majority of assets with which DOD is commonly concerned. These assets include—

- Aircraft and components at aviation facilities.
- Vehicle and carriage-mounted or -towed weapons systems and components at motor pools.
- Petroleum, oil, and lubricants (POL).
- Controlled medical substances and other medically sensitive items.
- Communication and electronics equipment; test, measurement, and diagnostic equipment (TMDE); night-vision devices (NVDs); and other high-value precision equipment and tool kits.
- Organizational clothing and individual equipment stored at central-issue facilities.
- Subsistence items at commissaries, commissary warehouses, and troop-issue facilities.
- Repair parts at installation-level supply activities and direct-support (DS) units with authorized stockage lists.
- Facilities-engineering supplies and construction materials.
- Audiovisual equipment, training devices, and subcaliber devices.
- Miscellaneous pilferable assets (not included above) and money.
- Mission-critical or high-risk personnel.
- General military and civilian populations.
- Industrial and utility equipment.
- Controlled cryptographic items.
- Sensitive information (included in TM 5-853-1, but not included in DA Pam 190-51).
- Arms, ammunition, and explosives (AA&E).
- Installation banks and finance offices.

RISK LEVELS

2-6. DA Pam 190-51 provides a procedure for determining risk levels—assessing the value of the assets to their users and the likelihood of compromise. These factors are assessed by answering a series of questions leading to value and likelihood ratings.

2-7. Asset value is determined by considering the following three elements:

- The criticality of the asset for its user and the Army as a whole.
- How easily the asset can be replaced.
- Some measure of the asset's relative value.

2-8. The relative value differs for each asset. For some assets, the relative value is measured in terms of monetary cost.

2-9. The likelihood of the threat is assessed for each applicable aggressor category by considering the asset's value to the aggressor, the history of or potential for aggressors attempting to compromise the asset, and the vulnerability of the asset based on existing or planned protective measures.

REGULATORY REQUIREMENTS

2-10. The risk level is the basis for determining the required protective measures for assets covered in AR 190-51. For each asset type, there may be physical protective measures, procedural security measures, and terrorism counteraction measures. These measures are specified by risk level. The measures identified in AR 190-51 are the minimum regulatory measures that must be applied for the identified threat level. The minimum regulatory measures for AA&E are based on the risk category established in AR 190-11.

ANTITERRORISM/FORCE-PROTECTION CONSTRUCTION STANDARDS

2-11. In accordance with DOD Instruction 2000.16, the commanders in chief (CINCs) have developed standards for new construction and existing facilities to counter terrorism threat capabilities within the area of responsibility. These construction standards have specific requirements for such measures as standoff distance, perimeter barriers, building construction, and parking. The DOD construction standard provides for minimum standards that must be incorporated into all inhabited DOD structures regardless of the identified threat. These standards provide a degree of protection that will not preclude the direct effects of blast but will minimize collateral damage for buildings and people and will limit the progressive collapse of structures. These standards add relatively little cost, may facilitate future upgrades, and may deter acts of aggression. (All services have adopted common criteria and minimum standards to counter antiterrorism/force-protection [AT/FP] vulnerabilities and terrorism threats.) Protection to identified threat levels is described in the following paragraphs. Physical-security personnel must be familiar with the CINC and DOD AT/FP construction standards because these standards may affect elements of physical-security plans and how individual facilities are secured.

THREAT IDENTIFICATION

2-12. The threat must be described in specific terms to help determine the assets' vulnerabilities or to establish protective measures. This description should include the tactics that aggressors will use to compromise the asset (weapons, tools, and explosives are likely to be used in an attempt). For example, the threat might be described as a moving vehicle bomb consisting of a 4,000-pound vehicle containing a 500-pound explosive. Another example would be a forced-entry threat using specific hand, power, or thermal tools. These types of threat descriptions (called the design-basis threat) can be used to design detailed protective systems to mitigate the attacks. TM 5-853-1 and DA Pam 190-51 contain procedures for establishing design-basis threat descriptions in the format described above. These procedures can be used together or separately. Threats listed in the TM will be summarized later in this chapter. When using the TM as a lone source or in conjunction with DA Pam 190-51, the following actions occur:

- When the TM process is used alone, the user goes through an identical process to that in DA Pam 190-51 up to the point where the risk level would be determined. In TM 5-853-1, the value and likelihood ratings are used differently than in DA Pam 190-51. The likelihood rating is used to determine the weapons, tools, and explosives that will be used by a particular aggressor in carrying out a specific tactic. In this procedure, higher likelihood ratings result in more severe mixes of weapons, tools, and explosives. The assumption is that the more likely the attack, the more resources the aggressor is likely to use in carrying out the attack.
- When the procedure in TM 5-853-1 is used in conjunction with the results of the DA Pam 190-51 risk analysis, the likelihood rating is taken directly from the risk analysis and applied as described above.

LEVEL OF PROTECTION

2-13. The level of protection applies to the design of a protective system against a specified threat (for example, a bomb, breaking and entering, pilfering, and so forth). The level of protection is based on the asset's value rating from either DA Pam 190-51 or TM 5-853-1. The level increases as the asset's value rating increases. There are separate levels of protection for each tactic. TM 5-853-1 provides detailed guidance on how to achieve the levels of protection, and Chapter 3 of this manual provides a summary of the levels of protection as they apply to various tactics.

VULNERABILITIES

2-14. Vulnerabilities are gaps in the assets' protection. They are identified by considering the tactics associated with the threat and the levels of protection that are associated with those tactics. Some vulnerabilities can be identified by considering the general design strategies for each tactic described in TM 5-853-1 and as summarized in Chapter 3 of this manual. The general design strategies identify the basic approach to protecting assets against specific tactics. For example, the general design strategy for forced entry is to provide a way to detect attempted intrusion and to provide barriers to delay the aggressors until a response force arrives. Vulnerabilities may involve inadequacies in intrusion-detection systems (IDSs) and barriers. Similarly, the general design strategy for a moving vehicle bomb is to keep the vehicle as far from the facility as possible and to harden the facility to resist the explosive at that distance. Vulnerabilities may involve limited standoff distances, inadequate barriers, and building construction that cannot resist explosive effects at the applicable standoff distance.

PROTECTIVE MEASURES

2-15. Where vulnerabilities have been identified, protective measures must be identified to mitigate them. AR 190-13, AR 190-51, DA Pam 190-51, and TM 5-853-1 are effective tools for developing protective measures. The key to effective development of protective systems is a partnership between physical-security personnel and the installation engineers. Appendix E of this manual discusses information for office security, which should be listed in the physical-security plan (see Appendix F). Appendix G discusses personal-protection measures.

THE INTEGRATED PROTECTIVE SYSTEM

2-16. Protective systems integrate physical protective measures and security procedures to protect assets against a design-basis threat. The characteristics of integrated systems include deterrence, detection, defense, and defeat.

DETERRENCE

2-17. A potential aggressor who perceives a risk of being caught may be deterred from attacking an asset. The effectiveness of deterrence varies with the aggressor's sophistication, the asset's attractiveness, and the aggressor's objective. Although deterrence is not considered a direct design objective, it may be a result of the design.

DETECTION

2-18. A detection measure senses an act of aggression, assesses the validity of the detection, and communicates the appropriate information to a response force. A detection system must provide all three of these capabilities to be effective.

2-19. Detection measures may detect an aggressor's movement via an IDS, or they may detect weapons and tools via X-ray machines or metal and explosive detectors. Detection measures may also include access-control elements that assess the validity of identification (ID) credentials. These control elements may provide a programmed response (admission or denial), or they may relay information to a response force. Guards serve as detection elements, detecting intrusions and controlling access.

2-20. Nuclear, biological, and chemical (NBC) detection systems must be used to measure and validate acts of aggression involving WMD. NBC detection systems should also be used to communicate a warning.

DEFENSE

2-21. Defensive measures protect an asset from aggression by delaying or preventing an aggressor's movement toward the asset or by shielding the asset from weapons and explosives. Defensive measures—

- Delay aggressors from gaining access by using tools in a forced entry. These measures include barriers along with a response force.
- Prevent an aggressor's movement toward an asset. These measures provide barriers to movement and obscure lines of sight (LOSs) to assets.
- Protect the asset from the effects of tools, weapons, and explosives.

2-22. Defensive measures may be active or passive. Active defensive measures are manually or automatically activated in response to acts of aggression. Passive defensive measures do not depend on detection or a response. They include such measures as blast-resistant building components and fences. Guards may also be considered as a defensive measure.

DEFEAT

2-23. Most protective systems depend on response personnel to defeat an aggressor. Although defeat is not a design objective, defensive and detection systems must be designed to accommodate (or at least not interfere with) response-force activities.

SECURITY THREATS

2-24. Security threats are acts or conditions that may result in the compromise of sensitive information; loss of life; damage, loss, or destruction of property; or disruption of mission. Physical-security personnel and design teams must understand the threat to the assets they are to protect in order to develop effective security programs or design security systems. Historical patterns and trends in aggressor activity indicate general categories of aggressors and the common tactics they use against military assets. Aggressor tactics and their associated tools, weapons, and explosives are the basis for the threat to assets.

THREAT SOURCES

2-25. There are many potential sources of threat information. Threat assessment is normally a military-intelligence (MI) responsibility. MI personnel commonly focus on such security threats as terrorists and military forces. Within the US and its territories, the Federal Bureau of Investigation (FBI) has primary responsibility for both foreign and domestic terrorists. The FBI, the US Army Criminal Investigation Command (USACIDC [CID]), and local law-enforcement agencies are good sources for physical-security personnel to obtain criminal threat information. Coordinating with these elements on a regular basis is essential to maintaining an effective security program.

THREAT CATEGORIES

2-26. Security threats are classified as either human or natural. Human threats are carried out by a wide range of aggressors who may have one or more objectives toward assets such as equipment, personnel, and operations. Aggressors can be categorized and their objectives can be generalized as described below. (See DA Pam 190-51 and TM 5-853-1 for more information.)

Aggressor Objectives

2-27. Four major objectives describe an aggressor's behavior. Any one of the first three objectives can be used to realize the fourth. These objectives include—

- Inflicting injury or death on people.
- Destroying or damaging facilities, property, equipment, or resources.
- Stealing equipment, materiel, or information.
- Creating adverse publicity.

Aggressor Categories

2-28. Aggressors are grouped into five broad categories—criminals, vandals and activists, extremists, protest groups, and terrorists. Hostile acts performed by these aggressors range from crimes (such as burglary) to low-intensity conflict threats (such as unconventional warfare). Each of these categories describes predictable aggressors who pose threats to military assets and who share common objectives and tactics.

- Criminals can be characterized based on their degree of sophistication. They are classified as unsophisticated criminals, sophisticated criminals, and organized criminal groups. Their common objective is the theft of assets; however, the assets they target, the quantities they seek, their relative efficiency, and the sophistication of their actions vary significantly. Vandals and activists may also be included under this category.
- Vandals and activists are groups of protesters who are politically or issue oriented. They act out of frustration, discontent, or anger against the actions of other social or political groups. Their primary objectives commonly include destruction and publicity. Their selection of targets will vary based on the risk associated with attacking them. The degree of damage they seek to cause will vary with their sophistication.
- Extremists are radical in their political beliefs and may take extreme, violent actions to gain support for their beliefs or cause.
- Protesters are considered a threat only if they are violent. Lawful protesters have to be considered, but significant protective measures and procedures are not normally needed to control their actions. The presence of extremists or vandals/activists at a peaceful protest increases the chance of the protest becoming violent.
- Terrorists are ideologically, politically, or issue oriented. They commonly work in small, well-organized groups or cells. They are sophisticated, are skilled with tools and weapons, and possess an efficient planning capability. There are three types of terrorists—CONUS, OCONUS, and paramilitary OCONUS.
  - CONUS terrorists are typically right- or left-wing extremists operating in distinct areas of the US.
  - OCONUS terrorists generally are more organized than CONUS terrorists. They usually include ethnically or religiously oriented groups.

- Paramilitary OCONUS terrorist groups show some military capability with a broad range of military and improvised weapons. Attacks by OCONUS terrorists are typically more severe.

2-29. Natural threats are usually the consequence of natural phenomena. They are not preventable by physical-security measures, but they are likely to have significant effects on security systems and operations. They may require an increase in protective measures either to address new situations or to compensate for the loss of existing security measures. They may reduce the effectiveness of existing security measures by such occurrences as collapsed perimeter fences and barriers, inoperable protective lighting, damaged patrol vehicles, and poor visibility. Natural threats and their effects relative to security include the following:

- Floods may result in property damage, destruction of perimeter fences, and damage to IDSs. Heavy rains or snowfalls may have similar effects even if they do not result in flooding.
- Storms, tornadoes, high winds, or rain may cause nuisance alarms to activate and cause damage to IDSs. They may limit the visibility of security personnel and may affect closed-circuit television (CCTV) systems. Winds may also disrupt power or communication lines and cause safety hazards from flying debris.
- Earthquakes may cause nuisance alarms to activate or may disrupt IDSs. They may also cause broken water or gas mains, fallen electrical or communication lines, and weakened or collapsed buildings.
- Snow and ice can make travel on patrol roads difficult, may delay responses to alarms, may impede the performance of IDSs, and may freeze locks and alarm mechanisms. Heavy ice may also damage power and communication lines.
- Fires may damage or destroy perimeter barriers and buildings, possibly leaving assets susceptible to damage or theft.
- Fog can reduce the visibility of security forces, thereby requiring additional security personnel. It may also increase the response time to alarms and reduce the effectiveness of security equipment such as CCTV systems.

Aggressor Tactics

2-30. Aggressors have historically used a wide range of offensive strategies reflecting their capabilities and objectives. These offensive strategies are categorized into 15 tactics that are specific methods of achieving aggressor goals (see TM 5-853-1). Separating these tactics into categories allows facility planners and physical-security personnel to define threats in standardized terms usable as a basis for facility and security-system design. Common aggressor tactics include—

- **Moving vehicle bomb.** An aggressor drives an explosive-laden car or truck into a facility and detonates the explosives. His goal is to damage or destroy the facility or to kill people. This is a suicide attack.
- **Stationary vehicle bomb.** An aggressor covertly parks an explosive-laden car or truck near a facility. He then detonates the explosives either by time delay or remote control. His goal in this tactic is the

same as for the moving vehicle bomb with the additional goal of destroying assets within the blast area. This is commonly not a suicide attack. It is the most frequent application of vehicle bombings.

- **Exterior attack.** An aggressor attacks a facility's exterior or an exposed asset at close range. He uses weapons such as rocks, clubs, improvised incendiary or explosive devices, and hand grenades. Weapons (such as small arms) are not included in this tactic, but are considered in subsequent tactics. His goal is to damage the facility, to injure or kill its occupants, or to damage or destroy assets.
- **Standoff weapons.** An aggressor fires military weapons or improvised versions of military weapons at a facility from a significant distance. These weapons include direct (such as antitank [AT] weapons) and indirect LOS weapons (such as mortars). His goal is to damage the facility, to injure or kill its occupants, or to damage or destroy assets.
- **Ballistics.** The aggressor fires various small arms (such as pistols, submachine guns, shotguns, and rifles) from a distance. His goal is to injure or kill facility occupants or to damage or destroy assets.
- **Forced entry.** The aggressor forcibly enters a facility using forced-entry tools (such as hand, power, and thermal tools) and explosives. He uses the tools to create a man-passable opening or to operate a device in the facility's walls, doors, roof, windows, or utility openings. He may also use small arms to overpower guards. His goal is to steal or destroy assets, compromise information, injure or kill facility occupants, or disrupt operations.
- **Covert entry.** The aggressor attempts to enter a facility or a portion of a facility by using false credentials or stealth. He may try to carry weapons or explosives into the facility. His goals include those listed for forced entry.
- **Insider compromise.** A person authorized access to a facility (an insider) attempts to compromise assets by taking advantage of that accessibility. The aggressor may also try to carry weapons or explosives into the facility in this tactic. His goals are the same as those listed for forced entry.
- **Visual surveillance.** The aggressor uses ocular and photographic devices (such as binoculars and cameras with telephoto lenses) to monitor facility or installation operations or to see assets. His goal is to compromise information. As a precursor, he uses this tactic to determine information about the asset of interest.
- **Acoustic eavesdropping.** The aggressor uses listening devices to monitor voice communications or other audibly transmitted information. His goal is to compromise information.
- **Electronic-emanations eavesdropping.** The aggressor uses electronic-emanation surveillance equipment from outside a facility or its restricted area to monitor electronic emanations from computers, communications, and related equipment. His goal is to compromise information.

- **Mail-bomb delivery.** The aggressor delivers bombs or incendiary devices to the target in letters or packages. The bomb sizes involved are relatively small. His goal is to kill or injure people.
- **Supplies-bomb delivery.** The aggressor conceals bombs in various containers and delivers them to supply- and material-handling points such as loading docks. The bomb sizes in this tactic can be significantly larger that those in mail bombs. His goal is to damage the facility, kill or injure its occupants, or damage or destroy assets. Appendix H addresses the actions to take when a bomb is suspected.
- **Airborne contamination.** An aggressor contaminates a facility's air supply by introducing chemical or biological agents into it. His goal is to kill or injure people.
- **Waterborne contamination.** An aggressor contaminates a facility's water supply by introducing chemical, biological, or radiological agents into it. These agents can be introduced into the system at any location with varying effects, depending on the quantity of water and the contaminant involved. His goal is to kill or injure people.

2-31. The aforementioned tactics are typical threats to fixed facilities for which designers and physical-security personnel can provide protective measures. However, some common terrorist acts are beyond the protection that facility designers can provide. They cannot control kidnappings, hijackings, and assassinations that take place away from facilities or during travel between facilities. Protection against these threats is provided through operational security and personal measures (see Appendices G and I), which are covered in doctrine relative to those activities and are under the general responsibility of the CID.

TACTICAL ENVIRONMENT CONSIDERATIONS

2-32. When determining the assets and threats, the same considerations should be given to the systems approach in the tactical environment as when in the cantonment area. The same process of determining the assets, their risk level, and any regulatory guidance apply. Identifying potential threats and the level of protection required for the assets are necessary. Commanders and leaders must also identify additional vulnerabilities and other required protective measures. Commanders are not expected to have the same physical protective measures due to the impact of resources, budget, location, and situations.

2-33. Commanders must consider the various tactics used by aggressors and use their soldiers' abilities to counteract these tactics. Considerations for specific assets (such as military-working-dog [MWD] and explosive-ordnance-disposal [EOD] teams and their abilities to detect and disassemble a bomb) must be identified. Units must have the ability to improvise in a tactical environment. Their training and resourcefulness will compensate for shortcomings in the field.

2-34. The systems approach to security provides focus and integration of resources. Protective systems are mutually supporting and systematically developed to negate the threat. Commanders conduct an intelligence preparation of the battlefield (IPB) and vulnerability assessments (VAs) to determine risks. Security resources and measures are applied to mitigate risks and to deter, detect, defend, and defeat the threat.

Chapter 3

Design Approach

Developing protective systems to protect assets depends on an effective partnership between engineers and physical-security personnel. Physical-security personnel need to understand the basic approaches the engineers will take in laying out protective systems. Engineers must understand the issues involved with ensuring that anything they lay out is compatible with security operations and the operations of the asset users. The best way to ensure a viable design is through teamwork. This chapter provides a summary of the basic approaches to protecting assets against threats (the design strategies). Understanding these strategies is critical to being an effective team member in developing protective systems.

DESIGN STRATEGIES

3-1. There are separate design strategies for protecting assets from each tactic described in Chapter 2. There are two types of strategies associated with each tactic—the general-design and specific-design strategies. The general-design strategy is the general approach to protecting assets against tactics. The specific-design strategy refines the general-design strategy to focus the performance of the protective system on a particular level of protection. (See TM 5-853-1 for more information.)

PROTECTIVE MEASURES

3-2. Protective measures are developed as a result of the general- and specific-design strategies. These protective measures commonly take the form of site-work, building, detection, and procedural elements.

- Site-work elements include the area surrounding a facility or an asset. Technically, they are associated with everything beyond 5 feet from a building. They can include perimeter barriers, landforms, and standoff distances.
- Building elements are protective measures directly associated with buildings. These elements include walls, doors, windows, and roofs.
- Detection elements detect such things as intruders, weapons, or explosives. They include IDSs, CCTV systems used to assess intrusion alarms, and weapon and explosive detectors. These elements can also include the guards used to support this equipment or to perform similar functions.
- Procedural elements are the protective measures required by regulations, TMs, and standing operating procedures (SOPs). These elements provide the foundation for developing the other three elements.

VEHICLE BOMBS

3-3. Vehicle-bomb tactics include both moving and stationary vehicle bombs. In the case of a moving vehicle bomb, the aggressor drives the vehicle into the target. This is commonly known as a suicide attack. In a stationary vehicle bomb, he parks the vehicle and detonates the bomb remotely or on a timed delay.

GENERAL-DESIGN STRATEGY

3-4. Blast pressures near an exploding vehicle bomb are very high, but they decrease rapidly with distance from the explosion. The design strategy for these tactics is to maintain as much standoff distance as possible between the vehicle bomb and the facility and then, if necessary, to harden the facility for the resulting blast pressures. Barriers on the perimeter of the resulting standoff zone maintain the required standoff distance. The difference between moving and stationary vehicle-bomb tactics is that the aggressor using the moving vehicle bomb will attempt to crash through the vehicle barriers; the aggressor using the stationary vehicle bomb will not. Therefore, vehicle barriers for the moving vehicle bomb must be capable of stopping a moving vehicle at the perimeter of the standoff zone. For a stationary vehicle bomb, vehicle barriers must mark the perimeter of the standoff zone, but they are not required to stop the moving vehicle. They only need to make it obvious if an aggressor attempts to breach the perimeter.

LEVELS OF PROTECTION

3-5. There are three levels of protection for vehicle bombs—low, medium, and high. The primary differences between the levels are the degree of damage allowed to the facility protecting the assets and the resulting degree of damage or injury to the assets.

- **Low.** The facility or the protected space will sustain a high degree of damage but will not collapse. It may not be economically repairable. Although collapse is prevented, injuries may occur and assets may be damaged.
- **Medium.** The facility or the protected space will sustain a significant degree of damage, but the structure will be reusable. Occupants and other assets may sustain minor injuries or damage.
- **High.** The facility or the protected space will sustain only superficial damage. Occupants and other assets will also incur only superficial injury or damage.

SITE-WORK ELEMENTS

3-6. The two primary types of site-work elements for vehicle bombs are the standoff distance and vehicle barriers. The vehicle's speed must also be taken into consideration.

Standoff Distance

3-7. The standoff distance is the maintained distance between where a vehicle bomb is allowed and the target. The initial goal should be to make that distance as far from the target facility as practical. Figure 3-1 shows the distances required to limit building damage to particular levels (including the levels of protection described above) for a range of bomb weights. All bomb weights are given in terms of equivalent pounds of trinitrotoluene (TNT), which is a standard way of identifying all explosives regardless of their composition. The example in Figure 3-1 is a building of conventional construction (common, unhardened construction). Buildings built without any special construction at these standoff distances will probably withstand the explosive effects. Conventionally constructed buildings at standoff distances of less than those shown in Figure 3-1 will not adequately withstand blast effects. (Refer to TM 5-853-1 for information on hardening buildings to resist a blast.) Do not allow vehicles to park within the established standoff distances. Recognize that this restriction can result in significant operational and land-use problems.

3-8. **Exclusive Standoff Zone.** When an exclusive standoff zone is established, do not allow vehicles within the perimeter unless they have been searched or cleared for access. The zone's perimeter is established at the distance necessary to protect the facility against the highest threat explosive. All vehicles should be parked outside the exclusive standoff zone; only

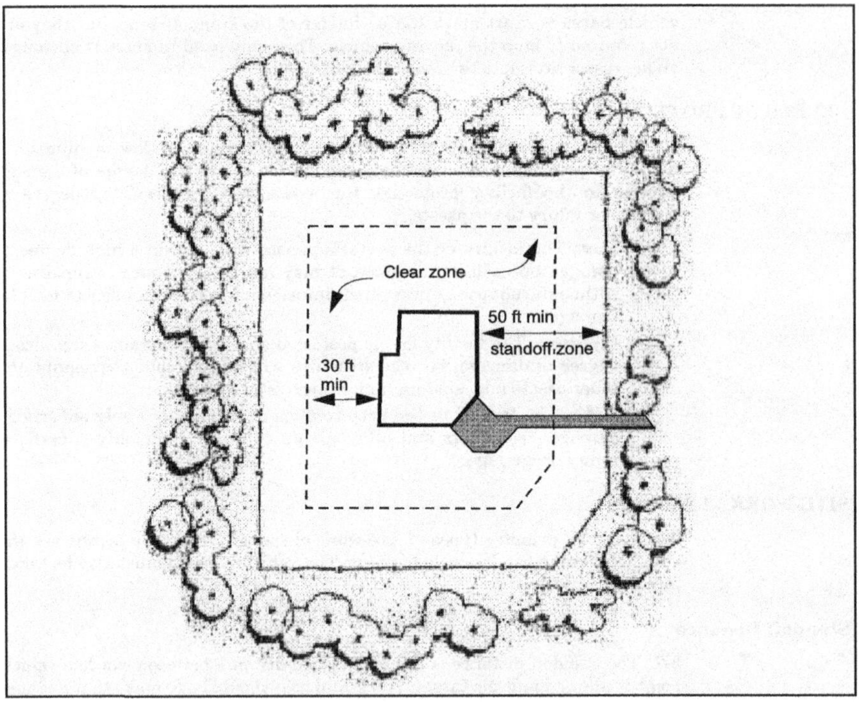

Figure 3-1. Standoff Distance

maintenance, emergency, and delivery vehicles should be allowed within the zone after being searched. Figure 3-2 shows an exclusive standoff zone.

3-9. **Nonexclusive Standoff Zone.** A nonexclusive standoff zone is established in a location having a mixture of cars and trucks (with relatively few trucks). A nonexclusive standoff zone takes advantage of aggressors being able to conceal a smaller quantity of explosives in a car than they can in a truck. Therefore, a nonexclusive standoff zone includes inner and outer perimeters. The inner perimeter is set at a distance corresponding to the weight of explosives that can be concealed in cars. The outer perimeter is set at a distance associated with the weight that can be placed in trucks (refer to TM 5-853-1). With these two perimeters, cars can enter the outer perimeter without being searched but they cannot enter the inner perimeter. Trucks cannot enter the outer perimeter, since it is established based on what they can carry. Figure 3-3 shows a nonexclusive standoff zone. The nonexclusive standoff zone provides the advantages of allowing better use of the parking areas and limiting the number of vehicles that need to be searched at the outer perimeter.

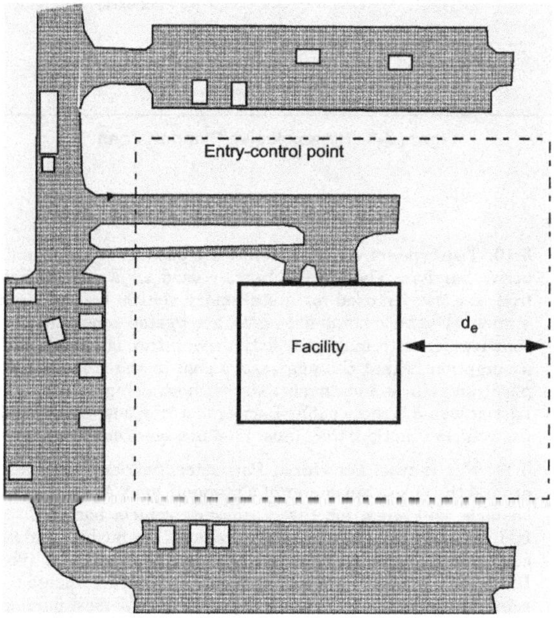

d_e = exclusive standoff-zone distance

Figure 3-2. Exclusive Standoff Zone

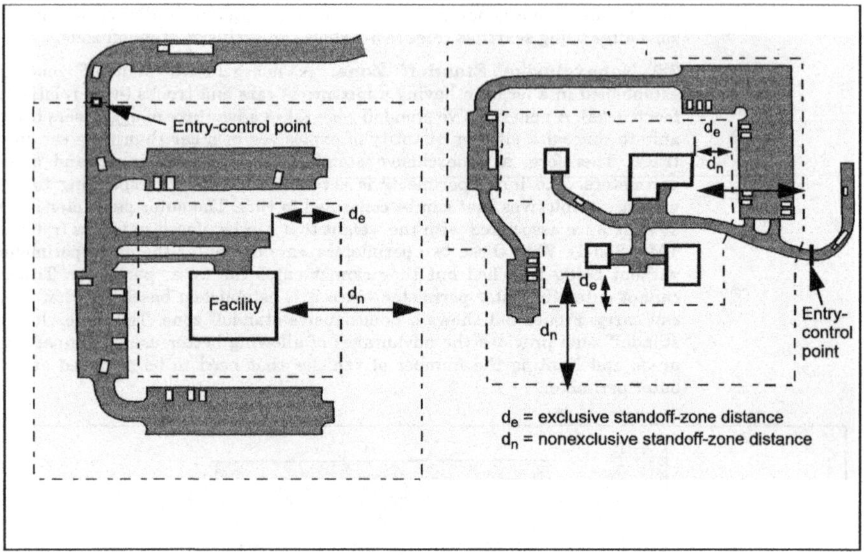

Figure 3-3. Nonexclusive Standoff Zone

Vehicle Barriers

3-10. Two types of vehicle barriers are used for vehicle bombs—perimeter and active barriers. The type of barrier used for a moving vehicle bomb differs from the barrier used for a stationary vehicle bomb. The barrier used for a stationary vehicle bomb does not have to stop a vehicle's motion. The goal for that barrier is to make anybody driving through the barrier noticeable. The assumption is that the aggressor's goal in the stationary vehicle bomb is to park the vehicle and sneak away without being noticed. Crashing through a barrier would be noticeable. Barriers for the moving vehicle bomb need to stop the vehicle's motion; they must be much more substantial.

3-11. **Perimeter Barriers.** Perimeter barriers are fixed barriers placed around the entire perimeter of a standoff zone. Anything that presents a fixed obstacle will work for the stationary vehicle bomb. Common applications include chain-link fences, hedges made of low bushes, and high (over 8 inches) curbs. Aggressors driving through such barriers are likely to be noticed. Barriers capable of stopping moving vehicles include chain-link fences reinforced with cable, reinforced concrete "Jersey barriers", pipe bollards, planters, ditches, and berms. When barriers such as the Jersey barriers and planters are used to stop moving vehicles, they must be anchored into the ground to be effective. The cables in the reinforced fence also have to be anchored into the ground or partially buried. Spaces between barriers should be no greater than 4 feet. Figure 3-4 shows common perimeter barriers for stationary or moving vehicle bombs. Refer also to TM 5-853-1.

3-12. **Active Barriers.** Active barriers are placed at openings in perimeters where vehicles need to enter or exit. These barriers must be able to be raised and lowered or moved aside. For the stationary vehicle bomb, barriers can be as simple as chain-link, pipe, or wooden gates that can be raised and lowered. Aggressors crashing through any of these or similar obstructions will likely draw attention. For the moving vehicle bomb, the barriers are heavy structures and have many construction and operations considerations

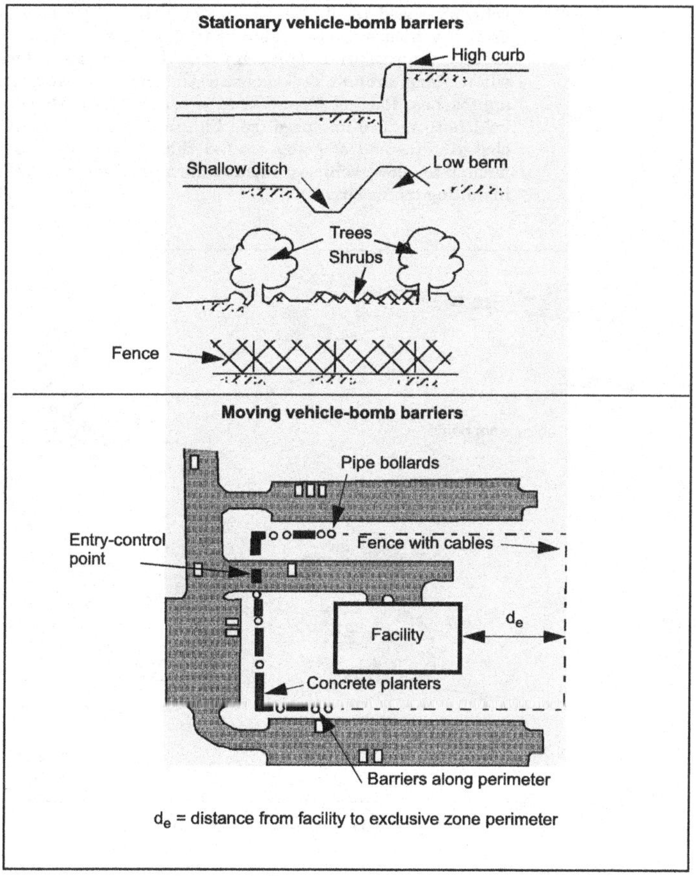

Figure 3-4. Perimeter-Barrier Application

associated with them. These barriers may stop vehicles weighing up to 15,000 pounds and travelling 50 miles per hour. They commonly cost tens of thousands of dollars (refer to TM 5-853-1). Some common active vehicle barriers are shown in Figure 3-5. For temporary or deployed conditions, park a vehicle across an opening and move it aside to grant access.

Speed Control

3-13. It is important to control the speed of a vehicle approaching a barrier used for a moving vehicle bomb. The energy from a vehicle that a barrier must stop increases as its speed increases. The energy also increases with more weight, but the effect of speed is much greater. Therefore, decreasing the vehicle's speed results in smaller and less costly barriers. The best way to limit a vehicle's approach speed to perimeter barriers is to place or retain obstacles in potential approach paths. The vehicles are forced to reduce speed when going around these obstacles. The same principle applies for road approaches. Placing obstacles in a serpentine pattern on the road forces a vehicle to reduce its speed (see Figure 3-6, page 3-8). If the vehicle hits the obstacles instead of going around them, they are still slowed down. Other means to slow vehicles include forcing them to make sharp turns and installing traffic circles.

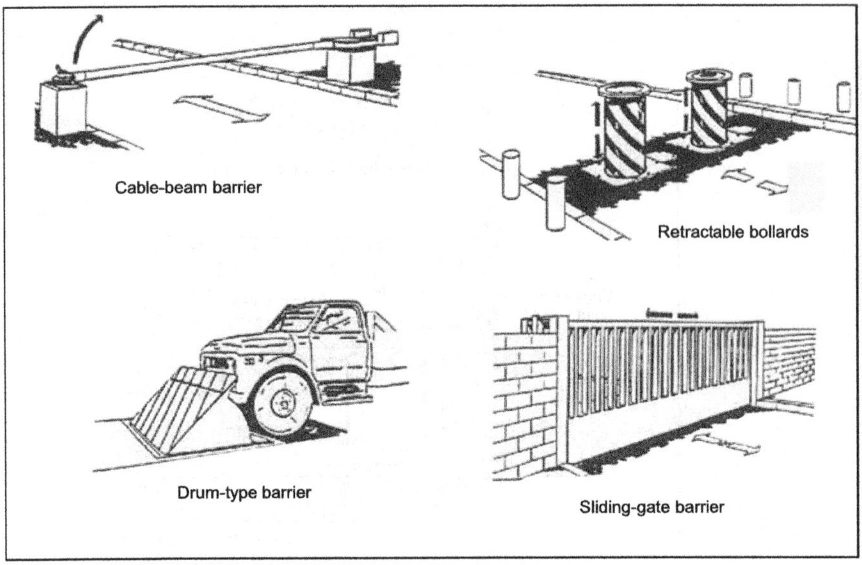

Figure 3-5. Active Vehicle Barriers

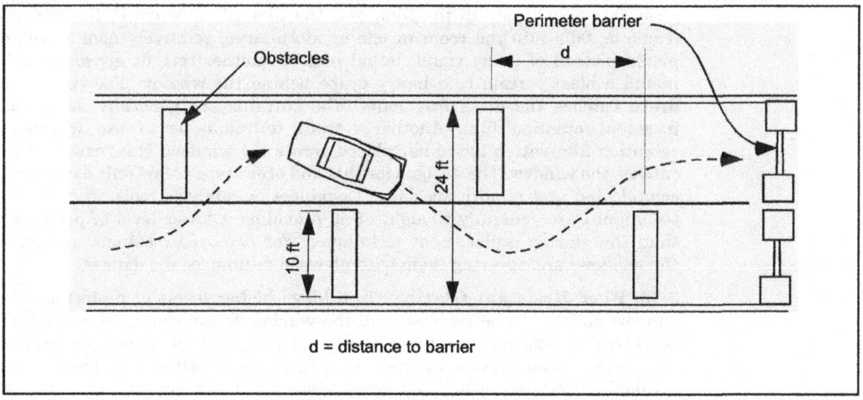

Figure 3-6. Serpentine Pattern

BUILDING ELEMENTS

3-14. Once the standoff distance is established and the site has been laid out, the designers can select the building components necessary to protect the assets against the threat explosives at the standoff distance. The building components include the walls, roofs, doors, and windows. Detailed design issues related to these building elements are covered in TM 5-853-1.

Walls and Roofs

3-15. If the distances shown for the desired damage levels in Figure 3-1, page 3-3, cannot be enforced, the building's walls and roofs will need to be strengthened. This can be achieved in new construction by using reinforced masonry or reinforced concrete in the walls and reinforced concrete in the roof. When the standoff distance is not available for existing construction, a more detailed analysis may be required to determine what the explosion's impact will be on the structure. When the construction is inadequate, more standoff distance should be investigated or the engineers should apply specialized techniques for retrofitting the construction to increase its strength.

Windows

3-16. Historically, glass fragments have caused about 85 percent of injuries and deaths in bomb blasts. There are two basic approaches to mitigating the effects of bomb blasts on glass—retrofitting the windows with film or curtains and using blast-resistant glazing.

3-17. **Retrofitting Windows.** One of the most common means of decreasing the hazards from broken glass is to install fragment-retention film on the glass. The film is a plastic (polyester) sheet that adheres to the window glass with a special adhesive. The film does not strengthen the glass; but when the glass breaks, it keeps the fragments from spreading throughout the room. The glass fragments stick to the film, and the film either stays in the window frame or falls into the room in one or more large, relatively nonhazardous pieces instead of many small, lethal pieces. Another retrofit approach is to install a blast curtain or a heavy drape behind the window. The curtain or drape catches the glass fragments. The curtains are generally used with fragment-retention film. Another retrofit technique is to use fragment-retention film with a metal bar placed across the window. This "catcher bar" catches the window. The designs for this and other types of retrofit devices are complicated and require specialized engineering-analysis tools. The retrofit techniques are generally thought of as providing a lower level of protection than the glazing replacement techniques. For deployed locations, removing the windows and covering them with plywood minimizes the danger.

3-18. **Blast-Resistant Glazing.** To achieve higher levels of protection, the window glass must be replaced and the window frame should be reinforced. Because of its expense, this procedure is generally limited to new construction and major renovations. Special blast-resistant glazing and frames are available that use either tempered glass or a plastic glazing (such as polycarbonate). Another promising type of blast-resistant glazing is laminated glass, in which several layers of common glass are adhered together with a special interlayer. The resulting laminated construction is usually stronger than common glass while retaining the same thickness. The interlayer acts similarly to fragment-retention film. For deployed locations, a means of minimizing the danger of windows is to remove them and replace them with plywood.

Doors

3-19. Doors are another building component particularly vulnerable to an explosive blast. Common metal and wood doors provide little resistance to a blast. The two ways to address the problem of doors is to install them in foyers or to replace them. Glass doors or doors containing windows should be avoided.

Foyers

3-20. Door hazards can be reduced by installing doors in foyers during construction or by adding foyers to existing buildings. When a door is located in a foyer and the outer door fails, the outer door flies into a wall instead of the building's interior (see Figure 3-7, page 3-10). The inner door then has a greater chance of remaining intact. This option generally provides a low level of protection.

3-21. Another option is to replace the doors with specially constructed blast-resistant doors and frames. These doors are commercially available and can provide a high level of protection, but they are very expensive and heavy. The doorframe must be made of the same type of material and provide the same level of protection as the door.

DETECTION ELEMENTS

3-22. Detection elements for vehicle bombs are limited to the use of guards to control access into standoff zones. The guards search vehicles seeking entry

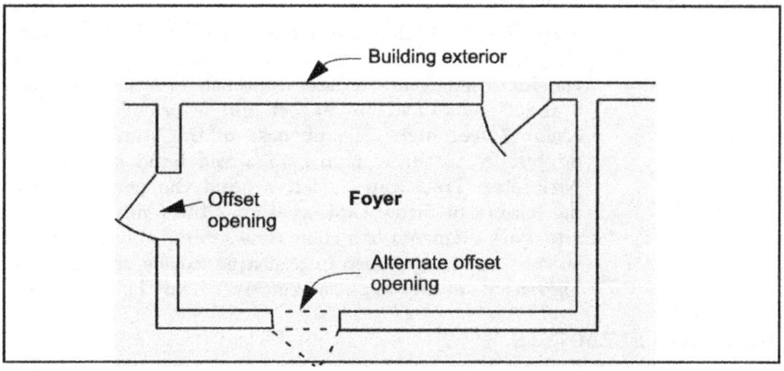

Figure 3-7. Door in Foyer

into the perimeter through an entry-control point. The recommended levels of searches depend on the required level of protection (see TM 5-853-1). Guards can be stationed at entry-control points continuously, or they can be summoned to an entry-control point when access is needed. The latter is commonly the case for the inner perimeter of exclusive standoff zones where only delivery and maintenance vehicles need access.

EXTERIOR ATTACK

3-23. An exterior attack is a physical attack using weapons such as rocks, clubs, improvised incendiary devices (IIDs) such as Molotov cocktails, explosives such as improvised explosive devices (IEDs), and hand grenades. The explosives can be thrown at or placed near a facility's exterior. Examples of IEDs for this tactic range from pipe bombs and hand grenades to briefcase-sized explosives.

GENERAL-DESIGN STRATEGY

3-24. Because the exterior attack is directed at a facility's exterior surfaces, the general-design strategy is to keep aggressors away from the facility (at a standoff distance) and, if necessary, to harden the facility's exterior components to resist the effects of weapons and explosives. A standoff distance from the facility reduces the degree of hardening required to resist weapons effects. When briefcase-sized bombs are a threat, an obstacle-free zone should be established around the facility and the explosives placed within should be detected and disarmed.

LEVELS OF PROTECTION

3-25. The levels of protection for exterior attacks are similar to those for vehicle bombs. Levels of protection vary based on the level of building damage and asset injury or damage allowed. However, due to the limited sizes of explosives involved in this tactic, the damage to the building will be much more localized and injuries or damage to assets will be confined to smaller areas.

SITE-WORK ELEMENTS

3-26. Site-work elements for exterior attacks are relatively limited because the explosive weights are more limited. Large standoff distances are not a consideration. The common approach to site-work elements is to lay out a standoff zone of about 50 feet and to provide a fence or perimeter barrier about 7 feet high. The purpose of the standoff is to make it harder for aggressors to throw pipe bombs and hand grenades at targets inside the perimeter. Trees can be left around the perimeter to make it harder for aggressors to throw explosives over the fence. The remaining component of site-work elements is a clear zone around the facility. A clear zone is applied so that anything placed in that area can be detected visually. This limits the aggressor's ability to place explosives near the target facility.

BUILDING ELEMENTS

3-27. Building elements for exterior attacks are similar to those for vehicle bombs. For small IEDs and IIDs, the building-element requirements do not increase the cost of the building significantly. For larger, briefcase-sized bombs, the measures are more significant than for incendiary devices but less than for vehicle bombs.

Walls and Roofs

3-28. Walls and roofs are not a problem with small explosives. Conventional construction normally provides adequate protection. Walls with 6-inch reinforced concrete or 8-inch, grout-filled, reinforced masonry will withstand the effects of typical pipe bombs or hand grenades. The corresponding roof construction is 6-inch reinforced concrete. In the case of briefcase-sized bombs, considerations similar to those discussed for vehicle bombs need to be employed.

Windows

3-29. A significant goal when constructing windows is to make them difficult to throw an explosive or incendiary device through, especially when considering smaller explosives. This is accomplished by constructing smaller windows or making narrow windows (see Figure 3-8, page 3-12). For existing windows, parts of the windows can be covered to achieve a narrow effect. These windows still may be susceptible to breakage due to explosive effects, even from the smaller explosives. This problem is solved by installing 3/4-inch-thick plastic (polycarbonate) glazing or by raising the windows over 6 feet high to develop a small standoff distance (as shown in Figure 3-9, page 3-12). A 3/4-inch glazing will also stop grenade fragments. Fragment-retention film, a blast curtain, or a heavy drape as described in vehicle-bomb tactics are also good applications for small bombs.

Conventionally Constructed Doors

3-30. Doors are not a significant problem with small bombs and incendiary devices. Generally, metal doors are adequate for incendiary devices, and doors placed in foyers (as shown in Figure 3-7) are adequate for pipe bombs and hand grenades. A similar application for briefcase-sized bombs would provide

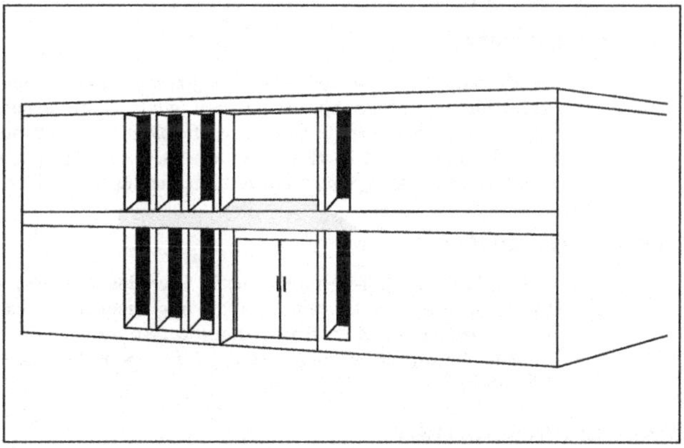

Figure 3-8. Narrow Recessed Windows

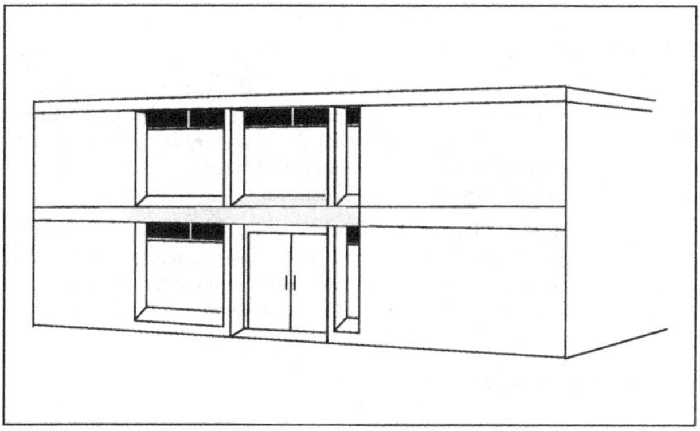

Figure 3-9. Narrow Raised Windows

only a low level of protection. To achieve higher levels of protection for briefcase-sized bombs, blast-resistant doors must be installed.

3-31. The requirements to meet the levels of protection for larger explosives are similar to those described for vehicle bombs, but they will not stop grenade fragments. Fragment-retention film and drapes or curtains can provide a low level of protection, but blast-resistant glazing is required to achieve a higher level of protection.

DETECTION ELEMENTS

3-32. Other than awareness of aggressor activity on or outside the site, detection is only a specific design goal where briefcase-sized bombs are anticipated. When that is the case, the clear zone around the building must be visually monitored so that any objects placed in it are detected. At higher levels of protection, visual surveillance is augmented by IDSs.

STANDOFF WEAPONS

3-33. The standoff-weapons tactic includes the use of AT weapons and mortars. In both of these tactics, the aggressor fires weapons at assets located in the protected facility from a distance. An AT-weapon attack requires a clear LOS to the target, while mortars can fire over obstacles and only need a clear line of flight.

GENERAL-DESIGN STRATEGY

3-34. Standoff-weapons attacks cannot be detected reliably before they occur. Protective design to resist these tactics relies on blocking LOSs to protected areas of a facility or hardening the facility to resist the particular weapon's effects. The approaches to protection against mortars and AT weapons differ from each other and will be discussed separately. Detection measures are not applicable for these tactics.

LEVELS OF PROTECTION

3-35. There are two levels of protection against both mortars and AT weapons. For AT weapons, the low level of protection depends on detonating the AT round before it hits the target facility. The high level of protection avoids the risk associated with that and hardens the building to resist the direct impact of the AT round.

3-36. For mortars, the low level of protection involves allowing some areas of the facility to be sacrificed. Those spaces provide a buffer to the assets to be protected. The assets within the sacrificial areas and the areas themselves may be destroyed. At the high level of protection, the building's exterior fully resists the mortar rounds and there are no sacrificial areas.

SITE-WORK ELEMENTS

3-37. The primary site-work element for standoff weapons is to obstruct LOSs from vantage points outside of the site. With AT weapons, the aggressor cannot hit what he cannot see. This is not true with mortars, but blocking LOSs from mortar firing points helps to make targeting more difficult. The LOSs are blocked by using trees, other buildings, vehicle parking areas, or fences. Another site-work element, a predetonation screen, applies only to an AT weapon. When using a predetonation screen, the AT round is detonated on the screen and its effects are dissipated in the distance between the screen and the target (see Figure 3-10, page 3-14). Any screen material (such as a

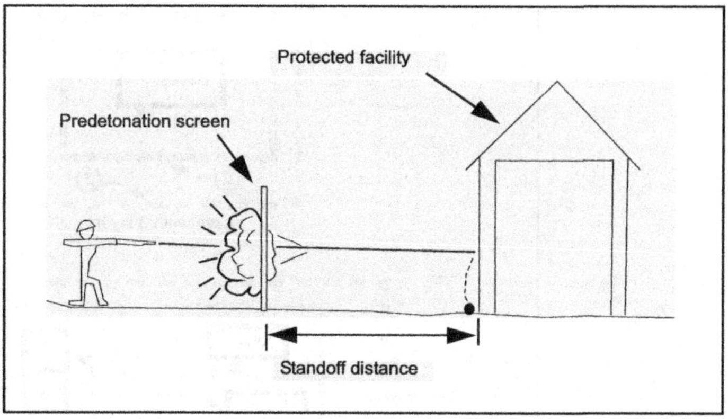

Figure 3-10. Predetonation Screen

wooden fence) will detonate the round unless it has spaces in it. The screen distances vary from less than 10 feet to almost 40 feet, depending on the building construction (see TM 5-853-1). This measure only applies to the low level of protection.

BUILDING ELEMENTS

3-38. Building elements for AT weapons and mortars involve the building's layout. This includes the materials used in the construction.

Layout

3-39. A building's interior layout is only an issue for the low level of protection against a mortar round. The layout issue involves designating sacrificial areas in which unimportant assets are located. The assets to be protected are located in a hardened interior layer. Figure 3-11 includes a plan view (from above). The sacrificial area has to be both around and above the protected area in case a mortar round comes from above. If such a layout is not feasible, other options include going to a higher level of protection and either hardening the entire building or building the facility underground (which are both very expensive).

Walls and Roofs

3-40. Walls and roofs must offer protection against both AT weapons and mortar rounds. The design of walls that protect against AT weapons varies with the level of protection. For the low level of protection where the round is predetonated, the walls can be of conventional construction, varying with the standoff distance from the predetonation screen to the wall. For higher levels of protection, the walls must resist the full effect of the round, requiring the walls to be 24-inch-thick reinforced concrete. Roofs are not an issue in

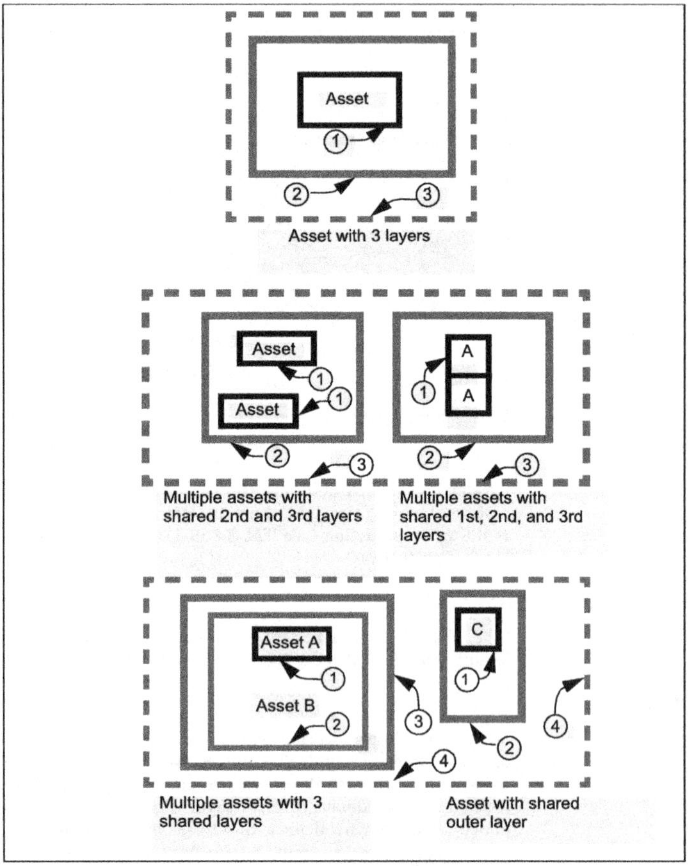

Figure 3-11. Assets Protected by Hardened Interior Layer

protecting against AT weapons because it is difficult to get direct LOSs to roofs. If such LOSs are possible, the roof should be designed like the walls.

3-41. To provide protection against mortar rounds, walls and roofs should be designed to resist the explosive effects in the rounds at the standoff distance that the sacrificial space provides. In the case of sacrificial areas, the walls can be of common construction. The interior protected-area walls are then designed of reinforced concrete or reinforced masonry for the standoff distance those sacrificial walls provide. When the walls must resist the full effect of the rounds (as in the higher level of protection), they are likely to be very thick (up to 30 inches of reinforced concrete for some improvised mortars). Similar considerations should be made for roofs. Roofs are designed to take the direct effects of the round or to take the round at the standoff distance provided by the sacrificial area.

Doors and Windows

3-42. It is impractical to provide doors and windows that are resistant to mortar rounds and AT weapons. Windows should only be used in sacrificial areas where there is a mortar threat. When there is an AT weapon threat, windows can only be used where the round is predetonated. The windows should be narrowed or raised to present a smaller target (see Figures 3-8 and 3-9, page 3-12). Doors should be placed in foyers (see Figure 3-7, page 3-10) for protection against AT rounds and to achieve a low level of protection against mortars. Blast-resistant doors are necessary to achieve a high level of protection against mortar rounds.

BALLISTICS

3-43. In a ballistics tactic, aggressors fire small arms at assets from vantage points outside of the target facility's control. Ballistic attacks cannot be detected reliably before they occur.

GENERAL-DESIGN STRATEGY

3-44. Protective measures to resist these tactics rely on blocking LOSs to protected areas of a facility or by hardening the facility to resist the ballistic effects. This strategy focuses on assets within buildings. Protecting people or property in the open is difficult and can only be addressed through operational measures. Detection measures are not applicable for this tactic.

LEVELS OF PROTECTION

3-45. There are only two levels of protection for this tactic. The low level of protection depends on blocking LOSs to assets. This strategy assumes that the aggressor cannot hit what he cannot see. The risk of an aggressor firing into a building randomly and hitting something is what makes this the low level of protection. The high level of protection involves hardening building components to resist the ballistic effects. These strategies can be thought of as either hardening or hiding.

SITE-WORK ELEMENTS

3-46. Site-work elements are of limited use for the ballistics tactic. When they are applied, they are used to obstruct LOSs from vantage points outside of the site, which is consistent with the low level of protection. The LOSs can be blocked using trees, other buildings, motor pools, or fences.

BUILDING ELEMENTS

3-47. Building elements are the principal means of protecting assets against a ballistics attack. They can be applied to achieve either the low or high level of protection.

Walls and Roofs

3-48. Walls and roofs are inherently opaque, so it is easy to achieve the low level of protection (hiding) with them. Achieving the high level of protection (hardening) for walls and roofs can be done within conventional construction using reinforced concrete, concrete-masonry units (CMUs), or clay brick. The material's required thickness is shown in Table 3-1. The thicknesses of CMUs and clay brick are nominal, meaning they do not represent the actual thickness of the material; they represent the thicknesses at which those materials are commercially available. Steel plates (mild steel and armor steel) and bullet-resistant fiberglass can be used to retrofit existing building components that would not provide the needed bullet resistance.

Table 3-1. Required Thicknesses, in Inches

Ballistics Type	Reinforced Concrete	Grouted CMU*	Clay Brick*	Steel Plate		Bullet-Resistant Fiberglass
				Mild	Armor	
.38 special	2	4	4	1/4	3/16	5/16
9 mm	2 1/2	4	4	5/16	1/4	7/16
7.62 and 5.56 mm	4	6	6	9/16	7/16	1 1/8
7.62-mm AP	6 1/2	8	8	13/16	11/16	N/A
*Nominal thicknesses						

Windows

3-49. Windows can include openings in walls and skylights, although skylights are only an issue where there are LOSs to them. When skylights require protection, treat them like windows. Achieving the low level of protection (hiding) for windows requires making it difficult to see through them, such as installing reflective film on the glass. An aggressor cannot see through the windows during daylight while it is lighter outside than inside, but he may see through them at night when the opposite might be true. Drapes or blinds that can be closed at night address that vulnerability. To achieve the high level of protection requires bullet-resistant window assemblies. These are commercially available for a wide range of ballistics types. They are purchased as manufactured-and-tested assemblies (including glazing and frames, both of which are equally bullet-resistant). The glazing materials and thicknesses and the framing details are proprietary to their manufacturers. The manufacturers make them according to industry test standards to ensure an effective product.

Doors

3-50. Doors without glass easily meet the requirements for the low level of protection. Meeting the high level of protection requires the installation of bullet-resistant door assemblies. Doors can be installed in foyers so that there is no direct LOS into assets within the building (see Figure 3-7, page 3-10).

FORCED ENTRY

3-51. In the forced-entry tactic, an aggressor tries to forcibly gain access to assets. He may use tools or explosives to breach building components or other barriers.

GENERAL-DESIGN STRATEGY

3-52. The general-design strategy for forced entry is to detect the aggressor early in the forced-entry attempt and delay him long enough for a response force to intercept him. The combination of detection and defensive measures must provide sufficient time for a response force to intercept the aggressor before he reaches the asset or before he escapes with it, depending on the protective goals for the asset. The first goal would apply where the asset is likely to be destroyed or where access to it is not acceptable. The second goal would be applied when the idea is to prevent it from being stolen.

LEVELS OF PROTECTION

3-53. Several levels of protection apply to forced entry. These levels vary in terms of system design, delay time, and response-force arrival time.

SITE-WORK ELEMENTS

3-54. Site-work elements do not normally play a major role in protecting against a forced entry. However, the site should be laid out and maintained so that an aggressor does not have a hiding place nearby that will conceal his attempts to break into the building. Another site-work element is the application of perimeter barriers, most commonly fences. Fences are effective at delineating a boundary and at keeping honest people honest, but they are ineffective for preventing a forced entry. The design strategy for forced entry is based on delaying the aggressor, and any serious aggressor could climb a fence in less than 4 seconds or can cut through a fence in less than 10 seconds. Therefore, fences are not used as delay elements, but they are used to establish boundaries and as platforms on which to hang sensors. The final site-work consideration is securing utility-access ports such as manholes. If there are utility tunnels through which aggressors can enter a building, those accesses should be locked using padlocks or locking bolts.

BUILDING ELEMENTS

3-55. Building elements are the principal construction elements of a system for protecting against a forced entry. The building elements are used to provide delay. The process for designing to resist forced entry involves laying out concentric "rings" of delay (called defensive layers). These defensive layers can include the facility's exterior, interior rooms within that layer, and containers within the interior rooms. The individual building components for each of the layers (walls, doors, windows, floors, ceilings, and roofs) provide the delay time (see TM 5-853-1).

DETECTION ELEMENTS

3-56. For a protective system to be effective against a forced entry, the aggressors must be detected at a point of adequate delay. Detection at that point can be achieved by using an IDS. Once a sensor detects an aggressor, the alarm annunciator communicates that event to security personnel, who then dispatch a response force. The alarm can be assessed through a guard response or via CCTV. Chapter 6 and TM 5-853-4 provide detailed discussion of IDSs, CCTV systems, and other elements of ESSs.

COVERT ENTRY AND INSIDER COMPROMISE

3-57. In the covert-entry tactic, an aggressor who is not authorized to be in the facility attempts to enter using false credentials. In the insider-compromise tactic, personnel with legitimate access to a facility try to compromise an asset. The insider may or may not have legitimate access to the asset itself. The purpose of the entry in either case can be to steal or otherwise compromise the asset or to destroy it. In the latter case, the aggressor may bring IEDs or IIDs.

GENERAL-DESIGN STRATEGY

3-58. The general-design strategy for both the insider-compromise and covert-entry tactics is to keep people from entering areas they are not authorized to enter. For covert entry, aggressors are denied access to controlled areas. For insider compromise, aggressors are denied access to assets within controlled areas based on their need to have access to them. The general-design strategy also includes detecting aggressors removing assets from protected areas and detecting aggressors carrying tools, weapons, and explosives into protected areas.

LEVELS OF PROTECTION

3-59. The levels of protection for these tactics address different issues, depending on whether the aggressor's goal is to steal or otherwise compromise an asset or to destroy it. When the goal is to steal or compromise an asset, the levels of protection vary with the number and sophistication of the access controls required to verify personnel access into a controlled area. When the goal is to destroy the assets, the levels of protection vary with the amount of damage the building (and the assets inside) are allowed to sustain and the sophistication of detecting weapons or explosives at entry points.

BUILDING ELEMENTS

3-60. Building elements vary with an aggressor's goal. To protect against theft or compromise of assets, building elements are used to establish and maintain controlled areas into which only authorized personnel can enter. For insider compromise, there may be an additional requirement that access be further limited among personnel otherwise authorized access to the controlled area. That access is based on the need to have access to a specific asset. The result is that the controlled area may be compartmentalized, and each compartmentalized area may have separate access requirements. There are no special construction requirements for these tactics if the goal is theft of compromise. The only requirement is that the building elements of controlled areas should provide enough resistance to require aggressors to force their way through them to gain entry and to provide evidence of the forced entry if it is attempted. Forcing entry would be contrary to the aggressor's assumed goal to be covert. In addition, a common design goal would be to limit the number of entrances into controlled areas because there will need to be access control at each entry.

3-61. To protect against the destruction of assets, building elements are used to shield assets from the effects of explosives going off at access-control points.

The basic approach is to lay out areas at access points in which guards can search for carried-in weapons, explosives, or incendiary devices. The construction of that area is designed to limit damage to the rest of the building if an explosive is detonated in that area. Those levels of damage are similar to those discussed in relation to vehicle bombs. The walls and doors between the access point and the protected area will be hardened, and the walls and doors to the outside will be of lightweight construction so that they may fail and vent the blast pressure away from the building. At the higher level of protection, the access-control area is located in a separate facility and the target building is hardened to resist an explosion in that separate facility.

DETECTION ELEMENTS

3-62. Detection elements for these tactics also vary based on the aggressor's goal. For theft, the detection elements are mainly related to access control. For destruction, the detection elements are used to detect weapons, explosives, or incendiary devices.

3-63. The main detection elements for theft or compromise are access-control devices. These can include procedural systems (such as guards checking ID), mechanical systems (such as keyed or combination locks), or electronic entry-control elements (such as electronic card readers, keypads, and biometric devices). Chapter 6 provides detailed discussion of electronic devices. The sophistication of these elements and the number used varies with the level of protection. For example, achieving the higher levels of protection requires the application of multiple forms of access-control elements such as a card reader and an electronic keypad for electronic-entry control or a badge check and badge exchange for a procedural system.

3-64. When destruction of the assets is the goal, detection is oriented toward detecting weapons, explosives, or incendiary devices. At the lower levels of protection, it is sufficient for guards to search for carried-in items. Achieving higher levels of protection requires the application of such equipment as metal detectors, X-ray machines, and explosive detectors.

SURVEILLANCE AND EAVESDROPPING

3-65. Surveillance and eavesdropping tactics include visual surveillance, acoustic eavesdropping, and electronic-emanations eavesdropping. In these tactics, aggressors remain outside of controlled areas and try to gather information from within those areas. The tools used for these tactics include ocular devices for the visual-surveillance tactic and listening devices and electronic-emanations-eavesdropping equipment for the eavesdropping tactic.

GENERAL-DESIGN STRATEGY

3-66. The general-design strategy for these tactics is to deny aggressors access to information assets. The kind of information (objects, operations, or files; secure conversations; or electronically processed data) and how it can be compromised differs for each tactic as do the specific protective strategies. Therefore, each tactic is addressed separately.

LEVELS OF PROTECTION

3-67. Each of these tactics has only one level of protection. Either one protects or fails to protect against these tactics.

SITE-WORK ELEMENTS

3-68. Site-work elements play a minor role in protecting assets from all surveillance or eavesdropping tactics. The main issue is to eliminate or control vantage points from which aggressors can surveil or eavesdrop on assets or operations. In addition, for the visual-surveillance tactic, a design goal can be to block LOSs from vantage points. Items used to block LOSs include trees, bushes, fences, and other buildings (see Figure 3-12).

BUILDING ELEMENTS

3-69. Building elements are the principal components of the protective strategies for surveillance and eavesdropping tactics. For visual surveillance,

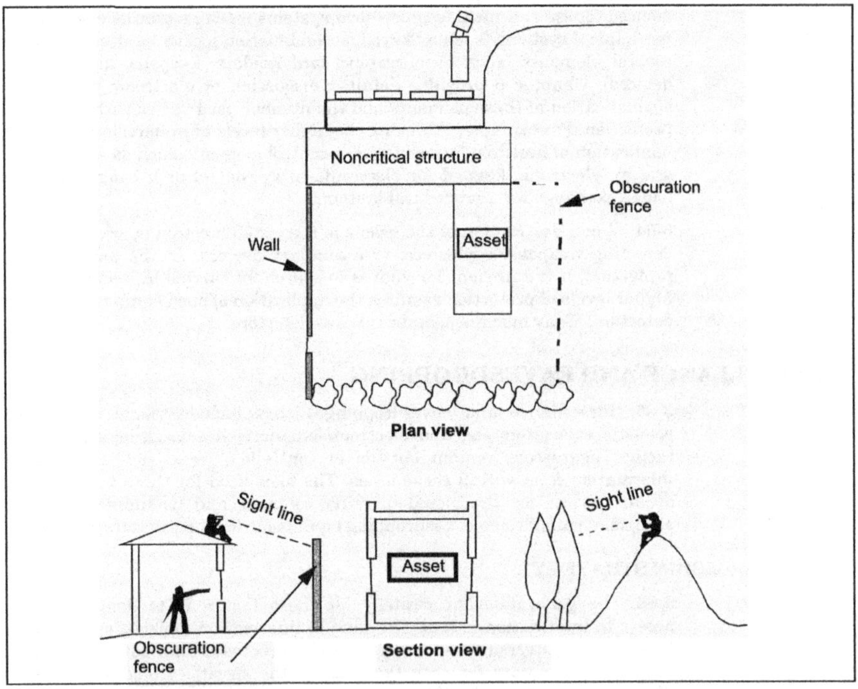

Figure 3-12. LOS Blocked From Potential Vantage Points

the building elements must block LOSs from outside the building. Walls and roofs perform this function effectively. Doors are only a problem when they have windows in them or are made of transparent materials. When this is the case, they can be treated like windows or they can be placed in foyers so that there are no direct LOSs through them. Windows can be treated with reflective film and drapes or blinds as described in the ballistics tactics. When there are LOSs through skylights, they should be treated like windows.

3-70. Building elements for acoustic eavesdropping relate to the construction of areas (preferably separated from the building exterior) that minimize the sound that can be transmitted through them. This requires specialized construction that has a sound-transmission-coefficient (STC) rating. Walls, floors, and ceilings can be constructed to achieve specific STC ratings using conventional construction materials as described in TM 5-853-1. Doors and windows that are STC rated are commonly manufactured and tested as assemblies. This type of design and construction can be expensive.

3-71. Protection against electronic-emanations eavesdropping involves the application of Terminal Electromagnetic-Pulse Emanation Standard (TEMPEST) guidance, most of which is classified. The protection is based on a TEMPEST assessment done for the Army by the US Army Intelligence and Security Command (INSCOM) and on guidance in AR 380-19. The results of a TEMPEST assessment will commonly lead to countermeasures from one or more of the following categories:

- Follow information security policies and procedures recommended during the assessments.
- Provide controlled space both inside and outside the facility.
- Provide TEMPEST-shielded equipment.
- Provide separation between electronic circuits that handle classified information and those that do not. This is commonly called red/black separation.
- Provide TEMPEST-shielded enclosures. This is specialized, metal-shielded construction that is very expensive.

MAIL AND SUPPLY BOMBS

3-72. In mail- and supply-bomb tactics, aggressors place bombs in materials delivered to a facility. Explosives used in supply bombs are significantly larger (briefcase size) than those in mail bombs (pipe bombs or smaller). Mail bombs are usually directed at individuals, while supply bombs may be used to target larger numbers of people. These tactics assume that the facility containing the asset has a mail-handling area or a supplies-handling and -receiving area. These tactics do not apply if mail or supplies are handled and screened in a different facility.

GENERAL-DESIGN STRATEGY

3-73. A bomb exploding within a building has more severe effects than the same size bomb exploding outside of the facility because the blast pressures cannot dissipate inside. Also, there is no standoff distance between the explosive and the facility to mitigate blast effects. The general-design strategy for mail and supply bombs is to detect delivered bombs before they explode and to harden the area where the explosion takes place. This minimizes the damage to the remainder of the facility. Occupants and contents within the mail room or supplies-handling area are likely to be killed or destroyed if an undetected bomb explodes.

LEVELS OF PROTECTION

3-74. The levels of protection for mail and supply bombs are based on the amount of damage allowed to the building and, therefore, the occupants of the building. They also vary based on the sophistication of the detection measures used.

BUILDING ELEMENTS

3-75. The purpose of building elements in relation to these bomb tactics is to shield assets from the effects of explosives going off at supply areas, receiving points, or mail rooms. The basic approach is to lay out either a mail room or a supplies-receiving area in which people can search suspicious packages for explosives or incendiary devices. Constructing this type of area will limit the damage to the rest of the building if an explosive is detonated there. Those levels of damage are similar to those discussed in relation to vehicle bombs.

Mail Rooms

3-76. Mail rooms should be located on the facility's exterior, away from any critical assets. The walls and ceiling between the mail room and the remainder of the building are hardened to keep the blast effects out of the facility. The exterior walls and doors should be of lightweight construction so that they may fail and vent the blast pressure away from the building. There may be an explosives container in the mail room where suspicious packages can be placed. If the package explodes, the container will keep its effects from causing damage or injury. The hardened construction will protect assets outside of the mail room if the explosion occurs outside of the container. Check with EOD personnel to determine the local policy for using explosive containers. At higher levels of protection, the mail room is constructed to completely contain the effects of an explosion either through hardened construction or by using a specialized construction called vented suppressive shielding. Mail rooms should not have windows into protected areas. Doors between the mail room and the rest of the building should be avoided, placed in foyers, or replaced with blast-resistant doors, depending on the desired level of protection.

Supplies-Handling Areas

3-77. Supplies-handling areas should also be on the building's exterior, away from critical areas of the facility. Walls and doors between the handling area and the protected area should be hardened, and the exterior walls and doors should be of lightweight construction so that they may fail and vent the blast pressure away from the building. There should be no windows between the handling area and the protected area. At the higher level of protection, the handling area is located in a separate facility and the target building is hardened to resist an explosion in that separate facility.

DETECTION ELEMENTS

3-78. Detection for these assets varies with the level of protection. At the lower levels of protection, bombs are detected by inspection. As the level of protection goes up, the sophistication of the detection increases. At the higher levels of protection, equipment such as X-ray examining devices, metal detectors, and explosives detectors can be used. Explosive-detection dogs are an alternative to explosive detectors.

CHEMICAL AND BIOLOGICAL CONTAMINATION

3-79. When using chemical- and biological-contamination tactics, aggressors introduce contaminants into the air or water supply to a facility or a group of facilities. Both airborne and waterborne contaminants include chemical, biological, and radiological agents. Aggressors may also forcibly enter a facility to contaminate water or air using the forced-entry tactic.

GENERAL-DESIGN STRATEGY

3-80. Both chemical and biological agents are difficult to detect in water and air supplies. Radiological agents are relatively easy to detect in water, but they are not commonly included in water-quality examinations. It is unlikely that all agents will be detected, so the general-design strategy for these tactics is to filter out suspected airborne contaminants or to shut off suspected waterborne contaminants. Also, because contaminants can easily be entered into the environment from inside a facility, the strategy includes limiting access to the facility (especially mechanical rooms, water intakes, and so forth).

LEVELS OF PROTECTION

3-81. The levels of protection for each of these tactics differ only in the frequency with which some protective measures are exercised. For the low level of protection, they are exercised only in response to a known threat. In the high level of protection, they are exercised continuously.

SITE-WORK ELEMENTS

3-82. Site-work elements are only significant for waterborne contamination. They include protecting water-treatment plants and water-storage structures. This protection may include constructing perimeter barriers (such as chain-link fences) and controlling access to the plant site. These measures are used because most contaminants require quantities on the order of truckloads to contaminate a water supply, so the focus of security is to keep such large vehicles under control. The perimeter barriers do not need to stop the vehicles because the assumption is that the aggressor wants to be covert. An overt act would alert people to avoid the water supply.

BUILDING ELEMENTS

3-83. Building elements for both tactics include controlling access so that aggressors cannot sneak in and plant devices in the building. Protection against airborne contamination at a facility involves making elements of the air-handling system (including air intakes) inaccessible and laying out toxin-free areas for people to be protected. A toxin-free area is an area in which the internal air pressure is higher than the external air pressure. Therefore, if a chemical, biological, or radiological device is set off outside, its contaminant will not be able to penetrate the protected area. Achieving that "net positive pressure" requires a significant air-handling system with air filters to filter contaminants out of the air. It also requires an air-lock entrance into the area so contaminants cannot enter through the door. At the low level of protection, the filters and the air-handling system are only used in response to a credible threat. At the high level of protection, that risk is not acceptable and the filters are run continuously.

3-84. The building-element issues for waterborne contamination are limited to providing protection against forced and covert entries into water-treatment plants and water-storage areas. These methods have been previously described. The only additional issue is the provision for alternative water sources. If it is suspected or detected that the water is contaminated, a backup water source should be in place (such as bottled water). For the high level of protection, bottled water should always be used for drinking.

Chapter 4

Protective Barriers

Protective barriers are used to define the physical limits of an installation, activity, or area. Barriers restrict, channel, or impede access and are fully integrated to form a continuous obstacle around the installation. They are designed to deter the worst-case threat. The barriers should be focused on providing assets with an acceptable level of protection against a threat.

OVERVIEW

4-1. Protective barriers form the perimeter of controlled, limited, and exclusion areas. Utility areas (such as water sources, transformer banks, commercial power and fuel connections, heating and power plants, or air-conditioning units) may require these barriers for safety standards. Protective barriers consist of two major categories—natural and structural.

- Natural protective barriers are mountains and deserts, cliffs and ditches, water obstacles, or other terrain features that are difficult to traverse.
- Structural protective barriers are man-made devices (such as fences, walls, floors, roofs, grills, bars, roadblocks, signs, or other construction) used to restrict, channel, or impede access.

4-2. Barriers offer important benefits to a physical-security posture. They create a psychological deterrent for anyone thinking of unauthorized entry. They may delay or even prevent passage through them. This is especially true of barriers against forced entry and vehicles. Barriers have a direct impact on the number of security posts needed and on the frequency of use for each post.

4-3. Barriers cannot be designed for all situations. Considerations for protective structural barriers include the following:

- Weighing the cost of completely enclosing large tracts of land with significant structural barriers against the threat and the cost of alternate security precautions (such as patrols, MWD teams, ground sensors, electronic surveillance, and airborne sensors).
- Sizing a restricted area based on the degree of compartmentalization required and the area's complexity. As a rule, size should be kept to a minimum consistent with operational efficiency. A restricted area's size may be driven by the likelihood of an aggressor's use of certain tactics. For example, protecting assets from a vehicle bomb often calls for a substantial explosives standoff distance. In these cases, mitigating the vehicle bomb would often be more important than minimizing the restricted area to the extent necessary for operational efficiency. Protective barriers should be established for—
  - Controlling vehicular and pedestrian traffic flow.

- Providing entry-control points where ID can be checked.
- Defining a buffer zone for more highly classified areas.
- Precluding visual compromise by unauthorized individuals.
- Delaying forced entry.
- Protecting individual assets.

4-4. If a secured area requires a limited or exclusion area on a temporary or infrequent basis, it may not be possible to use physical structural barriers. A temporary limited or exclusion area may be established where the lack of proper physical barriers is compensated for by additional security posts, patrols, and other security measures during the period of restriction. Temporary barriers (including temporary fences, coiled concertina wire, and vehicles) may be used. Barriers are not the only restrictive element, and they may not always be necessary. They may not be ideal when working with limited or exclusion areas or when integrated with other controls.

4-5. Because barriers can be compromised through breaching (cutting a hole through a fence) or by nature (berms eroded by the wind and rain), they should be inspected and maintained at least weekly. Guard-force personnel should look for deliberate breaches, holes in and under barriers, sand dunes building up against barriers, and the proper functioning of locks.

FENCING

4-6. Three types of fencing are authorized for use in protecting restricted areas—chain link, barbed wire, and barbed tape or concertina. The type used for construction depends primarily on the threat and the degree of permanence. It may also depend on the availability of materials and the time available for construction. Fencing may be erected for other uses besides impeding personnel access. It can impede observation, can serve as a means to defeat standoff-weapon systems (such as rocket-propelled grenades [RPGs]), and can serve as a barrier to hand-thrown weapons (such as grenades and firebombs).

4-7. Generally, chain-link fencing will be used for protecting permanent limited and exclusion areas. All three types of fencing may be used to augment or increase the security of existing fences that protect restricted areas. Examples would be to create an additional barrier line, to increase existing fence height, or to provide other methods that effectively add to physical security. It is important to recognize that fencing provides very little delay when it comes to motivated aggressors, but it can act as a psychological deterrent.

CHAIN LINK

4-8. Chain-link fence (including gates) must be constructed of 6-foot material, excluding the top guard. Fence heights for conventional arms and ammunition security must be 6 feet for standard chain-link, wire-mesh fencing. Chain-link fences must be constructed with 9-gauge or heavier wire. They must be galvanized with mesh openings not larger than 2 inches per side and have twisted and barbed selvages at the top and the bottom. The wire must be taut and securely fastened to rigid metal or reinforced-concrete posts set in concrete. It must reach within 2 inches of hard ground or pavement. On soft ground, it must reach below the surface deep enough to compensate for shifting soil or sand. Materials and construction must meet with the US Army Corps of Engineers (USACE) guide specifications shown in the USACE Standard (STD) 872-90 series. Weaknesses in the chain-link fence occur as a result of weather (rusting) or failure to keep it fastened to the post that affects the desired tightness. Damage to the fence and fence fabric may be the result of allowing vegetation and trees to grow on or near the fence. The interaction between the fence and the overgrowth often leads to fence damage and reduces the integrity and continuity of the fence as a perimeter boundary and barrier. The perimeter fence is the most obvious protective measure. A well-maintained fence indicates that the asset owner is dedicated to physical security.

BARBED WIRE

4-9. Standard barbed wire is twisted, double-strand, 13.5-gauge wire, with four-point barbs spaced an equal distance apart. Barbed-wire fencing (including gates) intended to prevent human trespassing should not be less than 6 feet high and must be affixed firmly to posts not more than 6 feet apart. The distance between strands should not exceed 6 inches, and at least one wire should be interlaced vertically and midway between posts. The ends must be staggered or fastened together, and the base wire must be picketed to the ground.

BARBED TAPE OR CONCERTINA

4-10. A barbed-taped obstacle (BTO) is fabricated from 0.025-inch stainless steel and is available in 24-, 30-, 40-, and 60-inch-diameter coils. The barbs shall have a minimum length of 1.2 inches, and the barb cluster's width shall be 1.21 inches. A BTO deploys tangle-free for fast installation. It may be recovered and used again. Fifty feet (plus or minus 2 inches) can be covered by 101 coil loops. Handling barbed tape requires the use of heavy barbed-tape gauntlets instead of standard barbed-wire gauntlets.

Barbed-Tape Concertina

4-11. Barbed-tape concertina (standard concertina barbed tape) is a commercially manufactured wire coil of high-strength-steel barbed wire that is clipped together at intervals to form a cylinder. When opened, it is 50 feet long and 3 feet in diameter. When used as the perimeter barrier for a restricted area, the concertina must be laid between poles with one roll on top of another or in a pyramid arrangement (with a minimum of three rolls).

4-12. Reinforced barbed-tape concertina consists of a single strand of spring-steel wire and a single strand of barbed tape. The sections between barbs of the barbed tape are securely clinched around the wire. Each coil is about 37 1/2 inches in diameter and consists of 55 spiral turns connected by steel clips to form a cylindrical diamond pattern when extended to a coil length of 50 feet. One end turn is fitted with four bundling wires for securing the coil when closed and each end turn is fitted with two steel carrying loops. The concertina extends to 50 feet without permanent distortion. When released, it can be retracted into a closed coil.

4-13. When possible, a top guard should be constructed on all perimeter fences and may be added on interior enclosures for additional protection. A top guard is an overhang of barbed wire or tape along the top of a fence, facing outward and upward at about a 45-degree angle. Placing barbed wire or tape above it can further enhance the top guard. Top-guard supporting arms will be permanently affixed to the top of fence posts to increase the overall height of the fence by at least 1 foot. (Due to liability issues in some locations, the top guards will not be allowed to face outward where the fence is adjacent to public areas.) Three strands of barbed wire spaced 6 inches apart must be installed on the supporting arms. The number of strands of wire or tape may be increased when required. The top guard of fencing adjoining gates may range from a vertical height of 18 inches to the normal 45-degree outward protection but only for sufficient distance along the fence to open the gates adequately. Bottom and top tension wires should be used in lieu of fence rails. A concrete sill may be cast at the bottom of the fence to protect against soil erosion. A bottom rail is used on high-security fences to prevent intruders from lifting the fence.

Gates and Entrances

4-14. The number of gates and perimeter entrances must be the minimum required for safe and efficient operation of the facility. Active perimeter entrances must be designed so that the guard force maintains full control. Semiactive entrances, such as infrequently used vehicular gates, must be locked on the inside when not in use. When closed, gates and entrances must provide a barrier structurally comparable to their associated barriers. Care must be afforded against the ability to crawl under gates. Top guards, which may be vertical, are required for all gates.

Triple-Standard Concertina (TSC) Wire

4-15. This type of fence uses three rolls of stacked concertina. One roll will be stacked on top of two rolls that run parallel to each other while resting on the ground, forming a pyramid. In many situations, this fence has been used effectively in place of a chain-link fence. (If perimeter fencing consists of TSC, a top guard is not feasible.)

Tangle-Foot Wire

4-16. Barbed wire or tape may be used in appropriate situations to construct a tangle-foot obstruction either outside a single perimeter fence or in the area between double fences to provide an additional deterrent to intruders. The wire or tape should be supported on short metal or wooden pickets spaced at irregular intervals of 3 to 10 feet and at heights between 6 and 12 inches. The wire or tape should be crisscrossed to provide a more effective obstacle. The space and materials available govern the depth of the field.

AIRCRAFT CABLE

4-17. Although not used very often, aircraft cable can be used as a temporary barrier. Refer to FM 5-34 for information required for determining the barrier's strength. The barrier is created using wire rope. Clips are spaced six times the diameter of the wire rope. Aircraft cable (deployed as described above or attached to a chain-link fence) can also be made to act as a barrier to moving vehicles. To do so, the cable must be anchored into the ground at both ends at about 200-foot intervals (see TM 5-853-1).

UTILITY OPENINGS

4-18. Sewers, air and water intakes and exhausts, and other utility openings of 10 inches or more in diameter that pass through perimeter barriers must have security measures equivalent to that of the perimeter (see TM 5-820-4). Specific requirements of various openings are discussed below:

- Manhole covers 10 inches or more in diameter must be secured to prevent unauthorized opening. They may be secured with locks and hasps, by welding them shut, or by bolting them to their frame. Ensure that hasps, locks, and bolts are made of materials that resist corrosion. Keyed bolts (which make removal by unauthorized personnel more difficult) are also available.
- Drainage ditches, culverts, vents, ducts, and other openings that pass through a perimeter and that have a cross-sectional area greater than 96 square inches and whose smallest dimension is greater than 6 inches will be protected by securely fastened welded bar grilles (refer to TM 5-853-3, Figure 8-1). As an alternative, drainage structures may be constructed of multiple pipes, with each pipe having a diameter of 10 inches or less. Multiple pipes of this diameter may also be placed and secured in the inflow end of a drainage culvert to prevent intrusion into the area. Ensure that any addition of grilles or pipes to culverts or other drainage structures is coordinated with the engineers so that they can compensate for the diminished flow capacity and additional maintenance that will result from the installation.

OTHER PERIMETER BARRIERS

4-19. Buildings less than two stories high that form part of a perimeter must have a top guard along the outside edge to deny access to the roof. When using masonry walls as part of a perimeter barrier, they must be at least 7 feet high and have a barbed-wire top guard. The top guard should be sloped outward at a 45-degree angle and carry at least three strands of barbed wire. This will increase the vertical height of the barrier by at least 1 foot.

4-20. Protect windows, active doors, and other designated openings by securely fastening bars, grilles, or chain-link screens. Fasten window barriers from the inside. If hinged, the hinges and locks must be on the inside. Building elements that provide delay against forced entry have stringent requirements. These elements should be designed according to TM 5-853-1.

SECURITY TOWERS

4-21. It is not acceptable to observe a perimeter from towers only. However, all towers should be located to provide maximum observation and should be constructed for protection from small-arms fire.

4-22. Mobile towers are useful in some temporary situations such as a large, open storage area where receiving and storing activities take place. All facilities using towers must have a support force available for emergencies. Tower personnel should be rotated at frequent intervals.

4-23. The height of a tower increases the range of observation during daylight hours and at night with artificial illumination. However, during inclement weather and during a blackout, towers lose this advantage and must be supplemented by on-ground observation.

4-24. The following considerations should be made when planning for the use of towers:

- Hardening the tower against small-arms effects by using sandbags, salvaged armor, or commercially fabricated bullet-resistant construction. This may require strengthening the tower supports, which should be performed only under the supervision of an engineer. The level of protection required must equate to the threat level identified during the IPB or the military decision-making process (MDMP). The best approach is to design for the worst identified threat rather than to try and modify the tower at a later date on short notice.
- Installing communications and alarm systems, both audible and visual (primary and alternate).
- Using appropriate surveillance, target-acquisition, and night-observation (STANO) equipment with the tower and perimeter barriers being surveilled. Infrared (IR) items may be especially valuable. Considerations for the selection and use of STANO equipment must be made while evaluating the effects of perimeter protective lighting.
- Providing security lighting for route protection to the tower. Security lighting also allows for support of the guard force entering or exiting the perimeter.
- Ensuring that the tower's height is determined according to the area of observation.
- Ensuring that towers have overlapping, mutually supporting fields of observation and fire.
- Providing towers with a backup fortified defensive fighting position, as appropriate.

INSTALLATION ENTRANCES

4-25. The number of installation or activity gates and perimeter entrances in active use should be limited to the minimum number required for safe and efficient operations. When necessary, install vehicle barriers in front of vehicle gates. Security lighting should be considered at entry points (see Chapter 5). Refer to TM 5-853-1 for the application and selection of these barriers.

4-26. Plans to use guards for controlling entry to an installation or activity must be predetermined based on the threat conditions (THREATCON). The construction of the guard post must be included in the security plan.

PERIMETER ENTRANCES

4-27. Active perimeter entrances should be designated so that security forces maintain full control without an unnecessary delay in traffic. This is accomplished by having sufficient entrances to accommodate the peak flow of pedestrian and vehicular traffic and having adequate lighting for rapid and efficient inspection. When gates are not operational during nonduty hours, they should be securely locked, illuminated during hours of darkness, and inspected periodically by a roving patrol. Additionally, warning signs should be used to warn drivers when gates are closed. Doors and windows on buildings that form a part of the perimeter should be locked, lighted, and inspected.

ENTRY-CONTROL STATIONS

4-28. Entry-control stations should be provided at main perimeter entrances where security personnel are present. Considerations for construction and use should be based on the information outlined in USACE STD 872-50-01.

4-29. Entry-control stations should be located as close as practical to the perimeter entrance to permit personnel inside the station to maintain constant surveillance over the entrance and its approaches. Additional considerations at entry-control stations include—

- Establishing a holding area for unauthorized vehicles or those to be inspected further. A turnaround area should be provided to keep from impeding other traffic.
- Establishing control measures such as displaying a decal on the window or having a specially marked vehicle.

4-30. Entry-control stations that are manned 24 hours each day should have interior and exterior lighting, interior heating (where appropriate), and a sufficient glassed area to afford adequate observation for personnel inside. Where appropriate, entry-control stations should be designed for optimum personnel ID and movement control. Each station should also include a telephone, a radio, and badge racks (if required).

4-31. Signs should be erected to assist in controlling authorized entry, to deter unauthorized entry, and to preclude accidental entry. Signs should be plainly displayed and be legible from any approach to the perimeter from a reasonable distance. The size and coloring of a sign, its letters, and the interval of posting must be appropriate to each situation.

4-32. Entry-control stations should be hardened against attacks according to the type of threat. The methods of hardening may include—

- Reinforced concrete or masonry.
- Steel plating.
- Bullet-resistant glass.
- Sandbags, two layers in depth.
- Commercially fabricated, bullet-resistant building components or assemblies.

WARNING SIGNS

4-33. A significant amount of warning signs should be erected to ensure that possible intruders are aware of entry into restricted areas. Warning signs augment control signs. They warn intruders that the area is restricted and that trespassing may result in the use of deadly force.

4-34. Warning signs should be installed along the limited area's physical barriers and at each entry point where they can be seen readily and understood by anyone approaching the perimeter. In areas where English is one of two or more languages commonly spoken, warning signs must contain the local language in addition to English. The wording on the signs will denote warning of a restricted area. The signs should be posted at intervals of no more than 100 feet. They must not be mounted on fences equipped with intrusion-detection equipment. Additionally, the warning signs prescribed in AR 190-13 should be posted at all entrances to limited, controlled, and exclusion areas. See Chapter 7 for more details.

OTHER SIGNS

4-35. Signs setting forth the conditions of entry to an installation or area should be plainly posted at all principal entrances. The signs should be legible under normal conditions at a distance not less than 50 feet from the point of entry. Such signs should inform the entrant of the provisions (search of the person, the vehicle, packages, and so forth) or prohibitions (such as against cameras, matches, and lighters and entry for reasons other than official business) that may be prescribed by the installation commander.

4-36. Signs or notices legibly setting forth the designation of restricted areas and provisions of entry should be plainly posted at all entrances and at other points along the perimeter line as necessary. The wording of these signs or notices is prescribed in AR 190-13.

INSTALLATION PERIMETER ROADS AND CLEAR ZONES

4-37. When the perimeter barrier encloses a large area, an interior all-weather perimeter road should be provided for security-patrol vehicles. Clear zones should be maintained on both sides of the perimeter barrier to provide an unobstructed view of the barrier and the ground adjacent to it. Roads within the clear zone should be as close to the perimeter barrier as possible without interfering with it. The roads should be constructed to allow effective road barriers to deter motor movement of unauthorized personnel during mobilization periods.

4-38. Clear zones should be kept clear of weeds, rubbish, or other material capable of offering concealment or assistance to an intruder attempting to breach the barrier. A clear zone of 20 feet or more should exist between the perimeter barrier and exterior structures, parking areas, and natural or man-made features. When possible, a clear zone of 50 feet or more should exist between the perimeter barrier and structures within the protected area, except when a building's wall constitutes part of the perimeter barrier. Ammunition supply points (ASPs) will have clear zones 12 feet outside of the ASP and 30 feet inside, and the vegetation will not exceed 8 inches (4 inches for high-threat and highly controlled areas). Refer to AR 190-11 and DOD 0-2000.12-H, Appendix EE, for further information.

4-39. When it is impossible to have adequate clear zones because of property lines or natural or man-made features, it may be necessary to increase the height of the perimeter barrier, increase security-patrol coverage, add more security lighting, or install an intrusion-detection device along that portion of the perimeter.

4-40. When considering the construction of a new site or perimeter, ensure that the plans include a fence located well inside the property line, thus permitting control of enough space outside the fence to maintain at least a minimal clear zone. The following considerations apply:

- On a large installation (such as a proving ground), it is unreasonable to construct an expensive perimeter fence and keep it under constant observation. Such an installation is usually established in a sparsely inhabited area. Its comparative isolation and the depth of the installation give reasonable perimeter protection. Under these circumstances, it is usually sufficient to post warning signs or notices, reduce access roads to a minimum, and periodically patrol the area between the outer perimeter and the conventionally protected vital area of the installation.
- An alternative to erecting new or replacing old chain-link fence involving an entire installation perimeter is to relocate or isolate the sensitive area or item by—
  - Relocating the item within a safe perimeter.
  - Consolidating the item with other items.
  - Erecting a chain-link fence (regulations permitting) around individual assets rather than the installation's perimeter.

ARMS-FACILITY STRUCTURAL STANDARDS

4-41. It is next to impossible to build a protective barrier that cannot be penetrated by a human or heavy armor. Therefore, as opposed to protecting a facility using only one barrier, enhance security by using a combination of barriers to increase delay. Multiple barriers also cause aggressors to expend more energy trying to breach all of the barriers. They also provide the appearance of additional security and may further deter some aggressors.

4-42. The interest of security must be kept in mind when constructing walls, ceilings, floors, and roofs. Facilities that house arms and ammunition are constructed as security barriers in the interest of deterring and delaying penetration. Construction guidelines for arms facilities are outlined in AR 190-11. AR 190-11 requires coordination with the engineer office, the safety office, the provost marshal office (PMO), or the security-force office when definitive drawings and specifications for new construction or upgrades or modifications of AA&E storage structures are proposed. This coordinated effort ensures that safety and physical-security requirements are met. AR 190-11 also addresses waivers and exceptions for AA&E storage structures, as well as the requirements for a tactical (training or operational) or shipboard environment. Waivers and exceptions are not discussed in this manual. The following guidelines are provided for securing AA&E in tactical and shipboard environments:

- The criteria and standards for protecting AA&E will be developed by the major Army command (MACOM) according to AR 190-11.
- The deploying commander will establish and enforce procedures for securing deployed AA&E based on the assessment of the threat, the objectives, the location, and the duration of the deployment.
- The AA&E in the tactical environment will be secured at all times.
- The AA&E will be under continuous positive control.
- Persons charged with the custody of AA&E will have the capability to sound the alarm if a forceful theft is attempted.
- A response force will be available to protect the AA&E.
- A system of supervisory checks will be established to ensure that all personnel comply with security measures. Supervisory checks of the AA&E holding area will be made to ensure that the AA&E being guarded have not been tampered with.
- All officers, noncommissioned officers (NCOs), or civilian equivalents will closely monitor the control of ammunition and explosives during field training or range firing.
- Selection of personnel to perform guard duties at AA&E holding areas will be closely monitored by commanders to ensure that only responsible individuals are assigned duties.

Chapter 5

Physical-Security Lighting

Security lighting allows security personnel to maintain visual-assessment capability during darkness. When security-lighting provisions are impractical, additional security posts, patrols, MWD patrols, NVDs, or other security means are necessary.

OVERVIEW

5-1. Security lighting should not be used as a psychological deterrent only. It should also be used along perimeter fences when the situation dictates that the fence be under continuous or periodic observation.

5-2. Lighting is relatively inexpensive to maintain and, when properly used, may reduce the need for security forces. It may also enhance personal protection for forces by reducing the advantages of concealment and surprise for a determined intruder.

5-3. Security lighting is desirable for those sensitive areas or structures within the perimeter that are under observation. Such areas or structures include pier and dock areas, vital buildings, storage areas, motor pools, and vulnerable control points in communication and power- and water-distribution systems. In interior areas where night operations are conducted, adequate lighting facilitates the detection of unauthorized persons approaching or attempting malicious acts within the area. Security lighting has considerable value as a deterrent to thieves and vandals and may make the job of the saboteur more difficult. It is an essential element of an integrated physical-security program.

5-4. A secure auxiliary power source and power-distribution system for the facility should be installed to provide redundancy to critical security lighting and other security equipment. During deployed operations, primary power may not exist or may be subject to constraints or interruptions due to poor infrastructure or hostile activity. Auxiliary power sources must be available for critical electrical loads and must be secured against direct and indirect fires as well as sabotage. If automatic-transfer switches are not installed, security procedures must designate the responsibility for the manual start of the source.

COMMANDER'S RESPONSIBILITY

5-5. Commanders determine perimeter lighting needs based on the threat, site conditions along the perimeter, surveillance capabilities, and available guard forces. Commanders ensure that security lighting is designed and used to discourage unauthorized entry and to facilitate the detection of intruders approaching or attempting to gain entry into protected areas.

PLANNING CONSIDERATIONS

5-6. Security lighting usually requires less intensity than working lights, except for ID and inspection at entry-control points. Each area of a facility presents its own unique set of considerations based on physical layout, terrain, atmospheric and climatic conditions, and security requirements. Information is available from the manufacturers of lighting equipment and from the installation's director of public works, who will assist in designing a lighting system. This information includes—

- Descriptions, characteristics, and specifications of various lighting fixtures, arc, and gaseous-discharge lamps.
- Lighting patterns of various fixtures.
- Typical layouts showing the most efficient height and spacing of equipment.
- Minimum levels of illumination and lighting uniformity required for various applications.

5-7. In planning a security-lighting system, the physical-security manager considers the—

- Cost of replacing lamps and cleaning fixtures, as well as the cost of providing the required equipment (such as ladders and mechanical buckets) to perform this maintenance.
- Provision of manual-override capability during a blackout, including photoelectric controls. These controls may be desirable in a peacetime situation but undesirable when a blackout is a possibility.
- Effects of local weather conditions on lighting systems.
- Fluctuating or erratic voltages in the primary power source.
- Grounding requirements.
- Provisions for rapid lamp replacement.
- Use of lighting to support a CCTV system.
- Limited and exclusion areas. Specific lighting requirements are referenced in AR 190-59 and TM 5-853-2. TM 5-853-4 provides guidance for facility applications that include CCTV cameras.
  - Lighting in these areas must be under the control of the guard force.
  - For critical areas (such as weapons storage areas), instantaneous lighting with a backup source is required. Any period without lighting in a critical area is unacceptable. Therefore, these areas generally have a requirement for backup power (such as diesel-engine generators, uninterrupted power supplies, and batteries) in case of power loss.
  - Security-lighting systems are operated continuously during hours of darkness.
  - Protective lights should be used so that the failure of one or more lights will not affect the operation of the remaining lights.
- Lighting requirements for adjoining properties and activities.
- Restrike time (the time required before the light will function properly after a brief power interruption).

- Color accuracy.
- Other facilities requiring lighting, such as parking areas.

PRINCIPLES OF SECURITY LIGHTING

5-8. Security lighting enables guard-force personnel to observe activities around or inside an installation while minimizing their presence. An adequate level of illumination for all approaches to an installation will not discourage unauthorized entry; however, adequate lighting improves the ability of security personnel to assess visually and intervene on attempts at unauthorized entry. Lighting is used with other security measures (such as fixed security posts or patrols, fences, and ESSs) and should never be used alone. Other principles of security lighting include the following:

- Optimum security lighting is achieved by adequate, even light on bordering areas; glaring lights in the eyes of an intruder; and little light on security-patrol routes. In addition to seeing long distances, security forces must be able to see low contrasts (such as indistinct outlines of silhouettes) and must be able to detect an intruder who may be exposed to view for only a few seconds. Higher levels of illumination improve these abilities.
- High brightness contrast between an intruder and the background should be the first consideration when planning for security lighting. With predominantly dark, dirty surfaces or camouflage-type painted surfaces, more light is needed to produce the same brightness around installations and buildings than when clean concrete, light brick, and grass predominate. When the same amount of light falls on an object and its background, the observer must depend on contrasts in the amount of light reflected. His ability to distinguish poor contrasts is significantly improved by increasing the illumination level.
- The observer primarily sees an outline or a silhouette when the intruder is darker than his background. Using light finishes on the lower parts of buildings and structures may expose an intruder who depends on dark clothing and darkened face and hands. Stripes on walls have also been used effectively, as they provide recognizable breaks in outlines or silhouettes. Providing broad-lighted areas around and within the installation against which intruders can be seen can also create good observation conditions.

5-9. To be effective, two basic systems or a combination of both may be used to provide practical and effective security lighting. The first method is to light the boundaries and approaches; the second is to light the area and structures within the property's general boundaries. Protective lighting should—

- Discourage or deter attempts at entry by intruders. Proper illumination may lead a potential intruder to believe detection is inevitable.
- Make detection likely if entry is attempted.
- Prevent glare that may temporarily blind the guards.

TYPES OF LIGHTING

5-10. The type of lighting system used depends on the installation's overall security requirements. Four types of lighting units are used for security-lighting systems—continuous, standby, movable (portable), and emergency.

5-11. Continuous lighting is the most common security-lighting system. It consists of a series of fixed lights arranged to flood a given area continuously during darkness with overlapping cones of light. Two primary methods of using continuous lighting are glare projection and controlled lighting.

- The glare security-lighting method is used when the glare of lights directed across the surrounding territory will not be annoying nor interfere with adjacent operations. It is a strong deterrent to a potential intruder because it makes it difficult to see inside of the area. Guards are protected by being kept in comparative darkness and being able to observe intruders at a considerable distance beyond the perimeter.
- Controlled lighting is best when it limits the width of the lighted strip outside the perimeter, such as along highways. In controlled lighting, the width of the lighted strip is controlled and adjusted to fit the particular need. This method of lighting may illuminate or silhouette security personnel.

5-12. Standby lighting has a layout similar to continuous lighting. However, the luminaries are not continuously lit but are either automatically or manually turned on when suspicious activity is detected or suspected by the security force or alarm systems.

5-13. Movable lighting consists of manually operated, movable searchlights that may be lit during hours of darkness or only as needed. The system normally is used to supplement continuous or standby lighting.

5-14. Emergency lighting is a system of lighting that may duplicate any or all of the above systems. Its use is limited to times of power failure or other emergencies that render the normal system inoperative. It depends on an alternative power source such as installed or portable generators or batteries.

FENCED PERIMETERS

5-15. Fenced perimeters require the lighting specifications indicated in TM 5-853-2. Specific lighting requirements are based on whether the perimeter is isolated, semi-isolated, or nonisolated.

- Isolated fenced perimeters are fence lines around areas where the fence is 100 feet or more from buildings or operating areas. The approach area is clear of obstruction for 100 or more feet outside of the fence. Other personnel do not use the area. Use glare projection for these perimeters and keep patrol routes unlit.
- Semi-isolated fenced perimeters are fence lines where approach areas are clear of obstruction for 60 to 100 feet outside of the fence. The general public or installation personnel seldom have reason to be in the area. Use controlled lighting for these perimeters and keep patrol routes in relative darkness.

- Nonisolated fenced perimeters are fence lines immediately adjacent to operating areas. These areas may be within an installation or public thoroughfares. Outsiders or installation personnel may move about freely in this approach area. The width of the lighted strip depends on the clear zones inside and outside the fence. Use controlled lighting for these perimeters. It may not be practical to keep the patrol area dark.

ENTRANCES

5-16. Entrances for pedestrians will have two or more lighting units providing adequate illumination for recognition of persons and examination of credentials. Vehicle entrances will have two lighting units located to facilitate the complete inspection of passenger cars, trucks, and freight cars as well as their contents and passengers. Semiactive and inactive entrances will have the same degree of continuous lighting as the remainder of the perimeter, with standby lighting to be used when the entrance becomes active. Gatehouses at entrances should have a low level of interior illumination, enabling guards to see approaching pedestrians and vehicles.

OTHER

5-17. Areas and structures within the installation's property line consist of yards; storage spaces; large, open working areas; piers; docks; and other sensitive areas and structures.

- Open yards (unoccupied land only) and outdoor storage spaces (material storage areas, railroad sidings, motor pools, and parking areas) should be illuminated. An open yard adjacent to a perimeter (between guards and fences) will be illuminated according to the perimeter's illumination requirements. Where lighting is necessary in other open yards, illumination will not be less than 0.2 foot-candle at any point.
- Lighting units are placed in outdoor storage spaces to provide an adequate distribution of light in aisles, passageways, and recesses to eliminate shadowed areas where unauthorized persons may hide.
- Illuminating both water approaches and the pier area safeguards piers and docks located on an installation. Decks on open piers will be illuminated to at least 1 foot-candle and the water approaches (extending to a distance of 100 feet from the pier) to at least 0.5 foot-candle. The area beneath the pier floor will be lit with small wattage floodlights arranged on the piling. Movable lighting is recommended as a part of the protective lighting system for piers and docks. The lighting must not in any way violate marine rules and regulations (it must not be glaring to pilots). Consult the US Coast Guard (USCG) for approval of protective lighting adjacent to navigable waters.

WIRING SYSTEMS

5-18. The wiring circuit should be arranged so that failure of any one lamp will not leave a large portion of the perimeter line or a major segment of a critical or vulnerable position in darkness. Feeder lines will be placed underground (or sufficiently inside the perimeter in the case of overhead wiring) to minimize the possibility of sabotage or vandalism from outside the perimeter. Another advantage to underground wiring is reduced effects from adverse weather conditions.

MAINTENANCE

5-19. Periodic inspections will be made of all electrical circuits to replace or repair worn parts, tighten connections, and check insulation. Keep fixtures clean and properly aimed.

POWER SOURCES

5-20. Primary and alternate power sources must be identified. The following is a partial list of considerations:

- The primary source is usually a local public utility.
- An alternate source (standby batteries or diesel-fuel-driven generators may be used) is provided where required and should—
  - Start automatically upon failure of primary power.
  - Be adequate to power the entire lighting system.
  - Be equipped with adequate fuel storage and supply.
  - Be tested under load to ensure efficiency and effectiveness.
  - Be located within a controlled area for additional security.

CCTV-CAMERA LIGHTING REQUIREMENTS

5-21. TM 5-853-4 provides a detailed discussion of CCTV-camera lighting requirements and guidelines for minimum lighting levels and lighting uniformity. The following considerations apply when lighting systems are intended to support CCTV assessment or surveillance:

- The camera's field of view.
- Lighting intensity levels.
- Maximum light-to-dark ratio.
- Scene reflectance.
- Daylight-to-darkness transitions.
- Camera mounting systems relative to lighting.
- The camera's spectral response.
- The cold-start time.
- The restrike time.

Chapter 6

Electronic Security Systems

An overall site-security system is comprised of three major subelements—detection, delay, and response. The detection subelement includes intrusion detection, assessment, and entry control. An ESS is an integrated system that encompasses interior and exterior sensors; CCTV systems for assessing alarm conditions; electronic entry-control systems (EECSs); data-transmission media (DTM); and alarm reporting systems for monitoring, controlling, and displaying various alarm and system information. Interior and exterior sensors and their associated communication and display subsystems are collectively called IDSs.

OVERVIEW

6-1. Many Army and DOD regulations specify protective measures, policies, and operations related to security. Although the regulations specify minimum requirements, it is possible that more stringent requirements will be necessary at specific sites. A designer will use a previously performed site survey to determine which regulations apply and to determine whether circumstances require more stringent measures. Refer to TM 5-853-4 for additional detailed information.

6-2. AR 190-13 requires the use of a standardized ESS, if practical and available. The receiving element must determine whether a standardized system can meet the requirements and whether it is available. After coordinating with the product manager for physical-security equipment to verify that a standardized system is available, the associated MACOM can issue approval to procure a commercial system in lieu of a standardized system.

USE OF ESS

6-3. An ESS is used to provide early warning of an intruder. This system consists of hardware and software elements operated by trained security personnel.

6-4. A system is configured to provide one or more layers of detection around an asset. Each layer is made up of a series of contiguous detection zones designed to isolate the asset and to control the entry and exit of authorized personnel and materials.

GENERAL ESS DESCRIPTION

6-5. An ESS consists of sensors interfaced with electronic entry-control devices, CCTV, alarm reporting displays (both visual and audible), and security lighting. The situation is assessed by sending guards to the alarm point or by using CCTV. Alarm reporting devices and video monitors are located in the security center. The asset's importance will determine whether multiple or redundant security centers are required and, ultimately, the required sophistication of all elements in the ESS. Digital and analog data are transmitted from local (field) interior and exterior locations to the security center for processing. Reliability and accuracy are important functional requirements of the data-transmission system.

ESS IMPLEMENTATION PROCESS

6-6. The ESS implementation process is shown in Figure 6-1. Implementing an ESS is based on general requirements tailored to a site-specific mission and physical profile. The process begins with a site survey that includes a top-down view of basic needs and classic configurations that are tailored to such site-specific characteristics as terrain, site geography, climatic conditions, the type of asset, and priorities. This data is used to determine the hardware and software requirements, taking into account the additional capacity that should be factored into the design system for future expansion. Once the requirements for an ESS have been identified, the user must determine whether an existing standardized system is suitable for the application. (AR 190-13 outlines the process for gaining approval to use nonstandard equipment.) The user must also secure funding for the equipment (refer to Appendix J). Depending on the current funding regulations, operation-and-maintenance, procurement, or other funds may be required. For example, operations and procurement, Army (OPA) funds may be required for IDS devices; and operations and maintenance, Army (OMA) funds may be required for installation items. A contract is normally awarded to procure and install the equipment. The procurement or installation must be overseen. This may be accomplished by reviewing submittals, inspecting the contractor's work, or responding to the contractor's requests for information. Once the equipment is installed, the acceptance-testing activities must be witnessed and verified. Site conditions during acceptance testing affect the demonstrated detection capability of an exterior IDS. As feasible, acceptance testing should be designed to determine a sensor system's probability of detection (PD) under a range of conditions. For some types of sensor systems, this may be as straightforward as conducting both daytime and nighttime trials to experience differences in temperature and solar heating. After the ESS has been accepted, it must be operated and maintained throughout the remainder of its life cycle. Planning for manpower to operate the system and forecasting the funding and personnel to properly maintain the system is critical for success.

ESS DESIGN CONSIDERATIONS

6-7. A facility may require interior and exterior ESS elements, depending on the level of protection required. The applicable regulations, threat, and design criteria will define the ESS's general requirements. For an existing ESS, hardware and software may need to be supplemented, upgraded, or completely replaced. A site layout (in which all assets are identified and located) is required. It is a useful design tool for such tasks as configuring the DTM.

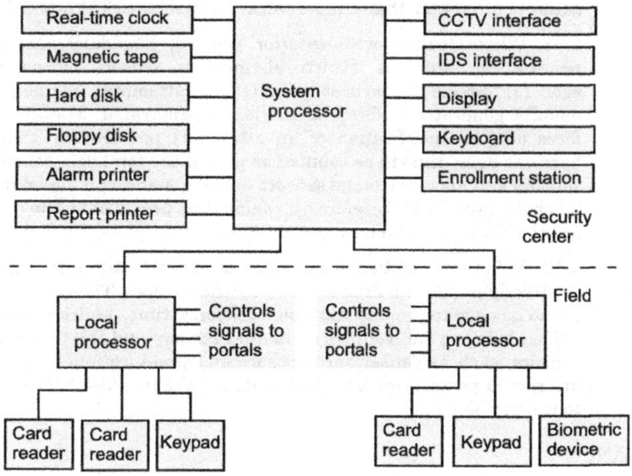

Figure 6-1. Entry-Control System Configured With Distributed Control

6-8. The exterior and interior IDSs should be configured as layers of unbroken rings concentrically surrounding the asset. These rings should correspond to defensive layers that constitute the delay system. The first detection layer is located at the outermost defensive layer necessary to provide the required delay. Detection layers can be on a defensive layer, in the area between two defensive layers, or on the asset itself, depending on the delay required. For example, if a wall of an interior room provides sufficient delay for effective response to aggression, detection layers could be between the facility exterior and interior-room wall or on the interior-room wall. These would detect the intruder before penetration of the interior wall is possible.

RESPONSE AND DELAY

6-9. When dealing with an ESS, the response time is defined as the time it takes the security force to arrive at the scene after an initial alarm is received at the security center. The total delay time is defined as the sum of all of the barriers' delay times, the time required to cross the areas between barriers after an intrusion alarm has been reported, and the time required to accomplish the mission and leave the protected area.

6-10. An ESS's basic function is to notify security personnel that an intruder is attempting to penetrate, or has penetrated, a protected area in sufficient time to allow the response force to intercept and apprehend him. To accomplish this, there must be sufficient physical delay between the point where the intruder is first detected and his objective. This provides delay time equal to or greater than the response time (refer to TM 5-853-1).

6-11. When dealing with interior sensors, boundary sensors that detect penetration (such as structural-vibration sensors or passive ultrasonic sensors) provide the earliest warning of an attempted intrusion. This alarm is usually generated before the barrier is penetrated. This gives the security force advance notification of an attempted penetration, thus allowing the barrier's delay time to be counted as part of the total delay time. Door-position sensors and glass-breakage sensors do not generate an alarm until the barrier has been breached; therefore, the delay time provided by the barrier cannot be counted as part of the total delay time.

6-12. Volumetric motion sensors do not generate an alarm until the intruder is already inside the area covered by the sensors. Therefore, if these sensors are to be used to provide additional response time, additional barriers must be placed between the volumetric motion sensors and the protected asset. Point sensors, such as capacitance sensors and pressure mats, provide warning of attempted penetration only if they detect the intruder before access is gained to the protected area.

BASIC GUIDANCE

6-13. An IDS is deployed in and around barriers (as detailed in TM 5-853-1). Voice communication links (radio, intercom, and telephone) with the response force are located in the security center. Security personnel will man the center and will alert and dispatch response forces in case of an alarm.

6-14. The barrier should always be deployed behind the IDS to ensure that integrity is maintained against intruders. An intruder will then activate the alarm sensor before penetrating or bypassing the barriers, thus providing delay for alarm assessment and response. The delay time is the determining factor in whether an assessment is conducted by dispatching a guard or by observing the CCTV. Normally, an intruder can climb a fence before a guard can be dispatched; therefore, a CCTV is usually required with an exterior IDS. Barriers can be located ahead of an alarm sensor as a boundary demarcation and can serve to keep people and animals from causing nuisance alarms by inadvertently straying into a controlled area. These barriers provide no additional response time because the barrier could be breached before the IDS sensors could be activated.

6-15. Data for monitoring and controlling an ESS are gathered and processed in the security center where the operator interacts with information from the ESS components located at remote facilities. The ESS's alarm-annunciation computer and its DTM line-termination equipment should be located in a controlled area and provided with tamper protection. Supervisory personnel should permit changes to software only, and these changes should be documented. If redundant DTM links connect the central computer to the local processor, diverse paths should be used to route these links.

6-16. The preferred medium for transmitting data in an ESS is a dedicated fiber-optics system. It provides for communications not susceptible to voltage transients, lightning, electromagnetic interference, and noise. Additionally, the fiber optics will provide a measure of communication-line security and wide bandwidth for video signals and increased data-transmission rates.

ESS EFFECTIVENESS

6-17. An ESS has a degree-of-protection effectiveness that is based on its probability of detecting intruders attempting to go over, under, around, or through the physical-security system. The intruder may use forced-entry, covert-entry, or insider-compromise tactics. A well-designed system will minimize the possibility of a successful penetration through covert entry or insider compromise. Interior and exterior alarm sensors have a PD based on the capability to detect an intruder passing through a sensing field. An intruder disturbs the steady-state quiescent condition of a sensor for a finite period. Sensors are designed to detect a person of minimum stature moving within a specific range of speeds and distances from the sensor, and any target outside of those parameters will probably not be detected. The PD for a specific sensor is usually specified at 0.9 or greater, but the designer must be aware that the PD is based on certain constraints and environmental conditions.

6-18. Manufacturer specifications usually do not discuss environmental or nuisance alarms that can be caused by climatic conditions (such as wind or rain) or by the intrusion of animals (including birds). The alarm annunciation is valid because the sensor's thresholds have been exceeded; however, the alarm does not represent a valid penetration attempt. If the assessment system is slow, the operator may not be able to determine the cause of the alarm and must, therefore, treat an environmental or nuisance alarm as real.

6-19. Another type of false alarm is caused by electronic-circuit tolerances being exceeded, resulting in the sensor's actuation. False alarms may also result from improper installation of the sensor or from effects of other equipment in the immediate area.

6-20. After an alarm is sensed and information is displayed in the security center, the console operator must determine the cause of the alarm (intrusion, nuisance, environmental, or false). Timely assessment is required when determining its cause. For example, if an intruder scales a fence in 10 seconds and runs 20 feet per second, the intruder will have overcome the barrier and be 2,200 feet from the point of penetration in 2 minutes. To conduct an accurate assessment of the alarm after 2 minutes, guards will have to search an area of about 200 acres. A fixed-television camera properly located and integrated with the alarm processor can assess the situation while the intruder is still in the controlled area.

6-21. For a CCTV camera to be effective, the area it views must be adequately lighted. To correlate the alarms and cameras in a large system (more than 10 cameras) in a timely manner, a computer-based processing system must be used to select and display alarms and camera scenes for the operator. A complex ESS has the following basic components:

- Intrusion-detection sensors.
- Electronic entry-control devices.
- CCTV.

- Alarm-annunciation system.
- DTM.

6-22. The intrusion-detection sensors are normally deployed in a series of concentric layers. The overall PD improves with each added layer of sensors. The layers (interior and exterior) should be functionally uniform; however, their overall effectiveness and cost are different. The exterior zones significantly differ from the interior zones due to the following considerations:

- The consistency of the PD.
- The PD.
- The cost per detection zone.
- The number of zones.
- The overall sensor coverage.

6-23. Exterior IDSs usually have PDs equal to those of interior IDSs. However, exterior sensors are more likely to experience weather-related situations that cause the system's PD to vary. Sensor phenomenology (passive infrared [PIR], microwave radar, and so forth) determines which environmental factors may alter the system's PD. The frequency of occurrence, severity, and duration of a weather event jointly determine whether it represents security vulnerability with the IDS in use. Typically, sophisticated intruders will attempt their penetration and challenge an ESS under conditions most favorable to themselves. Inclement weather (fog, snow, and rain) affects the usefulness of CCTVs and security lighting such that the capability for remote assessment of alarm events may be lost. Exterior IDSs are not necessarily less likely to detect a penetration attempt during fog, rain and snow; the effect of such site conditions on the IDS depends on sensor phenomenology. For example, fence motion caused by rain impact may drive the response of a fence-mounted sensor closer to satisfying the system's alarm criteria, with the result that the margin of disturbance available to the intruder is less. Also, certain buried sensors are more likely to detect an intruder when the ground is wet because of rain or melting snow. Since interior sensor systems are less influenced by environmental conditions, their PD is typically more consistent than that of some types of exterior sensor systems. Other considerations in comparing an interior and exterior ESS are the cost, the number and size of detection zones required, and the detection height.

- Because of environmental conditions, the exterior electronics must be designed and packaged for extremes of temperature, moisture, and wind. The result is that exterior electronic packages are more costly than equivalent packages for interior applications.
- State-of-the-art exterior sensors do not detect penetration attempts above the height of a fence (typically 8 feet). Fence-mounted sensors are usually limited to this height because the fence fabric or poles are used to support the sensor. For aboveground sensors in the controlled area between the fences, the sensor's mounting brackets and posts limit the detection height. In some applications of field sensors (especially buried sensors), the detection height is no more than 3 feet. For a facility, interior sensors can be deployed on walls, floors, or ceilings, thus permitting complete protection of the asset.

6-24. An interior ESS may be far less costly than that of a comparable exterior ESS. This comparison indicates to the designer the value of selecting and deploying a well-planned, well-designed, layered system. The basic rule in overall design of an ESS is to design from the inside out; that is, layered from the asset to the site boundary.

INTERIOR ESS CONSIDERATIONS

6-25. An interior ESS is typically deployed within a boundary in the immediate vicinity of the asset being protected. If the interior ESS operates in a controlled environment, its PD will be independent of any weather-induced variation in exterior conditions. Also, the physical-security system's effectiveness is enhanced by the interior barriers (walls, ceiling, and floor) that inherently impose a longer delay than exterior barriers (fences and gates).

6-26. Functionally, an interior asset should be viewed as being contained within a cube with sensors protecting all six faces. Interior sensors can be deployed at the cube's perimeter, in its interior space, or in the space immediately outside of the cube.

6-27. If an increased level of protection is dictated by the threat, and if the building is large enough, multiple layers of interior sensors may be deployed for a given asset. A multilayered interior IDS will improve the overall PD. Tamper protection and access-/secure-mode capabilities must be considered when planning and laying out interior sensors.

TAMPER PROTECTION

6-28. To minimize the possibility of someone tampering with circuitry and associated wiring, all sensor-related enclosures must be equipped with tamper switches. These switches must be positioned so that an alarm is generated before the cover has been moved enough to permit access to the circuitry of adjustment controls. In addition, several types of sensors should be equipped with tamper switches to protect against being repositioned or removed. Security screens containing grid-wire sensors and vibration sensors that can be easily removed from a wall are examples of sensors that require tamper switches.

ACCESS/SECURE MODE

6-29. During regular working hours, many of the interior sensors must be deactivated by placing the area in the access mode. For example, door-position sensors and volumetric sensors in occupied areas must be deactivated to prevent multiple nuisance alarms caused by the normal movement of people. This can be done locally or remotely. With local control, a switch is used to bypass or shunt alarm contacts when the sensor is placed in the access mode. When done remotely, the security-center operator usually enters a command that causes the processor software to ignore incoming alarms from those sensors placed in the access mode. However, when a sensor is placed in the access mode, its tamper-protection circuitry must remain in the activated or secure mode. During nonworking hours when the facility is unoccupied, all sensors must be placed in the secure mode. Certain devices (such as duress-alarm switches, tamper switches, grid-wire sensors covering vent openings, and glass-breakage sensors) should never be placed in the access mode. The designer must ensure that selected sensors can be placed in an access mode (if required) and that certain types of sensors (such as duress and tamper switches) are configured so that they cannot be put in the access mode under any condition.

EXTERIOR ESS CONSIDERATIONS

6-30. An exterior ESS is typically deployed at a site's boundary or some other significant boundary such as the demarcation fence for a group of bunkers. An exterior ESS has the advantage that it remains in the secure mode at all times.

6-31. The ideal configuration for an exterior ESS is a rectangle or a polygon, with all sides being straight. The ESS is located in and around barriers that typically include a dual fence. The outside fence is used for demarcation, and the interior fence is used to aid in detection and provide some delay. If dual fences are not used, the sensors should be deployed on the fence or inside it.

DESIGN GUIDELINES

6-32. The general-design criteria of a perimeter IDS involves primarily the selection and layout of exterior sensors that are compatible with the physical and operational characteristics of a specific site. Important factors to consider during the selection process include physical and environmental conditions at the site, the sensor's performance, and the overall cost of the system. Refer to TMs 5-853-1 and 5-853-2 for additional guidance on the requirements for and placement of exterior sensor systems. Since exterior barriers provide very little delay, exterior sensor systems generally do not provide a significant increase in the available response time.

Physical and Environmental Considerations

6-33. Physical and environmental considerations are often the determining factors for selecting exterior sensors. The site's characteristics can significantly affect a sensor's operational performance, both in terms of PD and the susceptibility to nuisance alarms. Exterior sensor systems should be selected on the basis of the frequency and duration of weather-related periods of poor detection capability. An exterior IDS may have an unacceptably low PD during a particular weather event or site condition, yet otherwise be superior to other IDSs in terms of good detection capability and a low nuisance-alarm rate. It may be appropriate to select that IDS in spite of its known vulnerability, precisely because the circumstances of its vulnerability are known and precautionary measures can be taken at those times. The overall performance of that IDS, together with its cost, may justify its selection.

6-34. Weather and climatic conditions at a specific site can significantly influence sensor selection. For example, IR detectors are not very effective in heavy rain, fog, dust, or snow. Deep snow can affect detection patterns and performance of both IR and microwave sensors. High winds can cause numerous false alarms in fence-mounted sensors. Electrical storms can cause alarms in many types of sensors and may also damage the equipment.

6-35. Vegetation can be a significant cause of nuisance alarms. Tall grass or weeds can disturb the energy pattern of microwave and both thermal IR and near-IR beam-break sensors. Vegetation growing near electric-field sensors and capacitance sensors can cause nuisance alarms. Large weeds or bushes rubbing against a fence can produce nuisance alarms from fence-mounted sensors. Large trees and bushes moving within the field of view of video motion sensors can cause nuisance or environmental alarms. A clear area must be established for exterior sensors. This area must be void of vegetation or contain vegetation of carefully controlled growth.

6-36. Topographic features are extremely important. Ideally, perimeter terrain should be flat, although gently sloping terrain is acceptable. Irregular terrain with steep slopes may preclude the use of LOS sensors and make CCTV assessment difficult. Gullies and ditches crossing the perimeter represent a vulnerability to LOS sensors and may be a source of false alarms (from flowing water) for buried line sensors. Large culverts can provide an intruder with an entry or exit route across the perimeter without causing an alarm. Likewise, overhead power and communication lines may permit an intruder to bridge the perimeter without causing an alarm.

6-37. Large animals (such as cows, horses, and deer) can cause nuisance alarms in both aboveground and buried sensors. Sensors sensitive enough to detect a crawling or rolling intruder are susceptible to nuisance alarms from small animals such as rabbits, squirrels, cats, and dogs. To minimize the interference from animals, a dual chain-link-fence configuration may be established around the site perimeter with the sensors installed between the fences.

Sensor Performance

6-38. Exterior sensors must have a high PD for all types of intrusion and have a low unwanted-alarm rate for all expected environmental and site conditions. Unfortunately, no single exterior sensor that is presently available meets both these criteria. All are limited in their detection capability, and all have high susceptibility to nuisance and environmental conditions. Table 6-1, page 6-10, provides estimates of PDs for various types of intrusions. Table 6-2, page 6-10, lists the relative susceptibility of various types of sensors to nuisance and environmental alarms.

Economic Considerations

6-39. Exterior sensor costs are usually given in cost per linear foot per detection zone (typically 300 feet). These costs include both equipment and installation. Fence-mounted sensors (such as strain-sensitive cable, electromechanical, and mechanical) are generally less costly than stand-alone and buried line sensors. Installation costs can vary significantly, depending on the type of sensor. Table 6-3, page 6-11, provides a comparison of relative costs for procuring and installing various types of exterior sensor systems. It should be remembered that the sensor system's cost is only a portion of the total cost for employing a perimeter IDS. Additional costs include fencing, site preparation, CCTV assessment, and perimeter lighting.

Table 6-1. Estimate of PD by Exterior Sensors

Type of Sensor	Slow walk	Walking	Running	Crawling	Rolling	Jumping	Tunnelling	Trenching	Bridging	Cutting	Climbing	Lifting
	Intruder Technique											
Fence mounted	N/A	N/A	N/A	N/A	N/A	VH	VL	L	VL	M/H	H	M/H
Taut wire	N/A	N/A	N/A	N/A	N/A	VH	VL	VL	VL	H	H	H
Electric field	VH	VH	VH	H	VH	VH	VL	L	L	N/A	N/A	N/A
Capacitance	VH	VH	VH	H	H	VH	VL	L	L	N/A	N/A	N/A
Ported cable	H	VH	VH	VH	VH	H	M	VH	L	N/A	N/A	N/A
Seismic	H	VH	H	M	M	M	L	M	L	N/A	N/A	N/A
Seismic/magnetic	H	VH	H	M	M	M	L	M	L	N/A	N/A	N/A
Microwave	H	VH	H	M/H	M/H	M/H	VL	L/M	L	N/A	N/A	N/A
IR	VH	VH	VH	M/H	M/H	H	VL	L	VL	N/A	N/A	N/A
Video motion	H	VH	VH	H	H	H	VL	L/M	M	N/A	N/A	N/A
VL = very low, L = low, M = medium, H = high, VH = very high, N/A = not applicable												

Table 6-2. Relative Susceptibility of Exterior Sensors to False Alarms

Type of Sensor	Wind	Rain	Standing water/runoff	Snow	Fog	Small animals	Large animals	Small birds	Large birds	Lightning	Overhead power lines	Buried power lines
	Intruder Technique											
Fence mounted	H	M	L	L	VL	L	M	L	L	L	VL	VL
Taut wire	VL	VL	VL	VL	VL	VL	L	VL	VL	VL	VL	VL
Electric field	M	L/H	VL	M	VL	M	VH	L	M	M	L	VL
Capacitance	M	M	VL	M	VL	M	VH	L	M	M	L	VL
Ported cable	VL	M	H	L	VL	VL	M	VL	VL	M	VL	L
Seismic	M	L	L	L	VL	L	VH	VL	VL	L	L	M
Seismic/magnetic	M	L	L	L	VL	L	VH	VL	VL	H	M	H
Microwave	L	L	M/H	L/M	L	M/H	VH	VL	M	L/M	L	VL
IR	L	L	L	M	M	M	VH	L	M	L	VL	VL
Video motion	M	L	L	L	M/H	L	VH	VL	M	L	L	VL
VL = very low, L = low, M = medium, H = high, VH = very high												

Table 6-3. Exterior IDS Sensor Cost Comparison

Type of Sensor	Equipment	Installation	Maintenance
Fence mounted	L	L	L
Taut wire	H	H	M
Electric field	H	M	M
Capacitance	M	L	M
Ported cable	H	M	M
Seismic	M	M	L
Seismic/magnetic	H	M	L
Microwave	M	M	L
IR	M	L	M
Video motion	M	L	M
L = low, M = medium, H = high			

PERIMETER LAYOUT AND ZONING

6-40. A protected area's perimeter is usually defined by an enclosing wall or fence or a natural barrier such as water. For exterior sensors to be effective, the perimeter around which they are to be deployed must be precisely defined. In most applications, a dual chain-link-fence configuration will be established around the perimeter. Typically, fences should be between 30 and 50 feet apart; as the distance increases, it is harder for an intruder to bridge the fences. If fence separation is less than 30 feet, some microwave and ported-coax sensors cannot be used. The area between fences (called the controlled area or isolation zone) may need to be cleared of vegetation and graded, depending on the type of sensor used. Proper drainage is required to preclude standing water and to prevent the formation of gullies caused by running water after a heavy rain or melting snow. Cleared areas are required inside and outside of the controlled area. These areas enhance routine observation, as well as sensor-alarm assessment, and minimize the protective cover available to a would-be intruder.

6-41. After the perimeter has been defined, the next step is to divide it into specific detection zones. The length of each detection zone is determined by evaluating the contour, the existing terrain, and the operational activities along the perimeter. Detection zones should be long and straight to minimize the number of sensors or cameras necessary and to aid guard assessment if cameras are not used. It may be more economical to straighten an existing fence line than to create numerous detection zones in accommodating a crooked fence line. If the perimeter is hilly and LOS sensors or CCTV assessment are used, the length of individual detection zones will be commensurate with sensor limitations. Entry points for personnel and vehicles must be configured as independent zones. This enables deactivation of the sensors in these zones; that is, placing them in the access mode during customary working hours (assuming the entry points are manned) without having to deactivate adjacent areas.

6-42. The specific length of individual zones can vary around the perimeter. Although specific manufacturers may advertise maximum zone lengths exceeding 1,000 feet, it is not practical to exceed a zone length of 300 feet. If the zone is longer, it will be difficult for an operator using CCTV assessment or for the response force to identify the location of an intrusion or the cause of a false alarm.

6-43. When establishing zones using multiple sensors, the designer should establish coincident zones where the length and location of each individual sensor will be identical for all sensors within a given zone. If an alarm occurs in a specific zone, the operator can readily determine its approximate location by referring to a map of the perimeter. This also minimizes the number of CCTV cameras required for assessment and simplifies the interface between the alarm-annunciation system and the CCTV switching system.

ESS ALARM-ANNUNCIATION SYSTEM

6-44. Status information from the various intrusion-detection sensors and entry-control terminal devices must be collected from the field and transmitted to the alarm-annunciation system in the security center, where it is processed, annunciated, and acted on by security personnel. The alarm-annunciation system may also interface with a CCTV system. There are typically two types of alarm-annunciation configurations available. The simplest configuration, which is suitable for small installations, is the point-to-point configuration. With this configuration, a separate transmission line is routed from the protected area to the security center (see Figure 6-2). The Joint-Service Interior Intrusion-Detection System (J-SIIDS) is typical of this type of configuration but will not be further discussed in this manual. The second, and more popular type, is a digital multiplexed configuration that allows multiple protected areas to communicate with the security center over a common data line. A block diagram of a typical multiplexed alarm-annunciation system is shown in Figure 6-3.

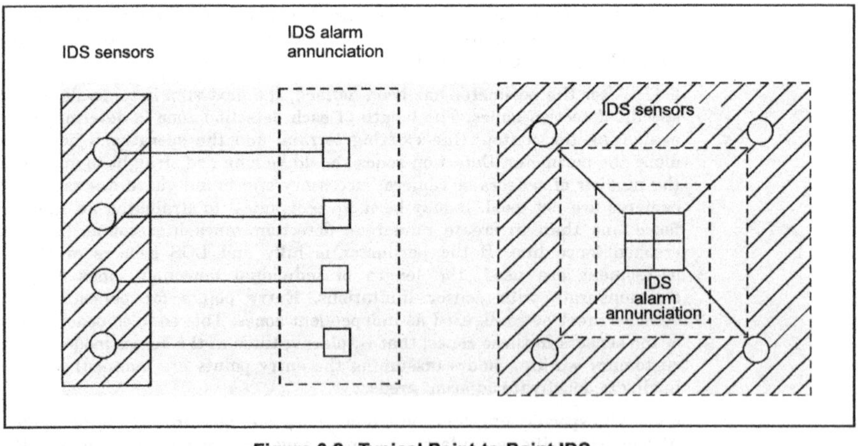

Figure 6-2. Typical Point-to-Point IDS

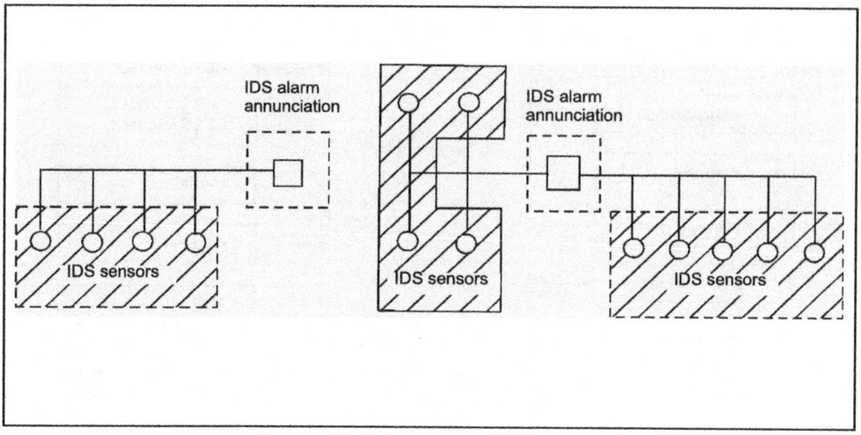

Figure 6-3. Typical Multiplexed IDS

ALARM-ANNUNCIATION CONFIGURATION

6-45. A block diagram of a typical alarm-annunciation system is shown in Figure 6-4, page 6-14. As shown in the figure, the central computer is the hub of the information flow. The central computer receives and displays alarm and device status information and sends operator-control commands to the ESS's local processors. It also interfaces with the CCTV system. For larger facilities, the management of the DTM communications tasks may be delegated to a separate communication processor so that the central computer can turn its full attention to interpreting the incoming information and updating the control and display devices located at the security console (display, logging, control, and storage devices).

6-46. The central computer may consist of one or more digital computers. The real-time clock is usually integral to the central computer and provides a time stamp for alarms and other events. It allows for time synchronization with the CCTV and other systems, if included. The console operator must be able to set the clock, which should include a battery backup. All system events must be properly time-correlated. For example, there will be an exact time correlation for an ESS alarm event reported on the alarm printer and the corresponding video scene recorded by the CCTV's video processor.

DATA STORAGE

6-47. Computer-based systems are required to store large amounts of information such as system software, application programs, data structures, and system events (alarm transactions and status changes). Therefore, a large amount of nonvolatile memory is required. The semiconductor memory provided with a central computer is designed for rapid storage and retrieval

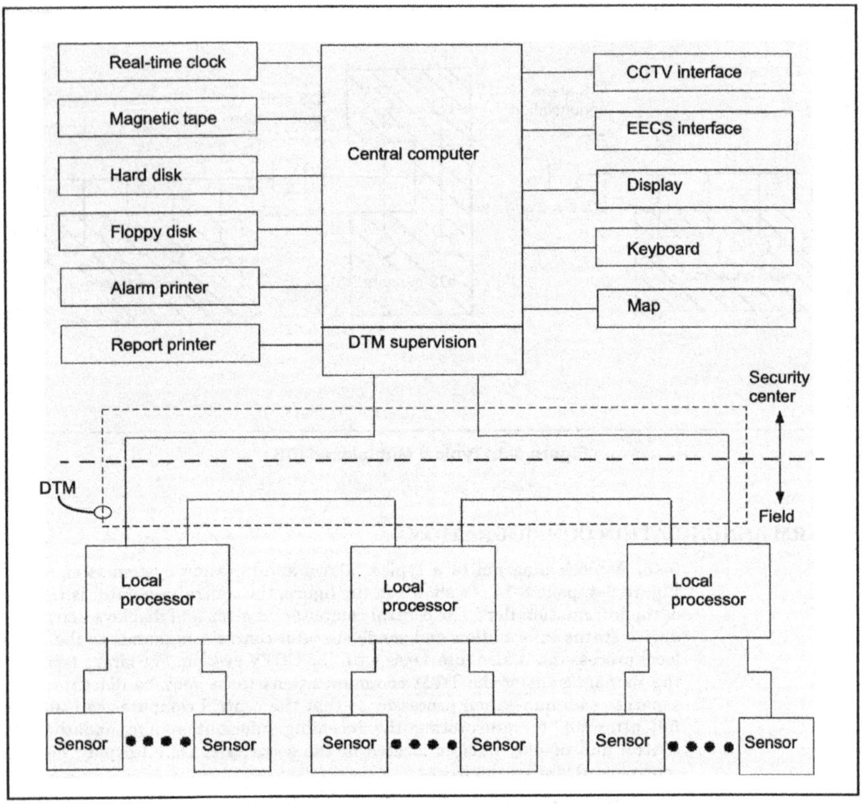

Figure 6-4. Typical IDS Alarm-Annunciation System

and possesses extremely fast access times. The most commonly used media for archival storage are magnetic tape; compact-disk, read-only memory (CD-ROM); and magnetic disk. These media are capable of economically storing large amounts of data.

OPERATOR INTERFACE

6-48. The operator interacts with the alarm-annunciation system through devices that can be seen, heard, or touched and manipulated. Visual displays and printers can be used to inform the operator of an alarm or the equipment's status. Audible devices are used to alert an operator to an alarm or the equipment's failure. Devices such as push buttons and keyboards permit an operator to acknowledge and reset alarms, as well as change operational parameters.

- **Visual displays.** The type of display used to inform the operator visually of the ESS's status is determined primarily by the system's complexity. Status information is usually displayed on monitors. Alphanumeric displays and map displays are seldom used. Monitors provide great flexibility in the type and format of alarm information that may be displayed. Both text and graphic information can be displayed in a variety of colors. Multiple alarms may also be displayed. If alarms are prioritized, higher-priority alarms may be highlighted by blinking, by using bold print or reverse video, or by changing colors. To assist the operator in determining the correct response, alarm-specific instructions may be displayed adjacent to the alarm information.
- **Audible alarm devices.** In conjunction with the visual display of an alarm, the alarm-annunciation system must also generate an audible alarm. The audible alarm may be produced by the ringing of a bell or by the generation of a steady or pulsating tone from an electronic device. In any case, the audible alarm serves to attract the operator's attention to the visual-alarm display. A silence switch is usually provided to allow the operator to silence the bell or tone before actually resetting the alarm.
- **Logging devices.** All alarm-system activity (such as a change of access/secure status, an alarm event, an entry-control transaction, or a trouble event) should be logged and recorded. Logged information is important not only for security personnel investigating an event, but also for maintenance personnel checking equipment performance for such causes as false and nuisance alarms. Most alarm-annunciation systems are equipped with logging and alarm printers.
- **Alarm printers.** Alarm printers are typically of the high-speed, continuous-feed variety. The printer provides a hard-copy record of all alarm events and system activity, as well as limited backup in case the visual display fails.
- **Report printers.** Most ESSs include a separate printer (report printer) for generating reports using information stored by the central computer. This printer will usually be typical of those found in modern office environments.
- **Operator control.** A means is required to transmit information from the operator to the system. The type of controls provided usually depends on the type of display provided. The following are consistent with the controls:
  - Keypads consist of a numeric display system that will generally be provided with a 12-digit keypad and several function keys such as access, secure, acknowledge, and reset. The keypad enables an operator to key in numeric requests for the status of specific zones.
  - Monitor-based systems are usually provided with a typewriter-type keyboard that enables an operator to enter more information using a combination of alphanumeric characters and function keys.
  - An ESS may be equipped with enhancement hardware/devices to help the operator enter information or execute commands quickly. A mouse or a trackball are typical examples.

FIELD-DATA COLLECTION

6-49. Sensor and terminal device data must be transmitted to the central alarm monitor located in the security center using a selected DTM. The following are DTM methods that may be used:

Local Processors

6-50. Multiplexing techniques can be used to minimize the number of data links needed to communicate field-device status to the security center. This is done through devices called local processors. The following is descriptive of a local processor's capabilities:

- A local processor may have very few device inputs, or it may have many (depending on the manufacturer). Rather than having a fixed number of inputs, many local processors are expandable. For example, a basic local processor may be provided with eight device inputs with additional blocks of eight inputs available by using plug-in modules.
- The local processor must provide line supervision for all communication links to sensors, terminal devices, and so forth. Usually, direct-current (DC) line supervision is supplied as the standard with more secure techniques available as options. The data communication links between the local processor and the central alarm monitor must also be supervised.
- Local processors can also provide output signals that can be used for such functions as activating sensor remote test features, light control, or portal control or activating a deterrent (such as a loud horn).
- The local processor contains a microprocessor, solid-state memory, and appropriate software. It has the capability to perform a number of functions locally (such as access-/secure-mode selection, alarm reset, card or keypad electronic-entry control, portal control, and device testing). If the communication link to the security center is temporarily lost, local processors can continue to operate in a stand-alone mode, storing data for transmission after the link is restored.
- The number of local processors required for a specific site depends on the number of protected areas and their proximity to each other and the number of sensors within a protected area. For example, a small building may require one local processor, whereas a large building may require one or more for each floor. An exterior IDS perimeter with two or three different sensors may require one local processor for every two perimeter zones. All local processors may be linked to the central computer using one common DTM link, or the DTM may consist of several links. The designer should note that the temporary loss of a DTM link would render all local processors on that link inactive for the duration of the loss.

Central Computer and Local-Processor Data Exchange

6-51. When the ESS is powered up or reset at the security center, the central computer will download all necessary operational information over the DTM to all local processors. After the download is complete, the central computer will automatically begin polling the local processors for ESS device status. In addition to alarm status, tamper indications, and local-processor status, the DTM may be required to convey security-center console-operator commands to field devices. Examples include security-area access-/secure-mode changes and initiation of the intrusion-sensor self test.

CCTV Interface

6-52. If a CCTV assessment system is deployed with the ESS, an interface between the two is required. This interface allows CCTV system alarms (such as loss of video) to be displayed by the ESS's alarm-annunciation system. The interface also provides IDS alarm signals to the CCTV's video switcher so that the correct CCTV camera will be displayed on the CCTV monitors to allow real-time alarm assessment and video recording as required.

ESS SOFTWARE

6-53. The software provided with computer-based ESS alarm-annunciation systems consists of three types—a standard operating system (such as the Microsoft®-disk operating system [MS-DOS]); vendor-developed application programs; and user-filled, site-specific data structures.

- **System software.** The designer will ensure that system software provided by the vendor conforms to accepted industry standards so that standard, follow-on maintenance and service contracts can be negotiated to maintain the central computer system.
- **Application software.** The vendor-developed application programs are typically proprietary and include ESS monitoring, display, and entry-control capabilities.
- **User-filled data structures.** These data structures are used to populate the site-specific database. Specific electronic address information, personnel access schedules, and normal duty hours are typically included in the site-specific database. The information may include preferred route descriptions for the response force, the phone number of the person responsible for the alarmed area, and any hazardous material that may be located in the alarmed area.

6-54. ESS software functions typically include the following:

- **Alarm monitoring and logging.** The software should provide for monitoring all sensors, local processors, and data communication links and notifying the operator of an alarm condition. All alarm messages should be printed on the alarm printer, archived, and displayed at the console. As a minimum, printed alarm data should include the date and time (to the nearest second) of the alarm and the location and type of alarm.
- **Alarm display.** The software should be structured to permit several alarms to be annunciated simultaneously. A buffer or alarm queue should be available to store additional alarms until they are annunciated and, subsequently, acted upon and reset by the console operator.
- **Alarm priority.** A minimum of five alarm-priority levels should be available. Higher-priority alarms should always be displayed before

lower-priority alarms. This feature permits an operator to respond quickly to the more important alarms before those of lesser importance. For example, the priority of alarm devices may be as follows:

- Duress.
- Intrusion detection.
- Electronic-entry control.
- Tamper.
- CCTV alarms and equipment-malfunction alarms.

- **Reports.** The application software should provide for generating, displaying, printing, and storing reports.

PASSWORDS

6-55. Software security will be provided by limiting access to personnel with authorized passwords assigned by a system manager. A minimum of three password levels shall be provided. Additional security can be provided by programmed restrictions that limit the keyboard actions of logged-in passwords to the user ranks of system managers, supervisors, and console operators, as appropriate.

OPERATOR INTERFACE

6-56. The software should enable an operator with the proper password to enter commands and to obtain displays of system information. As a minimum, an operator should be able to perform the following functions through the keyboard or the keypad:

- Log on by password to activate the keyboard.
- Log off to deactivate the keyboard.
- Request display of all keyboard commands that are authorized for the logged-in password.
- Request display of detailed instructions for any authorized keyboard command.
- Acknowledge and clear alarm messages.
- Display the current status of any device in the system.
- Command a status change for any controlled device in the system.
- Command a mode change for any access/secure device in the system.
- Command printouts of alarm summaries, status summaries, or system activity on a designated printer.
- Add or delete ESS devices or modify parameters associated with a device.

INTERIOR INTRUSION-DETECTION SENSORS

6-57. Interior intrusion-detection sensors are devices used to detect unauthorized entry into specific areas or volumetric spaces within a building. These sensors are usually not designed to be weatherproof or rugged enough to survive an outdoor environment. Therefore, this type of sensor should not be used outdoors unless described by the manufacturer as suitable for outdoor use.

6-58. Interior intrusion-detection sensors generally perform one of three detection functions—detection of an intruder penetrating the boundary of a protected area, detection of intruder motion within a protected area, and detection of an intruder touching or lifting an asset within a protected area. Therefore, interior sensors are commonly classified as boundary-penetration sensors, volumetric motion sensors, and point sensors. Although duress switches are not intrusion-detection sensors, they are included in this discussion because they are usually wired to the same equipment that monitors the interior intrusion-detection sensors.

BOUNDARY-PENETRATION SENSORS

6-59. Boundary-penetration sensors are designed to detect penetration or attempted penetration through perimeter barriers. These barriers include walls, ceilings, duct openings, doors, and windows.

Structural-Vibration Sensors

6-60. Structural-vibration sensors detect low-frequency energy generated in an attempted penetration of a physical barrier (such as a wall or a ceiling) by hammering, drilling, cutting, detonating explosives, or employing other forcible methods of entry. A piezoelectric transducer senses mechanical energy and converts it into electrical signals proportional in magnitude to the vibrations. To reduce false alarms from single accidental impacts on the barrier, most vibration sensors use a signal processor that has an adjustable pulse-counting accumulator in conjunction with a manual sensitivity adjustment. The count circuit can be set to count a specific number of pulses of specific magnitude within a predefined time interval before an alarm is generated. However, the circuitry is usually designed to respond immediately to large pulses, such as those caused by an explosion. The sensitivity adjustment is used to compensate for the type of barrier and the distance between transducers. Typically, several transducers can be connected together and monitored by one signal processor. Figure 6-5, page 6-20, shows an example of wall-mounted, structural-vibration sensors.

Glass-Breakage Sensors

6-61. Glass-breakage sensors detect the breaking of glass. The noise from breaking glass consists of frequencies in both the audible and ultrasonic range. Glass-breakage sensors use microphone transducers to detect the glass breakage. The sensors are designed to respond to specific frequencies only, thus minimizing such false alarms as may be caused by banging on the glass.

Passive Ultrasonic Sensors

6-62. Passive ultrasonic sensors detect acoustical energy in the ultrasonic frequency range, typically between 20 and 30 kilohertz (kHz). They are used to detect an attempted penetration through rigid barriers (such as metal or masonry walls, ceilings, and floors). They also detect penetration through windows and vents covered by metal grilles, shutters, or bars if these openings are properly sealed against outside sounds.

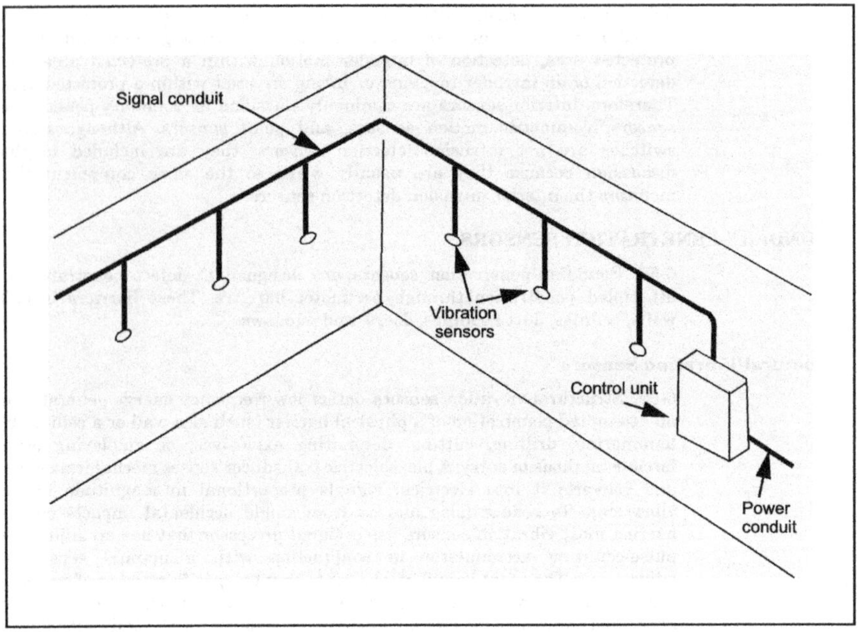

Figure 6-5. Wall-Mounted, Structural-Vibration Sensors

6-63. **Detection Transducer.** The detection transducer is a piezoelectric crystal that produces electrical signals proportional to the magnitude of the vibrations. A single transducer provides coverage of an area about 15 by 20 feet in a room with an 8- to 12-foot ceiling. A typical detection pattern is shown in Figure 6-6. Ten or more transducers can be connected to a signal processor. As with vibration sensors, the signal processor for a passive ultrasonic sensor has manual sensitivity adjustment and an adjustable pulse-counting accumulator.

6-64. **Sensors.** Passive ultrasonic sensors detect ultrasonic energy that results from the breaking of glass, the snipping of bolt cutters on metal barriers, the hissing of an acetylene torch, and the shattering of brittle materials (such as concrete or cinderblock). However, the sensors will not reliably detect drilling through most material nor attacks against soft material such as wallboard. Their effective detection range depends largely on the barrier material, the method of attempted penetration, and the sensitivity adjustment of the sensor. Examples of maximum detection distances for a typical sensor for different types of attempted penetration are shown in Table 6-4.

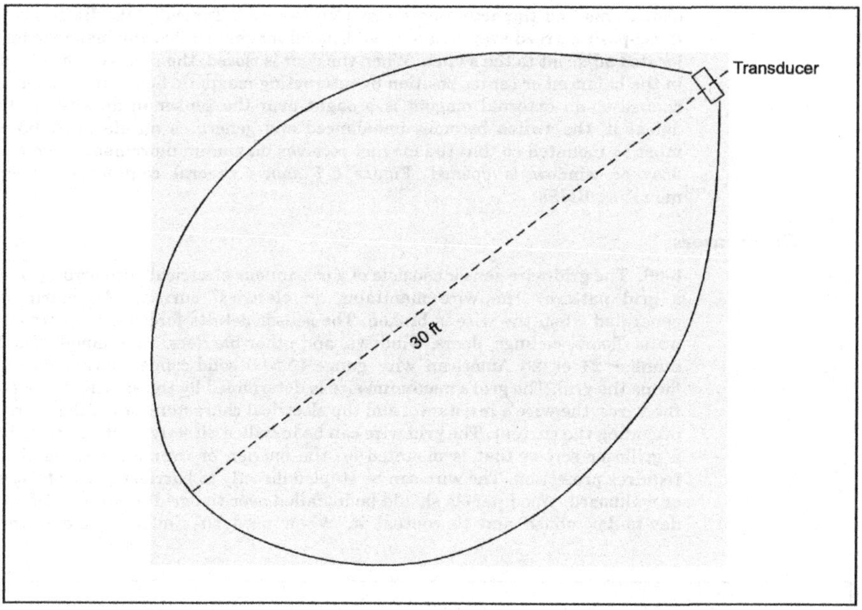

Figure 6-6. Typical Passive-Ultrasonic-Sensor Detection Pattern

Table 6-4. Detection Range for Passive Ultrasonic Sensors

Penetration	Distance (in Feet)
Cut 1/4-inch-thick expanded metal with bolt cutters	55
Cut 5/8-inch reinforcing bar with bolt cutters	45
Use acetylene cutting torch	39
Cut wood with circular saw	30
Cut 5/8-inch reinforcing bar with hacksaw	19
Drill through brick	15
Drill through 1/8-inch steel plate	6
Cut 1/8-inch steel plate with hacksaw	4
Drill through cinderblock	3

6-65. **Balanced Magnetic Switches.** Balanced magnetic switches (BMSs) are typically used to detect the opening of a door. These sensors can also be used on windows, hatches, gates, or other structural devices that can be opened to gain entry. When using a BMS, mount the switch mechanism on the door frame and the actuating magnet on the door. Typically, the BMS has a three-position reed switch and an additional magnet (called the bias magnet) located adjacent to the switch. When the door is closed, the reed switch is held in the balanced or center position by interacting magnetic fields. If the door is opened or an external magnet is brought near the sensor in an attempt to defeat it, the switch becomes unbalanced and generates an alarm. A BMS must be mounted so that the magnet receives maximum movement when the door or window is opened. Figure 6-7 shows several configurations for mounting BMSs.

Grid-Wire Sensors

6-66. The grid-wire sensor consists of a continuous electrical wire arranged in a grid pattern. The wire maintains an electrical current. An alarm is generated when the wire is broken. The sensor detects forced entry through walls, floors, ceilings, doors, windows, and other barriers. An enamel-coated number 24 or 26 American wire gauge (AWG) solid-copper wire typically forms the grid. The grid's maximum size is determined by the spacing between the wires, the wire's resistance, and the electrical characteristics of the source providing the current. The grid wire can be installed directly on the barrier, in a grille or screen that is mounted on the barrier, or over an opening that requires protection. The wire can be stapled directly to barriers made of wood or wallboard. Wood panels should be installed over the grid to protect it from day-to-day abuse and to conceal it. When used on cinder, concrete, and

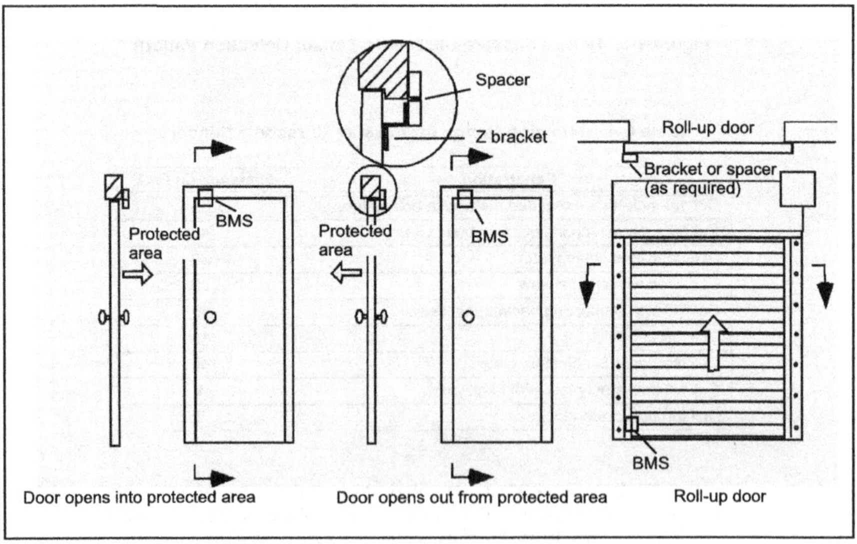

Figure 6-7. BMS Mounting Configurations

masonry surfaces, these surfaces must first be covered with plywood or other material to which the wire can be stapled. An alternative method is to staple the wire grid to the back side of a panel and install the panel over the surface.

VOLUMETRIC MOTION SENSORS

6-67. Volumetric motion sensors are designed to detect intruder motion within the interior of a protected volume. Volumetric sensors may be active or passive. Active sensors (such as microwave) fill the volume to be protected with an energy pattern and recognize a disturbance in the pattern when anything moves within the detection zone. Whereas active sensors generate their own energy pattern to detect an intruder, passive sensors (such as IR) detect energy generated by an intruder. Some sensors, known as dual-technology sensors, use a combination of two different technologies, usually one active and one passive, within the same unit. If CCTV assessment or surveillance cameras are installed, video motion sensors can be used to detect intruder movement within the area. Since ultrasonic motion sensors are seldom used, they will not be discussed here.

Microwave Motion Sensors

6-68. With microwave motion sensors, high-frequency electromagnetic energy is used to detect an intruder's motion within the protected area. Interior or sophisticated microwave motion sensors are normally used.

6-69. **Interior Microwave Motion Sensors.** Interior microwave motion sensors are typically monostatic; the transmitter and the receiver are housed in the same enclosure (transceiver). They may each be provided with a separate antenna or they may share a common antenna. The high-frequency signals produced by the transmitter are usually generated by a solid-state device, such as a gallium arsenide field-effect transistor. The power generated is usually less than 10 milliwatts, but it is sufficient to transmit the signal for distances up to about 100 feet. The shape of the transmitted beam is a function of the antenna configuration. The range of the transmitted beam can be controlled with a range adjustment. A variety of detection patterns can be generated (see Figure 6-8, page 6-24). The frequency of the transmitted signal is compared with the frequency of the signal reflected back from objects in the protected area. If there is no movement within the area, the transmitted and received frequencies will be equal and no alarm will be generated. Movement in the area will generate a Doppler frequency shift in the reflected signal and will produce an alarm if the signal satisfies the sensor's alarm criteria. The Doppler shift for a human intruder is typically between 20 and 120 hertz (Hz). Microwave energy can pass through glass doors and windows as well as lightweight walls or partitions constructed of plywood, plastic, or fiberboard. As a result, false alarms are possible because of the reflection of the microwave signals from the movement of people or vehicles outside of the protected area. The designer can sometimes take advantage of this when the protected area is large and contains a number of partitions, but this is not normally done.

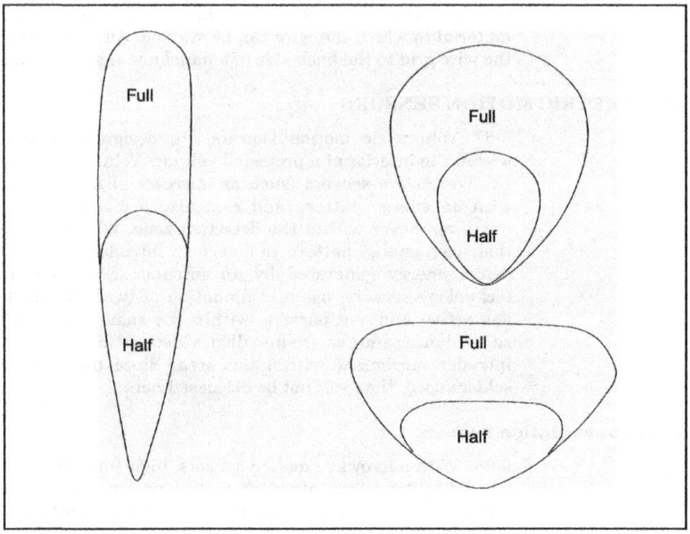

Figure 6-8. Typical Detection Patterns for Microwave Motion Sensors

6-70. **Sophisticated Microwave Motion Sensors.** Sophisticated microwave motion sensors may be equipped with electronic range gating. This feature allows the sensor to ignore the signals reflected beyond the settable detection range. Range gating may be used to effectively minimize unwanted alarms from activity outside the protected area.

PIR Motion Sensors

6-71. PIR motion sensors detect a change in the thermal energy pattern caused by a moving intruder and initiate an alarm when the change in energy satisfies the detector's alarm criteria. These sensors are passive devices because they do not transmit energy; they monitor the energy radiated by the surrounding environment.

6-72. All objects with temperatures above absolute zero radiate thermal energy. The wavelengths of the IR energy spectrum lie between 1 and 1,000 microns. Because the human body radiates thermal energy of between 7 and 14 microns, PIR motion sensors are typically designed to operate in the far IR wavelength range of 4 to 20 microns.

6-73. The IR energy must be focused onto a sensing element, somewhat as a camera lens focuses light onto a film. Two techniques are commonly used. One technique uses reflective focusing; parabolic mirrors focus the energy. The other uses an optical lens. Of the various types of optical lenses, Fresnel lenses are preferred because they can achieve short focal lengths with minimal thickness. Because IR energy is severely attenuated by glass, lenses are usually made of plastic.

6-74. The sensor's detection pattern is determined by the arrangement of lenses or reflectors. The pattern is not continuous but consists of a number of rays or fingers, one for each mirror or lens segment. Numerous detection patterns are available, several of which are shown in Figure 6-9. The PIR is not provided with a range adjustment, but the range can be adjusted

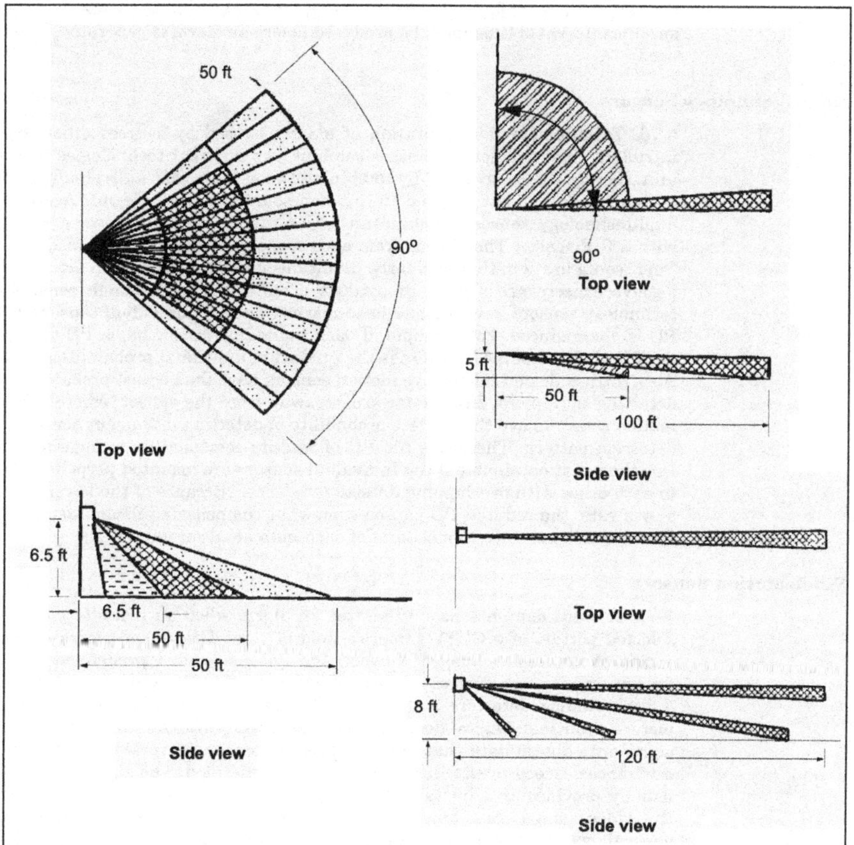

Figure 6-9. Typical Detection Patterns for a PIR Motion Detector

somewhat by manipulating the sensor's position; therefore, careful selection of the appropriate detection pattern is critical to proper sensor performance.

6-75. Most manufacturers use a pyroelectric material as the thermal sensing element. This material produces a change in electric charge when exposed to changes in temperature. To minimize false alarms caused by changes in ambient temperature, most manufacturers use a dual-element sensor. The sensing element is split into halves, one that produces a positive voltage pulse and the other a negative pulse when a change in temperature changes. An intruder entering one of the detection fingers produces an imbalance between the two halves, resulting in an alarm condition. Quadelement sensors that combine and compare two dual-element sensors are also in use. Pulse-count activation, a technique in which a predefined number of pulses within a specific interval of time must be produced before an alarm is generated, is also used.

Dual-Technology Sensors

6-76. To minimize the generation of alarms caused by sources other than intruders, dual-technology sensors combine two different technologies in one unit. Ideally, this is achieved by combining two sensors that individually have a high PD and do not respond to common sources of false alarms. Available dual-technology sensors combine an active ultrasonic or microwave sensor with a PIR sensor. The alarms from each sensor are logically combined in an "and" configuration; that is, nearly simultaneous alarms from both active and passive sensors are needed to produce a valid alarm. Although combined technology sensors have a lower false-alarm rate than individual sensors, the PD is also reduced. For example, if each individual sensor has a PD of 0.95, the PD of the combined sensors is the product of individual probabilities (0.9). Also, ultrasonic and microwave motion sensors have the highest probability of detecting movement directly toward or away from the sensor, whereas PIR motion sensors have the highest probability of detecting movement across the detection pattern. Therefore, the PD of sensors combined in a single unit is less than that obtainable if the individual sensors are mounted perpendicular to each other with overlapping detection patterns. Because of the lower false-alarm rate, the reduced PD can be somewhat compensated for by increasing the sensitivity or detection criteria of each individual sensor.

Video Motion Sensors

6-77. A video motion sensor generates an alarm when an intruder enters a selected portion of a CCTV camera's field of view. The sensor processes and compares successive images between the images against predefined alarm criteria. There are two categories of video motion detectors—analog and digital. Analog detectors generate an alarm in response to changes in a picture's contrast. Digital devices convert selected portions of the analog video signal into digital data that are compared with data converted previously; if differences exceed preset limits, an alarm is generated. The signal processor usually provides an adjustable window that can be positioned anywhere on the video image. Available adjustments permit changing horizontal and vertical window size, window position, and window sensitivity. More sophisticated units provide several adjustable windows that can be individually sized and positioned. Multiple windows permit concentrating on several specific areas of an image while ignoring others. For example, in a scene containing six doorways leading into a long hallway, the sensor can be set to monitor only two critical doorways.

POINT SENSORS

6-78. Point sensors are used to protect specific objects within a facility. These sensors (sometimes referred to as proximity sensors) detect an intruder coming in close proximity to, touching, or lifting an object. Several different types are available, including capacitance sensors, pressure mats, and pressure switches. Other types of sensors can also be used for object protection.

Capacitance Sensors

6-79. Capacitance sensors detect an intruder approaching or touching a metal object by sensing a change in capacitance between the object and the ground. A capacitor consists of two metallic plates separated by a dielectric medium. A change in the dielectric medium or electrical charge results in a change in capacitance. In practice, the metal object to be protected forms one plate of the capacitor and the ground plane surrounding the object forms the second plate. The sensor processor measures the capacitance between the metal object and the ground plane. An approaching intruder alters the dielectric value, thus changing the capacitance. If the net capacitance change satisfies the alarm criteria, an alarm is generated.

6-80. The maximum capacitance that can be monitored by this type of sensor is usually between 10,000 and 50,000 picofarads. The minimum detectable change in capacitance can be as low as 20 picofarads. The signal processor usually has a sensitivity adjustment that can be set to detect an approaching intruder several feet away or to require that the intruder touch the object before an alarm is generated.

6-81. Because air forms most of the dielectric of the capacitor, changes in relative humidity will affect the sensor's sensitivity. An increase in humidity causes the conductivity of the air to increase, lowering the capacitance. Moving a metal object (such as a file cabinet) closer to or away from the protected object can also affect the sensitivity of a capacitance sensor. Figure 6-10, page 6-28, illustrates a typical application using a capacitance sensor.

Pressure Mats

6-82. Pressure mats generate an alarm when pressure is applied to any part of the mat's surface, as when someone steps on the mat. One type of construction uses two layers of copper screening separated by soft-sponge rubber insulation with large holes in it. Another type uses parallel strips of ribbon switches made from two strips of metal separated by an insulating material and spaced several inches apart. When enough pressure is applied to the mat, either the screening or the metal strips make contact, generating an alarm. Pressure mats can be used to detect an intruder approaching a protected object, or they can be placed by doors or windows to detect entry. Because pressure mats are easy to bridge, they should be well concealed, such as placing them under a carpet.

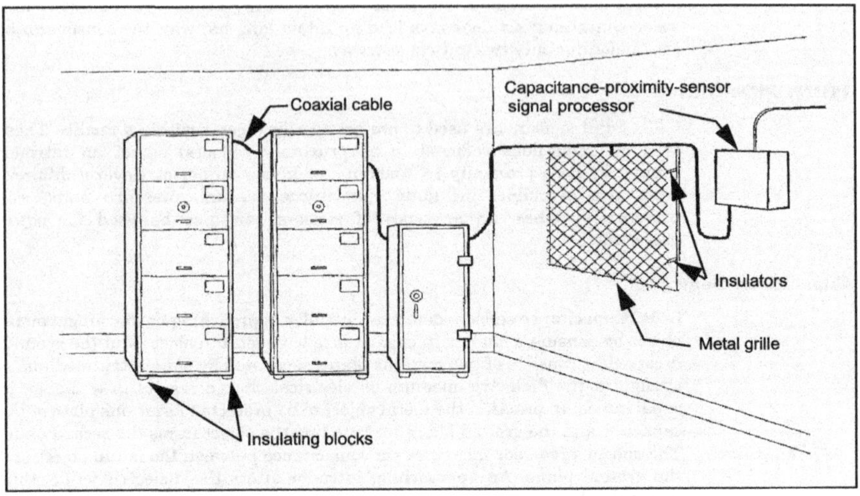

Figure 6-10. Capacitance-Proximity-Sensor Application

Pressure Switches

6-83. Mechanically activated contact switches or single ribbon switches can be used as pressure switches. Objects that require protection can be placed on top of the switch. When the object is moved, the switch actuates and generates an alarm. In this usage, the switch must be well concealed. The interface between the switch and the protected object should be designed so that an adversary cannot slide a thin piece of material under the object to override the switch while the object is removed.

DURESS-ALARM DEVICES

6-84. Duress-alarm devices may be fixed or portable. Operations and security personnel use them to signal a life-threatening emergency. Activation of a duress device will generate an alarm at the alarm-monitoring station. Because of the nature of the alarm, duress devices should never annunciate at the point of threat. These devices are customarily manually operated.

6-85. Fixed duress devices are mechanical switches permanently mounted in an inconspicuous location, such as under a counter or desk. They can be simple push-button switches activated by the touch of a finger or hand or foot-operated switches attached to the floor.

6-86. Portable duress devices are wireless units consisting of a transmitter and a receiver. The transmitter is portable and small enough to be conveniently carried by a person. The receiver is mounted in a fixed location within the facility. Either ultrasonic or RF energy can be used as the communication medium. When activated, the transmitter generates an alarm that is detected (within range) by the receiver. The receiver then activates a relay that is hardwired to the alarm-monitoring system.

EXTERIOR INTRUSION-DETECTION SENSORS

6-87. Exterior intrusion-detection sensors are customarily used to detect an intruder crossing the boundary of a protected area. They can also be used in clear zones between fences or around buildings, for protecting materials and equipment stored outdoors within a protected boundary, or in estimating the PD for buildings and other facilities.

6-88. Exterior sensors are designed to operate in outdoor environmental conditions. The detection function must be performed with a minimum of unwanted alarms such as those caused by wind, rain, ice, standing water, blowing debris, animals, and other sources. Important criteria for selecting an exterior sensor are the PD, the sensor's susceptibility to unwanted alarms, and the sensor's vulnerability to defeat.

6-89. The PD of an exterior sensor is much more vulnerable to the physical and environmental conditions of a site than that of an interior sensor. Many uncontrollable forces (such as wind, rain, ice, frozen soil, standing or running water, falling and accumulated snow, and blowing dust and debris) may affect an exterior sensor's performance. Although attention generally is directed to circumstances that cause a dramatic drop in the PD, environmental factors can also cause short-term increases in the PD. If controlled intrusions (intrusions by security personnel to verify the current detection capability of an IDS) are done while an IDS temporarily has a higher than usual PD as the result of current site conditions, the results may give a false indication of the general effectiveness of that IDS.

6-90. Because of the nature of the outdoor environment, exterior sensors are also more susceptible to nuisance and environmental alarms than interior sensors. Inclement weather conditions (heavy rain, hail, and high wind), vegetation, the natural variation of the temperature of objects in the detection zone, blowing debris, and animals are major sources of unwanted alarms.

6-91. As with interior sensors, tamper protection, signal-line supervision, self-test capability, and proper installation make exterior sensors less vulnerable to defeat. Because signal-processing circuitry for exterior sensors is generally more vulnerable to tampering and defeat than that for interior sensors, it is extremely important that enclosures are located and installed properly and that adequate physical protection is provided. Several different types of exterior intrusion-detection sensors are available. They can be categorized as—

- Fence sensors.
- Buried line sensors.
- LOS sensors.
- Video motion sensors.

FENCE SENSORS

6-92. Fence sensors detect attempts to penetrate a fence around a protected area. Penetration attempts (such as climbing, cutting, or lifting) generate mechanical vibrations and stresses in fence fabric and posts that are usually different than those caused by natural phenomena like wind and rain. The basic types of sensors used to detect these vibrations and stresses are strain-sensitive cable, taut wire, and fiber optics. Other types of fence sensors detect penetration attempts by sensing changes in an electric field or in capacitance. Mechanical and electromechanical fence sensors are seldom used and will not be discussed here.

Strain-Sensitive Cable

6-93. Strain-sensitive cables are transducers that are uniformly sensitive along their entire length. They generate an analog voltage when subject to mechanical distortions or stress resulting from fence motion. Strain-sensitive cables are sensitive to both low and high frequencies. The signal processor usually has a band-pass filter that passes only those signals characteristic of fence-penetration actions. An alarm is initiated when the signal's frequency, amplitude, and duration characteristics satisfy the processor's criteria. Because the cable acts like a microphone, some manufacturers offer an option that allows the operator to listen to fence noises causing the alarm. Operators can then determine whether the noises are naturally occurring sounds from wind or rain or are from an actual intrusion attempt. This feature is relatively costly to implement because it requires additional cable from each signal processor to the security center and, if CCTV is being used, it may be of limited benefit. Strain-sensitive cable is attached to a chain-link fence about halfway between the bottom and top of the fence fabric with plastic ties. One end of the cable is terminated at the signal processor and the other end with a resistive load. The DC through the cable provides line supervision against cutting or electrically shorting the cable or disconnecting it from the processor. A typical installation is shown in Figure 6-11.

Taut-Wire Sensor

6-94. A taut-wire sensor combines a physically taut-wire barrier with an intrusion-detection sensor network. The taut-wire sensor consists of a column of uniformly spaced horizontal wires up to several hundred feet in length and securely anchored at each end. Typically, the wires are spaced 4 to 8 inches apart. Each is individually tensioned and attached to a detector located in a sensor post. Two types of detectors are commonly used—mechanical switches and strain gauges.

- The mechanical switch consists of a specially designed switch mechanism that is normally open. The tensioned wires are mechanically attached to the switch, and movement of the wire beyond a preset limit causes the switch to close. To counteract small gradual movements of a wire (such as that caused by settling of the fence or by freezing or thawing of soil) switches are usually supported in their housing by a soft plastic material. This material allows the switch to self-adjust when acted upon by gradual external forces and wire effects

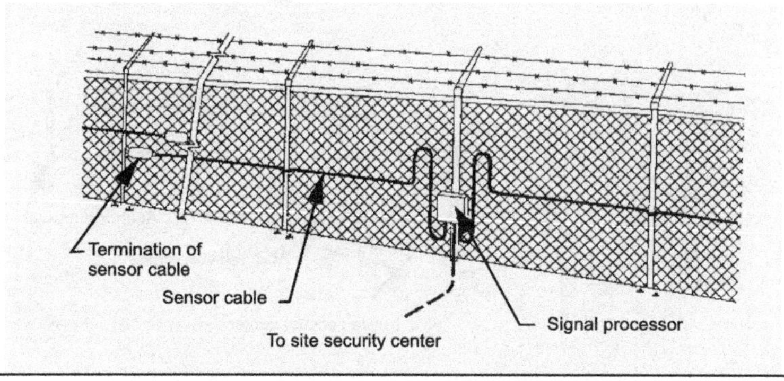

Figure 6-11. Typical Strain-Sensor Cable Installation

such as the relaxation of the wire with time and its thermal expansion or contraction.

- Strain-gauge detectors are attached to the taut wire with a nut on a threaded stud. When a force is applied to the taut wire, the resulting deflection is converted by the strain gauge into a change in electrical output that is monitored by a signal processor.

6-95. With sensors that use mechanical switches as detectors, the switches in a single sensor-post assembly are wired in parallel and are connected directly to the alarm-annunciation system. Pulse-count circuitry is not used because a single switch closure, such as that caused by an intruder moving or cutting one wire, is indicative of an intrusion attempt. Strain-gauge detectors in a sensor post are monitored by a signal processor. When the signal from one or more strain gauges satisfies the processor's criteria, an alarm is initiated.

6-96. The taut-wire sensor can be installed as a freestanding fence or can be mounted on an existing fence or wall. Figure 6-12, page 6-32, shows a freestanding configuration.

Fiber-Optic Cable Sensors

6-97. Fiber-optic cable sensors are functionally equivalent to the strain-sensitive cable sensors previously discussed. However, rather than electrical signals, modulated light is transmitted down the cable and the resulting received signals are processed to determine whether an alarm should be initiated. Since the cable contains no metal and no electrical signal is present, fiber-optic sensors are generally less susceptible to electrical interference from lightning or other sources.

Electric-Field Sensors

6-98. Electric-field sensors consist of an alternating-current (AC) field generator, one or more field wires, one or more sense wires, and a signal

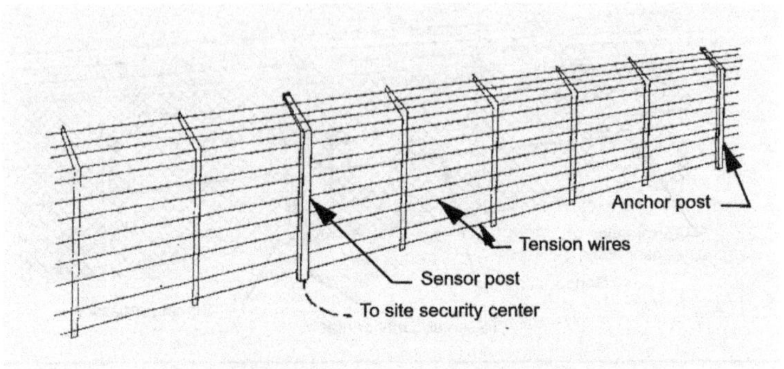

Figure 6-12. Typical Taut-Wire Installation

processor. The generator excites the field wires around which an electrostatic-field pattern is created. The electrostatic field induces electrical signals in the sense wires, which are monitored by the signal processor. Under normal operating conditions, the induced signals are constant. However, when an intruder approaches the sensor, the induced electrical signals are altered, causing the signal processor to generate an alarm.

6-99. Several different field- and sense-wire configurations are available. They range from one field wire and one sense wire to as many as four field wires and one sense wire or four field wires and four sense wires. Figure 6-13 shows the detection pattern produced by vertical three-wire (one field and two sense wires) configurations. The three-wire system has a wider detection envelope and is less costly (one less field wire and associated hardware). However, because of the tighter coupling between wires, the four-wire system is less susceptible to nuisance alarms caused by extraneous noise along the length of the zone.

6-100. A signal processor monitors the signals produced by the sense wires. The processor usually contains a band-pass filter that rejects high-frequency signals such as those caused by objects striking the wires. Additional criteria that must be satisfied before the processor initiates an alarm include signal amplitude and signal duration. By requiring the signal to be present for a preset amount of time, false alarms (such as those caused by birds flying through the detection pattern) can be minimized.

6-101. As with taut-wire sensors, electric-field sensors can be freestanding (mounted on their own posts) or attached by standoffs to an existing fence. They can also be configured to follow contours of the ground. The area under the sensor must be clear of vegetation, since vegetation near or touching sense wires can cause false alarms. These sensors can also be installed on the walls and roof of a building.

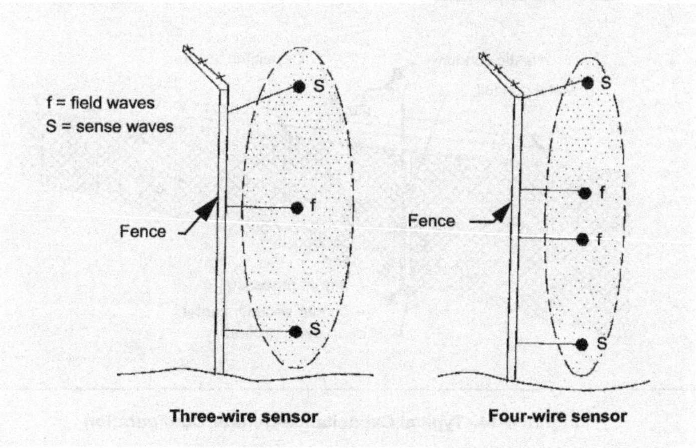

Figure 6-13. Typical Electric-Field-Sensor Detection Patterns

Capacitance Proximity Sensors

6-102. Capacitance proximity sensors measure the electrical capacitance between the ground and an array of sense wires. Any variations in capacitance, such as that caused by an intruder approaching or touching one of the sense wires, initiates an alarm. These sensors usually consist of two or three wires attached to outriggers along the top of an existing fence, wall, or roof edge. Figure 6-14, page 6-34, shows a typical capacitance sensor consisting of three sensor wires attached to the outrigger of a fence. To minimize environmental alarms, the capacitance sensor is divided into two arrays of equal length. The signal processor monitors the capacitance of each array. Changes in capacitance common to both arrays (such as produced by wind, rain, ice, fog, and lightning) are canceled within the processor. However, when changes occur in one array and not the other because of an intruder, the processor initiates an alarm.

BURIED-LINE SENSORS

6-103. A buried-line sensor system consists of detection probes or cable buried in the ground, typically between two fences that form an isolation zone. These devices are wired to an electronic processing unit. The processing unit generates an alarm if an intruder passes through the detection field. Buried-line sensors have several significant features:

- They are hidden, making them difficult to detect and circumvent.
- They follow the terrain's natural contour.
- They do not physically interfere with human activity, such as grass mowing or snow removal.

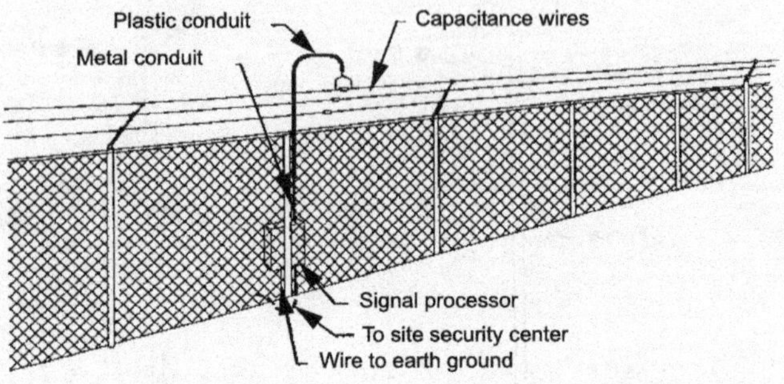

Figure 6-14. Typical Capacitance-Sensor Configuration

- They are affected by certain environmental conditions, such as running water and ground freeze/thaw cycles. (Seismic, seismic/magnetic, magnetic, and balanced pressure sensors are seldom used and will not be discussed here.)

6-104. The ported-coax cable sensor consists of two coax cables buried in the ground parallel to each other. An RF transmitter is connected to one cable and a receiver to the other. The outer conductor of each cable is ported (fabricated with small holes or gaps in the shield). The transmitter cable radiates RF energy into the medium surrounding the cables. A portion of this energy is coupled into the receiver cable through its ported shield. (Because of the ported shields, these cables are frequently referred to as leaky cables.) When an intruder enters the RF field, the coupling is disturbed, resulting in a change of signal monitored by the receiver, which then generates an alarm. Two basic types of ported-coax sensors are available—pulse and continuous wave.

- Pulse-type sensors transmit a pulse of RF energy down one cable and monitors the received signal on the other. The cables can be up to 10,000 feet long. The signal processor initiates an alarm when the electromagnetic field created by the pulse is disturbed and identifies the disturbance's approximate location.
- Continuous-wave sensors apply continuous RF energy to one cable. The signal received on the other cable is monitored for electromagnetic-field disturbances that indicate an intruder's presence. Cable lengths are limited to 300 to 500 feet. Additionally, the sensor is available in a single-cable configuration as well as two separate cables. The pattern typically extends 2 to 4 feet above the ground and can be 5 to 13 feet wide, depending on cable spacing and soil composition. Figure 6-15 represents a typical cross-section of a detection pattern created by a ported-cable sensor.

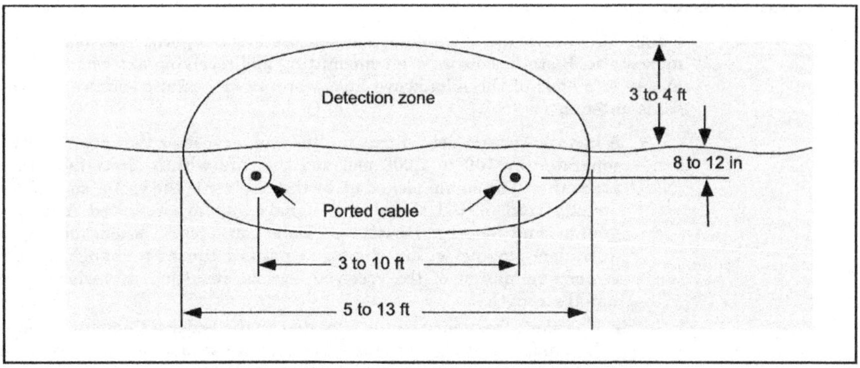

Figure 6-15. Typical Ported-Cable Detection Pattern

6-105. Sensor performance depends on properties of the medium surrounding the cables. Velocity and attenuation of the RF wave that propagates along the cables and the coupling between the cables are functions of the dielectric constant of the soil and its conductivity which, in turn, depends on its moisture content. For example, the velocity is greater and the attenuation is less for cables buried in dry, low-loss soil than in wet, conductive soil. Freeze/ thaw cycles in the soil also affect the sensor's performance. When wet soil freezes, the wave velocity and the cable coupling increase and the attenuation decreases, resulting in greater detection sensitivity. Seasonal sensitivity adjustments may be necessary to compensate for changing ground conditions.

6-106. Although usually buried in soil, ported cables can also be used with asphalt and concrete. If the asphalt or concrete pavement area is relatively small and only a few inches thick (such as a pedestrian pavement crossing the perimeter), the ported cables can be routed under the pavement. However, for the large and deep pavements, slots must be cut into the asphalt or concrete to accept the cable.

6-107. A portable ported-coax sensor is available that can be rapidly deployed and removed. The cables are placed on the surface of the ground rather than buried. This sensor is useful for temporary perimeter detection coverage for small areas or objects (such as vehicles or aircraft).

LOS SENSORS

6-108. The LOS sensors, which are mounted above ground, can be either active or passive. Active sensors generate a beam of energy and detect changes in the received energy that an intruder causes by penetrating the beam. Each sensor consists of a transmitter and a receiver and can be in a monostatic or bistatic configuration. Passive sensors generate no beam of energy; they simply look for changes in the thermal characteristics of their field of view. For effective detection, the terrain within the detection zone must be flat and free of obstacles and vegetation.

Microwave Sensors

6-109. Microwave intrusion-detection sensors are categorized as bistatic or monostatic. Bistatic sensors use transmitting and receiving antennas located at opposite ends of the microwave link, whereas monostatic sensors use the same antenna.

- A bistatic system uses a transmitter and a receiver that are typically separated by 100 to 1,200 feet and that are within direct LOS with each other. The signal picked up by the receiver is the vector sum of the directly transmitted signal and signals that are reflected from the ground and nearby structures. Detection occurs when an object (intruder) moving within the beam pattern causes a change in net-vector summation of the received signals, resulting in variations of signal strength.
    - The same frequency bands allocated by the Federal Communications Commission (FCC) for interior microwave sensors are also used for exterior sensors. Because high-frequency microwave beams are more directive than low-frequency beams and the beam pattern is less affected by blowing grass in the area between the transmitter and the receiver, most exterior sensors operate at the next to highest allowable frequency, 10.525 gigahertz (GHz).
    - The shape of the microwave beam and the maximum separation between the transmitter and the receiver are functions of antenna size and configuration. Various antenna configurations are available, including parabolic-dish arrays, strip-line arrays, and slotted arrays. The parabolic antenna uses a microwave-feed assembly located at the focal point of a metallic parabolic reflector. A conical beam pattern is produced (see Figure 6-16). A strip-line antenna configuration produces a nonsymmetrical beam that is higher than its height. Larger antenna configurations generally produce narrower beam patterns.

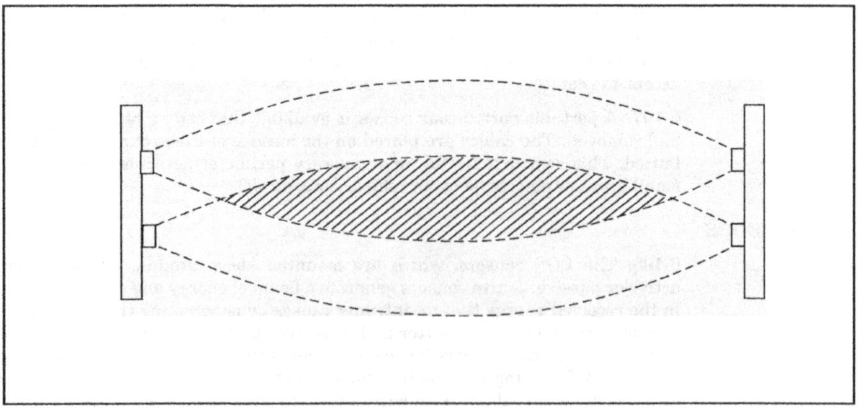

Figure 6-16. Stacked Microwave Configuration

- Monostatic microwave sensors use the same antenna or virtually coincident antenna arrays for the transmitter and receiver, which are usually combined into a single package. Two types of monostatic sensors are available. Amplitude-modulated (AM) sensors detect changes in the net-vector summation of reflected signals similar to bistatic sensors. Frequency-modulated (FM) sensors operate on the Doppler principle similar to interior microwave sensors. The detection pattern is typically shaped like a teardrop (see Figure 6-17). Monostatic sensors can provide volumetric coverage of localized areas, such as in corners or around the base of critical equipment.

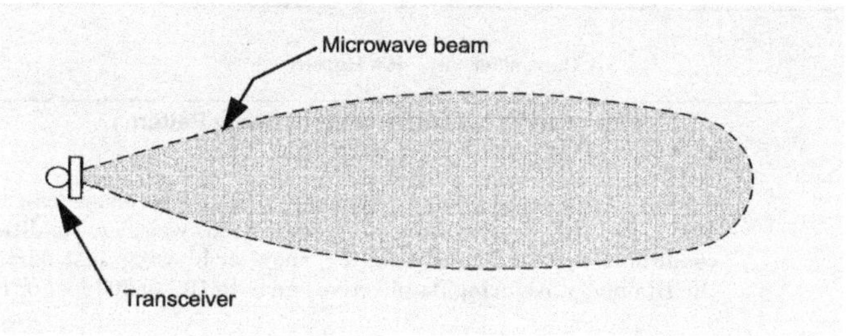

Figure 6-17. Typical Monostatic-Microwave-Sensor Detection Pattern

IR Sensors

6-110. The IR sensors are available in both active and passive models. An active sensor generates one or more near-IR beams that generate an alarm when interrupted. A passive sensor detects changes in thermal IR radiation from objects located within its field of view.

6-111. Active sensors consist of transmitter/receiver pairs. The transmitter contains an IR light source (such as a gallium arsenide light-emitting diode [LED]) that generates an IR beam. The light source is usually modulated to reduce the sensor's susceptibility to unwanted alarms resulting from sunlight or other IR light sources. The receiver detects changes in the signal power of the received beam. To minimize nuisance alarms from birds or blowing debris, the alarm criteria usually require that a high percentage of the beam be blocked for a specific interval of time.

6-112. Active sensors can be single- or multiple-beam systems. Because single-beam sensors can be easily bypassed, multiple-beam systems are generally used in perimeter applications. There are two basic types of multiple-beam configurations—one type uses all transmitters on one post and all receivers on the other post; the second type uses one transmitter and several receivers on each post. Both types are illustrated in Figure 6-18, page 6-38.

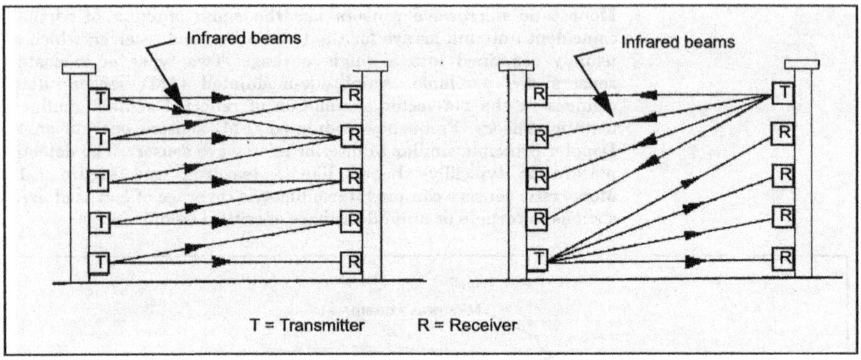

Figure 6-18. Typical IR-Sensor Beam Patterns

6-113. The spacing between transmitters and receivers can be as great as 1,000 feet when operation is under good weather conditions. However, conditions such as heavy rain, fog, snow, or blowing dust particles attenuate the IR energy, reducing its effective range to 100 to 200 feet or less.

VIDEO MOTION SENSORS

6-114. A video motion sensor generates an alarm whenever an intruder enters a selected portion of a CCTV camera's field of view. The sensor processes and compares successive images from the camera and generates an alarm if differences between the images satisfy predefined criteria. Digital devices convert selected portions of the analog video signal into digital data that are compared with data converted previously; if differences exceed preset limits, an alarm is generated.

6-115. The signal processor usually provides an adjustable window that can be positioned anywhere on the video image. Available adjustments permit changing the window's horizontal and vertical sizes, its position, and its sensitivity. More sophisticated units provide several adjustable windows that can be individually sized and positioned. Multiple windows permit concentrating on several specific areas of an image while ignoring others. For example, in a scene that contains several critical assets and multiple sources of nuisance alarms (such as large bushes or trees), the sensor can be adjusted to monitor only the assets and ignore the areas that contain the nuisance-alarm sources.

6-116. The use of video motion-detection systems for exterior applications has been limited, primarily because of difficulties with uncontrolled exterior environments. Lighting variations caused by cloud movement and shadows of slow-moving objects, birds and animals moving within the camera's field of view, camera motion and moving vegetation during windy conditions, and severe weather conditions have traditionally caused a multitude of unwanted alarms in this type of system. Systems using more advanced signal-processing algorithms have improved motion-detection capability and nuisance-alarm rejection; however, they are still subject to high unwanted-alarm rates under certain conditions and should be used with due caution and extreme care.

ELECTRONIC ENTRY CONTROL

6-117. The function of an entry-control system is to ensure that only authorized personnel are permitted into or out of a controlled area. Entry can be controlled by locked fence gates, locked doors to a building or rooms within a building, or specially designed portals.

6-118. These means of entry control can be applied manually by guards or automatically by using entry-control devices. In a manual system, guards verify that a person is authorized to enter an area, usually by comparing the photograph and personal characteristics of the individual requesting entry. In an automated system, the entry-control device verifies that a person is authorized to enter or exit. The automated system usually interfaces with locking mechanisms on doors or gates that open momentarily to permit passage. Mechanical hardware (such as locking mechanisms, electric door strikes, and specially designed portal hardware) and equipment used to detect contraband material (such as metal detectors, X-ray baggage-search systems, explosives detectors, and special nuclear-material monitors) are described in other documentation. Refer to TM 5-853-1 for additional information on determining entry-control requirements and integrating manual electronic-entry control into a cohesive system.

6-119. All entry-control systems control passage by using one or more of three basic techniques—something a person knows, something a person has, or something a person is or does. Automated entry-control devices based on these techniques are grouped into three categories—coded, credential, and biometric devices.

CODED DEVICES

6-120. Coded devices operate on the principle that a person has been issued a code to enter into an entry-control device. This code will match the code stored in the device and permit entry. Depending on the application, a single code can be used by all persons authorized to enter the controlled area or each authorized person can be assigned a unique code. Group codes are useful when the group is small and controls are primarily for keeping out the general public. Individual codes are usually required for control of entry to more critical areas. Coded devices verify the entered code's authenticity, and any person entering a correct code is authorized to enter the controlled area. Electronically coded devices include electronic and computer-controlled keypads.

Electronic Keypad Devices

6-121. The common telephone keypad (12 keys) is an example of an electronic keypad. This type of keypad consists of simple push-button switches that, when depressed, are decoded by digital logic circuits. When the correct sequence of buttons is pushed, an electric signal unlocks the door for a few seconds.

Computer-Controlled Keypad Devices

6-122. These devices are similar to electronic keypad devices, except they are equipped with a microprocessor in the keypad or in a separate enclosure at a different location. The microprocessor monitors the sequence in which the keys are depressed and may provide additional functions such as personal ID and digit scrambling. When the correct code is entered and all conditions are satisfied, an electric signal unlocks the door.

CREDENTIAL DEVICES

6-123. A credential device identifies a person having legitimate authority to enter a controlled area. A coded credential (plastic card or key) contains a prerecorded, machine-readable code. An electric signal unlocks the door if the prerecorded code matches the code stored in the system when the card is read. Like coded devices, credential devices only authenticate the credential; it assumes a user with an acceptable credential is authorized to enter. Various technologies are used to store the code upon or within a card. Hollerith, optically coded, magnetic-spot, capacitance, and electric-circuit cards are seldom used and will not be discussed here. The most commonly used types of cards are described as follows:

Magnetic-Stripe Card

6-124. A strip of magnetic material located along one edge of the card is encoded with data (sometimes encrypted). The data is read by moving the card past a magnetic read head.

Wiegand-Effect Card

6-125. The Wiegand-effect card contains a series of small-diameter, parallel wires about one-half inch long, embedded in the bottom half of the card. The wires are manufactured from ferromagnetic materials that produce a sharp change in magnetic flux when exposed to a slowly changing magnetic field. This type of card is impervious to accidental erasure. The card reader contains a small read head and a tiny magnet to supply the applied magnetic field. It usually does not require external power.

Proximity Card

6-126. A proximity card is not physically inserted into a reader; the coded pattern on the card is sensed when it is brought within several inches of the reader. Several techniques are used to code cards. One technique uses a number of electrically tuned circuits embedded in the card. Data are encoded by varying resonant frequencies of the tuned circuits. The reader contains a transmitter that continually sweeps through a specified range of frequencies and a receiver that senses the pattern of resonant frequencies contained in the card. Another technique uses an integrated circuit embedded in the card to generate a code that can be magnetically or electrostatically coupled to the reader. The power required to activate embedded circuitry can be provided by a small battery embedded in the card or by magnetically coupling power from the reader.

Laser Card

6-127. The optical memory card, commonly called the laser card, uses the same technology developed for recording video and audio disks for entertainment purposes. Data is recorded on the card by burning a microscopic hole (using a laser) in a thin film covering the card. Data is read by using a laser to sense the hole locations. The typical laser card can hold several megabytes of user data.

Smart Card

6-128. A smart card is embedded with a microprocessor, memory, communication circuitry, and a battery. The card contains edge contacts that enable a reader to communicate with the microprocessor. Entry-control information and other data may be stored in the microprocessor's memory.

Bar Code

6-129. A bar code consists of black bars printed on white paper or tape that can be easily read with an optical scanner. This type of coding is not widely used for entry-control applications because it can be easily duplicated. It is possible to conceal the code by applying an opaque mask over it. In this approach, an IR scanner is used to interpret the printed code. For low-level security areas, the use of bar codes can provide a cost-effective solution for entry control. Coded strips and opaque masks can be attached to existing ID badges, alleviating the need for complete badge replacement.

BIOMETRIC DEVICES

6-130. The third basic technique used to control entry is based on the measurement of one or more physical or personal characteristics of an individual. Because most entry-control devices based on this technique rely on measurements of biological characteristics, they have become commonly known as biometric devices. Characteristics such as fingerprints, hand geometry, voiceprints, handwriting, and retinal blood-vessel patterns have been used for controlling entry. Typically, in enrolling individuals, several reference measurements are made of the selected characteristic and then stored in the device's memory or on a card. From then on, when that person attempts entry, a scan of the characteristic is compared with the reference data template. If a match is found, entry is granted. Rather then verifying an artifact, such as a code or a credential, biometric devices verify a person's physical characteristic, thus providing a form of identity verification. Because of this, biometric devices are sometimes referred to as personnel identity-verification devices. The most common biometric devices are discussed below.

Fingerprints

6-131. Fingerprint-verification devices use one of two approaches. One is pattern recognition of the whorls, loops, and tilts of the referenced fingerprint, which is stored in a digitized representation of the image and compared with the fingerprint of the prospective entrant. The second approach is minutiae comparison, which means that the endings and branching points of ridges and valleys of the referenced fingerprint are compared with the fingerprint of the prospective entrant.

Hand Geometry

6-132. Several devices are available that use hand geometry for personnel verification. These devices use a variety of physical measurements of the hand, such as finger length, finger curvature, hand width, webbing between fingers, and light transmissivity through the skin to verify identity. Both two- and three-dimensional units are available.

Retinal Patterns

6-133. This type of technique is based on the premise that the pattern of blood vessels on the human eye's retina is unique to an individual. While the eye is focused on a visual target, a low-intensity IR light beam scans a circular area of the retina. The amount of light reflected from the eye is recorded as the beam progresses around the circular path. Reflected light is modulated by the difference in reflectivity between blood-vessel pattern and adjacent tissue. This information is processed and converted to a digital template that is stored as the eye's signature. Users are allowed to wear contact lenses; however, glasses should be removed.

Device Combinations

6-134. Frequently, an automated entry-control system uses combinations of the three types of entry-control devices. Combining two different devices can significantly enhance the system's security level. In some cases, combining devices results in reduced verification times.

APPLICATION GUIDELINES

6-135. The primary function of an automated entry-control system is to permit authorized personnel to enter or exit a controlled area. Features available to the designer are described below.

- **Enrollment.** All entry-control systems must provide a means of entering, updating, and deleting information about authorized individuals into the system's database files. This is usually accomplished with a dedicated enrollment station for enrolling and disenrolling purposes that is directly connected to the central-processing unit. When credential devices are used, all authorized users must be provided with an appropriate credential. A means should also be provided to disenroll a person quickly without having to retrieve a credential. When using biometric devices, additional enrollment equipment will be required.
- **Entry-control techniques.** Some entry-control functions require additional hardware, while others are accomplished with software. Those features accomplished with software require that the appropriate database be available for every portal affected by them. Typically, these techniques include—
  - Area zones.
  - Time zones.
  - Team zones.
  - Anti-pass back.

- Antitailgate.
- Guard tour.
- Elevator control.

- **Alarms.** Several types of alarms can be used with an entry-control system. These alarms must annunciate audibly and visually in the security center.
- **Entry denial.** Most entry-control devices are configured to permit the user three entry attempts. If more than three unsuccessful entry attempts are made within a specified period, the device generates an alarm. An alarm is also generated if an invalid credential is used or attempted entries are detected that violate specified area, time, or team zoning requirements.
- **Communication failure.** This alarm is generated when a loss of communication between the central processor and the local equipment is detected.
- **Portal open.** If a portal door remains open longer than a predefined time, an alarm is generated.
- **Duress.** This alarm is generated when a special duress code is entered at a keypad.
- **Guard overdue.** This duress alarm is generated when a security guard is determined to be overdue at a checkpoint during a predefined guard tour.
- **Software tamper.** This type of alarm is generated when unauthorized persons are detected attempting to invoke certain system commands or modify database files.

PERFORMANCE CRITERIA

6-136. The overall performance of an entry-control system can be evaluated by examining the verification error rate and the throughput rate. An entry-control system can produce two types of errors—denial of admission of a person who should be admitted or admission of a person who should not be admitted. These are commonly referred to as false-reject errors (type I errors) and false-accept errors (type II errors). Although a false-reject error does not constitute a breach of security, it does create an operational problem that must be handled by an alternative method. False-accept errors constitute a breach of security. Ideally, both false-reject and false-accept error rates should be zero; in practice, however, they are not. In fact, they tend to act in opposition to each other. When the system is adjusted to minimize the false-accept error rate, the false-reject error rate usually increases. Verification error rates are typically measured in percent (number of errors/number of attempts x 100 percent). These error rates are typically very low for coded and credential devices, but many become significant if biometric devices are used.

6-137. The throughput rate is the number of persons that can pass through an entry point in a given unit of time and is usually expressed in persons per minute. It is the time required to approach the entry-control device and for the device to verify information (verification time) and the time required passing through the entry point. Typically, an individual can approach the device and pass through in 3 to 5 seconds. Verification time depends on the type of device and may vary from 3 to 15 seconds. Table 6-5 provides a list of typical verification times for different types of entry-control devices.

Table 6-5. Typical Verification Times of Entry-Control Devices

Device	Verification Time
Keypad	3 seconds
Card reader	3 seconds
Keypad/card reader	6 seconds
Biometric/keypad	6 to 15 seconds
Biometric/card reader	6 to 15 seconds
Biometric	2 minutes

DATA TRANSMISSION

6-138. A critical element in an integrated ESS is the DTM that transmits information from sensors, entry-control devices, and video components to display and assessment equipment. A DTM link is a path for transmitting data between two or more components (such as a sensor and alarm reporting system, a card reader and controller, a CCTV camera and monitor, or a transmitter and receiver). The DTM links connect remote ESS components to the security center. An effective DTM link ensures rapid and reliable transmission media, transmission technique, associated transmission hardware, and degree of security to be provided for the communication system.

6-139. A number of different media are used in transmitting data between elements of an IDS, an EECS, and a CCTV system. These include wire lines, coaxial cable, fiber-optic cable, and RF transmission.

- **Wire line.** Wire lines are twisted pairs that consist of two insulated conductors twisted together to minimize interference by unwanted signals.
- **Coaxial cable.** Coaxial cable consists of a center conductor surrounded by a shield. The center conductor is separated from the shield by a dielectric. The shield protects against electromagnetic interference.
- **Fiber optics.** Fiber optics uses the wide bandwidth properties of light traveling through transparent fibers. Fiber optics is a reliable communication medium best suited for point-to-point, high-speed data transmission. Fiber optics is immune to RF electromagnetic interference and does not produce electromagnetic radiation emission. The preferred DTM for an ESS is fiber-optic cables unless there are justifiable economic or technical reasons for using other types of media.
- **RF transmission.** Modulated RF can be used as a DTM with the installation of radio receivers and transmitters. An RF transmission system does not require a direct physical link between the points of communication, and it is useful for communicating over barriers such as bodies of water and heavily forested terrain. A disadvantage is that the signal power received depends on many factors (including transmission power, antenna pattern, path length, physical

obstructions, and climatic conditions). Also, RF transmission is susceptible to jamming and an adversary with an appropriately tuned receiver has access to it. The use of RF will be coordinated with the communications officer to avoid interference with other existing or planned facility RF systems.

6-140. There are two basic types of communication links—point-to-point and multiplex lines. A point-to-point link is characterized by a separate path for each pair of components. This approach is cost effective for several component pairs or when a number of scattered remote areas communicate with a single central location. The multiplex link, commonly referred to as a multidrop or multipoint link, is a path shared by a number of components. Depending on the number and location of components, this type of configuration can reduce the amount of cabling required. However, the cost reduction from reduced cabling is somewhat offset by costs of equipment required to multiplex and demultiplex data.

6-141. Data links used to communicate the status of ESS devices or other sensitive information to the security center must be protected from possible compromise. Attempts to defeat the security system may range from simple efforts to cut or short the transmission line to more sophisticated undertakings, such as tapping and substituting bogus signals. Data links can be made more secure by physical protection, tamper protection, line supervision, and encryption.

CCTV FOR ALARM ASSESSMENT AND SURVEILLANCE

6-142. A properly integrated CCTV assessment system provides a rapid and cost-effective method for determining the cause of intrusion alarms. For surveillance, a properly designed CCTV system provides a cost-effective supplement to guard patrols. For large facilities, the cost of a CCTV system is more easily justified. It is important to recognize that CCTV alarm-assessment systems and CCTV surveillance systems perform separate and distinct functions. The alarm-assessment system is designed to respond rapidly, automatically, and predictably to the receipt of ESS alarms at the security center. The surveillance system is designed to be used at the discretion of and under the control of the security center's console operator. When the primary function of the CCTV system is to provide real-time alarm assessment, the design should incorporate a video-processing system that can communicate with the alarm-processing system.

6-143. A candidate site for a CCTV assessment system will typically have the following characteristics:

- Assets requiring ESS protection.
- A need for real-time alarm assessment.
- Protected assets spaced some distance apart.

6-144. Figure 6-19, page 6-46, shows a typical CCTV system configuration. A typical site will locate CCTV cameras—

- Outdoors, along site-perimeter isolation zones.
- Outdoors, at controlled access points (sally ports).

- Outdoors, within the protected area, and at viewing approaches to selected assets.
- Indoors, at selected assets within the protected area.

6-145. The security console is centrally located in the security center. The CCTV monitors and the ancillary video equipment will be located at this console, as will the ESS alarm-processing and -annunciation equipment.

CCTV CAMERA COMPONENTS

6-146. An optical-lens system that captures and focuses reflected light from the scene being viewed onto an image target is common to all CCTV cameras. The image target converts reflected light energy into electrical impulses in a two-dimensional array of height and width. An electronic scanning system (reading these impulses in a predetermined order) creates a time-sensitive voltage signal that is a replica of optical information captured by the lens and focused on the target. This voltage signal is then transmitted to a location where it is viewed and possibly recorded. For components and technical information regarding CCTV cameras, see the appropriate TMs.

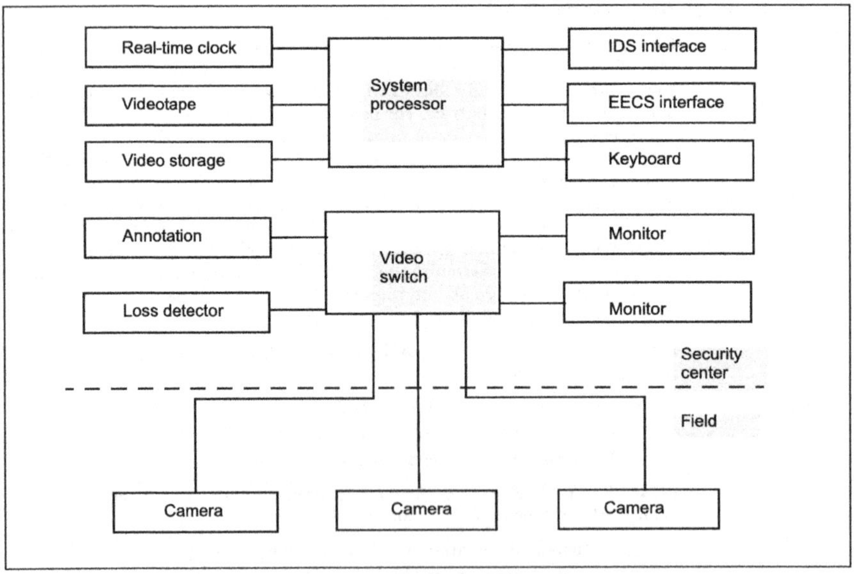

Figure 6-19. Typical CCTV System

VIDEO SIGNAL AND CONTROL LINKS

6-147. A CCTV transmission system is needed to convey video signals from various facility cameras to the security center and to carry commands from the security center to the cameras. Information may be sent via metallic cable, RF, or optical transmission.

Metallic Cable

6-148. Metallic video cables are electrical conductors manufactured specifically for the transmission of frequencies associated with video components. Coaxial cable is a primary example of this type of transmission media. Devices such as video-equalization amplifiers, ground loop correctors, and video-distribution amplifiers may be required.

RF Transmission

6-149. For a system that has widely separated nodes, RF transmission may be a good alternative to metallic cable and associated amplifiers. The information can be transmitted over a microwave link. A microwave link can be used for distances of about 50 miles, as long as the receiver and the transmitter are in the LOS.

Fiber-Optic Cable

6-150. In fiber-optic cable systems, electrical video signals are converted to optical light signals that are transmitted down the optical fiber. The signal is received and reconverted into electrical energy. An optic driver and a receiver are required per fiber. The fiber-optic transmission method provides a low-loss, high-resolution transmission system with usable length three to ten times that of traditional metallic in cable systems. Fiber-optic cable is the transmission media favored by DA.

CCTV-SYSTEM SYNCHRONIZATION

6-151. Timing signals are processed within the image-scan section of the CCTV camera. These signals may be generated internally from a crystal clock, derived from the camera's AC power source, or supplied by an external signal source. The camera should be capable of automatic switchover to its internal clock in case of external signal loss. When CCTV cameras are supplied by a common external (master) signal source or are all powered from the same AC power source, all cameras scan in synchronism. In this case, a console CCTV monitor will display a smooth transition when switched from one video source to another. Without this feature, the monitor display breaks up or rolls when switched between video sources. The rolling occurs for as long as it takes the monitor to synchronize its scan with that of the new video source, typically one second. The resynchronization delay will be experienced by all system components that receive video information, including recorders. To avoid this delay, the designer must specify that all cameras are powered from the AC power phase or must specify master synchronization for the design.

VIDEO PROCESSING AND DISPLAY COMPONENTS

6-152. As shown in Figure 6-19, page 6-46, CCTV camera signals propagate through the video transmission system and through coverage at the security center. In very simple configurations with only a few cameras and monitors, a hardwired connection between each camera and console monitor is adequate. As the number of cameras increases, the need to manage and add supplemental information to camera signals also increases. Psychological testing has demonstrated that the efficiency of console-operator assessment improves as the number of console monitors is reduced, with the optimum number being four to six monitors. Effectiveness is also enhanced by the use of alarm-correlated video. Major components of the video-processor system are the video switcher, the video-loss detector, the alarm-processor communication path, the master video-sync generator, video recorders, and monitors.

- **Video switchers.** Video switchers are required when the number of cameras exceeds the number of console monitors or when a monitor must be capable of selecting video from one of many sources. Video switchers are capable of presenting any of multiple video images to various monitors, recorders, and so forth.
- **Video-loss detector.** Video-loss detectors sense the continued integrity of incoming camera signals.
- **ESS interface and communication path.** There must be a means of rapid communication between the ESS alarm-annunciation and video-processor systems. The alarm processor must send commands that cause the video switcher to select the camera appropriate for the sensor reporting an alarm. The video-processor system must report system tampering or failures (such as loss of video) to the alarm processor. The path should also pass date-and-time synchronizing information between processors so that recorded video scenes and printed alarm logs are properly correlated.
- **Master video-sync generation and distribution.** Master video sync includes a crystal-controlled timing generator, distribution amplifiers, and a transmission link to each camera.
- **Video recorders.** Video recorders provide the means to record alarm-event scenes in real time for later analysis. A recorder typically receives its input through dedicated video-switcher outputs. To support recorder playback, the recorder output is connected to a dedicated switcher input and must be compatible with the switcher-signal format. In addition, the recorder receives start commands from the switcher, and compatibility must exist at this interface. Videocassette recorders should be used when alarm events are to be recorded for later playback and analysis. The cassettes can record in time lapse for up to 240 hours (depending on the user-selected speed) and will change to real-time recording on command. The cassettes can be erased and reused or archived if required.
- **Monitors.** Monitors are required to display the individual scenes transmitted from the cameras or from the video switcher. In alarm-assessment applications, the monitors are driven by dedicated outputs of the video switcher and the monitors display video sources selected by the switcher. For security-console operations, the 9-inch monitor is the

smallest screen that should be used for operator recognition of small objects in a camera's field of view. Two 9-inch monitors can be housed side by side in a standard 19-inch console. If the monitors are to be mounted in freestanding racks behind the security console, larger units will be used.

6-153. Video-processor equipment will be specified to append the following alphanumeric information so that it appears on both monitors and recordings. The equipment must allow the operator to program the annotated information and dictate its position on the screen. This information includes—

- Time and date information.
- Video-source or alarm-zone identification.
- Programmable titles.

CCTV APPLICATION GUIDELINES

6-154. Site-specific factors must be taken into consideration in selecting components that comprise a particular CCTV system. The first is the system's size in terms of the number of cameras fielded, which is the minimum number needed to view all ESS sensor-detection fields and surveillance cameras. Another factor is that some CCTV cameras may require artificial light sources. Finally, there are CCTV-system performance criteria and physical, environmental, and economic considerations. Each is discussed in detail in TM 5-853-4.

Scene Resolution

6-155. The level to which video details can be determined in a CCTV scene is referred to as resolving ability or resolution. It is generally accepted that for assessment purposes, three resolution requirements can be defined. In order of increasing resolution requirements, they are detection, recognition, and identification.

- Detection is the ability to detect the presence of an object in a CCTV scene.
- Recognition is the ability to determine the type of object in a CCTV scene (animal, blowing debris, or crawling human).
- Identification is the ability to determine object details (a particular person, a large rabbit, a small deer, or tumbleweed).

6-156. A CCTV assessment system should provide sufficient resolution to recognize human presence and to detect small animals or blowing debris. Given an alarmed intrusion sensor, it is crucial that the console operator be able to determine if the sensor detected an intruder or if it is simply responding to a nuisance condition. (Refer to TM 5-853-4 for detailed design applications.)

Illumination Levels

6-157. For interior applications where the same camera type is used in several different areas and the scene illumination in each area is constant, specify the manually adjustable iris. This allows a manual iris adjustment appropriate for each particular area's illumination level at the time of installation. If the camera must operate in an area subject to a wide dynamic range of illumination levels (such as would be found outdoors), specify the automatically adjusted iris feature.

Cost Considerations

6-158. The cost of a CCTV system is usually quoted as cost-per-assessment zone. When estimating the total system cost, video-processor equipment costs and the video-transmission system's costs must be included. Other potentially significant costs are outdoor lighting system upgrades and the site preparation required to support the CCTV cameras. The CCTV systems are expensive compared to other electronic security subsystems and should be specified with discretion.

DESIGN GUIDELINES

6-159. The design and application of CCTV systems are quite complex and should be left to professionals who are abreast of the current state-of-the-art systems. Some of the general design guidelines include the following:

- **System familiarity.** Before designing an effective CCTV assessment system, the designer must be familiar with the ESS's sensor placement and the detection field's shape.
- **CCTV camera placement and lighting.** The placement of exterior cameras requires more attention than that of interior cameras because of weather and illumination extremes. The field-of-view alignment, illumination range, and balanced lighting are major design factors. Exterior CCTV design considerations include environmental housings, camera mounting heights, system types, and so forth. Indoor design considerations include the mounting location and tamper detection. The layout for indoor alarm-assessment cameras is subject to three constraints—
  - The camera's location should enclose the complete sensor detection field in the camera's field of view.
  - Lighting that is adequate to support alarm assessment will be provided.
  - Protection from tampering and inadvertent damage by collision during normal area operations will be provided.

Chapter 7

Access Control

Perimeter barriers, intrusion-detection devices, and protective lighting provide physical-security safeguards; however, they alone are not enough. An access-control system must be established and maintained to preclude unauthorized entry. Effective access-control procedures prevent the introduction of harmful devices, materiel, and components. They minimize the misappropriation, pilferage, or compromise of materiel or recorded information by controlling packages, materiel, and property movement. Access-control rosters, personal recognition, ID cards, badge-exchange procedures, and personnel escorts all contribute to an effective access-control system.

DESIGNATED RESTRICTED AREAS

7-1. The installation commander is responsible for designating and establishing restricted areas. A restricted area is any area that is subject to special restrictions or controls for security reasons. This does not include areas over which aircraft flight is restricted. Restricted areas may be established for the following:

- The enforcement of security measures and the exclusion of unauthorized personnel.
- Intensified controls in areas requiring special protection.
- The protection of classified information or critical equipment or materials.

DEGREE OF SECURITY

7-2. The degree of security and control required depends on the nature, sensitivity, or importance of the security interest. Restricted areas are classified as controlled, limited, or exclusion areas.

- A controlled area is that portion of a restricted area usually near or surrounding a limited or exclusion area. Entry to the controlled area is restricted to personnel with a need for access. Movement of authorized personnel within this area is not necessarily controlled since mere entry to the area does not provide access to the security interest. The controlled area is provided for administrative control, for safety, or as a buffer zone for in-depth security for the limited or exclusion area. The commander establishes the control of movement.
- A limited area is a restricted area within close proximity of a security interest. Uncontrolled movement may permit access to the item. Escorts and other internal restrictions may prevent access within limited areas.

- An exclusion area is a restricted area containing a security interest. Uncontrolled movement permits direct access to the item.

7-3. The security protection afforded by a restricted area pertains particularly to subversive-activity control; that is, protection against espionage, sabotage, or any such action adversely affecting national defense. Within this context, the designation "restricted area" is not applicable to an area solely for protection against common pilferage or misappropriation of property or material that is not classified or not essential to national defense. For example, an area devoted to the storage or use of classified documents, equipment, or materials should be designated as a restricted area to safeguard against espionage. An installation communications center should also be so designated to safeguard against sabotage. On the other hand, a cashier's cage or an ordinary mechanic's tool room should not be so designated, although the commander may impose controls to access. This may be a simple matter of posting an "off limits to unauthorized personnel" sign. The PM or the physical-security manager acts as an advisor to the commander. In his recommendations, he must consider evaluating the purpose of designating a restricted area and coordinating with the intelligence officer and the staff judge advocate (SJA).

7-4. A restricted area must be designated in writing by the commander and must be posted with warning signs according to AR 190-13. In areas where English is one of two or more languages commonly spoken, warning signs will be posted in English and in the local language (see Figure 7-1).

7-5. An installation may have varying degrees of security. It may be designated in its entirety as a restricted area, with no further restrictions; or it may be subdivided into controlled, limited, or exclusion areas with restrictions of movement and specific clear zones. Figure 7-2 depicts a simplified restricted area and the degrees of security.

CONSIDERATIONS

7-6. There are other important considerations concerning restricted areas and their lines of division. These considerations include the following:

- A survey and analysis of the installation, its missions, and its security interests. This can determine immediate and anticipated needs that require protection. Anticipated needs are determined from plans for the future.
- The size and nature of the security interest being protected. Safes may provide adequate protection for classified documents and small items; however, large items may have to be placed within guarded enclosures.
- Some security interests are more sensitive to compromise than others. Brief observation or a simple act by an untrained person may constitute a compromise in some cases. In others, detailed study and planned action by an expert may be required.
- All security interests should be evaluated according to their importance. This may be indicated by a security classification such as confidential, secret, or top secret.

RESTRICTED AREA

THIS (INSTALLATION OR ACTIVITY) HAS BEEN DECLARED A RESTRICTED AREA BY AUTHORITY OF (TITLE, COMMANDING GENERAL OR COMMANDING OFFICER) IN ACCORDANCE WITH THE PROVISIONS OF THE DIRECTIVE ISSUED BY THE SECRETARY OF DEFENSE ON 20 AUGUST 1954, PURSUANT TO THE PROVISIONS OF SECTION 21, INTERNAL SECURITY ACT OF 1950. UNATHORIZED ENTRY IS PROHIBITED.

ALL PERSONS AND VEHICLES ENTERING HEREIN ARE LIABLE TO SEARCH. PHOTOGRAPHING OR MAKING NOTES, DRAWINGS, MAPS, OR GRAPHIC REPRESENTATIVES OF THIS AREA OR ITS ACTIVITIES ARE PROHIBITED UNLESS SPECIFICALLY AUTHORIZED BY THE COMMANDER. ANY SUCH MATERIAL FOUND IN THE POSSESSION OF UNAUTHORIZED PERSONS WILL BE CONFISCATED.

Figure 7-1. Sample Restricted-Area Warning

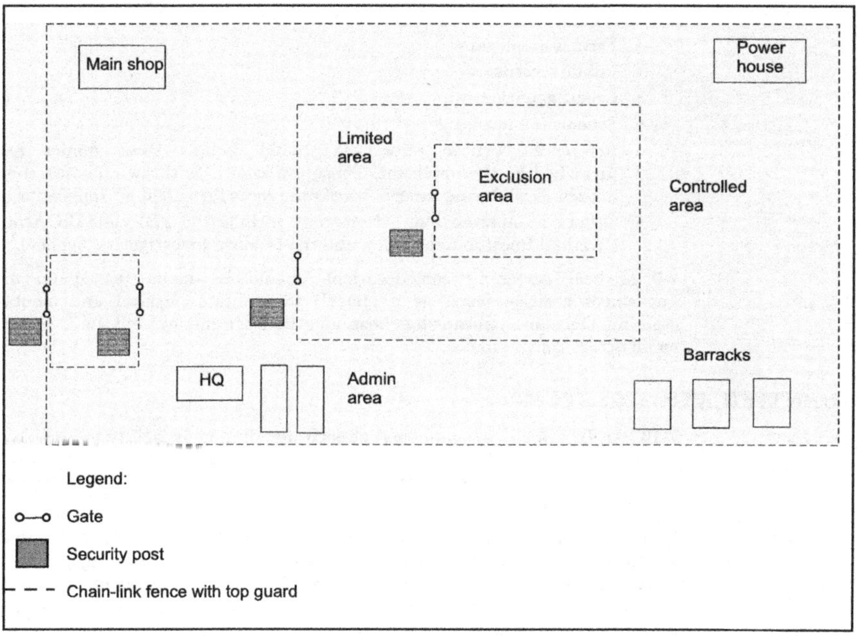

Figure 7-2. Schematic Diagram of a Simplified Restricted Area and the Degrees of Security

- Parking areas for privately owned vehicles (POVs) are established outside of restricted areas. Vehicle entrances must be kept at a minimum for safe and efficient control.
- Physical protective measures (such as fences, gates, and window bars) must be installed.

EMPLOYEE SCREENING

7-7. Screening job applicants to eliminate potential acts of espionage and sabotage and other security risks is important in peacetime and is critical during a national emergency. Personnel screenings must be incorporated into standard personnel policies.

7-8. An applicant should be required to complete a personnel security questionnaire, which is then screened for completeness and used to eliminate undesirable applicants. A careful investigation should be conducted to ensure that the applicant's character, associations, and suitability for employment are satisfactory. The following sources may be helpful in securing employment investigative data:

- State and local police (including national and local police in overseas areas).
- Former employers.
- Public records.
- Credit agencies.
- Schools (all levels).
- References. (These references should include those names not furnished by the applicant. These are known as throw offs, and they are obtained during interviews of references furnished by applicants.)
- Others as appropriate. (These may include the FBI, the US Army Criminal Records Repository, and the Defense Investigative Agency).

7-9. Medical screening considerations should be made (based on an applicant's position [such as a guard]) to evaluate physical and mental stamina. Once an applicant has been identified for employment, he is placed on an access-control roster.

IDENTIFICATION SYSTEM

7-10. An ID system is established at each installation or facility to provide a method of identifying personnel. The system provides for personal recognition and the use of security ID cards or badges to aid in the control and movement of personnel activities.

7-11. Standard ID cards are generally acceptable for access into areas that are unrestricted and have no security interest. Personnel requiring access to restricted areas should be issued a security ID card or badge as prescribed in AR 600-8-14. The card's/badge's design must be simple and provide for adequate control of personnel.

7-12. A security ID card/badge system must be established for restricted areas with 30 or more employees per shift. Commanders may (at their discretion) authorize a card/badge system in restricted areas for less than 30 people.

ID METHODS

7-13. Four of the most commonly used access-control ID methods are the personal-recognition system, the single-card or -badge system, the card- or badge-exchange system, and the multiple-card or -badge system.

Personal-Recognition System

7-14. The personal-recognition system is the simplest of all systems. A member of the security force providing access control visually checks the person requesting entry. Entry is granted based on—

- The individual being recognized.
- The need to enter has been established.
- The person is on an access-control roster.

Single-Card or -Badge System

7-15. This system reflects permission to enter specific areas by the badge depicting specific letters, numbers, or particular colors. This system lends to comparatively loose control and is not recommended for high-security areas. Permission to enter specific areas does not always go with the need to know. Because the ID cards/badges frequently remain in the bearer's possession while off duty, it affords the opportunity for alteration or duplication.

Card- or Badge-Exchange System

7-16. In this system, two cards/badges contain identical photographs. Each card/badge has a different background color, or one card/badge has an overprint. One card/badge is presented at the entrance to a specific area and exchanged for the second card/badge, which is worn or carried while in that area. Individual possession of the second card/badge occurs only while the bearer is in the area for which it was issued. When leaving the area, the second card/badge is returned and maintained in the security area. This method provides a greater degree of security and decreases the possibility of forgery, alteration, or duplication of the card/badge. The levels of protection described in TM 5-853-1 require multiple access-control elements as the levels of protection increase. In the case of the badge exchange, this system counts as two access-control elements.

Multiple-Card or -Badge System

7-17. This system provides the greatest degree of security. Instead of having specific markings on the cards/badges denoting permission to enter various restricted areas, the multiple card/badge system makes an exchange at the entrance to each security area. The card/badge information is identical and allows for comparisons. Exchange cards/badges are maintained at each area only for individuals who have access to the specific area.

MECHANIZED/AUTOMATED SYSTEMS

7-18. An alternative to using guards or military police (MP) to visually check cards/badges and access rosters is to use building card-access systems or biometric-access readers. These systems can control the flow of personnel entering and exiting a complex. Included in these systems are—

- Coded devices such as mechanical or electronic keypads or combination locks.
- Credential devices such as magnetic-strip or proximity card readers.
- Biometric devices such as fingerprint readers or retina scanners.

7-19. Access-control and ID systems base their judgment factor on a remote capability through a routine discriminating device for positive ID. These systems do not require guards at entry points; they identify an individual in the following manner:

- The system receives physical ID data from an individual.
- The data is encoded and compared to stored information.
- The system determines whether access is authorized.
- The information is translated into readable results.

7-20. Specialized mechanical systems are ideal for highly sensitive situations because they use a controlled process in a controlled environment to establish the required database and accuracy. One innovative technique applied to ID and admittance procedures involves dimension comparisons. The dimension of a person's full hand is compared to previously stored data to determine entry authorization. Other specialized machine readers can scan a single fingerprint or an eye retina and provide positive ID of anyone attempting entry.

7-21. An all-inclusive automated ID and access-control system reinforces the security in-depth ring through its easy and rapid change capability. The computer is able to do this through its memory. Changes can be made quickly by the system's administrator.

7-22. The commercial security market has a wide range of mechanized and automated hardware and software systems. Automated equipment is chosen only after considering the security needs and the environment in which it operates. These considerations include whether the equipment is outdoors or indoors, the temperature range, and weather conditions. Assessment of security needs and the use of planning, programming, and budgeting procedures greatly assist a security manager in improving the security posture.

CARD/BADGE SPECIFICATIONS

7-23. Security cards/badges should be designed and constructed to meet the requirements of AR 600-8-14. Upon issuing a card/badge, security personnel must explain to the bearer the wear required and the authorizations allowed with the card/badge. This includes—

- Designation of the areas where an ID card/badge is required.

- A description of the type of card/badge in use and the authorizations and limitations placed on the bearer.
- The required presentation of the card/badge when entering or leaving each area during all hours of the day.
- Details of when, where, and how the card/badge should be worn, displayed, or carried.
- Procedures to follow in case of loss or damage of the card.
- The disposition of the card/badge upon termination of employment, investigations, or personnel actions.
- Prerequisites for reissuing the card/badge.

VISITOR IDENTIFICATION AND CONTROL

7-24. Procedures must be implemented to properly identify and control personnel. This includes visitors presenting their cards/badges to guards at entrances of restricted areas. Visitors are required to stay with their assigned escort. Guards must ensure that visitors stay in areas relating to their visit; an uncontrolled visitor, although conspicuously identified, could acquire information for which he is not authorized. Foreign-national visitors should be escorted at all times.

7-25. Approval for visitors should be obtained at least 24 hours in advance (if possible). Where appropriate, the installation should prepare an agenda for the visitor and designate an escort officer. Measures must be in place to recover visitor cards/badges on the visit's expiration or when they are no longer required.

7-26. Physical-security precautions against pilferage, espionage, and sabotage require the screening, ID, and control of visitors. Further information about visiting requirements and procedures are found in ARs 12-15 and 381-20. Visitors are generally classed in the following categories:

- Persons with whom every installation or facility has business (such as suppliers, customers, insurance inspectors, and government inspectors).
- Individuals or groups who desire to visit an installation or facility for personal or educational reasons. Such visits may be desired by educational, technical, or scientific organizations.
- Individuals or groups specifically sponsored by the government (such as foreign nationals visiting under technical cooperation programs and similar visits by US nationals). Requests for visits by foreign nationals must be processed according to AR 380-10.
- Guided tours to selected portions of the installation in the interest of public relations.

7-27. The ID and control mechanisms for visitors must be in place. They may include the following:

- Methods of establishing the authority for admitting visitors and any limitations relative to access.

- Positive ID of visitors by personal recognition, visitor permit, or other identifying credentials. Contact the employer, supervisor, or officer in charge to validate the visit.
- The use of visitor registration forms. These forms provide a record of the visitor and the time, location, and duration of his visit.
- The use of visitor ID cards/badges. The cards/badges bear serial numbers, the area or areas to which access is authorized, the bearer's name, and escort requirements.

7-28. Individual groups entering a restricted area must meet specific prerequisites before being granted access. The following guidance is for group access into a restricted area:

Visitors

7-29. Before allowing visitors into a restricted area, contact the person or activity being visited. After verifying the visitor's identity, issue a badge, complete the registration forms, and assign an escort (if required). Visitors may include public-utility and commercial-service representatives.

Very Important Persons

7-30. The procedures for admitting very important persons (VIPs) and foreign nationals into restricted areas are contained in AR 12-15. Special considerations and coordination with the protocol office are necessary. A 24-hour advance notice is desirable for these requests, along with an agenda for the visit and the designation of an escort, if appropriate.

Civilians Working on Jobs Under Government Contract

7-31. To allow these personnel to conduct business in restricted areas, the security manager must coordinate with the procurement office. The security manager must also identify movement-control procedures for these employees.

Cleaning Teams

7-32. Supervisors using cleaning teams must seek technical advice from the physical-security office on internal controls for each specific building. This may include providing escorts.

DOD Employees in Work Areas After Normal Operating Hours

7-33. Supervisors establish internal controls based on coordination with the security manager. They also notify security personnel of the workers' presence, type, and duration of work.

ENFORCEMENT MEASURES

7-34. The most vulnerable link in any ID system is its enforcement. Security forces must be proactive in performing their duties. A routine performance of duty will adversely effect even the most elaborate system. Positive enforcement measures must be prescribed to enhance security. Some of these measures may include—

- Designating alert and tactful security personnel at entry control points.
- Ensuring that personnel possess quick perception and good judgment.
- Requiring entry-control personnel to conduct frequent irregular checks of their assigned areas.
- Formalizing standard procedures for conducting guard mounts and posting and relieving security personnel. These measures will prevent posting of unqualified personnel and a routine performance of duty.
- Prescribing a uniform method of handling or wearing security ID cards/badges. If carried on the person, the card must be removed from the wallet (or other holder) and handed to security personnel. When worn, the badge will be worn in a conspicuous position to expedite inspection and recognition from a distance.
- Designing entry and exit control points of restricted areas to force personnel to pass in a single file in front of security personnel. In some instances, the use of turnstiles may be advisable to assist in maintaining positive control.
- Providing lighting at control points. The lighting must illuminate the area to enable security personnel to compare the bearer with the ID card/badge.
- Enforcing access-control measures by educating security forces and employees. Enforcement of access-control systems rests primarily with the security forces; however, it is essential that they have the full cooperation of the employees. Employees must be instructed to consider each unidentified or improperly identified individual as a trespasser. In restricted areas where access is limited to a particular zone, employees must report unauthorized individuals to the security force.
- Positioning ID card/badge racks or containers at entry control points so that they are accessible only to guard-force personnel.
- Appointing a responsible custodian to accomplish control procedures of cards/badges according to AR 600-8-14. The custodian is responsible for the issue, turn in, recovery, and renewal of security ID cards/badges.

7-35. The degree of compromise tolerable in the ID system is in direct proportion to the degree of security required. The following control procedures are recommended for preserving the integrity of a card/badge system:

- Maintenance of an accurate written record or log listing (by serial number) all cards and badges and showing those on hand, to whom they are issued, and their disposition (lost, mutilated, or destroyed).
- Authentication of records and logs by the custodian.
- A periodic inventory of records by a commissioned officer.
- The prompt invalidation of lost cards/badges.
- The conspicuous posting at security control points of current lists of lost or invalidated cards/badges.
- The establishment of controls within restricted areas to enable security personnel to determine the number of persons within the area.
- The establishment of the two-person rule (when required).

- The establishment of procedures to control the movement of visitors. A visitor-control record will be maintained and located at entry control points.

SIGN/COUNTERSIGN AND CODE WORD

7-36. This method of verifying identity is primarily used in a tactical environment. According to the local SOP, the sign/countersign or code-word procedures must be changed immediately if compromised.

DURESS CODE

7-37. The duress code is a simple word or phrase used during normal conversation to alert other security personnel that an authorized person is under duress. A duress code requires planning and rehearsal to ensure an appropriate response. This code is changed frequently to minimize compromise.

ACCESS-CONTROL ROSTERS

7-38. Admission of personnel to a restricted area is granted to those identified and listed on an access-control roster. Pen-and-ink changes may be made to the roster. Changes are published in the same manner as the original roster.

7-39. Rosters are maintained at access control points. They are kept current, verified, and accounted for by an individual designated by the commander. Commanders or their designated representatives authenticate the rosters. Admission of persons other than those on the rosters is subject to specific approval by the security manager. These personnel may require an escort according to the local SOP.

METHODS OF CONTROL

7-40. There are a number of methods available to assist in the movement and control of personnel in limited, controlled, and restricted areas. The following paragraphs discuss the use of escorts and the two-person rule:

ESCORTS

7-41. Escorts are chosen because of their ability to accomplish tasks effectively and properly. They possess knowledge of the area being visited. Escorts may be guard-force personnel, but they are normally personnel from the area being visited. Local regulations and SOPs determine if a visitor requires an escort while in the restricted area. Personnel on the access list may be admitted to restricted areas without an escort.

TWO-PERSON RULE

7-42. The two-person rule is designed to prohibit access to sensitive areas or equipment by a lone individual. Two authorized persons are considered present when they are in a physical position from which they can positively detect incorrect or unauthorized procedures with respect to the task or operation being performed. The team is familiar with applicable safety and security requirements, and they are present during any operation that affords access to sensitive areas or equipment that requires the two-person rule. When application of the two-person rule is required, it is enforced constantly by the personnel who constitute the team.

7-43. The two-person rule is applied in many other aspects of physical-security operations, such as the following:

- When uncontrolled access to vital machinery, equipment, or materiel might provide opportunity for intentional or unintentional damage that could affect the installation's mission or operation.
- When uncontrolled access to funds could provide opportunity for diversion by falsification of accounts.
- When uncontrolled delivery or receipt for materials could provide opportunity for pilferage through "short" deliveries and false receipts.
- When access to an arms or ammunition storage room could provide an opportunity for theft. Keys should be issued so that at least two people must be present to unlock the locks required under the provisions of AR 190-11.

7-44. The two-person rule is limited to the creativity of the PM and the physical-security manager. They should explore every aspect of physical-security operations in which the two-person rule would provide additional security and assurance and include all appropriate recommendations and provisions of the physical-security plan. An electronic-entry control system may be used to enforce the two-person rule. The system can be programmed to deny access until two authorized people have successfully entered codes or swiped cards.

SECURITY CONTROLS OF PACKAGES, PERSONAL PROPERTY, AND VEHICLES

7-45. A good package-control system helps prevent or minimize pilferage, sabotage, and espionage. The local SOP may allow the entry of packages with proper authorization into restricted areas without inspection. A package-checking system is used at the entrance gate. When practical, inspect all outgoing packages except those properly authorized for removal. When a 100 percent inspection is impractical, conduct frequent unannounced spot checks. A good package-control system assists in the movement of authorized packages, material, and property.

7-46. Property controls are not limited to packages carried openly, but they include the control of anything that could be used to conceal property or material. Personnel should not be routinely searched except in unusual situations. Searches must be performed according to the local SOP.

7-47. All POVs on the installation should be registered with the PM or the installation's physical-security office. Security personnel should assign a temporary decal or other temporary ID tag to visitors' vehicles to permit ready recognition. The decal or the tag should be distinctly different from that of permanent-party personnel.

7-48. When authorized vehicles enter or exit a restricted area, they undergo a systematic search, including (but not limited to) the—

- Vehicle's interior.
- Engine compartment.
- External air breathers.
- Top of the vehicle.
- Battery compartment.
- Cargo compartment.
- Undercarriage.

7-49. The movement of trucks and railroad cars into and out of restricted areas should be supervised and inspected. Truck and railroad entrances are controlled by locked gates when not in use and are manned by security personnel when unlocked. The ID cards/badges are issued to operators to ensure proper ID and registration for access to specific loading and unloading areas.

7-50. All conveyances entering or leaving a protected area are required to pass through a service gate manned by security forces. Drivers, helpers, passengers, and vehicle contents must be carefully examined. The examination may include—

- Appropriate entries in the security log (including the date, operator's name, load description, and time entered and departed).
- A check of the operator's license.
- Verification of the seal number with the shipping document and examination of the seal for tampering.

7-51. Incoming trucks and railroad cars must be assigned escorts before they are permitted to enter designated limited or exclusion areas. Commanders should establish published procedures to control the movement of trucks and railroad cars that enter designated restricted areas to discharge or pick up cargo (escorts will be provided when necessary).

7-52. The best control is provided when all of these elements are incorporated into access-control procedures. Simple, understandable, and workable access-control procedures are used to achieve security objectives without impeding operations. When properly organized and administered, access-control procedures provide a method of positively identifying personnel who have the need to enter or leave an area.

TACTICAL-ENVIRONMENT CONSIDERATIONS

7-53. Access-control procedures during tactical operations may establish additional challenges for the commander. In some instances, the commander cannot provide a perimeter barrier (such as a fence) based on METT-TC. Commanders are still required to provide security measures for restricted areas, although they may not always have the necessary assets. Early-warning systems and the use of guards become crucial. A restricted area may become a requirement without prior notice during an operation. Figure 7-3 and Figure 7-4, page 7-14, are examples of temporary tactical restricted and exclusion areas.

7-54. Commanders must plan for these considerations when developing their budget. Funding must be requested and set aside to support physical-security requirements during tactical operations. Resources will not always be available; therefore, commanders must implement procedures that support access-control measures. Improvising will become common practice to overcome shortfalls concerning access-control equipment in the field.

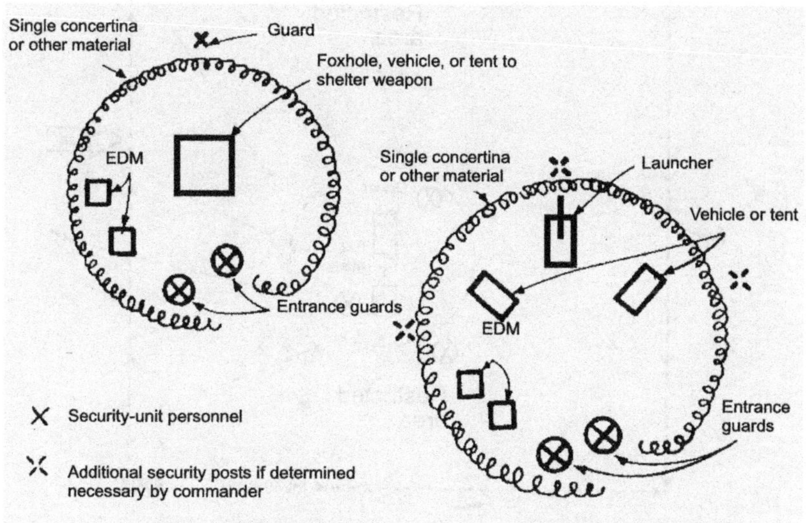

NOTE: Equipment, personnel, and positioning is according to the unit's SOP.

Figure 7-3. Sample Layout of Temporary Tactical Restricted Areas

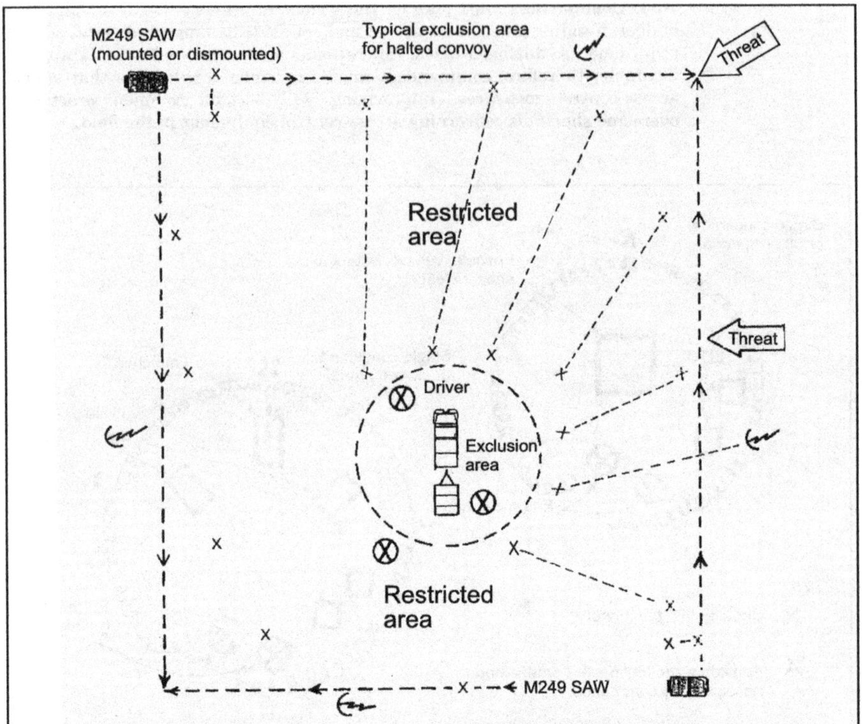

NOTE: Other vehicles and personnel, to include the LPs, form a defensive perimeter inside the primary directions of fire of the M60 machine guns. During attack, alternate machine-gun crews should be placed in the perimeter away from primary machine guns and be spaced so that they can be used as replacements and so that large gaps do not appear in the perimeter.

Legend:

X — MP rifleman

◄ – – — Principal direction of fire

( ) — Temporary exclusion area (size depends on terrain and threat force)

— Listening post

Figure 7-4. Sample Layout for a Temporary Tactical Exclusion Area

Chapter 8

Lock and Key Systems

Locks are the most acceptable and widely used security devices for protecting facilities, classified materials, and property. All containers, rooms, and facilities must be locked when not in actual use. Regardless of their quality or cost, locks are considered delay devices only. Some locks require considerable time and expert manipulation to open, but all locks can be defeated by force and with the proper tools. Locks must never be considered as a stand-alone method of security.

INSTALLATION AND MAINTENANCE

8-1. The USACE is responsible for installing locking devices in newly constructed facilities. Installation-level engineers are responsible for maintaining the locking devices. Physical-security personnel must work closely with engineer personnel to ensure that locks meet the standards and are installed according to applicable regulations. One source of assistance and information is the DOD Lock Program Technical Support Hotline at the Naval Facilities Engineering Services Center, Port Hueneme, California.

TYPES OF LOCKING DEVICES

8-2. The degree of protection afforded by a vault, a safe, or a filing cabinet may be measured in terms of the lock's resistance. Locking devices are listed in TM 5-805-8. Types of locking devices include key and combination locks.

8-3. ARs 190-11, 190-51, 50-5, and 50-6 prescribe specific types of locks for specific types of facilities. AR 380-5 prescribes standard facilities for storing classified material and contains guidance for different storage requirements.

KEY LOCKS

8-4. Key locks consist of, but are not limited to, the following:

- Cylindrical locksets are often called key-in-knob or key-in-lever locks. They are normally used to secure offices and storerooms. The locking cylinder located in the center of the doorknob distinguishes these locks. Some cylindrical locksets have keyways in each of the opposing knobs that require a key on either side to lock and unlock them. Others unlock with a key, but may be locked by pushing or rotating a button on the inside knob. These locks are suitable only for very low-security applications. Using these locks may require compensatory measures in the form of additional locks on containers within the room.
- Dead-bolt locks are sometimes called tubular dead bolts. They are mounted on the door in a manner similar to cylindrical locksets. The primary difference is in the bolt. When the bolt is extended (locked),

the dead bolt projects into the doorframe at least one inch, and it cannot be forced back (unlocked) by applying pressure to the end of the bolt. The dead-bolt lock has the potential for providing acceptable levels of protection for storerooms and other areas where more security is desired. It is recommended for use in military housing as an effective security measure in the installation's crime-prevention program. In situations where there is a window in or adjacent to the door, a double-cylinder dead-bolt lock (one that requires a key to open from either side) should be used.

- Mortise locks are so named because the lock case is mortised or recessed into the edge of the door. The most common variety of mortise locks has a doorknob on each side of the door. Entrance doors often have an exterior thumb latch rather than a doorknob. The mortise lock can be locked from inside by means of a thumb turn or by a button. Mortise locks are considered low-security devices since they weaken the door in the mortised area.
- Drop-bolt locks (often referred to as jimmy-proof locks) are normally used as auxiliary locks similar to dead bolts. Both the drop-bolt lock body and the strike have interlocking leaves similar to a door hinge. When closed, locking pins in the lock body drop down into the holes provided in the strike and secure the locking system. Since the lock body and the strike are interconnected with locking pins when closed, the lock essentially becomes a single unit and is extremely difficult to separate.
- Rim-cylinder locks are mounted to the door's inside surface and are secured by screws in the door face. These locks are generally used with drop-bolt and other surface-mounted locks and latches. They consist of an outer barrel, a cylinder and ring, a tailpiece, a back mounting plate, and two mounting screws. The tailpiece screws are usually scored so that the lock can be tailored to fit varying door thicknesses.
- Unit locks are ideal in heavily traveled facilities (such as hospitals or institutional buildings). These locks are a complete, one-piece unit that slides into a notch cut into the door's latch edge. The one-size cutout of the door edge simplifies the door preparation for the lock.
- Mechanical, push-button combination locks are digital (push buttons numbered 1 through 9) combination door-locking devices used to deny area access to any individual not authorized or cleared for a specific area. These locks are normally used for access control and should be backed up by door locking devices when the facility is unoccupied.
- Padlocks are detachable locks that are typically used with a hasp. Low-security padlocks, sometimes called secondary padlocks, are used to deter unauthorized access, and they provide only minimal resistance to force. Low-security locks are made with hardened steel shackles. Precautions must be taken to avoid confusing these locks with similar brass or bronze locks. The brass or bronze locks are commonly used but do not meet the security requirements of the hardened shackled locks. High-security padlocks may be used to secure AA&E. They provide the maximum resistance to unauthorized entry when used with a high-security hasp.

8-5. Some locks have interchangeable cores, which allow the same key system to include a variety of locks. Padlocks, door locks, cabinet locks, and electrical key switches can all be operated by the same key system. Because these cores are removable by a special key, this system allows for rapid rekeying of locks in the event that the key is compromised.

8-6. Locks are keyed in several different ways. When several locks are keyed differently, each is operated by its own key. When they are keyed alike, one key will open them all. Locks that are master-keyed are keyed differently, yet have one key that will open them all. Master-keying is done for convenience and represents the controlled loss of security. Master-keying is not used unless permitted by regulation.

COMBINATION LOCKS

8-7. Combination locks are available as padlocks or as mounted locks. They are low-security padlocks with combinations that are either fixed or changeable. Combination locks may be either mechanical or electronic. They are operated by entering a particular sequence of numbers. When the correct combination is entered, the lock's bolt is retracted. Combination locks used for securing classified material must meet Federal Specification FF-L-2740.

8-8. Although the lock is the most accepted and widely used security device, it is only a delay device and should never be considered as a positive bar to entry. A lock can (and will) be defeated. The best defense for locking devices is a good key-control program. Refer to AR 190-51, Appendix D, for standard key and lock procedures. Additional key and lock procedures for AA&E can be found in AR 190-11, Chapter 3.

Chapter 9

Security Forces

The security force for an installation or a facility provides the enforcement element in the physical-security program. This force consists of personnel specifically organized, trained, and equipped to protect the command's physical-security interests. It is a commander's most effective tool in a comprehensive, integrated, physical-security program. Vulnerability tests are periodically conducted to determine and ensure the state of readiness of security forces (see Appendix K).

TYPES OF SECURITY FORCES

9-1. On installations, security forces may be MP forces, security police, DOD civil-service security guards, or contract guards. Interior guard duties are performed by installation unit troops on a roster basis. MP forces normally perform security duties that require higher degrees of training and experience. These include—

- Security of restricted areas.
- Security of specific sensitive gates.
- Supervisory or coordinated roles with other military or DOD civil-service security guards.
- Responsibility for monitoring and responding to intrusion alarms.

9-2. An MP unit may perform the entire physical-security function alone based on METT-TC, the area, and the facilities. When an MP unit cannot assume responsibilities for all of the physical-security requirements in the command, other forces may be required. Additional forces may consist of the following:

- Personnel furnished by units of the installation's command on a daily or weekly basis. While this method has the single advantage of providing additional manpower, it has the disadvantages of rapid turnover and the lack of training. If this manpower is used, personnel should be assigned the least sensitive posts or patrols. For extended augmentation, units may be attached to MP units. The MP unit may also be augmented by reserve units or units in rotation.
- The combat-arms branches (especially the infantry) may attach their forces to MP units and may be designated as security guards assisting in the required operations.
- Military or paramilitary units of the host country may also be attached to or operate in coordination with MP forces. They may also be supplemented with national police of their own country.

- The installation's band may be a source of military force during wartime. (The band is assigned enemy-prisoner-of-war [EPW] duty as a wartime duty.) The band is doctrinally capable of providing security at the division tactical operations center (DTOC) and ASPs, assisting in the perimeter defense of the command post (CP), and operating the dismount point for the CP. It is capable of providing access control at the DTOC and the ASPs and augmenting or relieving security personnel on the defensive perimeter.

9-3. Civil-service security guards are uniformed civilian employees from a government agency. They are customarily trained and organized along military lines. The organization may be completely civil service or may be composed of civil-service personnel under military supervision. In either case, they are under the operational control of the PM or the security officer.

9-4. Labor-service personnel (local civilian personnel) have been organized and used successfully in theaters of operation. These types of units were organized after World War II and since that time have established enviable records in the physical-security field. They are distinctively uniformed, organized, and equipped. They have set and maintained the highest security standards, resulting in a minimal loss of property. While not military organizations, these units have successfully developed a high sense of duty and esprit de corps that has been reflected in their outstanding contributions to the physical security of installations in overseas commands.

AUTHORITY AND JURISDICTION

9-5. It is most important that the PM or the security officer determine (and instruct his security force in) the extent and limitations of the commander's jurisdiction in the field of law enforcement and investigations. Those jurisdictions include—

- Jurisdiction of place.
  - Military installations and facilities. Whether state or federal law or both are applicable on a military installation or facility depends largely on the nature of jurisdiction over the land involved. The amount of federal jurisdiction may vary between different areas of the installation or facility. The legal formalities of acquiring jurisdiction over land under the control of the Secretary of the Army are accomplished at DA level and according to the provisions of AR 405-20. Information and advice relating to jurisdictional questions should be referred to the local SJA.
  - Areas outside of military installations. Areas outside of military installations are generally subject to state and local laws; however, there are exceptions. Information and advice in this regard should be obtained through the local SJA.
  - Overseas areas. In overseas areas, jurisdiction varies according to the military situation and existing international treaties, contracts, and agreements. Guidance should be obtained in each instance from the commander and the SJA and set forth in appropriate command directives.

- Jurisdiction of personnel.
  - Jurisdiction of personnel generally follows the limitations of jurisdiction of the installation.
  - MP forces have jurisdiction and authority over personnel as described in AR 190-14 and related publications.
  - Authority for federal civilian employees assigned to security, police, and guard duties is derived from the installation's commanding officer. These personnel can have no more authority than he possesses and are subject to any limitations imposed thereon.
  - Security-force personnel may enforce all offenses under the Uniform Code of Military Justice (UCMJ), military regulations, federal laws and regulations, and state laws where applicable.
  - Security-force personnel may be given the same authority as MP forces over all personnel subject to military jurisdiction, including apprehension, detention, and search.
  - Civilian security-force personnel have no specific grant of authority over civilians other than the right of citizen's arrest.
  - The commander is the source of jurisdiction and authority for all other personnel assigned to security-force duties.

PERSONNEL SELECTION

9-6. Regardless of the use of structural, mechanical, or electronic equipment, the human element in security operations makes the difference between success or failure. Commanders and supervisors have a responsibility to ensure that security personnel who control access to restricted areas and classified activities are qualified based on criteria in AR 380-67. Personnel who perform physical-security duties must be disciplined and alert, have sound judgment, be confident and physically fit, and possess good interpersonal communication skills.

SECURITY CLEARANCE

9-7. Security-clearance criteria for security positions must be based on the security classifications of the information to which access will be granted. Security positions are normally designated as sensitive and require a secret security clearance. ARs 381-20 and 380-67 describe criteria and procedures governing security clearances. Appropriate civilian-personnel regulations should also be consulted when civilians are involved.

9-8. Positive evaluation of the reliability of all personnel must be made before they are entrusted with classified or sensitive information. (The Individual Reliability Program is prescribed in AR 190-56.) Follow-up action must be made on all personnel who are granted a security clearance to ensure that they continue to meet the criteria for their clearance. Personnel not meeting or adhering to the prescribed standards must have their security clearances revoked and thereby lose their access to areas containing classified information or material (see AR 380-67).

ORGANIZATION AND EMPLOYMENT OF FORCES

9-9. The organization of a security force will vary, depending on circumstances and the forces available. Forces consist of—

- Mobile patrols. A mobile detachment of ground, sea, or air forces dispatched to gather information or carry out a security mission.
- The response force. A mobile force with appropriate fire support (usually designated by the area commander) to deal with Level II threats in the rear area (Army). This is normally an MP function.
- Reserves. That portion of a force withheld from action or uncommitted to a specific course of action so as to be available for commitment at the decisive moment. Its primary purpose is to retain flexibility throughout an offensive action.
- Any combination of these three.

9-10. Instructions to the security force should be issued in writing. These instructions are normally in the form of general, special, or temporary orders. They should be carefully and clearly worded and include all phases of each assignment. They should be reviewed at least monthly to ensure that they are current. Categories of instructions of each are as follows:

- General orders are those orders that concern the security force as a whole and are applicable at all posts and patrols.
- Special orders pertain to a permanent post or patrol. Each permanent post or patrol should have special orders issued concerning the location, duties, hours manned, arms, ammunition, and other equipment required and the instructions for using force in enforcement and apprehension activities.
- Temporary orders are issued for a short period and cover a special or temporary situation. If it can be predetermined, such orders should indicate the period of time for which they are valid.

9-11. A security-force SOP that outlines policies, organization, authority, functions, and other required information should be prepared for required reading. Each security-force member should be held responsible for full knowledge and understanding of the contents of the SOP. Each installation PM, physical-security officer, or chief of a guard force should conduct periodic inspections and tests to determine each individual's degree of understanding of these instructions. Instructions should be provided in writing regarding the safeguarding and control of the SOP. Its contents may not be classified; however, the information could assist an intruder in breaching security.

HEADQUARTERS AND SHELTERS

9-12. The location of the security force's headquarters will depend on the size and layout of the installation or facility. The objectives are the efficient control of the security force and the adequate security of vital activities. On a small installation, there is frequently only one full-time entrance that may be supplemented by several part-time entrances. At these installations, the logical location of the headquarters would be at or near the main entrance. On larger installations, it might be better to locate the headquarters near the center of the cantonment area.

9-13. The security force's headquarters should be the control point for all physical-security matters for the installation and the monitoring point for protective alarm and communication systems. This office should have a reliable and independent means to contact nearby civil authorities. A list of key telephone numbers should be available for use in emergency operations.

9-14. Personnel shelters should be available to protect the guards from the elements. The design can be temporary or hardened and include adequate space for guard-force personnel only. The facility should have heat, ventilation, storage space for essential accessories, lighting that will not expose the occupant, and good visibility in all directions.

EXECUTION OF SECURITY ACTIVITIES

9-15. Security personnel must exercise good interpersonal communication skills when carrying out their duties with other employees. Bad employee relations can result if security personnel become impertinent and assume powers not rightfully theirs. Security personnel must understand the methods and techniques that will detect security hazards and assist in identifying violators and intruders.

9-16. Written reports or journals are recommended for security activities. These should be prepared by either the security force's supervisor or the personnel at the security post. These reports should record all activities, actions, and visits at the security post.

9-17. It must be strongly emphasized that security personnel will be used for security duties only and should not be given other routine functions except as directed by the commander or his representative. Security personnel should have no fire-fighting or similar duties regularly assigned. Such emergencies offer an excellent diversion to cover an intruder's entrance. Consequently, during such times, security personnel must be exceptionally alert when performing their duties. However, the security force may be cross-trained in other areas (such as fire fighting) so that they may be used when required and when circumstances permit (such as when they are off duty).

9-18. Personnel who are assigned to fixed posts should have a designated method of relief. The security force's shift supervisor should establish a relief schedule (about every two hours) according to local policies and the SOP. A simple but effective plan of operation should be worked out for the security force to meet every foreseeable emergency. Practice alarms should be conducted frequently to test the plan's effectiveness. Such plans should be designed to prevent a diversion at one point on the installation, drawing off the guards or distracting their attention from another section of the installation where unauthorized entry may be made. Routes and times for security patrols should also be varied at frequent intervals to preclude establishing a routine that may be observed by potential intruders.

TRAINING REQUIREMENTS

9-19. The extent and type of training required for security forces will vary according to the importance, vulnerability, size, and other factors affecting a particular installation or facility. The training program's objective is to ensure that all personnel are able to perform routine and emergency duties competently and efficiently.

BENEFITS OF PROPER TRAINING

9-20. Efficient and continuing training is the most effective means of obtaining and maintaining maximum proficiency of security-force personnel. Regardless of the selection process, new personnel seldom have all of the qualifications and experience necessary to do the job. In addition, new or revised job requirements frequently mean that personnel must be retrained. Training can bridge the void between ability and job requirement.

9-21. Supervisors need to remember that all personnel do not have the same training needs. It is a waste of valuable time to train an individual in a subject that he has already mastered. Past experience, training, acquired skills, and duty assignments should be evaluated for each person as an aid in planning an effective training program.

9-22. A good training program benefits both the installation and the security force. The task of supervising the security force is made easier, there is much less wasted time, fewer mistakes are made, and there is less friction with other agencies. A good training program helps to instill confidence through developing increased skill proficiency. The training program provides for more flexibility and better physical protection, fewer required personnel, and less time to learn duties. Training establishes systematic and uniform work habits.

BASIC TRAINING

9-23. As a minimum, personnel (including civil-service security personnel) who have not had security training should receive training in their security duties. This training includes—

- The care and use of weapons, if required. No person should be placed on security duty unless weapons training has occurred within the past 12 months. Weapons training must be according to AR 190-14.
- Areas of responsibility and authority of security personnel, particularly on apprehension, search and seizure, and the use of force
- The location and use of first aid and fire-control equipment and electrical switches.
- Duties in case of emergencies such as alerts, fires, explosions, and civil disturbances.
- Common forms of sabotage and espionage activity.
- The location of hazardous and vulnerable equipment and material.

IN-SERVICE TRAINING

9-24. All newly assigned individuals are given special instructions for each post. When possible, their first assignment should be with an experienced person. Additional in-service training and periodic retraining to review basic material and procedures are continuous requirements.

9-25. Scheduling in-service training and classes to enable all of the security force or a complete shift to participate is often difficult. Therefore, the supervisor must exercise good judgment when scheduling training to ensure that each person has the opportunity to receive the training.

EVALUATION OF TRAINING

9-26. Testing designed to evaluate performance is a necessary step in the training program. These tests may be oral or written or may be a type of performance test. They should be administered annually to ensure that the entire force maintains high standards of proficiency. A testing program also helps to improve training by—

- Discovering gaps in learning.
- Emphasizing main points.
- Evaluating instructional methods.

9-27. Security training received by personnel at their units must be entered in unit training charts or records. The record serves to—

- Indicate individual degrees of skill.
- Establish priorities of instruction.
- Present a consolidated picture of the security force's training status.
- Help certify guard personnel.

SUPERVISION

9-28. A security supervisor is tasked with overseeing and directing the work and behavior of other members of the security force. To obtain maximum performance from each member of his force, the supervisor must have a complete understanding of leadership principles and be capable of applying them.

9-29. The supervisor is responsible for understanding the operations of all posts. Additionally, he is often responsible for selecting, inducting, training, and ensuring the productivity, safety, morale, and advancement of guard-force members.

9-30. To ensure an alert, presentable, and efficient security force, the leadership must provide consistent and intelligent supervision. To earn the respect and cooperation of the guard force, supervisors must be professional in their conduct. The security force's morale and efficiency is a direct reflection of the quality of its supervision.

9-31. The ratio of supervisory personnel to security personnel should be determined by the individual characteristics of each installation. At small installations, the ratio may be higher than at large installations.

9-32. There must be sufficient supervision to enable the inspection of each post and patrol. It is also essential that supervisors be in contact with security headquarters to control emergencies that may arise. Specific duties of a supervisor include the inspection and briefing of the relief shift and the inspection of posts, vehicles, and equipment during visits to posts and patrols.

SUPPLEMENTS TO SUPERVISION

9-33. Various means and devices may be used as supplements to personnel supervision. These include the following:

- **Recorded tour systems.** Personnel record their presence at strategic points throughout an installation by using portable watch clocks or similar devices. These are effective means of ensuring that such points are regularly covered. This system provides an after-the-fact type of supervision.
- **Supervisory tour systems.** A signal is transmitted to a manned central headquarters at the time the post is visited. These systems provide instantaneous supervision and a means of detecting interference with normal security activities and initiating an investigation or other appropriate action.

9-34. All personnel on security duty should be required to report periodically to headquarters by the usual means of communication. The frequency of such reports will vary, depending on a number of factors. Regularity should be avoided to preclude setting a pattern by which an intruder can gauge an appropriate time for entrance.

MANAGEMENT

9-35. The physical-security supervisor is responsible for managing and developing the security organization. A physical-security program is greatly enhanced by a well-developed educational program.

9-36. The physical-security supervisor acts as an advisor and assists in formulating policies for the installation's physical-security measures. The goal should be the best security within the restrictions of the commander's budget guidance. Physical-security planners must remember that anyone can provide adequate security with unlimited funds; however, this is not a realistic approach. There must be a constant endeavor to effect justifiable economy where possible without jeopardizing the physical-security program.

UNIFORMS

9-37. All security-force personnel are required to wear the complete prescribed uniform as outlined in their special orders. Deviations from the prescribed uniform should not be made except for items to protect the guard force's health, comfort, and safety. The duty uniform will be worn during all tours of duty and may be worn during off-duty hours only between the place of residence and the place of duty. Each member of the security force is required to maintain high standards of appearance.

VEHICLES

9-38. The security force should be furnished with sufficient and reliable vehicles to maintain patrol standards established by the installation commander. Vehicles assigned to the force should be equipped with two-way radios to obtain the greatest possible use of all personnel and vehicles.

FIREARMS

9-39. Before issuing weapons, the security force will be briefed on the use of force. Security-force personnel will be issued weapons as prescribed by AR 190-11 and the unit's SOP. The commander may prescribe other weapons for the security force based on needs and requirements. Weapons normally are loaded with live ammunition, except where prohibited for safety reasons. The use of privately owned weapons while on duty is not authorized. Weapons and ammunition issued to security-force personnel will not be removed from the installation except in the course of official duty. When not in use, weapons are secured in arm racks in storage rooms as prescribed by AR 190-11.

9-40. Weapons are inspected as necessary to ensure proper maintenance. A written report is prepared and filed on the discharge of any weapon except for authorized and supervised training. The patrol supervisor or an MP investigator prepares the report (DA Form 3975).

9-41. Ammunition supplies for the security force's use must be maintained in secured storage containers according to AR 190-11. Ammunition must be issued only under proper supervision for authorized purposes. Ammunition issued to members of the security force must be accounted for by individual members immediately upon completion of duty. Any ammunition unaccounted for will be the subject of a report of its disposition by the individual.

COMMUNICATIONS

9-42. The security force should be equipped with two-way radios. These may be vehicle-mounted and portable, or they may be telephones. A secure-voice capability should be used where possible. This equipment is considered essential for the efficient operation of the security force and the accomplishment of its assigned mission. Proper use and care by security personnel will enhance the equipment's usefulness and capability.

MISCELLANEOUS EQUIPMENT

9-43. Security managers or supervisors should obtain other equipment necessary to accomplish their security mission. Items in this category may include (but are not limited to) warning lights; sirens; spotlights; portable lights; flashlights; first aid kits; traffic-control devices; and items of wear for the health, comfort, and safety of security personnel. Some of this equipment may require local purchase.

MILITARY WORKING DOGS

9-44. The requirements for physical protection of installations or facilities within the US and overseas theaters of operation continue to increase. Manpower available for this purpose has always been (and probably will continue to be) limited. The MWD, properly trained and properly used, can enhance a physical-security program. See AR 190-12 and DA Pam 190-12 for information regarding the use of MWDs.

SUMMARY

9-45. A security force is the critical element of a successful physical-security program. It is as strong as its weakest member. A comprehensive training program is essential to a knowledgeable, disciplined, and alert security force. A well-trained security force will be prepared to respond to a security breach.

Chapter 10

In-Transit Security

In-transit security subjects the movement of cargo to different, and frequently, more demanding aspects of physical security. Cargoes may be moved via port, rail, pipeline, or convoy. Regardless of the mode of movement, commanders must aggressively apply the principles of physical security to their protection. Security forces must be provided at the most vulnerable areas of each cargo movement.

IN-PORT CARGO

10-1. Ports and harbors are prime targets for enemy and criminal activities. Perimeter areas of these facilities are more vulnerable because of the extensive distance and exposed beach or pier areas. Terminal areas may include fully developed piers and warehouses or may be an unimproved beach where logistics-over-the-shore (LOTS) or roll-on/roll-off (RORO) operations are conducted.

10-2. If a Theater Army Area Command (TAACOM) MP unit must provide security for cargo in a port, the main effort is to provide security from the perimeter of the port outward. Security measures focus on aggressive patrolling to detect, report and, if need be, combat enemy threats. Measures may include—

- Conducting route and area reconnaissance patrols.
- Developing police intelligence in the area of operations (AO).
- Controlling traffic in the area surrounding the port.
- Conducting mounted or dismounted patrols (with MWDs, if available) around the port's perimeter.
- Establishing an access-control/ID section.
- Watching for diversions of supplies out of the port.
- Providing a response force to react to incidents inside the port's perimeter.
- Providing observation and early warning of threat ground and air attacks.

10-3. When providing security for cargo, the focus is on providing a security overwatch for the cargo as it moves from the port to the combat area. Inside a port's perimeter, access to cargo is limited by—

- Operating random mounted or dismounted patrols (with MWDs, if available).
- Using combined patrols as a response force for incidents inside the perimeter.

- Controlling access to the most restricted areas.

10-4. On occasion, the MP may have to safeguard highly critical cargo inside a port's perimeter. The type and degree of security provided is based on logistical security information. Some examples are the—

- Types and values of the cargo stored.
- Vulnerability of the cargo to a land threat.
- Likelihood of theft, diversion, pilferage, or sabotage by military personnel, local workers, black marketers, or enemy agents.
- Location and nature of the port facilities.
- HN agreements.
- Degree of entrance and exit controls.

10-5. Safeguarding the most critical cargo waiting to be transferred to land transport is the priority. The following measures help to safeguard stored cargo:

- Establishing access-control procedures.
- Searching bundles and packages being taken from the area.
- Examining trip tickets and documentation of cargo vehicles.

10-6. If the restricted area is a pier or other maritime environment, access from the water must be controlled as well as from the land. Entry on the landward side of a pier can be limited with fencing, pass control, and aggressive patrolling; but the part of the pier that protrudes over the water is accessible from the sides and from below. Methods for securing the pier along its water boundaries include—

- Patrols (both walking on the pier and in small boats).
- Protective lighting.
- Log booms.
- Nets.
- Buoys or floats.
- Anchored or pile-mounted navigational aids and signaling devices.
- Barges.

10-7. While most of the barriers described above will stop or impede access to facilities from boats or swimmers, nets are among the most effective. Well-marked, partially submerged objects are also effective; however, there may be legal prohibitions against placing barriers that may constitute a hazard to navigation. These barriers should be placed only after coordination with and approval by the appropriate legal and HN authorities. Sometimes it is best to close off the waterside of a pier. A floating boom will keep small boats out. Suspending a cable or a chain-link net from the bottom of the boom will deny access underwater.

10-8. At least two security zones must be established on a facility's waterside—the reaction zone and the keep-out (exclusion) zone. Security forces in these zones notify vessels, craft, and swimmers that they are entering restricted waters and should alter their course. Security forces may stop and search intruders if necessary. Security zones should extend at least 1,000 meters from the nearest protected asset; however, in some port areas this large security zone is not possible. In such cases, other measures (such as boat patrols) must be increased to mitigate the possibility of attack.

10-9. A reaction zone extends from the high-water mark to a distance beyond the maximum range of anticipated waterborne threats. Security forces will stop and challenge intruders inside the reaction zone.

10-10. The keep-out zone is the zone closest to the protected assets. It extends from the asset to the maximum range of anticipated threat weapons. Security forces should prevent the entry of all unauthorized craft or vessels into this zone. The tactical response force (in this case, a boat) may be used. In addition to organic security, forces may be provided by HN or contracted personnel.

10-11. To keep the cargo secured while transferring from one transport method to another, the traffic moving in and out of cargo-handling areas must be controlled. MP forces can—

- Set up a single access-control point.
- Erect field-expedient barriers. Truck trailers or other large vehicles can be used to constrict the traffic flow if permanent barriers are not in place.
- Limit entry to mission-essential personnel, vehicles, and equipment (as designated by the port authority).

10-12. A holding area should be provided if gates are used by vehicles other than cargo vehicles. Cargo vehicles can pull into the holding area while they are being checked. The holding area should be large enough to handle the volume and size of traffic. A wooden deck or platform at, or slightly higher than, the level of the truck bed can be used to facilitate checking. The platform must be at least as long as the vehicle (such as an empty flatbed trailer). Such a platform makes it quicker and easier to observe and check cargo.

10-13. Cargo is less likely to be diverted if a close watch is kept on cargo documentation and container safety. Containerized cargo is less likely to be stolen or sabotaged. However, containers must be watched closely as they are filled and sealed. Cargo can be pilfered before the seal is applied. An unsealed container can be moved to a stacking area; or someone may apply a false seal, break the seal later, remove the cargo, and then apply a legitimate seal.

10-14. At access-control points—

- Inbound and outbound containers should be inspected. Signs of damage or unserviceability should be observed.
- Containers must be inspected for the presence of seals or locks and hinges. Their serviceability should also be checked.
- The document's transport number, container number, and seal number should be checked to ensure that they match those numbers on the transportation control-and-movement document. (Check the seals by handling them, not simply by a visual check.)
- Containers with valid documents only should be allowed to pass inbound or outbound through the control point.

RAIL CARGO

10-15. Because a train's movement is determined directly by the condition of the tracks, cargo moving by rail is particularly vulnerable to attack. The destruction of switches, signals, or the track may be a delaying harassment; or it could trigger a major catastrophe. Since railroads can be such high-value targets, the commander may task MP or other US forces to provide on-board security for critical cargo.

10-16. Most train crews consist of four or five people who control the train—the engineer, a conductor, a fireman, a senior brakeman, and a brakeman or a flagman. The conductor is the train commander unless a transportation railway service officer is assigned to the train. The train commander is responsible for the train's operation and security. He makes all decisions affecting the train. The security force's commander is responsible for the cargo's security. The train crew and the security force watch for and report any discrepancies or interruptions to normal procedures at any time during the movement. Information about the movement is usually sent along the movement route by the chief dispatcher through a telephone circuit.

10-17. A four- to six-person security force is usually enough to secure railway shipments of sensitive freight, but additional security forces may be needed for moving critical cargo. In addition to a military security force, the shipper or loading agency may send specially trained personnel with highly sensitive cargo. The number of MP in a train security force depends on the—

- Sensitivity of the freight.
- Priority of need for the freight.
- Terrain over which the train will pass.
- Length of the train.
- Duration of the trip.
- Degree of threat.

10-18. Security forces prepare and maintain a record (by car number) of guarded cars in the train. Security forces can ride in—

- A specific car that requires protection.
- The caboose.
- A security-force car. (If only one security car is used, it should be near the center of the train; if more than one is used, cars should be spaced to provide the best protection for the train.)

10-19. The security force on a train must keep a constant check on car doors, seals, wires, and locks to detect tampering. The following instances must be noted and reported immediately:

- Irregularities in procedures.
- The presence or actions of unauthorized persons.
- Deficiencies or incidents that occur.

10-20. When planning rail-cargo security, the time schedule for the rail movement must be obtained. A map reconnaissance of the route should be provided, detailing bridges and tunnels that are especially vulnerable.

10-21. Security-force actions should be planned at scheduled stops or relief points, and forces should be deployed according to these plans. Locations of MP units and other friendly forces should be plotted along the route, and their radio frequencies and call signs should be noted. An intelligence report covering the route should also be obtained. This report should indicate sites where sabotage may occur, attacks may be expected, or thefts and pilferage are likely.

10-22. The shipper is responsible for the security of all carload freight until it is turned over to the Transportation Railway Service and the loaded cars are coupled to a locomotive for movement. The shipper or field transportation officer should complete the freight waybill or the government bill of lading. This report shows the car number, a brief description of contents, the weight of the load, the consignor, the consignee, the origin, and the destination. In addition, it may show special instructions for the movement or security of the car and its contents. Careful documentation is essential for—

- Securing the shipment.
- Locating cars with critical cargo.
- Ensuring that priority movement is authorized.

10-23. Transportation officers are responsible for the completeness, correctness, and proper handling of waybills. Each car must have a waybill; this allows cars to be detached or left behind should they become defective en route. If this occurs, a team from the security force must remain with the cargo until they are relieved.

10-24. Railway cars are sealed after loading. A seal shows that a car has been inventoried and inspected. The standard method of sealing a railway boxcar door (in addition to padlocks or wires) is with a soft metal strap or a cable seal that contains a serial number. Maintaining rigid accountability of all seals is necessary to prevent the undetected replacement of an original seal with another. While sealing does not prevent pilferage, a broken seal is a good indicator that the car and its contents have been tampered with. Train security forces or operating crews can easily check the seals on cars when the train stops. Broken seals should be reported immediately to help pinpoint the time and place of a possible theft. When vehicles are shipped by railcar, sensitive and high-value items must not be secured in the vehicles. Container-express (CONEX) and military-van (MILVAN) containers are ideal for shipping these and other small items on flatcars since they greatly reduce the chance of pilferage. These containers must be locked and sealed and, if possible, placed door to door for additional security.

10-25. When operations permit, cars containing highly pilferable freight, high-priority cargo, or special shipments are grouped in the train to permit the most economical use of security forces. When flatcars or gondolas are used to transport sensitive or easily pilfered freight, security forces should be placed where they can continuously observe and protect these cars.

10-26. When the train is stopped, security forces should dismount and check both sides of the train, verifying that seals, locks, and wires are intact. They must report a broken seal immediately to help pinpoint the time and place of the theft.

10-27. If the security force is relieved by another security force while en route, a joint inspection of the cars is conducted. The relief force signs the record being kept on the guarded cars. Consignees assume responsibility for the security of loaded freight cars at the time they arrive at their destination. When the trip is complete, the receiver or his agent will inspect the cars. The security force obtains a receipt for the cars, which is then attached to the trip report. The trip report should include—

- Dates and times the trip started and ended.
- Any additional information required by the local SOP or command directive.
- Recommendations for correcting deficiencies or for improving future security on trains.

10-28. Because unloading points are highly vulnerable to pilferage and sabotage, cars should be unloaded as soon as possible to reduce the opportunity for loss. MP forces are normally not available for the security of freight in railway yards. For more information regarding rail cargo, see FM 55-20.

PIPELINE CARGO

10-29. Pipeline systems are widely used in a theater of operation to transport bulk petroleum products or other liquids. Such systems are open to a number of security threats from the point of entry to the point of final delivery. Pipeline systems are composed of storage and dispersing facilities, pump stations, and extended pipelines. They also include discharging facilities for tankers at ports or other water terminals.

10-30. The type and extent of risk to a pipeline varies with the level of conflict in the AO. In a communications zone, the chief hazard is likely to be pilferage. Pipelines can be tapped by loosening the flange bolts that join sections of pipe or by cutting holes in the hose line. The risk rises if gasoline is scarce and expensive on the civilian market. Sabotage is a security hazard during all levels of conflict. It is committed by any method such as simply opening pipe flanges, cutting a hose line, or setting fires and causing explosions to destroy portions of the line.

10-31. In areas of conflict, the likelihood of sabotage and interdiction increases. Pipeline systems are vulnerable to air attacks, especially aboveground sections of the pipeline, pump stations, and storage facilities.

10-32. Security forces should be deployed in the best manner to provide coverage to the most vulnerable portions of the pipeline that are at the greatest risk to enemy, terrorist, partisan, and ground attack. Patrols should be set up to screen isolated areas and remote pumping stations. Sensors should also be considered, along with aerial security. Security patrols will—

- Detect, report, and respond to attacks on or sabotages of the pipeline.
- Monitor critical parts of the pipeline on a routine but random basis.
- Monitor ground sensors and other intrusion-detection devices. These are often used at pump stations and elsewhere along the pipeline to detect and identify threats to the system.

- Check line-pressure devices in pipeline and pumping facilities. These devices monitor the flow and detect breaks in the line, which may indicate pilferage of gasoline (or other petroleum products).

10-33. Dedicated security forces are rarely sufficient in number for the surveillance of an entire pipeline system. All available supporting forces (in the course of their normal duties) should observe and report items of intelligence for further investigation. Examples of suspicious activities in the pipeline area might include the unusual presence of commercial tanker trucks, the appearance of gasoline drums or cans, or an increased use of motor vehicles in fuel-scarce areas. Other resources available to the commander for coordination and support include HN and MP elements responsible for the AO, as well as the security officer of the petroleum group or battalion.

CONVOY MOVEMENT

10-34. As convoy movements are tactical in nature and are discussed in detail in FM 19-4, they will be briefly discussed here. When moving by convoy, consideration should be made for the following:

- Congested traffic areas.
- Travel during night hours when traffic is reduced rather than travel during daylight hours when traffic congestion is heaviest.
- National holidays. Traffic may be three times heavier than on a normal day. Also, if you are moving a convoy overseas on a national holiday, the HN people may not be receptive to your action and the result may be unwanted reactions on their part.
- The use of a marked HN police vehicle in conjunction with the convoy. The HN people are more receptive to an activity when it is represented by one of their own. Additionally, the HN police may be able to diffuse a potential crisis.
- Security of the convoy. Security of the convoy is foremost important both during movement and stops. During extended or overnight stops, special consideration must be given to securing the loaded vehicles.

Chapter 11

Inspections and Surveys

Inspections and surveys are valuable tools to a commander's physical-security program. These tools collectively measure and identify the readiness of a commander's physical-security program. The survey provides the installation commander with an overall security posture of the installation.

INSPECTIONS

11-1. Physical-security inspections are conducted at DA installations, activities, and facilities by trained physical-security inspectors. Some facilities on an installation may be exempt from inspection due to their mission. These facilities are inspected under the guidance of regulations and directives unique to those activities. Inspection personnel will be trained and will conduct inspections according to AR 190-13.

11-2. A physical-security inspection is a recorded assessment of physical-security procedures and measures implemented by a unit or an activity to protect its assets. The inspection is recorded on DA Form 2806-1-R (see AR 190-13 for using the form).

COORDINATION

11-3. Liaison and coordination should be established with other agencies on the installation before an inspection. The director of facility engineers can provide information to benefit the overall security program. Other agencies, such as MI (threat analysis) and local law-enforcement agencies, may have input essential to the security program.

SECURITY LIBRARY

11-4. A security library is necessary to help personnel prepare for and conduct an inspection. This library may include—

- The mission and history of each activity to be inspected.
- Previous inspection reports.
- A copy of the most current risk analysis.
- The SOPs and ARs specific to physical security.

ENTRANCE INTERVIEWS

11-5. Entrance interviews are usually required before the actual inspection. During the interview, the inspector establishes a rapport with the unit representative. The inspector identifies the following during the interview:

- All members of the inspection team.

- An overview of the last inspection.
- Areas to be inspected and the order of inspection.
- A review of waivers, work orders, and exceptions.
- Changes to the unit's mission (if any).

CONDUCTING INSPECTIONS

11-6. The inspection should be conducted from the outside to the inside of the facility, activity, or area with regard to the following:

- Observation of the facility will be conducted during all hours of the day.
- Interviews of managerial and operational personnel will be performed.
- Security forces should be inspected so as not to disrupt the mission (if possible).
- An assessment should be made of security-force training, especially if security-force knowledge proves inadequate.
- Inspection of entry and movement control by the guard force should not hinder operations.
- All communications systems used by the guard force should be thoroughly inspected. The guard force should have two reliable and efficient means of communication, one of which is a radio.
- Inspections should be conducted according to regulations appropriate for the facility.

EXIT INTERVIEWS

11-7. Exit interviews should be conducted as soon as possible after the inspection. The commander should be informed of any deficiencies or compliments noted. A rating on the inspection's results will not be provided during the exit interview. The approving authority, not the inspector, will determine the inspection rating. The rating will be forwarded to the unit along with the final report.

11-8. Recommendations will be made according to regulations. Written reports should be forwarded through channels in a timely manner according to the PM's SOP. The commander's report of actions taken will be required and reviewed by the PM's staff.

SURVEYS

11-9. A physical-security survey differs from an inspection in that a survey covers a formal assessment of an installation's physical-security program. Each survey includes a complete reconnaissance, study, and analysis of installation property and operations. The survey provides the commander with an assessment of the installation's overall security posture. It consists of the threat and the mission, and it advises the commander on the installation's physical-security program's strengths and weaknesses.

PHYSICAL-SECURITY SURVEY

11-10. The physical-security survey is a formal recorded assessment of an installation's physical-security program. See AR 190-13 for further information on this type of survey.

SECURITY-ENGINEERING SURVEYS

11-11. While a security-engineering survey is largely an engineer function, it must be coordinated with physical-security personnel to be successful. A security-engineering survey is the process of identifying (by means of an on-site survey) engineering requirements associated with facility enhancements for physical security and antiterrorism, including an IDS installation. This type of survey should be conducted when planning new construction, renovations, or upgrades to existing facilities where there are likely to be physical-security requirements. A security-engineering survey may also be requested by the PM or an equivalent security officer to evaluate existing security. This survey—

- Identifies assets to be protected.
- Identifies threats to these assets and the level of protection required to protect them.
- Identifies the protective measures.
- Determines the cost of the protective measures.

Appendix A

Metric Conversion Chart

This appendix complies with current Army directives which state that the metric system will be incorporated into all new publications. Table A-1 is a conversion chart.

Table A-1. Metric Conversion Chart

Metric to English			English to Metric		
Multiply	**By**	**To Obtain**	**Multiply**	**By**	**To Obtain**
Length					
Centimeters	0.0394	Inches	Inches	2.54	Centimeters
Meters	3.28	Feet	Feet	0.0305	Meters
Meters	1.094	Yards	Yards	0.9144	Meters
Kilometers	0.621	Miles (stat)	Miles (stat)	1.5609	Kilometers
Kilometers	0.540	Miles (naut)	Miles (naut)	1.853	Kilometers
Millimeters	0.039	Inches	Inches	25.40	Millimeters
Area					
Square centimeters	0.1550	Square inches	Square inches	6.45	Square centimeters
Square meters	10.76	Square feet	Square feet	0.0929	Square meters
Square meters	1.196	Square yards	Square yards	0.836	Square meters
Volume					
Cubic centimeters	0.610	Cubic inches	Cubic inches	16.39	Cubic centimeters
Cubic meters	35.3	Cubic feet	Cubic feet	0.0283	Cubic meters
Cubic meters	1.308	Cubic yards	Cubic yards	0.765	Cubic meters
Milliliters	0.0338	US liq ounces	US liq ounces	29.6	Milliliters
Liters	1.057	US liq quarts	US liq quarts	0.946	Liters
Liters	0.264	US liq gallons	US liq gallons	3.79	Liters
Weight					
Grams	0.0353	Ounces	Ounces	28.4	Grams
Kilograms	2.20	Pounds	Pounds	0.454	Kilograms
Metric tons	1.102	Short tons	Short tons	0.907	Metric tons
Metric tons	0.984	Long tons	Long tons	1.016	Metric tons

Appendix B

Sample Installation Crime-Prevention Handbook

This appendix provides guidance on planning, organizing, directing, and controlling installation crime-prevention programs. It provides guidance on developing an installation program, criminal analyses to identify crimes, guidance on which crimes to address, command and individual countermeasures for particular crimes, and program-evaluation procedures.

SECTION I — INSTALLATION CRIME-PREVENTION PROGRAMS

B-1. In the past few years, the Army has shifted an increasingly larger percentage of its manpower from combat-service-support activities to combat organizations. This change means that fewer MP personnel are available to support a larger number of units. To meet this challenge, it is necessary to reevaluate the way we do business and to emphasize those programs or procedures that have the greatest impact on our installation crime rates. Crime prevention is one program that can have a major impact on installation crime rates at a relatively minor cost in both dollars and manpower. It takes less effort to discourage a criminal from perpetrating a crime or to teach a soldier to avoid becoming a victim than it does to investigate a crime, identify the offender, prosecute him, and punish him. In addition, a proactive approach to law enforcement can help maintain the high quality of service life that can improve the retention of first-term soldiers.

B-2. The Army is a large organization that performs a variety of activities in many different environments. Crimes that are major problems on one installation may be totally absent from others. For example, most military installations have a significant number of robberies while most depots have none. Because of this, any rigid, centrally controlled program—no matter how carefully thought out—is bound to be inappropriate in many locations. Therefore, DA has elected to provide only the most general guidance and to allow commanders to develop crime-prevention programs that address their local problems.

CRIME-PREVENTION WORKING GROUPS

B-3. The installation is the smallest practical level for implementing crime-prevention programs. If these programs are developed and implemented at a lower level, then crime is often not eliminated but is merely displaced from units with good programs to units with less effective programs. Also, crime does not affect personnel only when they are in their place of duty. In many cases, a company commander's troops are victimized in areas over which he has little control. Unit commanders are responsible for implementing many anticrime measures; however, the selection of overall program goals, the ID of appropriate countermeasures, and quality control should be done at the installation level.

B-4. Crime prevention must always be recognized as a commander's program rather than as an MP program. MP personnel have the expertise to analyze data, identify major problems, and develop lists of possible countermeasures. They should perform these functions in support of an installation crime-prevention council appointed by the installation commander and composed of representatives of all of the installation's major organizations and activities. The advantages of using this type of system are—

- It provides representatives of all major segments of the post population with a forum where they can identify criminal problems that are of the greatest concern to them.
- It allows the representatives of all major commands to review the available options to counter a crime and to select the level of resource commitment that is compatible with their missions and internal priorities.
- It helps ensure that the resources of the entire community, rather than only those of the MP force, are mobilized to attack the problem.
- It is easier to obtain the support of the whole population if its representatives are instrumental in the development of the program.

CRIME-PREVENTION OFFICERS

B-5. The installation's crime-prevention officer is normally a senior NCO or an officer who has a solid background as an MP investigator or a physical-security inspector (PSI). He supports the installation council by performing a crime-data analysis to identify problem areas, drafting programs for the council's consideration, inspecting the implementation of council-mandated measures, and coordinating the efforts of unit crime-prevention officers in the implementation of the crime-prevention program.

B-6. As a member of the PM's staff, the crime-prevention officer develops the law-enforcement section of the crime-prevention program, develops and maintains the written crime-prevention plan, and coordinates crime-prevention programs with civilian police agencies and community groups.

B-7. Crime-prevention officers are also appointed in each organization down to the company level. At this level, written crime-prevention plans are not required; however, SOPs are established. The crime-prevention officers serve as their organizations' focal points for coordinating installation crime-prevention plans; they supervise the implementation of the installation's program within their organizations.

CRIME-PREVENTION PROGRAM DEVELOPMENT

B-8. The starting point for developing a crime-prevention program must be a thorough analysis of criminal activity on the installation. This identifies significant criminal problems that are susceptible to crime-prevention efforts. Crimes that are most susceptible to crime-prevention measures are those for which a high probability of reoccurrence exists. Crimes such as murder normally are not repetitive and are poor candidates for inclusion in the crime-prevention program. Since it is seldom practical to attack all criminal problems simultaneously, they should be prioritized based on their impact on the command's ability to perform its mission and their impact on installation personnel. Next, the whole range of countermeasures that can be used to combat each problem must be identified (see Figure B-1). Table B-1, page B-4, identifies (by offense) programs that have been successful in countering specific criminal problems. Sections III and V of this appendix contain discussions of the strengths, weaknesses, and applicability of the countermeasures listed in Table B-1. Once developed and prioritized, the list of criminal problems and possible countermeasures must be presented to the installation crime-prevention council for action. The council should decide which crimes will be addressed and which countermeasures will be used for each crime. The council must then identify specific objectives for its anticrime campaigns.

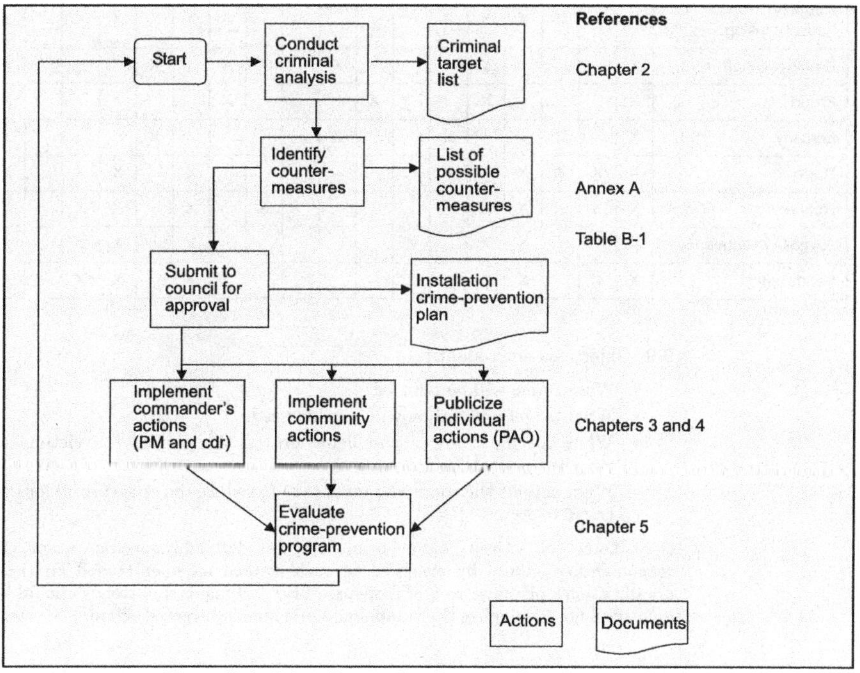

Figure B-1. Crime-Prevention Program-Development Cycle

Table B-1. Offenses Countermeasure Matrix

Crime	Command/Law-Enforcement Countermeasures												Community Programs					
	Crime Hot line	Lighting	Environmental Changes	Patrol/Surveillance	Publicity Campaigns	Residential Security Survey	Juvenile Programs	Fraud Programs	Employee Theft	Arson Programs	CCTV	Warning Signs	Neighborhood Watch	Operation ID	Neighborhood Walks/Escorts	Mobile Patrols	Project Lock	Education
Arson				X	X					X			X			X		X
Auto theft	X	X			X								X		X	X	X	X
Burglary/ housebreaking	X	X	X	X	X	X							X	X	X	X		X
Employee theft	X				X				X		X	X		X				X
Fraud	X				X			X				X						X
Larceny	X		X			X			X					X				X
Rape	X	X	X	X	X	X							X		X	X		X
Robbery	X	X	X	X	X						X		X		X	X		X
Juvenile delinquency	X			X	X		X						X		X	X		X
Vandalism	X	X		X	X		X						X		X	X		X

B-9. Objectives must identify—

- What crime will be reduced.
- What target population will be addressed.
- What specific changes and behaviors on the part of the victims or perpetrators will be encouraged.
- What actions the command must take to reduce the opportunity for the crime to occur.

B-10. Once objectives have been clearly defined, specific areas of responsibility should be assigned to each council member (based on their organization's primary area of responsibility) and major milestones should be identified for developing the campaign against each targeted crime.

TRAINING

B-11. The prerequisite skills for successful performance as an installation crime-prevention officer are best developed through on-the-job experience as the supervisor of MP investigations or physical-security inspections. More important than any technical skill is the cultivation of a frame of mind that instinctively examines each case to determine not only what occurred, but also how the crime could have been prevented. Technical skills (such as criminal-data analysis) that may not have been developed as an MP investigator or physical-security supervisor are presented in courses taught by several civilian agencies. These classes should be used to the fullest extent possible.

CIVILIAN CRIME-PREVENTION ORGANIZATIONS

B-12. There are many civilian crime-prevention organizations at the national, state, and local levels. Many of these organizations have produced crime-prevention material (including posters, radio spots, and leaflets). Material and programs sponsored by civilian agencies should be used to support Army crime-prevention efforts. However, when material from a source outside of DOD is used, a copyright release must be obtained. Normally, it is necessary to get a release for each separate item that is used. If there is any doubt as to the necessity of securing a copyright release, the crime-prevention officer should refer the matter to the local SJA.

SECTION II — CRIMINAL ANALYSIS

B-13. Criminal analysis is a system for identifying trends and patterns where they may exist. It is a routine, ongoing function for the PM and battalion- and brigade-level staffs. Criminal analysis is the foundation upon which the installation force-protection program is based. Moreover, criminal analysis is an integral component of the police intelligence-operations function and is applicable across the operational continuum. An effective criminal analysis establishes the following:

- Crimes having a significant impact on the installation.
- The segments of the population being victimized.
- The ID of criminals/perpetrators.
- The most common time of occurrence.
- The areas that experience the highest number of incidents.
- Offense information (such as types of weapons or victims' actions that contribute to the offense).
- Information critical to an installation's VA.
- Information essential in formulating a successful patrol-distribution plan.
- Police information and criminal intelligence fused with tactical intelligence.

B-14. With this type of information, specific countermeasures are developed to reduce the opportunity for a crime to occur or to remove the incentives for perpetrators. Without an effective criminal analysis, the overall security effort is unfocused. Moreover, the installation/base patrol-distribution plan may be skewed. When this occurs, broad countermeasures are implemented and meaningful results are lost.

B-15. A professional analysis of criminal data is essential for protecting soldiers, units, and installations. The Military Police Management Information System (MPMIS) provided the initial software capable of assisting in the development of criminal-analysis information. As this system is modernized and replaced by the Military Police Automated Control System (MPACS), the MP Corps and the CID will have interface with the Army Battle Command System (ABCS). The analysis of police, criminal, and tactical information results in the development of police intelligence (PI). The PI, which is coordinated with the Assistant Chief of Staff, G2 (Intelligence) (G2) or the Intelligence Directorate (Joint Command) (J2), provides relevant information and intelligence (RII) that greatly contribute to the commander's critical information requirements (CCIR). The RII provided vertically and horizontally ensures that the common operational picture is properly shared.

SOURCES OF INFORMATION

B-16. The basic source for identifying crimes that warrant examination is the installation's Law Enforcement and Discipline Report (DA Form 2819). The information in this report can be used to plot the amount of and seasonal variations for each major type of crime. This will tell little about conditions that produce the crime, but it is useful in identifying crimes that can be eliminated from detailed analyses due to low frequency rates.

B-17. The Law Enforcement and Discipline Report identifies which major category of crime should be targeted, but it does not tell which type of crime is causing the most problems. For example, It may indicate that robbery is a problem, but it will not discriminate between robbery of commercial establishments (such as the post exchange [PX], the Class VI store, or banks) and muggings of individuals. Since countermeasures for these two types of robbery are different, it is necessary to collect additional information. The best sources for this information are the military-police report (MPR) and the CID report of investigation (ROI). If there are fewer than 200 cases in the past year for a particular type of crime, they should all be examined. With a larger annual caseload, a random sample that is large enough to give results of plus or minus 5 percent accuracy should be examined.

B-18. General factors must be identified for any type of crime. These are the types of victims, the perpetrators, geographical data, and chronological data.

VICTIMS

B-19. It is essential to determine (as precisely as possible) the segment of a post population that is being victimized. Junior soldiers live in different areas on post than officers and senior NCOs. They patronize different clubs and, for the most part, work in different areas. Information programs must use multiple sources to ensure that the same message reaches the different portions of the population. If the specific population segment that is being victimized is not identified, it is possible to spend a large amount of resources and have little impact on the targeted crime because the message is not getting to the people who need it.

PERPETRATORS

B-20. As with the victim profile, data on perpetrators is essential to ensure that countermeasures are targeted against the correct population. When on-post auto thefts are committed by civilians with legitimate reasons for entering the installation, it may not make sense to increase security (as a countermeasure) at post entrances. Countermeasures are developed based on specific profiles identified by case type and the information described above. Successful countermeasures depend on correct criminal analyses. An analysis and profile of perpetrators will help to focus MP/CID countermeasures.

GEOGRAPHICAL DATA

B-21. The MPR and the ROI should contain a street address or a description of the incident's location (such as "in a parking lot on the east side of building 1409"). This type of description fulfills requirements for identifying the location of the incident in court. It is further enhanced by identifying the normal duty hours and the type of activity that takes place in the location (for example, "in a parking lot on the east side of the NCO club (building 1401), normal operating hours 1800-0400" or "in a parking lot on the east side of building 1408, a troop billets for Company A, 15th Cav"). This type of information may help to develop a list of specific types of areas where a particular crime is occurring, making it easier for the PM to provide increased MP patrols in these areas. For example, if there have been several incidents in which vehicles have been broken into and stereos removed, it would help to know that 90 percent of them occurred in the parking lot by troop billets.

CHRONOLOGICAL DATA

B-22. As with geographical data, the more specific the time-of-occurrence pattern is for a crime, the easier it is to apply sufficient resources to affect the crime rate. For each crime having a significant impact, determine the following:

- Major seasonal variations.
- Monthly variations. Is there a concentration of crimes immediately before or after payday?
- Weekend occurrences. Is there a concentration of incidents on weekends? Each day represents about 14.25 percent of the week. Concentrations of crime higher than that for any particular day may be significant.
- Time of day. Is there a particular period that accounts for a disproportionate share of the incidents?

CRIME-SPECIFIC FACTORS

B-23. In addition to the factors that should be examined for all crimes, the following crime-specific factors are useful in analyzing specific offenses:

- Housebreaking and burglary.

  - The type of building that was attacked (family-housing unit, troop barracks, PX, and so forth).
  - Whether the facility was occupied or unoccupied.
  - The point of entry (door, window, and so forth).
  - The method of entry (unsecured door, forced door, forced window, and so forth).
  - The property that was stolen. (Was it marked?)
- Robbery.
  - The number of perpetrators.
  - The perpetrator's method of operation.
  - The type of weapon used.
  - The type of robbery (street mugging, residential robbery, or commercial robbery).
  - The victim's injuries.
  - Actions by the victim that contributed to his being targeted.
- Larceny.
  - The type of property taken.
  - Whether the property was secured or unsecured.
  - The perpetrator's method of operation.
- Auto theft.
  - The type of vehicle stolen (POV, motorcycle, and so forth).
  - Whether the vehicle was secured or unsecured.
  - Whether the vehicle was recovered and where it was recovered.
  - Whether the vehicle was stripped of parts.
  - The perpetrator's method of operation.
- Forgery.
  - The type of document that was forged.
  - How the document was obtained.
  - The type of ID used in passing the forged document.
  - The con games or techniques used.
- Rape and sex offenses.
  - The perpetrator's method of operation.
  - The relationship between the victim and the perpetrator (blood relatives, acquaintances, or strangers).
  - The degree of force used.
- Aggravated assault or murder.
  - The relationship between the victim and the perpetrator.
  - The motivation.
  - The weapon used.

B-24. Personnel who have tried to use MPRs and ROIs as a sole source of data know that in many cases the required information is not contained in sufficient detail to be useful. To correct this situation, investigators and inspectors must be trained to recognize conditions that contributed to the crime as well as the information needed to identify and prosecute the offender. As a minimum, the PM ensures that all law-enforcement personnel know general and specific crime factors that are required to analyze each type of crime. He must also provide feedback through the law-enforcement operations staff when incomplete reports are received. If this is done, high-quality data can be collected without generating additional reports to collect criminal information for analysis.

B-25. The MPRs and ROIs are not the only sources of information on which to base criminal analyses. Physical-security inspection results, summaries of common deficiencies noted by the inspector general, data from the SJA claims section, and information summaries from reports of survey on lost government material can all produce worthwhile information on conditions that lead to the commission of crimes. Additionally, a review of civilian criminal statistics and civilian police information may contribute to the pattern analysis. The reviewing authority must understand which information is relevant. The MP commanders ensure that the CCIR are published in operations orders (OPORDs), operation plans (OPLANs), and SOPs. This effort ensures that MP forces at every level recognize relevant information.

INDIVIDUAL CRIMINAL ANALYSIS

B-26. A criminal analysis is most effective when it is applied to the class of criminal offenses with a high probability of recurrence. Single-incident crimes do not lend themselves to analyses. Most crimes against persons do not usually benefit from analyses, with the notable exceptions of rape, robbery, and related combinations of offenses (such as kidnapping and rape; robbery and attempted murder; burglary and rape; and burglary, robbery, and kidnapping). Analyzing isolated criminal offenses has some value (such as gaining knowledge of where these offenses are most likely to reoccur). However, this knowledge is usually difficult to use effectively for preventive purposes.

B-27. There are universal factors available for analysis for most crimes. The availability of these factors varies greatly between criminal types and specific reported offenses.

B-28. In addition to the universal factors, there are an almost infinite number of factors that may be considered specific to a particular criminal class or type. These crime-specific factors are data elements that are usually recorded during the reporting of a particular type of offense and are used for analysis purposes.

B-29. Crime-specific factors provide information that can be used by the analyst to connect crimes with similar characteristics. Information regarding physical evidence may have considerable value in the analysis of several criminal types (such as burglary and auto theft). Therefore, the suitability of different crimes to analyses depends on the general factors, specific modus operandi (MO) factors, and physical evidence. With these considerations in mind, different criminal types and classifications are examined to determine their applicability.

RESIDENTIAL HOUSEBREAKING AND BURGLARY

B-30. Although housebreaking and burglary are two of the most difficult crimes to prevent, they are the most suitable for analyses. The typical housebreaker or burglar establishes an MO pattern based on successful past offenses. Usually, a burglar will continue to commit similar crimes until he is apprehended.

B-31. The available information for evaluation must consider the specificity, accuracy, and value of the information received. Informational factors are presented in the order of relative importance. It should be emphasized that the factors listed in Figure B-2 are for analysis purposes; the order for investigative solutions may vary.

COMMERCIAL HOUSEBREAKING

B-32. The analysis of commercial housebreaking is in many respects easier than that of residential housebreaking or burglaries. More pertinent information is available to the analyst for analyzing commercial housebreaking (which is more specialized and will exhibit more specific MO characteristics). The analyst normally has specific information regarding the point and method of entry, the victim/target description, and the property-loss description. The time factors for commercial housebreaking may be of less value in a commercial-housebreaking analysis than in a residential-burglary analysis since most commercial housebreakings occur at night or during weekends.

B-33. The commercial housebreaker is generally more mobile than the residential burglar. An analysis of commercial housebreaking is not as

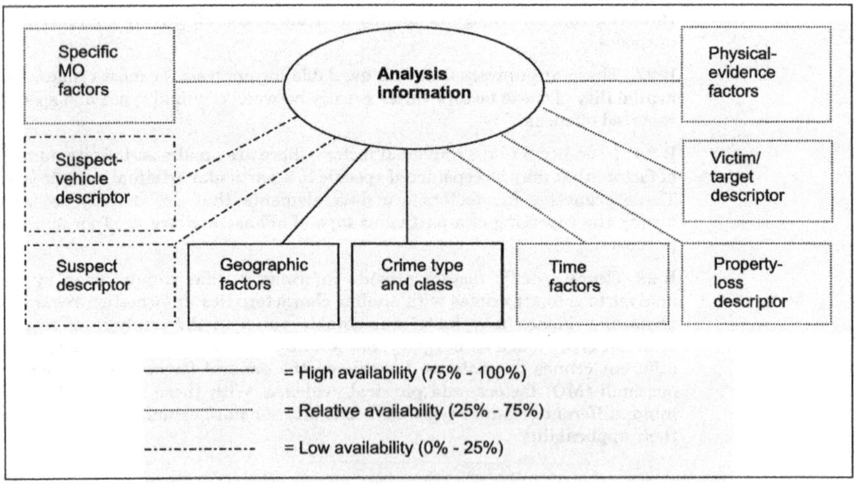

Figure B-2. Residential Housebreaking and Burglary

restricted in geographical area as an analysis of residential burglaries. The commercial burglar may travel a considerable distance to attack a particular type of business. The residential burglar is less discriminating because he has a greater number of potential targets.

B-34. The analysis of commercial housebreaking should be systematically directed to examining the information factors shown in Figure B-3. A commercial housebreaker may be a juvenile criminal, an addict, a professional thief, a soldier, or an enemy agent. Each of these types provides a different set of factors to analyze.

ROBBERY

B-35. Robberies are well suited for analysis. This class of criminal offender usually operates in a given geographical area. The commercial robber generally seeks a sizeable amount of cash; while targets of the street robber frequently include credit cards, checks, and other valuables in addition to cash.

B-36. The presence of physical evidence is more probable in street robberies than in commercial robberies because the criminal offender frequently discards evidence (such as purses and wallets) after completing the offense. The mugger frequently uses surprise as part of his MO in order to reduce his chance of apprehension by physical ID.

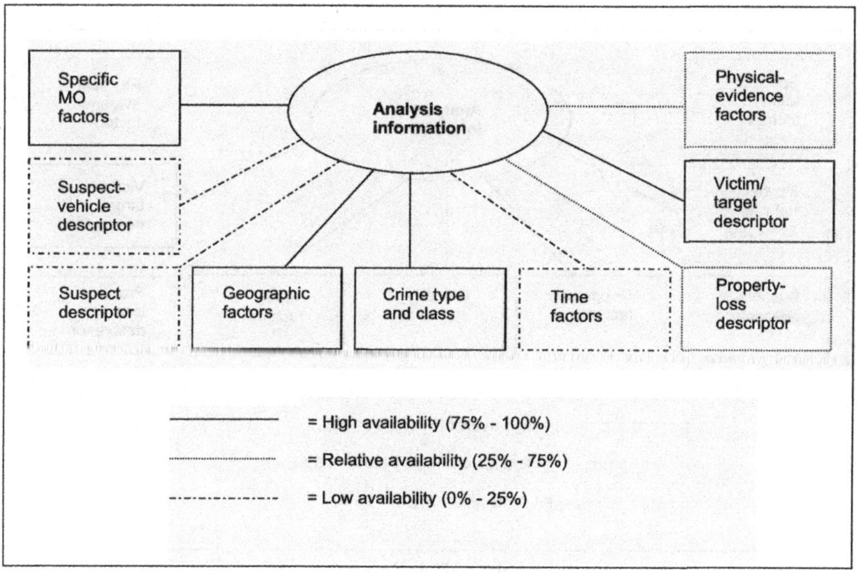

Figure B-3. Commercial Housebreaking

B-37. The probability of universal and crime-specific informational factors for a typical robbery is a systematic method for examining the case. See Figure B-4 for the analysis of a robbery case.

AUTO THEFT

B-38. Auto theft is probably the best-suited crime for analysis. The analyst frequently has more information available on any given auto theft than any other crime. Suspect descriptive information is usually lacking in auto larceny cases. However, the greater availability of crime-specific information makes auto theft extremely receptive to an analysis.

B-39. Auto-theft offenders include joyriders, professional car dismantlers, and wholesalers of stolen vehicles. In many cases, these offenders will establish and maintain a particular MO until apprehended.

B-40. Helpful elements in analyzing auto-theft cases are the availability of information, the presence of physical evidence, and the ability of the stolen property to be traced. Of special importance are the geographic factors, suspect-vehicle descriptors, and property-loss descriptors. In analyzing an auto theft, the analyst usually has information concerning two geographic locations for analysis—where the vehicle was stolen and where it was recovered. The suspect-vehicle descriptors, property-loss descriptors, and victim/target descriptors are the same (see Figure B-5). This greatly augments the analysis of auto thefts. In an examination of specific MO factors in an auto

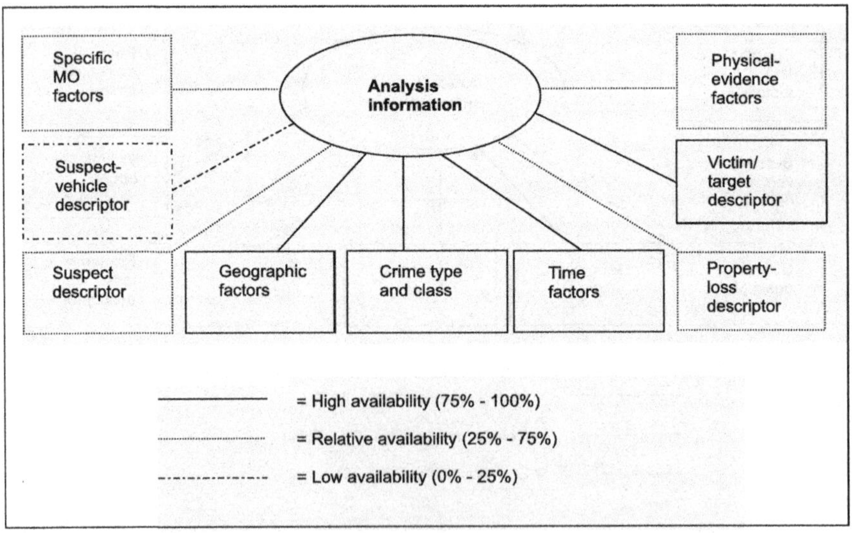

Figure B-4. Robbery

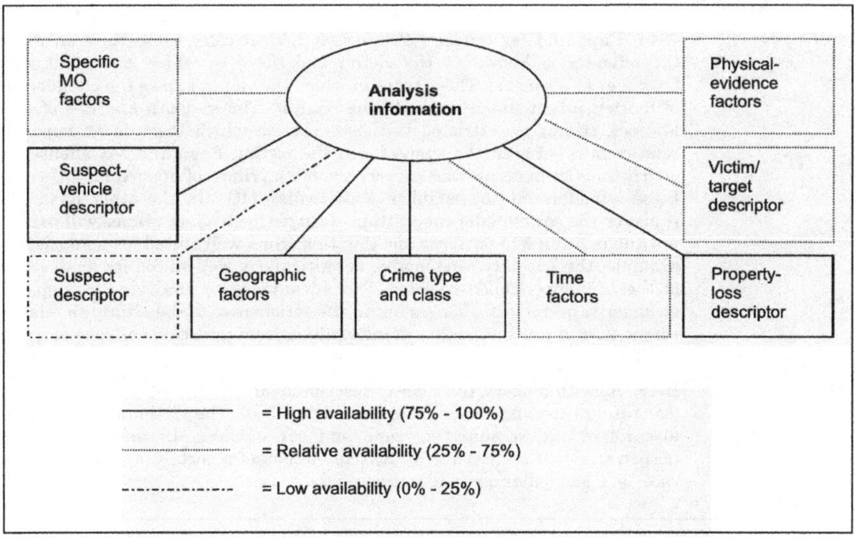

Figure B-5. Auto Theft

theft, the analyst will place emphasis on the vehicle's condition when recovered. Factors of primary importance in analyzing auto-theft cases are the vehicle's recovery location, the vehicle's make and model, and the degree of stripping. Coordination with local civilian police or HN police may reveal potential buyers of stolen automobiles.

B-41. Larceny and theft cases are not usually well suited for analysis. However, benefits are derived from analyzing larceny offenses when specific factors are selected, such as auto accession thefts.

B-42. The problem with trying to analyze general theft cases results from the large volume of reported offenses, the large number of possible crime types, and the many possible MO patterns. To analyze theft cases, the analysis operation must first restrict the number of cases and crime classification to a workable level. This may be done by classifying general theft cases into special categories and analyzing them by considering only special MO factors. Logical classifications include thefts of autos, auto accessories, bicycles, items on shipping docks, and tools and equipment.

B-43. Many of the general theft cases are helpful in analyzing other types of crimes. For example, an increasing trend in stolen automobile parts may have common perpetrators involved in auto theft and stripping. When particular patterns appear, a specific in-depth analysis can be conducted. It is essential that information developed during the analysis process be shared with the G2/ J2, the SJA, the CID, and others (as authorized).

RAPE AND SEX CRIMES

B-44. Rape and sex crimes fall into two distinct categories—those in which the offender is known to the victim and those in which he is unknown (stranger to stranger). Those cases in which the victim knows the offender are of limited informational value to the analyst. The in-depth analysis of rape and sex crimes is restricted to those cases in which there is no apparent relationship between the suspect and the victim. Rape and sex offenses in which the victim knows the suspect are often crimes of opportunity. Usually, these offenders do not establish a particular MO. On the other hand, the rapist or the sex offender committing stranger-to-stranger crimes will usually provide definite MO patterns, making this crime well suited for analysis. For example, the burglary/rape motive is particularly well-suited for analysis, as is the kidnapper/child molester. The advantages to analyzing stranger-to-stranger rape and sex offenses lies in the seriousness of the crime, its relative rareness, and the availability of information for a particular offense or group of offenses.

B-45. As with robbery, the suspect descriptors are important to the analyst for examining rape and sex offenses (see Figure B-6). The victim descriptors are also important in analyzing rape and sex offenses. In many cases, the perpetrator will restrict his attacks to victims of a certain age, a particular race, or a particular occupation grouping.

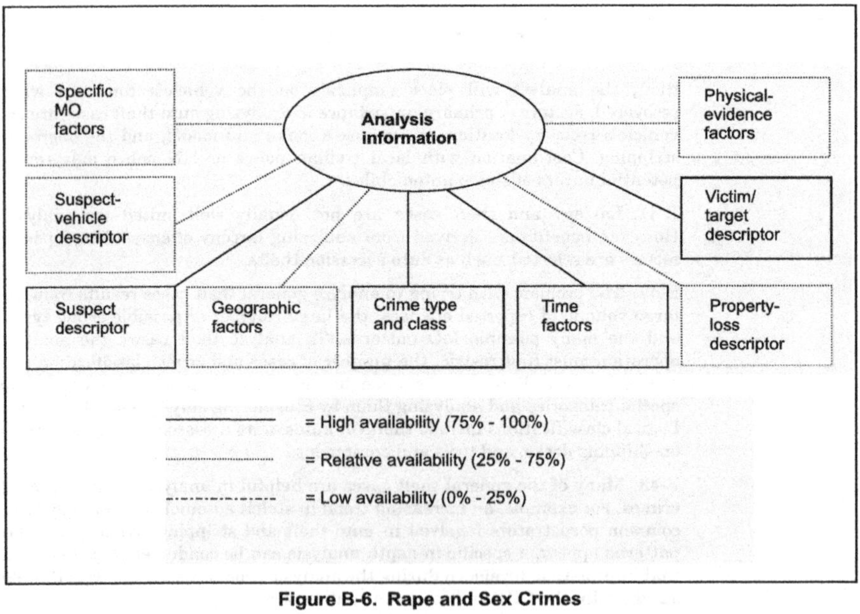

Figure B-6. Rape and Sex Crimes

FORGERY AND FRAUD

B-46. Forgery and fraud cases can provide correlative information for other reported criminal offenses such as burglary and strong-armed robbery. The analysis of forgeries is also aided by the fact that most forgers are repeaters who have established definite MO patterns.

B-47. An analysis of forgery and fraud cases can generate information that is disseminated to area merchants in the form of check-warning bulletins. The use of these bulletins can enhance community relations.

B-48. The primary problem in analyzing forgery and fraud cases is the immediate availability of information. In some cases, a delay of between three or four days to several months occurs between the time the offense was committed and the time it was reported.

ASSAULT AND MURDER

B-49. Most criminal offenses constituting assault and murder do not lend themselves to analysis. The comparative rarity of these crimes involving complete strangers makes it difficult to plot and predict assaults and murders. A tremendous amount of time and research is necessary to predict assaults and murders effectively. This data may identify specific areas with high rates of violent crime. This allows the commander to identify unsafe areas and warn soldiers to stay away from them.

CRIMINAL-ANALYSIS PROCEDURES

B-50. To organize the flow of information within a PMO and to facilitate analyses in support of crime-prevention efforts, the staff reviews a list of offenses. This list contains offenses that lend to analyses. The analyses provide the azimuth for crime suppression and countermeasures (for example, focused investigation/surveillance and patrol distribution). When final action on an MPR or an ROI covering one of these offenses is completed, it is forwarded to the staff for review.

B-51. The staff uses preprinted work sheets containing the factors of particular interest for an offense. They annotate and review the specific criminal factors. If some of the information is not available, the MP or CID investigator continues the investigation. The period covered by each work sheet will depend on the volume of cases at the particular installation. For a low-incidence-rate crime (such as robbery), one work sheet will suffice for the entire calendar year. For a more frequent crime (such as larceny), it may be necessary to use a different work sheet for each month or quarter. A summary sheet can be used to keep track of the data from all of the reporting periods in the year.

B-52. The data-collection sheets give specific information on individual types of crime. This information, together with general crime trends that are identified by comparing the data from DA Form 2819 for the current period to earlier periods and the geographic data from the crime-occurrence map, gives most of the information that is required to develop tightly targeted crime-prevention programs. While this system requires additional forms (the data-collection sheets and the crime-occurrence map) to be developed for use by the

MP Intelligence Officer (US Army) (S2), it does not increase the patrol's workload.

B-53. In addition to the information recorded in the forms, handwritten notes should be kept on items or trends that become apparent to the analyst during the review. For example, as a result of reading the cases on robbery to extract information for the collection form, the analyst may notice that a high proportion of the victims were attacked while crossing an unlighted athletic field near a troop billet area. This should be recorded so that it may be used in publicity campaigns and so that it may be passed along to patrols who routinely patrol the troop billet areas.

B-54. A geographic analysis is performed to determine information not available in the original data-element source (for example, a crime report). When a pin is used to indicate a crime location on a map, the relationship to other reported crimes of the same type becomes apparent to the analyst. Mapping is usually the source for a geographic analysis.

B-55. Mapping-analysis techniques involve the use of a map to depict the actual geographical relationships between particular criminal events according to prescribed data elements. A pin map displaying the actual locations of burglaries over a period of time is an example of such a technique.

B-56. In selecting a mapping technique, several things must be considered. These include the number of data elements that are to be recorded in the mapping, the retrievability of stored data, and the number of maps to be maintained.

B-57. The number of data elements to be recorded on a map can dictate the technique to be used. For example, if the case number is to be recorded in conjunction with the location of the offense, a dot (color-coded, self-adhesive paper disk) map or a flag-pin map can be used, whereas a typical color-coded pin map cannot. If particular MO factors are recorded, various color-coded and marked pins can be applied to identify specific MO patterns.

B-58. In addition, the length of time a particular map or set of maps is maintained should be considered. No set rules are established regarding a map's maintenance time; however, a number of agencies develop annual statistics based upon maps. Generally, maps are maintained for short-term (quarterly) and long-term (annually) periods for each type of crime. This decision should be based on the volume of criminal activity, physical-space limitations, the specific mapping technique adopted, and the number of maps to be maintained. As geospatial-terrain data becomes available, mapping techniques can be fully integrated into the MPACS.

B-59. Stored map data can be easily retrieved for comparing data. Pin maps can be photographically recorded and digitized for software applications. Computers and scanning equipment can be used to store the data for later retrieval. Color-coded paper dots placed on acetate overlays covering area maps facilitate the retrieval of recorded data. Digitized photographs of the maps are excellent sources of evidence for trials and court-martials. Moreover, the developed maps may contribute to the CCIR.

B-60. The number of maps to be maintained must be based on physical-space limitations, staffing, the types of crimes selected for analysis, and the maintenance period for each map. There are five types of crime suited to geographical analysis (burglary, robbery, housebreaking, auto theft, and sex crimes). Other crimes (such as theft of government property) may be depicted as a geographical analysis and contribute to RII. Housebreaking and burglaries can be placed on the same map by using different colored pins for each type of crime. The incorporation of numbered pins of different colors can be used to designate the type of premise attacked. Auto-theft and recovery maps may be kept on the same map with recorded information on thefts of vehicle parts.

B-61. The number of maps maintained should reflect the needs of the agency according to the volume of reported crimes and the different data elements or informational factors to be recorded. Therefore, the more detailed the analysis, the more maps are required. The actual number of maps maintained must be left to the discretion of the PM or the commander.

B-62. Periodically, a summary of the information that is collected, countermeasures, and operational/tactical procedures are briefed to the installation commander. Whether this is done monthly, quarterly, or semiannually will depend on the volume of reported crime and the commander's preference.

CRIMINAL-ANALYSIS SUMMARY

B-63. Criminal analysis is an important element to the MP function of PI. The techniques mentioned in this appendix are not all-inclusive. The PM/CID commander establishes formal procedures for integrating data and developing PI. It may require the formation of an ad hoc team, or it may be accomplished with computers and the MPACS. Criminal analysis requires close coordination with the G2/J2, the public affairs office (PAO), civilian police, and others based on METT-TC.

SECTION III — COMMAND AND LAW-ENFORCEMENT COUNTERMEASURES

B-64. This section discusses actions that commanders or PMs can take to reduce crime on Army installations. In many cases, the actions described have applications other than strictly crime prevention. However, only the crime-prevention aspects are discussed.

B-65. Most of the data available on the effectiveness of law-enforcement measures in reducing crime comes from studies conducted by civilians. While most of the findings of these studies are applicable to crime-prevention programs on Army installations, differences in population, the degree of control that can be exercised by authorities, and other environmental factors may dictate that the civilian recommendation be modified before implementation on military installations.

CRIME HOT LINES

B-66. A crime hot line is a dedicated crime-reporting telephone number located at the MP or security desk. This hot line allows anyone in the community to make an immediate report of an observed crime or suspicious activity. An effective crime-reporting program is a deterrent to crime and enhances law-enforcement responses to such incidents. Considerations in implementing a reporting program include the following:

- It should be publicized that personnel reporting incidents are allowed to remain anonymous if they desire. Some individuals will report their observations only if they know they can remain anonymous. If deemed feasible, neighborhood-watch blocks could be provided designated block numbers that could be used when reporting crimes or other suspicious activities. This system would allow neighborhoods to receive feedback on the disposition of the reported incident.
- The hot line's phone number should be easy to remember. This could include a number where extension digits are all the same, are in ascending or descending order, or spell out a word (such as 4357 [HELP]). Sticker labels listing the number could be placed on the phone with other emergency numbers for quick reference.
- The program should be well-publicized.
- Members on the installation should be educated on the desired procedures for reporting incidents.
- Each call should be documented. Records should be maintained concerning the results of these calls to evaluate the program's effectiveness.

CRIME PREVENTION THROUGH ENVIRONMENTAL DESIGN

B-67. Crime-prevention practitioners are recognizing the importance of considering design and physical planning in crime reduction. The crime-prevention officer has an opportunity to influence the design of facilities through the installation planning board. However, to be effective, he must understand a number of concepts about the relationship between the physical design of buildings and crime occurrences. These include the concepts of territoriality, natural surveillance, and defensible space.

TERRITORIALITY

B-68. Historically, a single-family home on its own piece of land and somewhat isolated from its neighbors (but often by as little as a few feet) has been considered to be the family's territory. The single-family home sits on a piece of land buffered from neighbors and the public street by intervening grounds. At times, symbolic shrubs or fences reinforce a boundary. The positioning of lights in windows that look out on the grounds also act to reinforce the claim.

B-69. Unfortunately, as the population has grown and the need for housing has increased, the trend toward developing single-family units has been paralleled, if not surpassed, by the development of row houses, apartment buildings, and various high-rise structures. Architects, planners, and designers involved in developing structures have not paid a great deal of attention to crime control or the need for an individual or a family group to identify with its home in a manner that might affect crime. Therefore, most families living in apartment buildings consider the space outside their apartment door to be distinctly public. In effect, they relegate responsibility for all activity outside the immediate confines of their apartment to public authorities. A question is whether environmental design can be used to extend the boundaries of these private realms, subdividing public space outside quarters so that more of the common space comes under the resident's influence and responsibility.

B-70. Through extensive research of efficiently functioning housing developments, a number of mechanisms have been identified that may be used in the design process (or may be added after construction). These mechanisms encourage the residents of multifamily dwellings to identify more with the ground or area around their immediate home site and to assume responsibility for its protection. Presented below is a brief discussion of a number of the mechanisms.

- **Site design.** If the grounds around a set of quarters can be directly identified with a particular building and the residents of that building take a personal interest in the use or upkeep of that area, they will play a role in protecting it. Through proper site design, a recreational area adjoining a building may be used as a buffer zone by providing play equipment for young children and seating areas for adults. The fact that children play and adults sit in these areas serves to increase the residents' concerns with the activities taking place there. Strangers are usually recognized and their activities come under observation and immediate questioning.
- **Street design.** Research has shown that by the placement, enclosure, or rerouting of streets and traffic the nature of a particular area can be changed and the crime rate reduced. For example, a particular portion of a street might be closed to vehicular traffic, and play equipment and seats may be added. In a number of areas where this technique has been used, it has been found that most residents know or at least recognize people up and down the block and strangers on the street are identified. Similar approaches that involve rerouting traffic, using one-way streets, or blocking off streets has lowered the crime rate in some areas.
- **Symbolic barriers.** The types of barriers that planners may use in laying out an area include open gateways, light standards, low walls, and plantings. Both physical and symbolic barriers serve the same purpose—to inform an individual that he is passing from a public to a private space. Symbolic barriers identified by residents as boundary lines serve as defining areas of comparative safety. Many places warrant the use of symbolic barriers, including transition points between a public street and the semipublic grounds of a building; an area between a building's lobby and its corridors; or hallways on particular floors of a building.
- **Internal design.** Although economics may sometimes enter the picture, a building's interior may be designed for specific groupings of apartment units and shared entrances. These factors may cause the residents of these apartments to develop a concern for the space immediately adjacent to their dwelling. For example, on each floor of an apartment building, two to four families might be required to share

a common corridor area. The apartment doors would be grouped around that common corridor, and access to elevators or stairs might be screened by a glazed partition. The net effect would be that the floor's residents would adopt the corridor as a collective extension of their dwelling unit and would take an increased interest in its maintenance and use.

- **Facilities and amenities.** The location of particular facilities (such as play and sitting areas and laundry facilities) will tend to give an area a high intensity of use and support the idea of territoriality. The presence of residents involved in various activities (children at play and people chatting or engaged in other types of activities) allows for casual surveillance by concerned members of the family and screens out possible intruders.

B-71. Reducing the number of apartment units grouped together to share a collectively defined area and limiting the number of buildings that comprise a housing project are important factors for creating an environment that residents will help to protect. Research has documented the fact that housing projects comprised of fewer high-rise buildings (two to four) have lower crime rates than projects containing a larger number of buildings. Based on this finding, it is argued that there appears to be much less freedom of movement in the public spaces of the smaller high-rise projects. Unlike buildings and large developments, every building of a small grouping usually has an entrance directly off a public street. These dwellings more closely resemble middle-income, high-rise developments and look more private.

B-72. As a crime-prevention officer, you may not be in a position to directly use these techniques. However, your familiarity with these approaches and the value of their use in the crime-prevention process are important elements in your arsenal of tools to create public involvement in reducing crime. In particular, the purpose of outlining these tools is not to equip you to be a designer, but rather to equip you to communicate with those who are involved in that profession. The discussion that follows will further enhance your ability to converse with designers.

NATURAL SURVEILLANCE

B-73. Experience has shown that the ability to observe criminal activity may not be adequate to stimulate an observer to respond with assistance to the person or property being victimized. The decision to act depends on the presence of motivational conditions, including—

- The degree to which the observer has developed a sense of personal and property rights that are being violated by the criminal act.
- The degree to which the observer feels that the event is within his area of influence.
- The observer's ability to clearly identify whether the act is unusual for the particular area.
- The observer's identification with either the victim or the property being vandalized.
- The degree to which the observer believes he can effectively alter the course of events he is observing.

B-74. Based on these conditions, a number of mechanisms have been identified that can be used to design the grounds and internal areas of apartment units, housing developments, and other residential areas to facilitate natural monitoring of activities taking place. By providing opportunities for surveillance through the positioning of windows in relation to stairs, corridors, or outside areas, continual natural observation will be maintained and crime will be deterred. If such steps are taken, the security of observed areas will be understood by the potential criminal, making him think twice before committing a crime.

B-75. The first of these natural surveillance mechanisms involves the positioning of service areas and access paths leading to apartment buildings to facilitate surveillance by residents and authorities. For example, buildings might be designed so that their entries face and are within 50 feet of a street, so that well-lit paths lead to the front door or the lobby, and so that the lobby is arranged to afford good visibility from the street. Other related steps focus on the strategic placement of windows, fire stairwells, lobby lights, and mailboxes so that they can be easily viewed from the street. Elevator waiting areas on each floor can also be designed so that they can be seen from the street level. Research has proven that if steps such as these are taken, residents will be more likely to become involved with protecting the facility, MP patrols will be in a better position to observe what is going on, and criminals will be discouraged from vandalizing the site.

B-76. A second technique that might be used to increase surveillance is to design facilities so that people within them will naturally view commonly used paths, entries, and play and seating areas during their normal household activities. This concept also focuses on the strategic placement of windows, lighting, and open areas so that natural surveillance by residents is improved.

B-77. Another mechanism involves the subdivision of housing areas into small, recognizable, and identifiable groupings that improve visual surveillance possibilities. Research has shown that in housing developments where the surveillance of a neighbor's outside activities was possible, residents were found to be very familiar with everyone's comings and goings. The overall effect was to cement collective identity and responsibility through social pressure.

DEFENSIBLE SPACE

B-78. Defensible space is a term for a range of combined security measures that bring an environment more under the control of its residents. A defensible space is a residential environment that can be used by inhabitants for the enhancement of their lives while providing security for their families, neighbors, and friends. The physical mechanisms suggested to create safety and improve upkeep (as part of the defensible-space concept) are self-help tools wherein design catalyzes the natural impulses of residents rather than forcing them to surrender their shared social responsibilities to any formal authority.

B-79. Research has revealed investigative techniques that might be used to modify existing housing areas to make them more secure. The following methods may require alteration or adaptation to the particular situation on your installation:

- Widening major pathways and using colored decorative paving.
- Differentiating small private areas (front lawns) outside each dwelling unit from the public path with low, symbolic walls.
- Adding public-seating areas in the center of public paths far enough from private-dwelling units to eliminate conflicts of use but close enough to be under constant surveillance by residents.
- Designing play areas as an integral part of open space.
- Adding new and decorative lighting to highlight various paths and recreation areas at night and extending the residents' surveillance potential and feeling of security.
- Adding seats and path networks to recreational facilities where large, central court areas exist. This increases the interest and usability of the areas.
- Redesigning parking and play areas around buildings to create the illusion that the buildings are grouped where natural opportunities exist.
- Modernizing building entrances to create breezeways into building courts and to accommodate a telephone intercom for opening entry doors to the lobby.
- Providing video surveillance of public grounds and central paths by security of public monitors.
- Installing audio surveillance capabilities in elevators and at the doors of residences.

CRIME-PREVENTION MODEL

B-80. The model for crime prevention through environmental design is based on the theory that action must be taken to counter crime before it occurs. The critical element in this model is the environmental-engineering component. It provides both direct and indirect controls against criminal activity by reducing the opportunity for crime through science and technology and the use of various urban planning and design techniques. The model explains what environmental engineering is and how it supports crime prevention. With this information, you may be in a better position to understand and respond to questions and discussions on how urban design and planning can have an impact on the installation's criminal element.

THE ENVIRONMENTAL INFLUENCE ON CRIMINAL BEHAVIOR

B-81. The basic theory that supports crime prevention through environmental design is that urban environments can influence criminal behavior in two ways. First, the physical surroundings in which people live have an effect on each individual. These physical characteristics include noise, pollution, overcrowding, and the existence and unmonitored spreading of refuse and other unsightly waste. The second element that must be dealt with in the environmental-engineering formula concerns the social characteristics of the community that provide individuals with social relationships to which they must respond. Characteristics such as alienation, loneliness, anxiety, and dehumanization are seen as keys to criminal behavior.

B-82. In terms of these environmental characteristics, buildings are all too often constructed to be dangerous, with corridors and passageways hidden from public view. Elevators, basements, and storage and washroom areas are also laden with danger due to their design. Various large-scale housing developments are not secure in that they are often isolated from the main flow of traffic (both human and automobile) and are closed to public use and public view.

B-83. With regard to altering the social characteristics of the community and their relationship to criminal behavior, it should be recognized that behavior is future-oriented, not past-oriented. A man steals so that he can have a car or money in the future, not because in the past he experienced psychic trauma, a broken home, poverty, or delinquent associates. Criminal behavior can be explained directly in terms of the consequences of behavior and in terms of noncriminal variables such as poverty, race, or social class. Criminal behavior is viewed as a problem to be dealt with and not symptomatic of other problems (such as poverty, mental conflict, class conflict, unemployment, or undereducation). To change criminal behavior, it must be dealt with directly by removing the environmental reinforcement that maintains the behavior. The approach advocated is to change the environment to which the individual responds.

ACTION APPROACHES TO CRIME PREVENTION THROUGH PHYSICAL PLANNING

B-84. The primary focus of crime prevention has been on what architects, planners, and other nonpolice professionals can do in terms of various physical-planning strategies to reduce criminal opportunity. Experienced MP personnel have long recognized that certain physical conditions can contribute to the rate and nature of crime. They have also developed a capability to identify high crime-risk locations by noting such factors as poor lighting and weak points of entry as potential targets. The critical job is to identify specific areas concerning physical planning and design that can be responded to and actions can be taken against on the installation.

B-85. Attempting to reduce crime or the fear of crime by regulating physical environments is easier said than done. In fact, although crime prevention can be built into almost every aspect of community planning, it is often ignored for a number of reasons. For example, fragmentation of responsible agencies is a key problem. In addition, crime has historically been looked upon as the exclusive responsibility of MP forces; not of those in charge of education, housing, or health and welfare. Yet, with an understanding of crime prevention combined with the knowledge that design techniques can change the opportunity for criminal behavior, PSIs will be able to talk the language of the planner and the designer and to be able to advise them from a police perspective.

B-86. It is notable that a number of civilian police agencies have become involved in the physical-planning process and have achieved notable results from their work. For example, the Fremont, California, Police Department has been involved in a planning process and maintains that law enforcement should become an integral part of the master- or comprehensive-plan review to screen all redevelopment plans for safety and crime hazards. Working with other units of municipal government as well as architects and designers, the department drew up a set of model guidelines for the evaluation of projects. The model included evaluation criteria dealing with such subjects as the accessibility of buildings to patrol units; traffic flow and off-street parking provisions; and the location and regulation of cul-de-sacs, playgrounds, common greens, fences, and security entrances. In addition, working with agencies such as the American Institute of Architects, the National Public Works Association, the Association of Public Utilities, and others, the department identified a number of subjects that are of specific concern to police officers and that should be considered in the design and planning stage. As a result of these efforts, the following list of design concerns was developed by the department:

- Building setbacks (front, side, and rear).
- Wall construction, interior and exterior (industrial, commercial, and residential).
- Door construction, setbacks and security (industrial, commercial, and residential) (including carports, garages, and sliding-glass doors).
- Windows and skylights, setbacks, heights (from ground), show-window displays, and the type of frame or pane.
- Stairs (stairwells and staircases).
- Balconies.
- Utility boxes.
- Fences, walls, hedges, screens, setbacks, heights, and louvers.
- Parking (public and private).
- Lighting (industrial, commercial, and residential).
- Streets, sidewalks, and walkways (locations, slopes, curvature, grades, and the length of a block).
- Alleys (blind and through alleys).
- Visibility of valuables (people, safes, cash registers, and personal property).
- Signs (street signs and signals, traffic signs and signals, and advertising signs).
- Accessibility; approach, entrance, and exit (pedestrian, vehicular, services, residential, commercial, and industrial).
- Public utilities and easements (gas, water, telephone, and electrical).
- Public areas and facilities (public restrooms, parks, bus stops and shelters, playgrounds, recreation halls, and so forth).
- Street trees and shrubbery (types, heights, and locations).

B-87. With this information, you can improve the security aspects of the community's physical-planning process. There is a probability that the work and recommendations of the National Advisory Commission on Criminal Justice Goals and Standards may help you in your efforts to get involved in the design process. More specifically, the Commission noted that "every police agency should participate with local planning agencies and organizations, public and private, in community physical planning that affects the rate or nature of crime or the fear of crime."

B-88. The future role of MP forces in crime prevention through physical planning will depend on the PSI's initiative. The perspectives and knowledge of what is happening in this field, combined with a working knowledge of the language, should equip PSIs to sell this approach as part of the overall program. It is important to point out that others who represent professions other than crime prevention are aware of the relationships between urban planning and crime. Remember, they deal in concepts, approaches, and ideas that have not involved the realities that law-enforcement personnel face.

SPECIALIZED PATROL TACTICS AND SURVEILLANCE

B-89. Specialized patrol operations use a variety of tactics in attempting to control identified crime problems. The most common tactics include uniformed tactical patrols and suspect and area surveillances. The following paragraphs discuss these tactics in terms of their target crimes, operation requirements and characteristics, and established or perceived levels of effectiveness.

B-90. The appropriateness of a given tactic depends on the characteristics of a particular crime problem. The selection of specialized patrol tactics should be made on the basis of a careful and continuous analysis of crimes. Most crimes can be addressed by more than one tactic. Several tactics might be tried in an effort to find the best one, and it is quite possible that the most effective approach to a given crime problem will include the combination of several tactics.

UNIFORMED TACTICAL PATROLS

B-91. A uniformed tactical patrol is the most traditional and widely used form of specialized patrol. It is a simple, straightforward approach to specialized patrol that involves the same procedures and techniques used by MP officers on routine patrol. These include constant visible movement throughout an area to generate a sense of police presence, careful observation of street activity, vehicle and pedestrian stops, and citizen contacts. The difference between uniformed tactical patrols and routine patrols are that uniformed tactical patrols use these tactics in an intense, concentrated fashion. MP officers are relieved of the responsibility for responding to routine calls for services so that they can devote their full time and attention to patrol, thus intensifying its impact. In addition, uniformed tactical operations typically deploy a number of MP officers in target areas, thereby increasing the level of patrol in these areas.

B-92. Uniformed tactical patrols can be used to control virtually any type of suppressible crime (for example, crimes that can be viewed from locations where the police have a legitimate right to be and those that can be potentially affected by police operations). These suppressible crimes include street robberies, purse snatches, vehicle thefts, burglaries, and housebreakings. Uniformed tactical patrols can also have an impact on other types of crime as officers use observation, field interrogation, and citizen contacts to develop information on the locations, activities, vehicles, and associates of suspects.

B-93. The primary purpose of uniformed tactical patrol is deterrence. This tactic is based on the assumption that highly visible, active patrols will deter potential offenders. By increasing the perceived probability of apprehension, conspicuous patrol is thought to reduce the likelihood that crimes will occur. If the deterrence should fail, heightened patrol coverage is believed to increase the probability of the immediate apprehension of the suspects.

B-94. Uniformed tactical patrols are often used to saturate an area that is experiencing a particularly serious crime problem. Although it has been widely used for years, saturation patrol has never been clearly and adequately defined. Exactly what level and intensity of patrol constitutes saturation has never been determined, nor have the effects of different levels of patrol been clearly established. It is difficult to prescribe the level of uniformed tactical patrols that should be used to disrupt a crime pattern in a particular area. This should be determined through an analysis of the size and characteristics of the area of concentration of each potential target crime pattern coupled with an assessment of manpower availability.

B-95. Some patterns can be effectively handled by a very small increase in the patrol level. For example, in Portland, Oregon, Operation Crime Reduction Involving Many People (CRIMP) had a significant impact on street robberies and assaults in the city's skid-row area by initiating two-officer, uniformed foot beats in the area. This was sufficient to saturate the primary locations of the target crimes during the high-crime hours, and it led to a substantial reduction in these crimes with little apparent spillover into adjacent areas.

B-96. Other departments have used saturation patrols on a much larger scale. For example, in the mid-1950s, the New York City Police Department attempted to saturate an entire precinct by assigning over 200 additional officers to the precinct's patrol force. Foot beats covering extremely small areas were arranged in straight lines so that an officer could keep the entire street area of his beat in view at all times. The four-month experiment led to a dramatic reduction in crime in the precinct. Compared with the same period in the previous year, street muggings fell by 89.9 percent, burglaries where entry was made from the front of the building dropped by 78 percent, and so on. The only crime category that was not affected was the relatively private crime of murder. Since crime displacement was not examined in the experiment, its true impact remains unknown. The experiment strongly suggests that massive additions of police manpower can have a substantial effect on crime. The problem is that most departments do not have the ability to conduct even a limited version of this experiment.

B-97. The amount of resources required for saturation patrols can vary tremendously, and there is no definite way of determining how much additional patrol is needed. This can best be decided on a problem-by-problem basis, using experience and evaluations of past efforts as a guide. As a rule, patrol augmentation should be sufficient to affect rather quickly the perceptions of would-be offenders concerning the level of police activity in a particular area. It appears that for such a strategy to be effective, sufficient resources should be applied in a limited area to ensure a true saturation effect.

B-98. Uniformed tactical patrols can use various modes of transportation. Patrol cars are most often used; however, foot patrols can be effective in congested service districts and bikes have been used to good advantage in high-density residential areas. The mode of transportation should be selected based on visibility, mobility, and access to citizens.

B-99. Some specialized units deploy MP officers in unmarked police cars. This is done in an effort to effect a balance between overt and covert operations, hopefully gaining many of the advantages of both. Unmarked cars may also be readily available since in many departments investigators work primarily during the day, which leaves their vehicles free for specialized patrol in the evening and early morning watches. This approach has serious drawbacks. First, unmarked MP cars are somewhat less visible than marked cars, yet they are still easily recognizable as police vehicles to large segments of the public (especially when the officers in them are in uniform). Second, the use of these cars in uniformed tactical patrols could lead to the sacrifice of some of the deterrent effects of high visibility without realizing the apprehension benefits of truly covert patrols.

B-100. Several patrol techniques have been tried to increase the visibility of uniformed patrol and to enhance the sense of police presence. A tandem patrol uses two marked patrol cars that follow each other at intervals of one-half to several blocks. Two units can also patrol on parallel streets one block over or in an alley. Another approach combines foot and vehicle patrols in an effort to increase visibility. Officers park their marked cars in conspicuous locations in high-crime areas and then are transported to other high-crime areas where they patrol on foot. The frequent repetition of this procedure is seen as a way of multiplying patrol visibility.

B-101. Unless a target area is more or less completely saturated (as in the New York experiment), MP patrols should move in a random, unpredictable pattern. This will make it difficult for potential offenders to plan their crimes on the basis of observation of patrol activities. One department found that its uniformed-tactical-patrol operation was actually helping burglars to commit break-ins. Interviews with confessed burglars indicated that they were aware of a visible patrol passing through the neighborhoods at regular intervals and they planned their crimes accordingly.

B-102. In addition to increasing the level and visibility of patrol in high-crime areas, uniformed tactical patrols often use aggressive patrol tactics involving frequent vehicle and pedestrian stops. MP patrols stop, question, and perhaps search citizens when there are reasonable grounds for suspecting that the citizens may have committed, may be committing, or may be about to commit a crime. Since the concept of reasonable suspicion is vague, MP officers have a considerable amount of discretion in conducting field interviews. Field interviews that do not result in either immediate arrest of the citizen or in alleviation of the officer's suspicions are usually documented in field-interview cards or spot reports. Field interviews serve to generate information about the activities of probable suspects; more importantly, they make the suspects aware that the police know of their presence in a given area, regard them as suspicious, and are watching them closely. This is expected to reduce the likelihood that they will commit crimes, at least in the area in which the tactical force is working.

B-103. The extensive use of field interviews is often highly controversial. Tactical units that emphasize field interviews have been accused of being storm troopers who constantly harass citizens and abuse their rights. There is sometimes a certain amount of truth to the allegations of harassment, especially when the activities of a particular type or group of suspects are being closely monitored. Not surprisingly, individuals who are stopped and questioned frequently are likely to complain, particularly if they have reason to be concerned about close police scrutiny. A recent study of field interviews in San Diego found that when conducted at moderate levels, field-interview activities do not have a major effect on police and community relations. The majority of citizens in all three study areas accept field interviews as legitimate and properly conducted police activity. The majority of all citizens who were the subjects of contacts felt that the contact was justified and properly conducted. The study found that the suspension of field interviews in a particular area was associated with a significant increase in the level of suppressible crime and when reinstated in the area, these crimes declined. The results of the San Diego study provide confirmation for the widely held belief that field interviews can have an important deterring effect on suppressible crime without doing irreparable damage to police and community relations.

B-104. The potentially negative impact of field interviews on certain segments of the community can be held to a minimum if the interviews are conducted in a professional manner. Citizens should be informed of the reasons why they are being stopped. They should be detained for as little time as possible and should not be subjected to verbal or physical abuse. There is also no need to stop everything that moves.

B-105. While experience and a limited amount of research indicate that uniformed tactical patrols can have a positive impact on the level of suppressible crime in areas, the overall effectiveness of this tactic is a controversial and much-debated issue. The principal concern is that rather than reducing crime, uniformed tactical patrols may simply lead to its displacement from one area to another or from one period to another.

B-106. When conducting uniformed tactical patrols, a PM activity should carefully monitor crime trends for indications of possible displacement effects. This is an important aspect of evaluating the impact of tactical operations, and it will provide information to guide future deployments and tactical decisions.

AREA SURVEILLANCE

B-107. Covert patrol and surveillance of high-crime areas can be used to make apprehensions for crime problems. These problems include those for which there are no suspects who warrant personal surveillance, the suspects are too numerous to permit personal surveillance, and there are too many potential targets to conduct either physical or electronic stakeouts. Examples of these types of problems would be a rash of residential burglaries or auto thefts in a particular area.

B-108. This tactic simply involves the covert patrol of a particular area and the observation of suspicious or unusual activities and occurrences that might indicate the likelihood of a crime. Suspicious individuals are not stopped but are watched until they either commit an offense or the officers' suspicions are removed.

B-109. The list of various techniques that can be used in an area surveillance is virtually endless. The following are some techniques that have been effectively used by specialized patrol units:

- Mingling with citizens at the crime scene to pick up information on possible suspects.
- Maintaining rooftop surveillance of a shopping center's parking lot to locate larcenies from vehicles.
- Surveilling housing areas by posing as maintenance workers.
- Following likely crime victims such as elderly citizens leaving a bank.
- Surveilling rooftops for unusual activity from aircraft or higher buildings. Binoculars are used to facilitate surveillance, and rooftops are marked so that street units can be dispatched to check out suspicious circumstances.

B-110. As in all types of plain or uniformed patrol, care should be taken to ensure that area surveillance is truly covert. Rental vehicles that can be changed frequently provide an excellent, though expensive, means of covert transportation.

B-111. Police on covert patrol should be dressed to blend in with the environment in which they are working, and they should have apparently legitimate, nonpolice-related reasons for being where they are. Several specialized units have found that surveillance teams composed of one male and one female officer can work in many situations without arousing suspicion. An apparently married or romantically involved couple lingering in a park, meandering slowly down the street, or sitting together in a parked car would generally appear less suspicious than two male MP officers doing the same things. Finally, it should be noted that in some small neighborhoods where residents know each other well, covert surveillance may be difficult, if not impossible, since the presence of any stranger arouses immediate curiosity and suspicion.

SUMMARY

B-112. These tactics represent the basic approaches that specialized patrol operations normally take in trying to control suppressible crime. Some of the tactics (such as uniformed tactical patrols) are directed primarily at crime deterrence, while others (such as suspect surveillance) are used to achieve apprehensions for target crimes. The tactics are most commonly used independently of one another. However, there are some indications that the combined use of several tactics in an integrated operation might be an effective way of coping with particular types of crime. Especially promising is the coordinated use of highly visible and covert patrols. A visible patrol force could be deployed to a particular area to deter crime there and direct it toward other areas in which MP forces using covert tactics are working. To date, efforts to direct criminal activity to areas or targets where MP forces are set up to make apprehensions have only been tried on a sporadic basis. However, this appears to be a promising approach to crime control and warrants greater attention in the future. It can be viewed as the creative use of crime displacement.

B-113. Many of the tactics mentioned previously could not be effectively used by MP patrols that are responsible for handling routine calls for service. It is obvious that the effectiveness of decoy operations, physical stakeouts, and suspect surveillance would be destroyed if the MP officers had to interrupt these activities to handle calls for service. It is also generally unwise to have MP officers in civilian clothes respond to calls for service.

B-114. The importance of crime analyses in identifying crime problems and suggesting appropriate tactics deserves repetition. Specialized patrol units should be deployed to address clearly identified crime patterns, and they should choose their tactics on the basis of an analysis of these patterns and a comprehensive knowledge of the area of occurrence. The nature and characteristics of a particular crime pattern should be the driving force behind the selection of tactics. None of the tactics of specialized patrol can be effectively used unless crime patterns have been identified and analyzed.

PUBLICITY CAMPAIGNS

B-115. Public-information campaigns are an essential part of every crime-prevention effort. The installation PAO can provide assistance in—

- Informing the public of the magnitude of the local crime problem.
- Disseminating information on crime circumstances.
- Generating the interest and enthusiasm necessary to initiate and sustain community crime-prevention programs.

B-116. The PAO should be a member of the installation's crime-prevention council and must be involved from the start in planning crime-prevention efforts. There are many information vehicles at the installation level available to carry the message to the public, including—

- Digital marquees.
- Radio.
- CCTV.
- Installation and unit newspapers.
- Posters and leaflets.
- Commanders calls and similar meetings.
- Town-hall meetings.

B-117. Not all media are equally effective in reaching a particular segment of the post population. To ensure that the media campaign is effective, identify the segment of the population requiring the information and the exact message to be communicated. The more specific the target audience, the more effective the media campaign will be.

B-118. Whatever media is used, it is important to provide coordinated input to the PAO well in advance of the desired publicity campaign date to allow for writing, rewriting, and publishing. For material to be published in a magazine, a minimum lead time of six months is normally required. For installation newspapers, a lead time of a week may be sufficient. Check with the PAO to determine the correct lead times for local publications.

B-119. When addressing civic groups or school audiences, films are often more effective than straight lectures because of the animation and other special techniques used to illustrate critical points. Several Army films are available as are a large number of commercial films.

RESIDENTIAL-SECURITY SURVEYS

B-120. A security survey is an in-depth, on-site examination of a physical facility and its surrounding property. The survey is conducted to determine a residence's security status, to identify deficiencies or security risks, to define the protection needed, and to make recommendations to minimize criminal opportunity. Because of several common characteristics, one expert has likened the security survey to the traditional criminal investigation. This comparison hinges primarily on the facts that both techniques are systematic in nature; are aimed at identifying the method of a criminal act; and are, in effect, more an art than a science. It should be recognized, however, that the survey has two other advantages. First, it can be conducted before the commission of a crime; second, it can offer protection against, rather than just remedial action after, criminal victimization.

THE RESIDENTIAL-SECURITY SURVEY AS A CRIME-PREVENTION TOOL

B-121. A residential survey is not a substitute for an aggressive neighborhood watch. It supplements these efforts and should be established after Neighborhood Watch and Operation ID programs are established.

B-122. The security survey is the primary tool used in crime prevention to recognize, appraise, and anticipate potential loss in residential areas. It is often defined as the backbone of a local crime-prevention program. In practice, it combines the security experience, training, and education of the local crime-prevention officer and focuses on a single element—the analysis of a residential facility.

B-123. The inherent value of surveys has been proven by nearly 300 civilian police departments across the country that have established crime-prevention bureaus or units. An even broader endorsement of the survey technique, however, was provided by the National Advisory Commission on Criminal Justice Standards and Goals, which stated that, "Every police agency should conduct, upon request, security inspections of businesses and residences and recommend measures to avoid being victimized by crime."

B-124. In short, the security survey is a tool that informs a homeowner of the particular areas in which his home is susceptible to criminal victimization along with steps that can be taken to reduce and minimize that potential. Further, the survey is a tangible document that reflects the efforts of the MP force not only to be responsive to community needs, but to get the community more directly involved in the criminal-justice process.

THE CRIME-PREVENTION SURVEY—A PROCESS OF INVESTIGATION

B-125. An important factor in understanding the residential-security survey is that it must be considered as an ongoing process. While a particular survey will result in specific recommendations, each survey will provide a foundation for future action. In combination, these surveys will provide a database that can be used in the analysis of the community's crime problems and guide action toward the resolution or reduction of the problem on a community-wide basis.

B-126. As a starting point, five steps must be used in carrying out the actual survey, while four additional steps must be remembered afterward. These steps are—

- Analyzing the overall environment (neighborhood, block, and so forth).
- Assessing the general vulnerability of the premises.
- Defining the specific points of vulnerability.
- Recommending specific security procedures.
- Including specific remedial hardware recommendations.
- Urging the implementation of the recommendations.
- Conducting a follow-up to ensure that recommendations have been implemented.
- Keeping crime statistics to evaluate the survey's effect and the implementation of recommendations.
- Conducting a second survey of the premise's statistical analysis to determine the alteration in criminal activity in the areas surveyed.

B-127. This process is the continued involvement and participation of the police with the community. Once the survey is completed, the job is not finished. In fact, if this posture is assumed, it may later be learned that the recommendations were not implemented and that the work was done in vain. This can easily lead to a loss of community-wide public confidence in the programs. Similarly, even in those cases when recommendations are implemented, additional crime might be experienced. Attention to this fact, coupled with immediate follow-up, will be essential to avoid losing the confidence gained from the original survey. Through prompt actions, the proposal of additional tactics may become the final step needed to substantially reduce criminal opportunity. In short, the security-survey process is not a one-shot operation and it is not to be looked upon as a simple, straightforward task. It is a continuing, difficult, rigorous, yet effective approach to reducing crime that has not been systematically applied by police agencies in this country.

THE ROLE OF POLICE IN IMPLEMENTING CRIME-REDUCTION PROGRAMS THROUGH SECURITY SURVEYS

B-128. At a minimum, there are two ways to encourage people to improve their personal security. First, you can organize, conduct, and participate in broad-based public-information and -education programs that make use of such media as radio, television, and the press. Second, you can organize, undertake, and follow up on a series of person-to-person security surveys. Clearly, both of these techniques have their advantages. The security survey has a unique quality that does not exist in the public-education program. It provides an added incentive on the part of citizens to implement recommendations because of the personal relationships and respect established during the actual survey by a crime-prevention officer.

B-129. For example, consider a homeowner listening to someone on television recommending improvements of security by installing better locks or alarm systems. You might ask yourself these questions: How secure am I? Are the locks that I have on my doors adequate? If they are inadequate, what kind of locks should I install? Do I really need an alarm system? What can the MP officers do for me in terms of making recommendations about security? In short, you would be aware of the possible need for improved security, but you would not know enough of the specifics to warrant real action. Because home security is a complicated matter requiring careful analysis and the installation of carefully tailored systems, implementation must be approached in a personal way. Certainly, public-education programs can assist in some aspects of crime prevention (such as prevention against schemes, auto theft, and personal protection), but they are not as effective in causing improved security as is personal contact.

B-130. Moreover, for the first time, police are placed in a position where they can actually provide a different kind of advice or service that can be offered in an environment where a crisis has not yet occurred. This is a truly unique opportunity for crime-prevention officers.

CONDUCTING THE SURVEY

B-131. To develop a proper perspective of the types of crimes that a PSI will most frequently be trying to reduce, a review of the cases is necessary to get a broader feel for the actual conditions in your area. During this review process, pay specific attention to photographs in the files. Study crime scenes in an effort to identify the type of security device that was defeated. In particular, if a door was used as a point of entry, note whether it was of adequate construction; if the door frame was broken or separated; if hardware, such as strike plates and door trim was inadequate; and so forth. In addition, review photos to determine if lock cylinders were pulled or if door chains fastened to trim moldings were simply pulled away to permit easy entrance. If photographs of the crime scenes are not available, visit as many crime scenes as possible. While doing so, photograph security risks you can study later and use as examples in future presentations to community groups.

B-132. By becoming familiar with the MO of persons committing such crimes as burglary and larceny, you will be better-equipped to understand potential risk situations and to point them out to potential victims. Quite surprisingly, many of the cases investigated were invited by some obvious crime-risk hazard that was overlooked by a resident. Also, many additional crime risks existed at a particular crime scene that a burglar could have exploited. Such vulnerability might be an indication of other crime targets within the community to which you should pay particular attention in your survey work.

B-133. To be an effective security surveyor, it is necessary to develop an intimate knowledge of the crime factors in your community. You do not have to become a statistician; however, the more you know and understand about crime problems, the methods used in your community, and security devices or systems that were defeated, the better you will be equipped to analyze the potential crime when surveying a home.

B-134. In addition to the general types of crimes that occur in your community, it is necessary to develop an understanding of the details of particular types of offenses. For example, with regard to residential housebreaking, you should be familiar with the types of burglaries and approaches used in particular sections of the community.

B-135. On the surface, it might appear as if this would be a monumental task. However, in terms of housebreaking, you might pull the files for the last quarter and carry out the following steps:

- Tally the number of times entry was made through the front door, rear door, or through a window.
- Identify the approach used for entry (kicking the lock, throwing a shoulder through the door, jimmying the lock, and so forth).
- Determine how often a window was broken or removed or a mechanism was used to force the latch (when entry was made through a window).
- Attempt to determine whether security devices were used in residences (such as alarms, special lighting, or other systems).
- Identify the burglar's general escape route (down a back alley, through a schoolyard, and so forth).

B-136. In developing an understanding of the MO of crimes, statistics that illustrate exactly what is happening in your community will be valuable tools. Not only will you be able to use this information in explaining crime risks while you are surveying the site, but it will be invaluable in making public presentations.

B-137. Only when you have developed the ability to visualize the potential for criminal activity will you have become an effective crime-scene surveyor. It is important that when you arrive on a survey site, you are prepared to give a property owner sound advice on the type of security precautions he should consider.

B-138. In summary, to be a good crime-prevention surveyor, you have to be a good investigator. You must understand the criminal's method of operation and the limitations of standard security devices. In addition, you must be knowledgeable about the type of security hardware necessary to provide various degrees of protection. (Specific information on conducting surveys is contained in Chapter 11).

JUVENILE CRIME PREVENTION

B-139. Law enforcement has long recognized the importance of establishing personal contact with youth. This need was first met with the "Officer Friendly" concept. The goal of this concept was to enhance the image of police among younger children and attempt to negate the unfavorable image other segments of society provided children. This concept has evolved into today's Drug Abuse Resistance and Education (DARE) program which has spread to most law-enforcement agencies.

B-140. Crime-prevention officers can have an impact on juvenile crime by establishing a positive interaction with juveniles. Many areas suffer a high crime rate of which over 50 percent is attributable to juveniles. Nationally, it is known that youth 17 years of age and under (who make up just 16 percent of the population) commit 42 percent of the crimes that cause injury or loss of property. As the punitive aspects of the juvenile justice system today are of questionable value, it is logical that crime-prevention officers should develop programs aimed at juvenile crime.

B-141. The basic problem in establishing any juvenile crime-prevention program is the destruction of preconceived ideas by both youths and law-enforcement personnel. This obstacle can only be overcome by the building of an honest rapport with individual youths. Just as a crime-prevention officer must sell crime-prevention procedures to the public, he must also sell himself and his program to the youth.

B-142. One method of establishing the needed rapport with youth reverts to a similar system used in the Officer Friendly concept. Classroom presentations in high schools and junior high schools can be used to destroy the preconceived ideas held by youth regarding police and law enforcement in general. Once an MP officer enters the classroom, he discovers that the majority of students obtain their ideas of police one of two ways—

- From television shows that portray every officer as a slow-witted moron (such as the old "Car 54" show).
- From their peers at school who relate every negative aspect of law enforcement.

B-143. However the students gain their information, it seldom resembles fact and must be changed to effectively establish a juvenile crime-prevention program.

B-144. In making classroom presentations for crime-prevention purposes, the crime-prevention officer must display honesty at all times. When confronted by uncomfortable questions, the MP officer should render the law-enforcement's point of view and make a truthful explanation. Obviously, some answers will not always be well received, but they will establish the credibility of his presentation. Further, when making these presentations, many areas dealing with specific areas of crime prevention will arise (such as rape prevention, property engravement, residential and apartment security, vehicle theft, and burglary prevention).

B-145. In dealing with schools, the crime-prevention officer can establish many positive contacts with individuals inside the school system that will prove to be of great value. Individuals such as administrators, counselors, and faculty members offer the crime-prevention officer assistance in many different ways. Perhaps the most valuable assistance is in the area of establishing a juvenile crime-prevention program that is based on peer influence.

B-146. A crime-prevention program aimed at peer influence is essential to a crime-prevention unit. Peer influence is directly related to the Determination Theory of Delinquency Causation. Simply stated, the theory holds that delinquency occurs as a result of external influences on youth. While attempting to control all external influences on youths is an impossible task, the control of peer influence is possible and perhaps the most important external influence to be controlled.

PEER-INFLUENCE CRIME-PREVENTION PROGRAM

B-147. A peer-influence crime-prevention program can be established after the crime-prevention officer establishes himself in the junior/senior high schools as described earlier. The basic concept of the program is the negation of undesirable peer influence on predelinquent youths and the establishment of a positive peer influence.

B-148. The peer-influence program functions within the school district. A problem youth is detected at the junior high school level by his teacher, counselor, or the crime-prevention officer. When it is believed that the peer-influence program could serve a youth, his counselor arranges a conference with him. During the conference, the counselor attempts to determine the basic problem experienced by the youth and has him complete an application for admittance into the program.

B-149. When the application is complete, the counselor contacts the youth's parent or guardian and explains the program. If the parent agrees to allow participation, the application is sent to the parent for his signature.

B-150. When the completed application is returned to the counselor, it is forwarded to a senior high school counselor for review. If necessary, the senior high school counselor may contact the junior high school counselor personally to discuss the youth in more detail. The high school counselor then selects a suitable senior high school student who has been accepted for the peer-influence program and matches the student with the junior high school youth.

B-151. Once the match has been made, the senior high school student is given a briefing on the junior high school student by the crime-prevention officer. The senior youth is instructed to contact the junior youth (preferably at the junior youth's home) and introduce himself to the youth and to the youth's parent or guardian. After the initial contact has been made, the senior youth is instructed to plan activities with the junior youth and spend as much time as possible with him.

B-152. Each senior youth that volunteers for the program is required to complete an application. This application gives personal background information and references that are thoroughly checked by the crime-prevention officer. The youth's academic record is examined, but prime consideration is given to citizenship factors rather than grades. The crime-prevention officer personally interviews each applicant.

B-153. When a senior youth is selected for the program, he is given a basic outline to follow. He is warned not to preach to the junior youth but to lead him by allowing him to observe the senior youth's actions at different activities. The senior youth must devote the necessary amount of time to limit the junior youth's opportunity to continue the negative associations he had in the past.

B-154. While the peer-influence program is simple in nature, the results can be remarkable. During a three-year period in Irvin, Texas, approximately 600 students participated in the program. From this number, seven arrests were made from the group with one juvenile accounting for four of the seven arrests. As about 30 percent of the juveniles participating were legally judged delinquent, the relapse rate was officially established as 0.1 percent.

B-155. While juveniles judged as delinquent can participate in the program, the crime-prevention officer should concentrate his efforts on those juveniles identified as predelinquent by counselors or faculty members. Again, the personal contact of the crime-prevention officer and the school faculty is essential to the program's success. A personal rapport should be cultivated and maintained.

B-156. Operationally, the peer-influence program can be administered by the collective efforts of school personnel and MP officers who are committed to the program. Once the program is established, it will probably not require the services of an officer on a full-time basis (depending on the number of students involved). However, constant contact is necessary; the more time that can be devoted, the more success will be apt to follow.

CRIME-PREVENTION PRESENTATIONS FOR ELEMENTARY AND JUNIOR HIGH SCHOOLS

B-157. As the traditional American family structure has experienced a significant decline over recent years, more and more emphasis is being placed on schools to provide guidance for children. As a crime-prevention officer who is frequently invited to speak to students, you can play an important role in assisting the school to provide this guidance. To stress the importance of the above statement, many experts in child behavior believe that the school is second only to the family in molding a child's socialization and behavior. Many people feel that the school is increasingly replacing much of the family socialization process. For this reason, schools are the most logical environment to begin a juvenile crime-prevention program.

B-158. Crime-prevention officers can have an impact on the most common types of juvenile crime by preparing specific lectures for particular grade levels in the elementary and junior high schools. Classes discussing shoplifting, vandalism, and bicycle theft are well received by students and have a wide-reaching effect on the community as a whole due to the children relaying the information to their parents.

B-159. A number of films are currently on the market that assist in making crime-prevention presentations in the classroom. In addition to specifically addressing juvenile crimes, the films directly relate to the peer-influence causation of crime.

OVERVIEW OF SELECTED JUVENILE CRIME-PREVENTION PROGRAMS

B-160. In addition to the programs described above, there are a variety of other juvenile programs. You will notice that some of the programs are designed to offer activities for the juvenile while at the same time attempting to provide one or more of the following:

- Positive peer influence.
- Education.

- Police/juvenile cooperation.
- An understanding of law-enforcement and MP duties.

B-161. One program deals with the enforcement aspect, with the goal being truancy reduction. Also mentioned are programs designed for child safety and victimization reduction.

DARE Program

B-162. DARE is a drug-abuse educational program that gives children and parents the skills to recognize and resist the subtle and overt pressures that cause them to experiment with drugs, alcohol, and violence. This unique program uses uniformed law-enforcement officers to teach a formal curriculum to students ranging from kindergarten to twelfth grade. DARE gives special attention to fifth through ninth graders to prepare them for entry into middle/junior high and high school where they are most likely to encounter pressures to use drugs or alcohol.

Police Explorer Programs

B-163. Explorer programs are run in conjunction with the Boy Scouts of America and a sponsoring police agency. The programs offer a wide range of flexibility in that they may be structured along the lines of a police cadet program with all activities centered around law-enforcement training. This type of explorer program usually seeks juveniles who have an active interest in pursuing a law-enforcement career. Being an explorer in this type of program will offer the student the opportunity to view the duties of a police officer and to take part in minor police activities (such as traffic direction and observation riding in patrol cars). The selection and training of the students for this type of program strictly depend on the sponsoring agency.

B-164. Other police-sponsored explorer programs may be designed to provide the members with other nonpolice activities (such as hiking, camping, and canoeing). At the same time, this type of program will offer the student the chance to meet with police officers and learn more about police functions, though this is a secondary purpose. Selection for this program would not be as stringent as the above-mentioned program. A local council office of the Boy Scouts can provide additional information on this program.

Alateen

B-165. The Alateen program is run in cooperation with the Alcoholics Anonymous organization. This program's intended purpose is to provide assistance for a child's emotional needs if he or she is faced with family problems caused by alcoholism of one or both parents. Your local chapter of Alcoholics Anonymous can provide more information on this program.

Truancy-Enforcement Program

B-166. Various cities have successfully adopted truancy-enforcement programs that focus attention on identifying and apprehending youths that are on unexcused absence from school. The primary goal behind this program is to prevent or reduce the incidence of daytime residential burglary. The San Angelo, Texas, Truancy Enforcement Program selected a target area in an effort to see how effective the program would be in reducing daytime burglaries. The initial evaluation of the program showed positive results. The San Angelo program uses uniformed crime-prevention officers in marked police cars patrolling around the school and known hangouts. All street personnel in the patrol division are also encouraged to check school-age individuals. Those students under 17 years of age who are suspected of truancy are taken to the school office while those over 17 years of age are detained only long enough to obtain their name and other identifying information, which is then turned into the school office. The assistant principal in charge of attendance completes the investigation and takes the appropriate action. The action varies (depending on the circumstances) but can include counseling with the student, the teacher, and the parents or, in extreme cases, suspension for a short period.

B-167. The emphasis behind these programs is to inform youths of the facts of the law. It was shown that many of the problems with the youthful offenders would be eliminated if they were presented the facts of the law to which they are subject. The courses can be adapted to fit the needs and desires of the individual school system, and the extent of the areas covered will depend on the commitment and availability of the local MP officers. Various civilian school districts have incorporated these law-enforcement courses as part of their regular curriculum.

Criminal Victimization Reduction

B-168. Programs designed toward this goal normally are structured to reduce the possibility of a child becoming a victim of a child molester. Most commonly, officers give a verbal presentation to groups of children using visual aids that help to act as a positive reinforcement to learning. Some effective methods have been the use of "talking traffic lights," "talking bicycles," and "talking motorcycles." These items are usually equipped with a tape-recorded message that can be controlled by the officer conducting the program. The talking traffic light uses a full-size traffic light that flashes red or green according to the type of person that is being described.

VANDALISM

B-169. Today's vandals often attack their own territory. School vandalism—the illegal and deliberate destruction of school property—is committed by students themselves. They break so many windows that in large districts the funds spent annually on replacing broken windows could pay for a new school. Vandals destroy about $3 million worth of school bus seats annually, and they commit enough arson to account for 40 percent of all vandalism costs.

B-170. School vandalism outranks all other assaults on private and public property. At the end of the 1973 school year, the average cost of damages from vandalism was estimated at $63,031 per school district. That figure could have paid the salaries of eight reading specialists or could have financed a school breakfast program for 133 children for one year. A typical school's chance of being vandalized in a month are greater than one in four, and the average cost of each act of vandalism is $81. Yet, these figures do not include the hidden costs of school vandalism—increased expenses for fencing, intrusion and fire detectors, special lighting, emergency communications equipment, and vandal-resistant windows. Every dollar spent in replacing a window or installing an alarm cannot be spent on education.

B-171. School vandalism can also have enormous social cost. The impact of an 89-cent can of spray paint used to cover a wall with racial epithets far exceeds the monetary cost of removing the paint. An abusive word scrawled across a hallway wall can destroy student morale, disrupt intergroup relations, undermine the authority of an administration, or even close the school. Incidents with high social costs damage the educational process as much as those with high monetary costs. Today's vandal is not a hardened, war-scarred veteran. Instead of grizzled whiskers, he sports peach fuzz. He is almost literally the boy next door. In fact, the typical vandal differs quite dramatically from the typical juvenile delinquent.

B-172. It is significant that vandals fall into a well-defined and relatively narrow age group. They are usually early adolescent males who are highly subject to group pressures and transitory impulses. It is not at all unusual for adolescents to act out whatever is controlling them at the moment—rage, boredom, pent-up energy, or the sheer joy of "wreckreation." While there are conditions that may predispose or provoke a youth toward vandalism, the problem seems to be almost human nature. Few among us have never written on a sidewalk or scrawled initials on a school desk. Vandalism cuts across all strata of society, all geographic regions, and all racial lines.

B-173. The causes of vandalism remain obscure. Though research addressing the "why" of vandalism is growing, it has yet to yield clear-cut answers. Among the motivating factors often cited are anger, frustration, hostility, bitterness, alienation, futility, inequality, restricted opportunity, emotional pain, failure, prejudice, revenge, and the need for attention. Although much of the research is convincing, the fact remains that many vandals do not appear to be among the most angry, frustrated, hostile, alienated, or needy youth. Only a small fraction of the youngsters who fall into that category actually commit acts of vandalism. So, while most experts agree that vandalism is not totally senseless, they do not claim to fully understand its causes. In fact, vandalism is often not understood by vandals themselves. Many vandals report that they do not know why they did it. Many others, according to case reports, offer the unsolicited observation that destruction is fun. Still others express satisfaction and exhilaration. Few consider themselves criminals. For the time being, we can conclude only that motives for vandalism are diverse. But the whys notwithstanding, the vandal profile suggests that our task is, in large part, to anticipate and redirect the impulses of young teenagers.

B-174. Schools are by no means the helpless victims of early adolescence. Many school factors, most of which are amenable to change, influence the amount of vandalism that schools experience. The following characteristics are typical of schools that suffer high property damage or loss:

- Vandalism is higher when there is poor communication between the faculty and the administration (such as when the principal fails to define policy or makes policy decisions unilaterally).
- Hostility and authoritarian attitudes on the part of teachers toward students often result in students "taking it out" on the school.

- Limited contact between teachers and students reduces student involvement with the school and increases the likelihood of vandalism.
- Schools characterized by intense competition for leadership positions suffer greater property damage and loss.
- The chances for vandalism increase when the students do not value their teachers' opinions of them.
- Schools at which students strive to get good grades experience more vandalism.
- Parents of students in high-damage schools express less favorable attitudes toward their schools than do other parents.
- The school is a convenient target for vandalism when it is close to students' homes.
- The chances for vandalism are increased when grades are used as a disciplinary tool.
- Damage is greater in larger schools where there is more property to destroy. This correlation between school size and vandalism prevails regardless of whether the school is located in an urban, suburban, or rural setting.
- Fewer offenses occur when rules are well understood by students and are consistently and firmly enforced by teachers and administrators.

Vandalism Prevention

B-175. If the special problems of early adolescence, often intensified by social or personal pressures, interact with school conditions to produce vandalism, then preventive measures must address the nature of both the child and the school. Furthermore, prevention must include both physical and human responses. At present, most vandalism-prevention or -reduction programs rely on physical security—bigger and better electronic alarm systems, patrol guards, dogs, tamperproof locks, and window grilles. These techniques help, but they address only 20 percent of the problem—those incidents involving breakage. These incidents usually occur when school is not in session and in the absence of witnesses. The techniques have little effect on the day-to-day trashing of the school or on the disruptive acts aimed at the school's routine (bomb threats, the setting of fires, and false fire alarms) that are committed during school hours. The most sophisticated physical and electronic barriers are not sufficient to keep vandals from what they consider an attractive target. In fact, it has been argued that alarms and armed guards, besides lowering student and staff morale, often themselves become a challenge, inviting rather than deterring vandals. Vandalism prevention requires not a narrow or piecemeal approach, but a varied and comprehensive effort that includes both physical and human components geared to the school's specific problems. Furthermore, an effective long-term program must involve the community, parents, neighbors, police, and civic groups as well as students, teachers, and school administrators.

B-176. Schools are an easy target for vandals. Most are public, secular, and often unoccupied. Most will remain public and secular; but they need not remain unoccupied, unprotected, or unobserved. The following are techniques that have made some schools less vulnerable to vandals. These are especially effective against problems occurring during nonschool hours.

B-177. **Occupy the School.** Employ a custodial force around the clock. In most schools, the entire custodial force works at one time, leaving the school at night. As an alternative, custodians can be assigned staggered shifts so that the school is always occupied. Twenty-four-hour custodians are particularly appropriate in schools suffering sporadic property damage that demand more than a roving patrol but less than permanent security guards.

B-178. Invite police to use the school buildings at night. Police can be issued keys to the schools in their patrol areas so that they can use school offices to write their reports. This places a police officer in the school when it might otherwise be unoccupied, and it places a police car in front of the school.

B-179. Bring the community into the school. The school is an excellent place for recreational programs; health clinics; adult-education classes; counseling centers; community gatherings; and Boy Scout, Girl Scout, and Parent-Teacher Organization (PTO) meetings. The presence of people in the school building not only reduces the opportunity for vandalism, but also stimulates community and student interest in the school.

B-180. **Watch the School.** Use school neighbors as eyes and ears. Ask nearby homeowners to watch the school and report suspicious activities. Emphasize careful observation and rapid reporting, but discourage direct involvement in any situation observed. Such programs work best if they are organized but based on informal involvement, if they are accompanied by overall involvement of parents and community with the school, and if they offer some sort of prestige to participants.

B-181. **Employ Roving Patrols.** A uniformed patrol used instead of or in conjunction with an alarm system can deter vandalism. The individuals who patrol should establish rapport with neighborhood youths and open communication with community leaders. They should also vary their patrol patterns.

B-182. Hire student patrols during the summer and on weekends. The school district or community can provide its youth with part-time or summer employment and, at the same time, curb vandalism by paying students to patrol the school grounds during weekends, holidays, and summer vacations. These students should be paid an adequate wage and considered an integral part of the school's security force.

B-183. **Control Access to the School by Using an Alarm System.** Alarms are the most expensive vandalism control measure a school can use. While they can detect vandals, they cannot apprehend them; they can merely signal the alarm-system monitor, which may be miles away. They can, however, be an efficient deterrent and should be considered as part of any comprehensive plan to control vandalism. If alarm systems are linked with a surveillance camera, the chances of apprehending intruders are greatly increased.

B-184. **Design the School With Vandalism Prevention in Mind.** The following designs for preventing vandalism can be implemented when building a school:

- Limit ground-to-roof access.
- Eliminate low, overhanging roofs.
- Avoid unnecessary exterior fixtures.
- Plant trees that cannot be climbed near buildings.
- Consider raising as much of the school building as possible from ground level.
- Build the school at some distance from residential areas. While it should be located near the homes of most of those it serves, it will suffer less property damage if there is a buffer zone between it and surrounding residential areas.
- Design the school with plenty of defensible space so that the normal flow of school traffic allows continuing, casual surveillance of the premises.
- Use vandal-resistant surfaces. Use harder surfaces in damage-prone areas. For walls, use epoxy paint or glazed tiles that are easily and inexpensively replaced or repaired; use small wall panels and keep replacement panels in stock; and place permanent signs, building names, and decorative hardware at a level that cannot be reached from the ground. Replacing damaged areas immediately shows a sense of pride in the appearance and helps to eliminate copycat acts of vandalism.
- Plan windows carefully. Avoid windows that are vulnerably placed. Use small panes of glass to simplify replacement; use thick, tempered glass, thick acrylic, or Plexiglas® for windows in heavily traveled or hangout areas. Avoid useless windows in student stores, administrative offices, and industrial-arts storage areas.
- Plan entries with multiple uses in mind. Install flexible internal gates to block off specific areas or corridors when necessary. Provide separate exterior entries for community and student use. Inside the building, create areas for informal gatherings near entrances and exits by installing soft-drink machines and benches.
- Locate or relocate the playground's access roads to provide better surveillance by roving patrols.
- Consider outdoor lighting. Opinions on this issue are divided. Many schools report a decline in vandalism after installing hardened exterior night lighting. Others report that elimination of all night lighting reduces vandalism, presumably because young adolescents are afraid of the dark. If lighting is used, it should be directed away from windows to keep vandals from seeing the process of destruction or its outcome.
- Channel graffiti. Graffiti artists will usually select light, smooth surfaces rather than dark, rough surfaces. Therefore, school officials can channel graffiti onto one or two walls designed to withstand such treatment. Students or maintenance staff can paint most walls at regular but not too frequent intervals. One wall can be officially designated a "legitimate" graffiti wall, though this approach removes some of the challenge inherent in informal graffiti.

B-185. **Make the Target Less Attractive.** The school is not only an easy target but also an attractive one. It represents enforced learning, discipline, and mandatory attendance to students who are, simply by virtue of their age, rejecting adult standards and striving to achieve independence from adult control. Students are additionally provoked if the school functions in an impersonal, undemocratic, repressive, or manipulative manner, further increasing the likelihood of vandalism.

B-186. **Revise the Curriculum.** The list of characteristics associated with vandals and vandalized schools indicates that property damage and loss are higher when competition for rewards is intense, the availability of rewards is limited, or the distribution of rewards is unfair. All of this suggests that school policy and atmosphere have a direct bearing on school vandalism. Below are changes in school governance that can help remove the features that make a school an attractive target for vandals. These procedures, especially in combination, have proven effective against all forms of vandalism, including those that most commonly occur while school is in session.

- **Alternative schools.** Though originally designed to perform educational functions, alternative schools have proven effective in reducing school violence and vandalism. They provide an option to students who are not benefiting from the regular academic classroom. These schools operate within or alongside the traditional school. They are characterized by low student-to-teacher ratios, intense individual counseling, and extensive use of the community as a learning resource. They offer an alternative to suspension, personalize the learning environment, and provide successful experiences to students who have performed poorly in the past.
- **Law-related education and police-school liaison programs.** In many communities, the police department has assigned school resource or liaison officers to local junior and senior high schools. These officers may on occasion assume policing duties, but their primary function is to counsel students and teach law-related courses. These activities acquaint students with the positive role that law plays in our society and personalize the relationship between the "cop on the beat" and "the kid on the corner."
- **Action learning.** This term refers to apprenticeship programs as well as training in practical aspects of adult life. The former allows students academic credit for community work (such as tutoring and assisting physicians, lawyers, or other professionals). The latter provides instruction in skills such as checkbook balancing, comparative shopping, and applying for a job. Both address the boredom and frustration that are linked to truancy, violence, and vandalism.
- **Reward distribution.** The school's reward structure is related to school crime. Although the school may offer attractive incentives, most students receive very little in the way of rewards. Many of those who lose out still care about the competition. They become frustrated, and they vent their anger on the apparent source of their problems, the school. It should be possible to spread the rewards around without compromising performance standards, perhaps by recognizing improvement as well as achievement or by acknowledging forms of achievement other than scholastic, athletic, and social.

B-187. **Change Administrative Policy and Organizational Structure.** According to the National Institute of Education's Safe-School Study, the principal's leadership is a critical factor in reducing or preventing school crime. Strong principals who are visible, available, and responsive to students and staff appear to be most successful in eliminating violence and vandalism. Those who are less successful are often described as unavailable and ineffective.

B-188. The Safe-School Study also found that the exercise of discipline through clear enunciation and even-handed enforcement of rules is associated with a low incidence of school crime. To increase the likelihood that students will find school a fulfilling experience, many districts are establishing minischools—schools small enough to allow the individual student to feel significant. Similarly, several large schools are currently functioning on a house basis—the school is divided programmatically into several smaller units with which students can more easily identify.

B-189. **Involve the Students.** Establishing a vandalism fund uses peer pressure to the school's advantage. The community or the school district puts aside a certain amount of money and announces that the funds will be used to cover the costs of vandalism. Any money left over reverts to the students to be used as they choose. This plan works because it educates students about the costs of vandalism, allows them to see the positive results of curbing property damage and, most important, gives them full responsibility for the problem.

B-190. Several school districts have established voluntary student security advisory councils that conduct workshops and small group discussions focusing on vandalism and violence prevention. These committees increase awareness of the school's problems, generate recommendations for action, and give students opportunities to participate in school decision making.

B-191. Older students can be helpful in influencing younger students. In several communities, junior and senior high school students visit fourth- and sixth-grade classrooms where they show a film about vandalism and then lead a discussion on the causes and consequences of vandalism.

B-192. School beautification projects involve students in the care of the school building and grounds in an attempt to increase their pride in the responsibility for the school. The more effective projects are selected by the students themselves and continue throughout the year. The projects focus on marginal students rather than school leaders.

B-193. **Involve the Parents.** An open-door policy allows parents to go directly to their child's classroom whenever they wish without securing a special visitor's permit from the office. This simple policy offers parents concrete evidence that they are indeed welcome at school.

B-194. In some schools, parents serve on antitruancy committees along with teachers and students. They visit youngsters and their families in an effort to resolve truancy problems. In other schools, parents serve as hall monitors, supervise extracurricular activities, and help in classrooms. Parents are also the school's best source of guest speakers and contacts for work-study or apprenticeship programs. Some school districts have initiated faculty men's clubs to acquaint fathers with the male teaching staff and to encourage them to assume more responsibility for their children's progress in school.

B-195. **Involve the Community.** In some communities, students and law-enforcement, school-district, and city personnel sponsor daylong forums on vandalism. The forums introduce citizens to the causes and costs of vandalism and give them an opportunity to voice their concerns and initiate preventive programs.

B-196. Police departments can initiate public-relations programs within schools and youth-service agencies. In addition, they can enlist the help of youth in preventing vandalism through police-sponsored groups, such as the Police-Youth Service Corps, which recruits adolescents from high-crime areas to work as public-safety aides. Similarly, law students from neighboring universities can be brought into the schools at minimal or no cost to discuss the legal implications of vandalism.

Picking up the Pieces

B-197. The preventive measures listed on the preceding pages can function as part of a long-range, proactive response plan. However, they do not address the question of immediate response. What should the school do, right away, about 20 broken windows, a cherry bomb in the toilet, or recurring racial graffiti on the wall? The answer is to repair the damage immediately. Quick repair keeps perpetrators from admiring their handiwork, lessens the epidemic effect of vandalism, and minimizes any social impact the act may have.

B-198. Initiate formal record-keeping procedures, and ensure that they are followed. Schools faced with serious problems should begin recording all acts of vandalism. They should also consult law-enforcement personnel about when police should and should not be called. When a school begins to have problems, it should work with the juvenile justice system so that the two institutions can coordinate their efforts with regard to school-age offenders.

B-199. Careful record keeping allows a school to plot the incidence of vandalism to find out precisely where and when each type of offense is occurring. For example, using incident analyses, the National Institute for Education's Safe-School Study found that—

- Fire and bomb threats most often occur on Tuesdays.
- School-property offenses tend to occur with greater frequency toward the end of each semester, especially in November and December.
- Break-ins and school-property offenses occur most often on weekends and Mondays.

B-200. This type of information is invaluable in planning a vandalism reduction and prevention plan.

Restitution

B-201. Restitution is a set of legal and administrative procedures through which the school receives payment from vandals for damages they cause. While it seems reasonable to require payment for damages, restitution does not appear to be worth the effort. In the first place, most vandals are not caught. In the Los Angeles School District (which has an aggressive restitution program), only 30 percent of the offenders are ever identified.

From this 30 percent, most restitution is paid before matters get to court. Going through lengthy legal processes to obtain the rest is simply not cost-effective. However, a parent faced with the possibility of a court case may make a greater effort to keep his or her child out of trouble.

CONCLUSION

B-202. The goal of crime prevention is to reduce crime through public awareness and education. The skills that crime-prevention officers acquire can eventually benefit all segments of our communities. It is logical then for crime-prevention officers to attack one of the major crime problems in our society—juvenile delinquency. Efforts of crime-prevention units will certainly be lacking if the juvenile-crime problem is not given a high priority and if juvenile-crime programs are not established.

FRAUD

B-203. Two types of fraud can be affected by crime-prevention efforts—fraud against the government and fraud against individuals in the Army. Since the countermeasures for each of these categories of fraud are different, they will be discussed separately.

FRAUD AGAINST THE GOVERNMENT

B-204. Fraud of this type is a loss to the Army due to manipulation of systems from within the government with criminal intent. Typical examples of fraud are—

- The diversion or theft of government property by falsifying documents such as purchase orders, shipping documents, and so forth.
- False claims for temporary-duty (TDY) pay, for reimbursement for losses due to the movement of household goods, or for reimbursement for material reported as stolen.
- Overcharges or underproduction on contracts with the Army.
- Bribery.
- Kickbacks to secure purchase orders.
- Use of one's official position for personal gain.

B-205. This list is not all-inclusive. There are many more ways to defraud the government. Some of the factors that make it easy to perpetrate a fraud are—

- The lack of independent verification of records, transactions, and reports.
- The lack of adequate supervision.
- Unrealistic budgeting and acquisition requirements.
- Failure to correct deficiencies identified by existing systems.
- Concentration, at the operational level, of responsibility and authority for an entire process in one individual.

B-206. There are built-in measures to discourage fraud for most Army material-control systems. These programs are routinely examined by the inspector general (IG) along with other command inspectors. A crime-prevention officer's main function is to encourage the reporting of fraud by ensuring that the community understands what type of activities should be reported and by ensuring that the numbers are posted in work areas where personnel are in a position to detect fraud.

FRAUD AGAINST INDIVIDUALS

B-207. The second type of fraud is con games or consumer fraud. This crime is perpetrated by professional con men, unscrupulous businesses, and amateurs who see a chance to make a fast buck. Most of the time, the operator offers a "something for nothing" deal. By the time the victim realizes that he has been duped, the con man is long gone. Frequently, the victim is so embarrassed that he has been duped that he does not report the crime. The only way that a crime-prevention officer can help prevent this crime is to publicize common con games and illegal business practices and to encourage victims to report them. When a case of fraud occurs, the specific con technique that was used should be advertised widely to prevent others from being fooled by the same ploy.

Repair Fraud

B-208. Repair frauds are simple to execute but difficult to detect. Some unscrupulous repairmen do not fix the problem, but they charge you anyway. Some use inferior parts; others charge you for work that you did not expect or need. Some also do "insurance" work; they will repair one thing but ensure that something else will soon go wrong. The following are methods that you can use to protect yourself from repair fraud:

- Shop around. Ask friends, neighbors, or coworkers for references. When you find repairmen you trust, stick with them.
- Do not try to diagnose the problem yourself unless you are an expert. The mechanic may take your advice, even if it is wrong. If you know exactly what the problem is, do not tell the mechanics; wait and see if their recommendations agree with your diagnosis. That way you will know whether needless repairs are suggested.
- Try to get several detailed written estimates before any work is done. Compare job descriptions and materials to be used. Be sure to ask if there is a charge for an estimate.
- Ask for the old parts to ensure that replacements were really installed.
- Ensure that you get a guarantee on any work done.
- Make sure that the work was done before you pay. Take your car for a test drive. Plug in the refrigerator. Test the television.
- Ask your local consumer affairs office about the laws in your state and what specific protection it gives regarding professional services to be licensed or certified.

Home-Improvement Fraud

B-209. Home repairs and improvements can be costly. Be cautious if somebody offers to do an expensive job for an unusually low price, if a firm offers to make a free inspection, or if the workers "just happened to be in the neighborhood." These are the favorite tricks of dishonest home-repair firms.

Some firms offer a price you just cannot resist. Once you sign the contract, you learn why—they never deliver the service. Others send door-to-door inspectors to do a free roof, termite, or furnace inspection. These free inspections will turn up plenty of expensive repairs. Some fly-by-night companies will offer to do the work on the spot. When they leave, you may be left with a large bill and a faulty repair job. To avoid home-improvement fraud, try the following:

- Get several estimates for every repair job. Compare prices and terms. Check to see if there is a charge for estimates.
- Ask your friends for recommendations, or ask the firm for references and check them.
- Check the ID of all inspectors.
- Call the local consumer affairs office or the Better Business Bureau to check the company's reputation before you authorize any work.
- Be suspicious of high-pressure sales tactics.
- Pay by check, never with cash. Arrange to make payments in installments—one-third at the beginning of the job, one-third when the work is nearly completed, and one-third after the job is done.

Land Fraud

B-210. Real estate can be a great investment. The enterprising real estate salesperson knows how anxious you are to find just the right property, especially for an investment or a retirement home—a nice warm climate, not too crowded, or a new development. Some dishonest agents will promise you anything—a swimming pool, country club, or private lake—to get your name on the contract. Even if the sales agent promises you luxury, they may not guarantee the basics such as water, energy sources, and sewage disposal.

B-211. Most land developers offering 50 or more lots (of less than 5 acres each) for sale or lease through the mail or by interstate commerce are required by law to file a Statement of Record with the Housing and Urban Development Administration (HUD). This document tells you almost everything you need to know about your future home—legal title, facilities available in the area (such as schools and transportation), availability of utilities and water, plans for sewage disposal, and local regulations and development plans. All of this information must be given to you in a property report prepared by the developer. Always ask to see this report before you sign anything.

B-212. If the developer does not give you a copy of the property report for the lot you are considering, you can obtain it from HUD for a fee. Send requests to: Department of Housing and Urban Development, Office of Interstate Land Sales Registration, 451 Seventh Street South West, Washington, DC 20410.

Door-to-Door Sales

B-213. Beware of the following door-to-door sales tactics:

- "Nothing like it in the stores!" This is a true statement. The vacuum cleaners in the stores are probably of better quality and come with a better warranty.

- "Won't find this price anywhere." This is also a true statement. The prices in the stores are probably lower.
- "Easy credit!" This is another true statement. They do not care what your credit rating looks like. Once you sign for the purchase, paying for it is your problem. Be wary of low monthly payments. Find out the total amount you will pay over the life of the loan, then subtract the actual cost of the item itself. The difference is what you will pay in interest. Your bank, your credit union, or a local legal-aid society can tell you if the interest rate is fair.

B-214. Be cautious of these words, and be firm if the salesperson pressures you to buy. If you do get trapped, you are protected by a Federal Trade Commission regulation. When you make a purchase in your home totaling $25 or more, the salesperson must give you a written contract and two Notice of Cancellation forms. You have three days to change your mind and use one of these forms to cancel your contract.

Charity Fraud

B-215. Charity fraud does a lot of harm. The swindler takes advantage of a person's good will and takes his cash—money that was meant for people in need. You can ensure that any money you give gets into the right hands. Remember these pointers when somebody asks you for a donation:

- Ask for ID, either the organization's or the solicitor's. Find out what the charity's purpose is and how the funds are used. Ask if contributions are tax deductible. If you are not satisfied with the answers, do not give money.
- Give to charities that you know. Check out those you have never heard of or others whose names are similar to a well-known charity.
- Do not fall for high-pressure tactics. If solicitors will not take no for an answer, tell them no anyway, but do not give them your money.
- Be suspicious of charities that only accept cash. Always send a check made out to the organization, not to an individual.
- If a solicitor reaches you by telephone, offer to mail your donation. Shady solicitors usually want to collect quickly.

Self-Improvement Fraud

B-216. Con artists know that everyone wants to look better, feel better, and be a better person. Selling worthless plans and cures is one of the easiest ways for them to make a "quick buck." The following ads can look tempting:

- "Miracle reducing plan."
- "Look like a model in only five days."
- "Learn to speak Spanish while you sleep."
- "You can have a new, dynamic personality."

B-217. What can you do? Be careful! Read the small print. Know what the product contains. You should check with your doctor before you embark on any diet or exercise program.

Medical and Health Fraud

B-218. Most of us do not know much about medicine; that is why we go to doctors. It is also why we fall for miracle cures and other phony health products and services. Patent medicines, health spas, and mail-in lab tests should be warning signs for the potential consumer. A laboratory in Texas advertised nationally that it had perfected a fail-safe urine test for cancer. More than 15,000 tests were made at $10 each before authorities stopped this fraudulent company.

Unsolicited Merchandise

B-219. Cagey con artists will send you a gift in the mail—a tie, a good luck charm, or a key chain. What do you do with it if you did not order it? If you are the kind of person they are looking for, you will feel guilty and pay for it, but you are not obligated to. If you—

- Have not opened the package, mark it "return to sender." The post office will send it back at no charge to you.
- Open the package and do not like what you find, throw it away.
- Open the package and like what you find, keep it, free of charge. This is a rare instance where the rule of "finders, keepers" applies unconditionally.

B-220. Whatever you do, do not pay for it. Look at your gift as an honest-to-goodness way of getting something for nothing. Do not get conned if the giver follows up with a phone call or a visit—by law the gift is yours to keep.

Mail Fraud

B-221. The following are examples of mail fraud:

- **The contest winner.** "You've won! This beautiful brand-name sewing machine is yours for a song. To claim your prize, come to our store and select one of these attractive cabinets for your new machine. Bring this letter with you and go home with a new sewing machine for next to nothing." Treat an offer like this carefully. Shop around before you claim your prize. Chances are, the cost of the cabinet will be more than the machine and cabinet are worth.
- **The missing heir.** You have just received a very official looking document. The sender is looking for the rightful heirs to the estate of someone with your last name. It could be you. To find out, just send $10 for more information. There may be thousands of people with your last name, and letters like these are often mailed nationwide. Even if there really was claimed estate, it is highly unlikely that you would be an heir. Save your money; why help a swindler get rich?

B-222. These are just two examples of mail fraud. Many of the other frauds described in this section can be handled through the mail. When they are, the US Postal Service can launch a full-scale investigation. If you think you have been cheated in a mail-fraud scheme—

- Save all letters, including envelopes.

- See if your neighbors or business associates received the same material.
- Contact your local postmaster who can direct you to your regional Postal Inspector's Office.

SUMMARY

B-223. The following are recommendations to avoid being duped:

- Do not believe something-for-nothing offers. You get what you pay for.
- Be suspicious of high-pressure sales efforts.
- Take your time. Think about the deal before you part with your money.
- Get all agreements in writing. Insist that agreements are made in "plain English," not "legalese."
- Read all contracts and agreements before signing. Have a lawyer examine all major contracts.
- Compare services, prices, and credit offers before agreeing to a deal. Ask friends what their experiences have been with the firm or service in question.
- Check the firm's reputation with your local consumer affairs office or the Better Business Bureau.

INTERNAL THEFT

B-224. Employee theft is a major problem in most large organizations. It is estimated that one-third of all business bankruptcies are a result of theft. Of course, in the Army there is no danger of the organization going out of business; however, waste and internal theft can divert significant amounts of critical resources from mission-essential activities. Unlike many businesses, the Army has recognized that internal losses can be a problem, and most areas have adequate controls mandated by regulation. As crime-prevention officers, we should monitor reports of survey, physical-security inspection reports, and results of crime-prevention inspections to ensure that control measures are being followed. When it is apparent that the correct measures are not being followed, the problem must be identified for the senior commander's action.

B-225. Research has shown that an organization's atmosphere is just as important as management controls and physical security in preventing employee theft. To control this problem, the following items are essential:

- Employees, both civilians and soldiers, must believe they are part of a professional organization that expects superior performance from all of its members. When second-rate work is accepted, the lax attitude carries over into property-control procedures and losses increase. Standards must be high, but fair.
- Leadership must set the example. If leaders, supervisors, and managers take advantage of their positions to use government material or services for their own benefit (even in minor ways), their employees will also feel justified in diverting Army resources to their own use.

- The organization must show a genuine concern for the problems of its personnel. If an employee feels he has been treated unfairly, it is easy for him to justify stealing from the organization. He believes that he is only taking what he would be getting if the organization was fair.
- The organization must take appropriate disciplinary or administrative actions in cases of theft. There can be no acceptable level of internal theft. If employees believe that an activity considers a certain level of loss to be acceptable, then material theft will grow rapidly. Everyone will consider the material he takes to be within the 2 percent that the activity expects to lose.
- Policies on internal thefts must be enforced. Frequently, a thief is a long-term employee whose honesty has always been above question. In these cases, the temptation is very strong to let the offender go with a "slap on the hand." However, this is a clear signal to other employees that diversion of Army resources for personal use is acceptable to management. If an employee chooses to steal and is caught, he should be prosecuted.
- Publicity campaigns using posters and other media should be used to disseminate command policies on internal theft. Crime hot lines are also useful in increasing reports of employee thefts.

PILFERAGE

B-226. The protection of property, including the prevention of pilferage of government supplies and equipment, is one of the primary functions of the MP and civil service security forces. This function may include protecting supplies and equipment while in storage areas, during the issue process, or while they are in transit.

B-227. Pilferage is probably the most common and annoying hazard with which security personnel are concerned. It can become such a financial menace and detriment to operations that a large portion of the security force's efforts may have to be devoted to its control. Pilferage, particularly petty pilferage, is frequently difficult to detect, hard to prove, and dangerous to ignore.

B-228. Military property loss throughout the world would increase millions of dollars each year if subjected to uncontrolled pilferage. However, the risks incurred cannot be measured in terms of dollars alone. Loss of critical supplies for tactical units could result in loss of life or a danger to national defense. In some areas, losses could assume such proportions as to jeopardize an installation's mission. All installations and facilities can anticipate loss from pilferage. Actual losses will depend on such variable factors as the type and amount of materials, equipment, and supplies produced, processed, and stored at the facility; the number of persons employed; social and economic conditions in surrounding communities; command attitude; and physical-security measures used. Because these factors differ greatly in various types of installations and in different geographical locations, each must be considered separately.

B-229. To determine the severity of this hazard at any given installation or facility, it is necessary to determine the amount of loss that may be occurring.

Unfortunately, this is not always an easy task. Accounting methods may not be designed to pinpoint thefts; consequently, such losses remain undisclosed or they are lumped together with other shrinkages, thus effectively camouflaging them.

B-230. Inventory losses may be inaccurately labeled as pilferage for the following reasons:

- Failure to detect shortages in incoming shipments.
- Improper stock usage.
- Poor stock accounting.
- Poor warehousing.
- Improper handling and recording of defective and damaged stock.
- Inaccurate inventories.

B-231. In some cases, inventory losses may be impossible to detect because of the nature and quantities of materials involved. Stock inventory records may not be locally maintained, or there may be no method for spot checks or running inventories to discover the shortages.

PROFILE OF PILFERERS

B-232. Physical-security personnel must be able to recognize and counteract two types of pilferers—casual and systematic. A casual pilferer is one who steals primarily because he is unable to resist the temptation of an unexpected opportunity and has little fear of detection. There is usually little or no planning or premeditation involved in casual pilferage, and the pilferer normally acts alone. He may take items for which he has no immediate need or foreseeable use, or he may take small quantities of supplies for the use of family or friends or for use around his home. The degree of risk involved in casual pilferage is normally slight unless a very large number of persons are involved.

B-233. Casual pilferage occurs when the individual feels the need or desire for a certain article, and the opportunity to take it is provided by poor security measures. Though it involves unsystematic theft of small articles, casual pilferage is nevertheless very serious. It may have a great cumulative effect if permitted to become widespread, especially if the stolen items have a high cash or potential value.

B-234. There is always the possibility that casual pilferers, encouraged by successful theft, may turn to systematic pilferage. Casual pilferers are normally employees of the installation and usually are the most difficult to detect and apprehend.

B-235. A systematic pilferer is one who steals according to preconceived plans. He steals any and all types of supplies to sell for cash or to barter for other valuable or desirable commodities. He may work with another person or with a well-organized group of people, some of whom may be members of a cleaning team or even be in an advantageous position to locate or administratively control desired items or remove them from storage areas or transit facilities.

B-236. The act of pilferage may be a one-time occurrence, or such acts may extend over a period of months or even years. A large quantity of supplies with great value may be lost to groups of persons engaged in elaborately planned and carefully executed systematic-pilferage activities. Systematic pilferers may or may not be employees of the installation; if they are not, they frequently conspire with employees.

TARGETS FOR PILFERAGE

B-237. Both casual and systematic pilferers have certain problems to overcome to accomplish pilferage objectives. These problems include the following:

- A pilferer's first requirement is to locate the item or items to be stolen. For the casual pilferer, this may be accomplished through individual search or even accidental discovery. In systematic pilferage, more extensive means are generally used. These may consist of surveilling by members of the group or checking shopping and storage areas or documents by those who have access to them.
- The second requirement is to determine the manner in which he can gain access to and possession of the desired item. This may involve something as simple as breaking open a box, or it may be as complex as surveying security factors (such as physical safeguards or security procedures) for weaknesses. It may also involve attempting to bribe security forces, altering or forging shipping documents or passes, or creating disturbances to divert the attention of security personnel while the actual theft is taking place.
- The third requirement is to remove the stolen items to a place where the thief may benefit from his act. Articles of clothing may be worn to accomplish this. Small items may be concealed in many possible places on the thief's body or in vehicles. Through falsification of documents, whole truckloads of supplies may be removed from their proper place without immediate discovery.
- Finally, to derive any benefit from his act, the pilferer must use the item himself or dispose of it in some way. The casual pilferage of supplies is intended primarily to satisfy the need or desires of the thief. The systematic pilferer usually attempts to sell the material through "fences," pawnbrokers, or black-market operations.

B-238. Detection of use or disposal can help prevent similar pilferage through investigation and discovery of the means used to accomplish the original theft. Similarly, each of the problems faced by would-be pilferers offers an opportunity for constructive preventive measures. Careful study of the possible opportunities for the pilferer to solve his problem is essential in security work.

B-239. The primary concern of a systematic pilferer in selecting a target is its monetary value. Since he steals for personal profit, the systematic pilferer looks for items from which he can realize the greatest financial gain. This means he must also have or be able to find a ready market for items he may be able to steal. He pilfers small items of relatively high value (such as drugs, valuable metals, or electronic items). However, we cannot discount the possibility that a systematic pilferer may, if the profit is substantial, select a target of great size and weight. As a rule, bulk-storage areas contain most of the material that may be selected by systematic pilferers.

B-240. The casual pilferer is likely to take any item easily accessible to him. Since he normally will remove the item from the installation by concealing it on his person or in his POV, size is also an important consideration. Monetary value and available markets are not of any great concern to the casual pilferer, because he usually does not have any idea of selling the property he steals. Storage areas containing loose items are more likely to tempt casual pilferers than bulk-storage areas.

METHODS OF PILFERAGE

B-241. There are many ways that pilfered items may be removed from installations. Because the motives and targets likely to be selected by systematic and casual pilferers are very different, the methods of operation for each are very different.

B-242. As stated above, the casual pilferer steals whatever is available to him and generally removes it from the installation by concealing it on his person or in his automobile. The methods of the systematic pilferer are much more varied and complex. The means he may use are limited only by his ingenuity.

B-243. Shipping and receiving operations are extremely vulnerable to systematic pilferage. It is here that installation personnel and truck drivers have direct contact with each other and readily available means of conveyance. This offers a tempting opportunity for collusion.

B-244. One individual must not have control of all shipping and receiving transactions. Obviously, this procedure invites manipulation of government bills of lading and inaccurate storage and movement procedures through failure of one activity to compare bills and invoices with another activity. The opportunities for monetary kickbacks increase without a sound system of checks and balances.

B-245. Railway employees assigned to switching duties on the installation can operate in a similar manner. However, this operation is more difficult because a railway car normally cannot be directed to a location where stolen property can be easily and safely removed.

B-246. Tanker trucks used for shipping petroleum products may be altered to permit pilferage of the product. Trash- and salvage-disposal activities offer excellent opportunities to the systematic pilferer to gain access to valuable material. Property may be hidden in waste materials to be recovered by a accomplice who removes trash from the installation.

B-247. Other methods that may be used by systematic pilfers to remove property from military installations include—

- Throwing items over fences to be retrieved later by themselves or by accomplices.
- Packaging property and sending it to outside addresses through mail channels.
- Conspiring with security personnel.

- Wearing loose-fitting clothing to conceal small items.
- Removing items by using vehicles belonging to outside contractors and vendors.

CONTROL MEASURES FOR CASUAL PILFERAGE

B-248. Specific measures for preventing pilferage must be based on careful analyses of the conditions at each installation. The most practical and effective method for controlling casual pilferage is to establish psychological deterrents. This may be accomplished in a number of ways, such as—

- Searching individuals and vehicles leaving the installation at unannounced times and places.
- Conducting spot searches occasionally to detect attempts of theft.
- Making employees aware that they may be apprehended if they attempt to illegally remove property.

B-249. Care must be taken to ensure that personnel are not demoralized nor their legal rights violated by oppressive physical control or unethical security practices. An aggressive security-education program is an effective means of convincing employees that they have much more to lose than to gain by engaging in acts of theft. Case histories may be cited where employees were discharged or prosecuted for pilferage. Care must be taken in discussing these cases to preclude the identification of individuals because of possible civil suits for defamation of character. Also, it is generally poor policy to publicize derogatory information pertaining to specific individuals. It is important for all employees to realize that pilferage is morally wrong no matter how insignificant the value of the item taken.

B-250. It is particularly important for supervisory personnel to set a proper example and maintain a desirable moral climate for all employees. All employees must understand that they have a responsibility to report any loss to the proper authorities. Adequate inventory and control measures should be instituted to account for all material, supplies, and equipment. Poor accountability, if it is commonly known, provides one of the greatest sources of temptations to the casual pilferer.

B-251. Identifying all tools and equipment by some mark or code (where feasible) is necessary so that government property can be identified. Installation tools and equipment have counterparts on the civilian economy and cannot otherwise be identified as government property. Another control method requires individuals to sign for tools and equipment. The use of the signature control method reduces the temptation to pocket the item.

B-252. In establishing any deterrent to casual pilferage, physical-security officers must not lose sight of the fact that most employees are honest and disapprove of theft. Mutual respect between security personnel and other installation employees must be maintained if the facility is to be protected from other more dangerous forms of human hazards. Any security measure that infringes on the human rights or dignity of others will jeopardize rather than enhance the overall protection of the installation.

CONTROL MEASURES FOR SYSTEMATIC PILFERAGE

B-253. Unlike the casual pilferer, the systematic thief is not discouraged by psychological controls. Nothing short of active physical-security measures are effective in eliminating loss from this source. Some of these measures include—

- Establishing security surveillance at all exits from the installation.
- Establishing an effective package- and material-control system.
- Locating parking areas for POVs outside of the activity's perimeter fencing.
- Eliminating potential thieves during the hiring procedure by careful screening and observation.
- Investigating all losses quickly and efficiently.
- Establishing an effective key-control system.
- Establishing adequate security patrols to check buildings, grounds, perimeters, and likely locations for clandestine storage of property removed from its proper location.
- Installing mechanical and electrical intrusion-detection devices where applicable and practical.
- Coordinating with supply personnel to establish customer ID to authenticate supply-release documents at warehouses and exit gates.
- Establishing appropriate perimeter fencing, lighting, and parking facilities and effective pedestrian, railway, and vehicle gate-security control.

HOW TO STOP EMPLOYEE THEFT

B-254. No matter what it is called—internal theft, peculation, embezzlement, pilferage, inventory shrinkage, stealing, or defalcation—thefts committed by employees are behind at least 60 percent of crime-related losses. So many employees are stealing so much that employee theft is the most critical crime problem facing businesses today.

B-255. Although employee theft results in part from factors beyond control, the extent of employee theft in any business is a reflection of its management—the more mismanagement, the more theft. An effective stop-employee-theft policy must include at least the following:

- Pre-employment screening.
- Analysis of opportunities for theft.
- Analysis of how employees steal.
- Management-employee communication.
- Prosecution of employees caught stealing.

B-256. Each employer must reduce losses as much as possible. A police state need not be created. Large monetary expenditures need not be made.

B-257. The best way to stop employee theft is to simply not hire those employees inclined to steal. The best way is also impossible. The employer must set up a screening process that will weed out obvious security risks. Many experts believe that personnel screening is the most vital safeguard against internal theft. The following are some basic guidelines for the employer:

- Have the applicant fill out a written application. Ensure that the written application does not discriminate and that it conforms to any applicable laws.
- Solicit references, but keep in mind that those contacted will give favorable opinions. Ask primary references for secondary references. In contacting the latter, make it clear that the applicant did not refer them.
- Interview. During an interview, assess the applicant's maturity and values. Observe his gestures.
- Use psychological deterrents. Inform the applicant that your business routinely runs a background security check or that fingerprints will be taken. The hope is that the dishonest applicant will not be back.
- Obtain credit-bureau reports, but only after following guidelines set forth in the Fair Credit Reporting Act.

Opportunities, Methods, and Control

B-258. Cases of employee theft have been documented in almost every conceivable phase of business operations, from theft of petty cash to theft of railroad cars. An infinite variety of methods have been used. Some of the areas that are most vulnerable are—

- Shipping and receiving.
- Inventory.
- Accounting and record keeping.
- Cash, check, and credit transactions.
- Accounts payable.
- Payroll.
- Facility storage units.

B-259. Some of the methods used include—

- Pilfering (one item at a time).
- Stealing from cash registers or altering cash-register records.
- Issuing false refunds.
- Using the back door or trash containers.
- Taking advantage as a supervisor.
- Avoiding package controls.
- Embezzling.
- Forging checks.
- Stealing credit cards.
- Manipulating computers and stealing computer time.
- Conspiring with night cleaning crews.
- Duplicating keys or using a master key that is not properly controlled.
- Conspiring with outsiders (such as inflating insurance claims).

B-260. Too many opportunities exist for employees to exploit. Reduce these opportunities and the losses will be reduced. Reduce opportunities by using the following controls:

- Perform random spot checks on all phases of business in addition to regular, comprehensive audits.
- Check the payroll. Make sure that you are not paying a fictitious or dead employee.
- Take physical inventory seriously.
- Know what you own. Be able to identify it.
- Do not allow one employee to perform all functions. Separate receiving, purchasing, and accounts payable. Separate accountants from cash.
- Control payment authorizations.
- Keep blank checks secured. Do not presign or use uncoded, unnumbered checks.
- Reconcile cancelled checks with original invoices or vouchers.
- Secure exits. Restrict employees to one exit, preventing exit from the rear of buildings.
- Establish strict package control.
- Inspect cash-register receipts daily. Inspect the tape and ensure that the employee is identified on the slip. Deposit money daily.
- Issue ID badges to decrease employee presence in unauthorized areas.
- Simplify red tape; make it harder for the employee to disguise theft.
- Locate employee parking away from the business area.
- Establish a usage schedule of supplies to isolate irregularities.

Management-Employee Communication

B-261. Leadership must be firm, yet reasonable. Most employees pattern their values after their leaders, so a good example must be set. If you expect employees to remain honest, do not take home office supplies or goods.

B-262. Train new employees, advising them of the company's values and the standards by which they will be expected to perform. Explain all security procedures, stressing their importance. Emphasize that any deviations will be thoroughly investigated.

B-263. Establish grievance procedures; give your employees an outlet for disagreement and be receptive to all grievances submitted. Ensure that employees are aware of grievance procedures and that no reprisals are taken.

B-264. Regularly evaluate employee performance and encourage employees to evaluate management. Unrealistic performance standards can lead either to desperation and anger (resulting in dishonesty) or to get-even attitudes. Regularly review salaries, wages, and benefits—do not force employees to steal from you.

B-265. Delegate responsibility. Unless decision making exists among lower and middle levels, there is a tendency for development of an "it's-us-against-them" attitude. Delegate accountability as well; no decision is valid if it is lost in a "pass-the-buck" routine.

SECTION IV — ARMY PROPERTY AT THE LOCAL LEVEL

B-266. Proper accountability by commanders and subordinate personnel cannot be overemphasized. To ensure accountability of property, commanders must establish, implement, and supervise an installation's, activity's, or organization's security program.

B-267. Weaknesses in security procedures at the installation, activity, or organizational levels involving military property create vulnerability supported by criminal activity. Criminal activity includes—

- Theft.
- Fraud.
- Property diversion.
- Property manipulation.

B-268. Commanders and subordinate personnel must conduct a risk analysis; identify military property and; in the interest of monetary value and mission accomplishment, design mandatory security measures for specific property.

B-269. Security doctrine (as outlined in this manual) should be used to the maximum extent in securing Army property vulnerable to theft, destruction, or manipulation. Certain categories of property (such as in Table B-2, page B-62) must be assessed for security vulnerability and protective treatment. Security protective measures addressing this military property should be documented in the installation's physical-security plan. If the security measures recommended in Table B-2 are implemented using established doctrine, they should eliminate or reduce the property's vulnerability. This will reduce the incidents of theft, pilferage, and manipulation at the installation.

MOTOR VEHICLES

B-270. Security of tactical vehicles should be based on a uniform and cost-effective approach. For example, to ensure that a tactical vehicle without a locking device is properly secured, install a clamp, chain, and locking device as illustrated in Figure B-7, page B-63. To install the security device properly while maintaining safety, refer to Technical Bulletin (TB) 9-2300-422-20. Army motor-vehicle security should also incorporate the use of the following:

- Key/lock security and accountability.
- Protective lighting.
- Fencing.
- Walking patrols, as appropriate.
- Frequent observation and visits by mobile patrols or unit personnel (such as the charge of quarters [CQ], the staff duty officer [SDO], and the staff duty noncommissioned officer [SDNCO]).

Table B-2. Recommended Security Measures

Property	Inventoried by—					Inspected by—							
	Security Plan	Hand Receipt/Property Book	Secured by IDS	PS Plan	Regulation	Unit Leaders	PS Officer	SDO	Maint Officer	Supply Officer	Dining-Facility Officer	CQ	SDNCO
Arms/ammunition	x	x	x	x	x	x	x	x	x	x			x
Small arms	x	x	x	x	x	x	x	x	x	x			x
Explosives	x	x	x	x	x	x	x	x	x	x			x
Communications/electronic equip	x	x	x	x	x	x	x	x	x	x			x
Hand tools, tool sets/kits, and shop equip	x	x	x	x	x	x	x		x	x			x
Subsistence items	x	x		x	x	x				x	x		x
Controlled substances, precious metals, and tax-free items	x	x	x	x	x	x	x	x		x			x
Accounts	x	x	x	x	x	x							
POL products	x	x		x	x	x	x	x	x	x		x	
Repair parts	x	x	x	x	x	x	x		x				
Aircraft	x	x	*	x	x	x	x	x	x				
Vehicles	x	x		x	x	x	x	x	x	x		x	x
Towed weapon systems/components	x	x	x	x	x	x	x	x	x				
Carriage-mounted weapon systems	x	x	x	x	x	x	x	x	x				x
Construction material	x		*	x		x	x			x			x
Special-issue clothing (CTA)	x	x	x	x	x	x	x	x			x		
Individual clothing and equipment	x	x	x	x	x	x		x		x			
Organizational equip/components	x	x	*	x	x	x	x	x	x	x		x	x
Compasses, binoculars, and flashlights	x	x		x	x	x	x			x			x
Medical-unique items	x	x	x	x	x	x	x	x		x			x
Housekeeping supplies and equipment	x	x	x	x	x	x				x		x	
Housing furniture	x	x		x	x	x				x		x	
Mess equipment	x	x	x	x	x	x	x			x	x		x
Office machines	x	x	x	x	x	x	x			x			x
Expendable/consumable supplies	x		x	x	**	x		x		x	x		x

*** Depends on facility availability and cost effectiveness**
**** Depends on local policy**

Figure B-7. Typical Clamp and Chain Installation

CONSUMER OUTLETS

B-271. The lack of initiative at the management level within operational consumer outlets does little to prevent or reduce pilferage. Such shortcomings are identified as—

- Failure to present a professional image.
  - Lack of continuing interest, motivation, and direction.
  - No alertness to internal control of pilferage.
- Failure to institute and implement methods of operational effectiveness and efficiency. These methods include—
  - A clearly defined delegation of responsibility.
  - Insistence on stringent accountability
  - Orientation and training programs for subordinate supervisors and current and new employees.
- Failure to emphasize and enforce established criteria for continual employment.
  - Rules of conduct.
  - Standards of job performance. (Officially request appropriate action for employees guilty of criminal acts or infractions conducive to criminal acts.)
  - Inattentive job attitudes of subordinate supervisors.
  - Inadequate personal checks of established accounting and inventory procedures. **NOTE: Checks on both a regular and unannounced basis tend to control access to official stock records and to ensure careful and organized storage or stocking of merchandise.**
  - Infrequent observation of an employee's job performance.

- Failure to report misconduct, criminal or otherwise, to superiors or responsible law-enforcement personnel in the activity.
- Failure to implement recommendations made during physical-security inspections or crime-prevention surveys.

PILFERAGE

B-272. Pilferage may be accomplished by individual employees, by a team of employees, or by employees and patrons in collusion. These actions can be greatly reduced by tightening supervision and security in the following areas:

- **Merchandise display or dispensing areas.** The following measures can reduce merchandise pilferage in display areas:
  - Detecting unauthorized price reductions.
  - Preventing or making it difficult to alter price tags.
  - Checking procedures for declaring merchandise old, shopworn, damaged, or salvage.
  - Providing more unpackaged items for personal consumption.
  - Discouraging the careless waste of foods and other perishable items.
- **Cash registers.** The following are methods of pilferage from cash registers:
  - Stealing directly from an unattended register.
  - Rerunning register tapes at lower figures (this is preventable if the reset key is maintained by the supervisor).
  - Clearing the register at a lower total figure than actual receipts for the operational period.
  - Reporting overrings and refunds falsely.
- **Theft of merchandise.** The following methods enable merchandise to be pilfered:
  - Underringing.
  - Reusing cash-register tapes. This occurs when employees fail to provide patrons with tapes or when patrons allow employees to retain tapes. The tapes allow employees to package merchandise and remove it from the premises.
  - Removing items from bags or containers carried out by employees.

SHOPLIFTING

B-273. Shoplifting is usually confined to sales areas and is committed by casual and systematic pilferers. Items that are most frequently pilfered—

- Are relatively small in size.
- Have a high degree of consumer desirability.
- Are easily carried in pocketbooks or secreted on the person.

B-274. Amateur, adult shoplifters share the following characteristics:

- Theft is from a sudden temptation (impulse theft). There is success in the initial theft, which leads to more temptation, stronger impulses, and more thefts.
- They rarely have a genuine need for the item.

- They usually have enough money to pay for the item.
- They display symptoms of nervousness and uneasiness.

B-275. Juvenile shoplifters have the following traits:

- They act on a dare or "to belong."
- They may be coached or directed by an adult.

B-276. Professional shoplifters share these characteristics. They—

- May be talkative and are usually polite and deliberate.
- Look for opportunities continually.
- Do not take many chances.
- Are very capable of spotting security personnel.
- Steal for resale.
- Have "fences" (usually).
- Steal "to order" (often). They may have a list describing the items to be pilfered.
- Use innovative techniques.

B-277. Genuine cases of kleptomania are rare. Kleptomaniacs share the following characteristics. They—

- Take items without regard to their value or use.
- Steal compulsively and often openly.
- Are nervous and shy.

B-278. Narcotic addicts as shoplifters have the following characteristics. Only MP/security personnel should attempt apprehension. These types of shoplifters—

- Are desperate for money and they fear imprisonment.
- Take big chances.
- Take the merchandise and exit the premises quickly.
- Steal when they are at their lowest physical and/or psychological ebb.
- Are dangerous if you try to apprehend them (sometimes violent).

B-279. Alcoholics and vagrants as shoplifters share the following traits. They—

- Usually steal because of need.
- Are often under the influence of liquor at the time of theft.
- Take the merchandise quickly, then exit the premises.
- Are less likely to steal regularly at a single location.

B-280. The most pilferage occurs when employee coverage is low or when employees are untrained, inexperienced, or indifferent to the issue. The ineffective use of floor space aids shoplifters by creating congestion in the patron's traffic flow. Allowing an emphasis on small rooms or partitioned areas causes congestion that clusters, isolates, or partially hides displays. Shoplifting involves the use of one or more of the following means to obtain items:

- Palming or placing an open hand on a small article, squeezing the muscles of the hand over the article to grasp it, and lifting the still open and apparently empty hand.
- Using fitting rooms to put on tight or close-fitting garments under clothing worn into the store.
- Trying on hats, gloves, sweaters, and jackets, then exiting the store.
- Stepping around counters and removing items from unlocked showcases.
- Handling several items at once and replacing all except the item pilfered.
- Using accomplices to create a diversion of employee attention when secreting items on the person. Such items include:
  - Clothing.
  - Pocketbooks or handbags.
  - Umbrellas.
  - Various items placed in packages or paper sacks containing merchandise paid for at other departments.

ARSON

B-281. On an installation there is little incentive for a professional arsonist to operate since the government owns the buildings and insurance fraud in collusion with the property owner is not possible. However, arson can still be a problem. In the civilian community, most deliberate fires are not set for profit. They are set to "get even" with the property owner or just for the excitement of watching something burn. Military installations are susceptible to this type of crime. On many Army posts, there is a large number of empty, wooden structures that are ideal targets for revenge seekers or vandals.

B-282. Arson is an easy crime to perpetrate, and it is relatively difficult to collect the information needed to convict an arsonist. However, this does not mean we are helpless in combating this crime. Several major cities, including Seattle, Denver, Houston, and Philadelphia, have developed successful programs to reduce the number of arsons. Some of the successful, proactive measures that have been developed are—

- Securing or disposing of materials that could be used to start fires. Ensuring that the regulations on the storage of gasoline, paint, and solvents are enforced and that paint lockers are locked.
- Enforcing command policies on the police of the post. Removing piles of trash, scrap lumber, and other material that burns easily.
- Securing empty buildings (especially empty wooden structures) and posting them as off-limits areas.
- Patrolling areas susceptible to arson.
- Encouraging participation in neighborhood watches, taxi patrols, and other community programs that increase surveillance.
- Encouraging the reporting of suspicious activities through the establishment of crime hot lines.

- Establishing a close working relationship between fire fighters and law-enforcement personnel to ensure that fires of suspicious origin are reported and thoroughly investigated.
- Offering rewards for information leading to the apprehension of the arsonist (if there is an outbreak of arson).

SECTION V — COMMUNITY CRIME-PREVENTION PROGRAMS

B-283. There are a number of crime-prevention programs set up in communities to help deter and detect crime. These programs are supported and often headed by members of the community in conjunction with local law-enforcement agencies. The programs are as diverse in nature as they are in number, yet they complement one another.

NEIGHBORHOOD WATCH PROGRAM

B-284. A neighborhood watch program is an organized network of citizens interacting with other neighbors and the police in preventing and detecting crime in their neighborhood. Law-enforcement efforts to reduce crime cannot be accomplished effectively without the support and cooperation of all citizens. A strong community involvement with neighbors helping themselves and other neighbors in becoming more alert to activities in the neighborhood, protecting their property, and reporting suspicious activities is essential to an effective crime-prevention program.

B-285. The Army Neighborhood Watch Program is designed to encourage Army service members and their families to actively participate in protecting their own property and the property of their neighbors, joining community crime-prevention programs, and reporting suspicious activities to MP officers. The program is designed to develop the following:

- The awareness of community crime trends and prevention efforts.
- The knowledge of quarters' security procedures.
- A cooperative system of surveillance over each neighbor's property.
- Accurate observation and reporting of suspicious activities.
- The establishment of a reliable, two-way information link between the community and MP forces.

B-286. Most neighbors know the routines of the other families that live near them. They know what cars are normally parked in the neighborhood and when families are on vacation or out of the area. Neighbors are in a very good position to recognize burglars and other intruders. Also, residents are in a good position to recognize safety hazards and crime-conducive conditions near their homes.

B-287. To capitalize on these advantages, neighborhood watch programs organize blocks in family-housing areas or floors in troop billets to improve police and community interaction. They also disseminate information on crime problems and countermeasures.

BLOCK CLUBS

B-288. Block clubs are the basic components of an installation-wide neighborhood watch program. The geographical size of a block club may vary widely depending on the population's density and the nature of the terrain. The key factor is that the terrain organized into a single block club should promote a feeling of unity and mutual assistance. In a troop billet area, block clubs could be organized along company lines, by individual barracks, or by floor. In a family-housing area, a block club could cover one high-rise apartment building, one block, or one or more streets that are so situated that the residents identify themselves as a subcommunity within the housing area.

Organization

B-289. Individual residents, community-service organizations, or MP units can initiate block clubs. Existing organizations (such as the PTO) have frequently established contacts within the community and have sponsored the organization of block clubs. Regardless of the approach used to organize the installation, every family or resident in the area should be contacted and encouraged to attend a block club (see Figure B-8).

B-290. A block captain and a deputy block captain should be elected at the initial block meeting. These individuals serve as "spark plugs" to sustain interest in their geographic area. They also represent their block club at district meetings. The effectiveness of the presentation at the initial block-club meeting is critical. A PMO representative should explain the crime problems on the installation and clearly outline the functions of a block club and how it can affect the crime problem. The main objective should be to generate enthusiasm and build a foundation upon which an effective neighborhood watch can be built. The use of camcorders or other audiovisual programs can generate interest. Association with programs in the civilian community may also help to generate interest.

B-291. The functions of a block club should be clearly identified. Block clubs—

- Serve as the "eyes and ears" of the police.
- Encourage the implementation of individual countermeasures, such as marking personal property.
- Disseminate crime-prevention information.
- Improve police and community relations

Block-Club Districts

B-292. A block-club district should be organized as an intermediate link between individual block clubs and the installation's crime-prevention council. These districts should cover at least one housing area and should be composed of the block captains and deputies from each block club in the district's area. District leaders should be elected from the block captains and should serve as members of the installation's crime-prevention council. The primary functions of districts are to transmit information from the crime-prevention council to the block clubs and to develop incentive awards to recognize effective participation by the clubs.

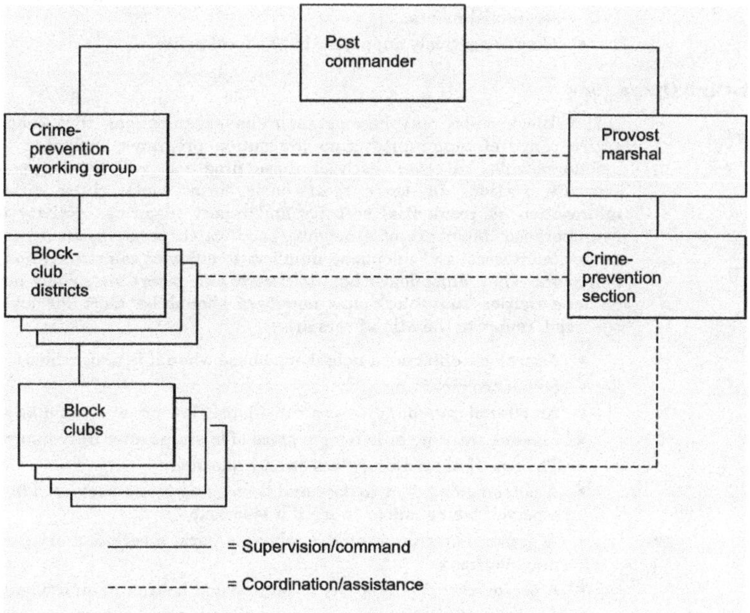

Figure B-8. Block-Club Organization

Maintaining Interest

B-293. Maintaining interest in neighborhood watch programs is a major problem. Typically, a particular incident generates the level of interest required to initially organize a system of block clubs and districts; but as the particular problem is overcome, interest wanes and the clubs gradually dissipate. A review of programs that have maintained their effectiveness for extended periods indicates that successful programs have the following characteristics:

- They are a formal organization with elected block captains and district leaders.
- They address a wide range of problems.
- They feature an incentive award program to recognize individuals and clubs that participate effectively.

- Their members attend periodic workshops to train leaders on various aspects of crime prevention.
- They have clear-cut, attainable objectives. The members need to see accomplishments.
- They are actively supported by the local police.

Block-Club Operations

B-294. Block clubs may become umbrella organizations that conduct the entire range of community crime-prevention programs, including foot and mobile patrols. However, each of these programs will be discussed in a separate section. In their most basic form, block clubs disseminate information on residential security and report suspicious activity in their neighborhood. Members of a neighborhood watch program should exchange names, addresses, and telephone numbers to enhance communication among neighbors. They must learn how to record and report suspicious activities. Some activities that block-club members should be alert for and, when observed, report to the MP officers are—

- A stranger entering a neighbor's house when it is unoccupied.
- Someone screaming.
- An offer of merchandise at a ridiculously low price (it could be stolen).
- Persons entering or leaving a place of business after duty hours.
- The sound of breaking glass or an explosion.
- A person going door to door and then going into a back or side yard, or a person trying a door to see if it is locked.
- A person loitering around a school, a park, a secluded area, or in the neighborhood.
- A person carrying property at an unusual hour or in an unusual place.
- A person exhibiting unusual mental or physical symptoms (he may be in need of medical help).
- A vehicle being loaded with valuables when parked by a closed business or untended residence.
- A business transaction conducted from a vehicle.

ARMY NEIGHBORHOOD WATCH MEETINGS

B-295. A PMO representative should attend the initial neighborhood block meeting to explain the relationships of crime prevention and the neighborhood watch programs. The meeting should be publicized with handout invitations announcing the time, location, and purpose of the meeting. If possible, the meeting should be held in the neighborhood (such as the training room in the troop billets, a local school, or a residence in family quarters). In addition to explaining the program concepts, the initial meeting could include the following:

- Installation and community criminal statistics concerning the nature and volume of housebreakings, larcenies, and other crimes.
- An exchange of names and telephone numbers (if applicable) of attendees. This information should also be placed on a neighborhood block sheet that is a geographical diagram of the block showing the

location of each room, apartment, and building address number in the block. The names and phone numbers of participants would be added to the address number of each residence drawn on the diagram. This block sheet should be distributed to block members at a subsequent meeting.
- Determination of the second meeting date and location.

B-296. Other agenda items and activities for subsequent meetings could include the following discussions:

- Residential security procedures and the conduct of quarters security inspections.
- Operation ID—marking and recording of personal property.
- The observation and report of suspicious activities.
- Other crime-prevention measures and procedures being implemented at the installation (such as rape prevention and citizen escort patrols).
- Fire prevention, personal safety, and other related activities.

OPERATION ID

B-297. Operation ID, which entails marking of property to make it identifiable and traceable to its owner if lost or stolen, was initiated by the Monterey Park, California, Police Department in 1963 and has been adopted by more than 80 percent of the police departments in the US. Operation ID is a low-cost, highly effective crime-prevention program. However, its success is contingent upon the willingness of individuals and communities to actively participate in marking and identifying their personal property.

B-298. Operation ID is designed to encourage Army service members and their families to mark their personal property with a standard Army-wide, owner-applied number. This numbering system permits the positive ID of the property and determines the location of the owner in case of theft or loss.

B-299. The principal advantages of Operation ID are theft deterrence and recovery of personal property. Marked stolen property is more difficult to dispose of, and illegal possession can result in the prosecution of a thief. Recovered lost or stolen property can only be returned if there is some means of identifying and locating the rightful owner.

METHODS OF IDENTIFYING PROPERTY

B-300. Various methods of establishing positive ID and ownership of property in case of loss are available for individuals. Each method has advantages and limitations, and a combination of these methods would be required to ensure the ID of all high-value personal property.

- Inscribing the owner's applied number with an etching or engraving tool would allow the recovering agency to visually identify the number inscribed on the property for notification and subsequent return of the property. Electrostatic markers are available for use at no cost to the individual. However, some personnel are reluctant to use this method since it can mar the property. The inscription should be made in a location that can be readily seen by the recovering agency but which would not deface the property's appearance or reduce its value. Some high-value personal items such as coins, jewelry, and silver cannot be inscribed with an owner's applied number. Another method of identifying these items would be required.

- Using invisible fluorescent ink, powder, or paste to mark the property can make it easier for the agency recovering the property to use an ultraviolet light and identify the owner. This marking will not mar the property. However, the fluorescent markings and ultraviolet light are an additional cost, and many agencies do not have ultraviolet lights to inspect recovered property.
- Using a laser photographic process to identify diamonds. Every diamond emits a unique reflection when penetrated by a low-level laser light. A laser photographic process has been developed to record a diamond's pattern of light reflection on film. Several jewelers throughout the country have the laser photographic equipment available. To register a diamond, two photographs are taken. One is provided to the owner and the other to a central registry. If a diamond is lost or stolen, the recovering agency can take the diamond to a jeweler that maintains the laser photographic equipment for print. This photograph would then be forwarded to the registry for owner ID.
- Photographing the personal item. Individuals can photograph personal high-value items that cannot be engraved. Although the agency recovering the lost or stolen items could not identify the owner, use of photographs could assist in verifying ownership if it is known that the item has been recovered. In addition, a photograph would assist in submitting claims against the government or private insurance companies, as appropriate.

RECORDING PERSONAL PROPERTY

B-301. Individuals should record identifying data (such as brand name, model, serial number, and value of the personal items), even if they use other methods of identifying property. This information would assist in determining what items may be lost, stolen, or damaged through fire, explosion, or other hazards. This information can also be used in claims against the government or private insurance companies, as appropriate.

IDENTIFICATION NUMBERING

B-302. There are various types of owner-applied numbering systems used to mark personal high-value items. Criteria that should be considered in determining which owner-applied numbering system to incorporate include—

- Uniqueness, where no two people have the same identifier.
- Permanence, so that the owner-applied number will not change.
- Ubiquity, so that an identifying number is available to any individual who desires one.
- Availability, where the identifying number can be easily obtained and remembered.
- Indispensable, so that there are incentives requiring an individual to have the number.

- Privacy, so that the number is not a means of infringing upon an individual's right to privacy.
- Uniformity, so that the owner-applied number would be readily recognized by law-enforcement agencies who handle or come into contact with the recovered property.
- Traceability, so that the property owner can be identified and located.

B-303. The most commonly used owner-applied numbering systems include—

- Driver's license number with the issuing state abbreviation prefix.
- Social security number.
- Personal numbers assigned to individuals by a local law-enforcement agency.
- Personal numbers with the marking agency's National Crime Information Center originating-agency ID number.
- A private numbering system maintained by a commercial organization.

STANDARD ARMY NUMBERING SYSTEM

B-304. To ensure that criteria for property ID numbering are met, the standard Army-wide, owner-applied numbering system is designated as the service member's social security number with a "USA" prefix. Upon recovery of lost or stolen property by other military or civilian law-enforcement agencies, the USA prefix, owner-applied number would alert the recovering agency that the property belongs to a member of the Army. The recovering agency can then contact the nearest Army installation's PMO/security office concerning the property and the owner-applied number inscribed. The service number can then be identified and located through the Army worldwide locator system. Since this locator system lists only social security numbers and locations of active Army service members, family members should also use the service member's social security number with the USA prefix when marking their personal items.

IDENTIFYING AND LOCATING OWNERS OF RECOVERED PROPERTY

B-305. The installation's PMO/security officer should be the initiator of tracer actions to identify and locate the owner of recovered property marked by the standard Army Operation ID numbering system. Continuous liaison should be maintained with local civilian law-enforcement agencies and other military installations to ensure that they are cognizant of the standard Army owner-applied numbering system and will contact the PMO/security office upon recovery of private property marked with this system. The PMO/security officer should readily accept custody of the private property if the recovering agency is willing to release the property and it is not required as evidence for criminal prosecution.

B-306. Upon notification of recovered property and the inscribed Army standard owner-applied number, the PMO/security officer should contact the servicing military personnel office (MILPO) for assistance in determining the name and location of the property owner by using the Army's worldwide locator microfiche.

B-307. If the owner-applied social security number is not listed on the Army's worldwide locator microfiche and the servicing MILPO is unable to provide the requested information, the PMO/security office should contact the following offices:

- For enlisted members: USA Enlisted Records and Evaluation Center, ATTN: PCRE-RF-L, Fort Benjamin Harrison, Indiana 46249.
- For officers: PERSCOM, ATTN: TAPC-MSR-S, 200 Stovall Street, Alexandria, Virginia 22332-0444.
- For warrant officers: PERSCOM, ATTN: TAPC-OPW, 200 Stovall Street, Alexandria, Virginia 22332-0444.

B-308. In case the owner-applied number cannot be identified by either the servicing MILPO or the USA Enlisted Records and Evaluation Center, the Army Reserve Personnel Command (AR-PERSCOM) should be contacted for assistance. The AR-PERSCOM retains files on separated, retired, and Reserve component Army members. Written requests should be forwarded to: Commander, AR-PERSCOM, ATTN: ARTC-PS, 1 Reserve Way, St. Louis, Missouri 63132.

B-309. When the name and location of the service member associated with the owner-applied number inscribed on the property is verified, the PMO/security office should notify the service member in writing that the property has been recovered. The notification should ascertain if the recovered property belongs to the service member, if the service member ever reported the property as lost or stolen, and if a claim was submitted to the SJA claims service for loss or theft of the property. Notification should also state where the property is located and a point of contact that the owner can deal with for the return of the property. If the property is to be retained as evidence for legal proceedings, the owner should be informed that the property will be returned upon completion of the proceeding. A copy of this letter should be provided to the US Army Claims Service, Fort Meade, Maryland 20755. The Army Claims Service will advise the PMO/security office if a claim was or was not submitted.

B-310. If the owner cannot be located, recovered property in the custody of the PM/security offices should be disposed of according to AR 190-22 and DOD 4160.21-M.

USE OF THE NATIONAL CRIME INFORMATION CENTER

B-311. Stolen articles may be entered into the National Crime Information Center (NCIC) file if a theft report has been made, if the item is valued at $500 or more, and if it has a unique manufacturer's assigned serial number and/or owner-applied number. Entering stolen personal property items meeting the above criteria into the NCIC or other police information systems as outlined in AR 190-27 is encouraged.

NEIGHBORHOOD WALKS

B-312. While most people are unwilling to participate more actively than as observers in a neighborhood watch, there are some individuals who want to become more actively involved in securing their neighborhoods. For this segment of the population, the organization of "neighborhood walks" provides a welcome opportunity to make a more active contribution. The basic idea is simple; residents patrol on foot through their own neighborhoods to observe and report crime. In practice, it is a little more complicated; however, neighborhood walks can have a dramatic impact on the crime rate, so the effort expended is worthwhile. The points that must be considered are—

- **Patrol composition.** Both adults and teenage children can volunteer to participate in neighborhood walks. However, there should be at least one adult in each party. There should be at least two people in each patrol group. Groups of four to six individuals are desirable since, by their numbers alone, they discourage attacks on the walkers. Larger groups are also more fun, and this is important when volunteers are providing the manpower.
- **Times/patrol duration.** As with other crime-prevention programs, maintaining a high level of interest can be a problem. A successful walk program in Philadelphia schedules groups for one 2-hour patrol per month. More frequent tours caused high drop-out rates among participants. Neighborhood walks should be conducted only during those times when the crime rate is the highest. Normally, there are not enough volunteers to conduct walks at times other than peak crime periods.
- **Functions.** Members of neighborhood-walk groups must understand that they are to observe only and not actively intervene in criminal acts. Participants and the government are legally liable for their actions during walks. When a crime or a suspicious activity is spotted, the neighborhood walkers should report it to the police. In Philadelphia, the walkers are equipped with horns. When a crime is spotted, they activate their horn and go to the nearest house to call the police. When residents hear the walker's horn, those in the immediate area turn on all of their lights and sound their own horns. The noise and increased lighting invariably causes the criminal to flee.
- **Neighborhood escorts.** In addition to observing and reporting criminal activity, neighborhood patrols can escort children and older persons between community-service facilities and residences. They can also request that owners secure property when they find it unsecured; for example, when there are unsecured bicycles parked on a front lawn.

VIGILANTISM

B-313. While active community participation is essential, vigilantism must be discouraged at all costs. Both formal law codes and US common law offer few protections for private citizens who take the law into their own hands. Block captains and installation crime-prevention officers must be alert for indications that neighborhood patrols are doing more than observing and take swift remedial action when required. Of course, all nonpolice participants must be prohibited from carrying weapons of any type while engaged in crime-prevention programs. Experience in neighborhoods having much higher violent crime rates than found on Army installations has demonstrated that passive devices like horns or whistles were adequate to discourage attacks.

These devices, plus the assignment of four to six individuals to a neighborhood patrol, provide sufficient protection.

MOBILE PATROLS

B-314. Some communities with high street-crime rates have been successful in organizing private citizens into mobile-patrol programs. Like neighborhood foot patrols, these mobile patrols serve as the "eyes and ears" of the police but do not actively intervene when they spot a crime in progress.

B-315. In a typical program, block captains or police crime-prevention officers assign specific patrol areas to each private mobile patrol. In addition, each patrol receives training on the patrol's functions, communications procedures, and emergency actions. Normally, patrols are instructed to blow their auto horns steadily when they observe a crime in progress. This is usually sufficient to drive off the criminal.

B-316. Most private citizen patrols use cellular telephones or citizen's band (CB) radios as direct communications links with the supporting police department. Installation CB radio clubs are often willing to sponsor anticrime patrols under police supervision. Commercial taxi companies that operate on military installations are also excellent candidates to organize into patrols. The cab drivers normally cover most of the high-crime-rate areas on the installation. Because of the frequency with which they cover them, they are familiar with the routine conditions in each area and are quick to spot suspicious activity.

B-317. The government cannot provide gasoline for POVs used for anticrime patrols; however, it may often provide magnetic signs to affix to the vehicle for identifying it as part of the police-sponsored, neighborhood-patrol program. As in the case of foot patrols, the installation crime-prevention officer must be alert for signs of vigilantism and must take positive action to discourage it if it appears.

PROJECT LOCK

B-318. Nearly one million automobiles are stolen in the US every year. The total value of cars stolen is around the billion dollar mark, making auto theft the nation's costliest crime involving property. Of even greater importance is the social impact of auto theft. For a growing number of young people each year, stealing cars represents the first step toward a life of crime.

B-319. Police agencies have been diligent in the apprehension of auto thieves and the recovery of stolen vehicles, but the auto-theft problem seems to be more amenable to improvement through prevention rather than punishment. Barring strict security, auto theft is one of the easier crimes to commit. All vehicles left unattended are vulnerable, and widespread prevention by police surveillance is a physical impossibility.

B-320. The problem has grown to serious proportions despite determined law enforcement because motorists continue to be negligent or unaware of their responsibility. As long as people invite theft by leaving their cars unlocked or leaving the key in the ignition, auto thefts will continue to climb. Almost half of all stolen cars each year had been left with keys in the ignition; nine out of ten of the stolen vehicles had been left unlocked.

B-321. If a significant reduction is to be made, the motorists themselves must make it. Widespread adoption of accepted and effective prevention practices by motorists presents the most logical and immediate improvement to this growing problem.

B-322. In 1963, the Boston police department conducted a broad information campaign with the assistance of the National Automobile Theft Bureau and the Insurance Information Institute. Since then, more than 525 "Lock Your Car" campaigns have been held in about 400 communities in 49 states.

B-323. In the months following these campaigns in such cities as Denver, Chicago, Atlanta, and San Francisco, significant reductions in the number of auto thefts (ranging from 9 to 54 percent) have been recorded. Undoubtedly, the "Lock Your Car" campaigns contributed to reducing auto-theft statistics.

B-324. Project Lock is designed to permit sponsoring groups to conduct one-day or one-week "Lock Your Car" campaigns. Its purpose is twofold; it—

- Alerts the public on the importance of locking cars and removing keys as a deterrent to auto theft.
- Contributes to the welfare of youths by preventing the commission of a first crime.

B-325. The following materials can be ordered to support Project Lock:

- Windshield flyers.
- Identifying insignia to be worn by inspectors.
- Tally cards for noting cars left with keys in ignitions or unlocked doors.

B-326. Groups considering sponsorship of this campaign should consider installation areas known to have car-theft problems. The PMO should assign patrolmen to accompany the teams on inspection day. Usually, the teams' routes can be arranged to fit the regular patrols of the MP officers. If uniformed police will not be available, the campaign should not be held.

COMMUNITY ORIENTATION

B-327. If the commander desires, a community orientation meeting might be held a month or so in advance of the campaign. This would be more desirable for a week-long rather than a day-long campaign. If such a meeting is planned, the PMO should issue invitations to representatives of the installation crime-prevention council, service clubs, women's clubs, PTOs, high schools, and churches.

B-328. The meeting should be opened with an introduction of the installation commander. After appropriate comments, the commander would read a proclamation setting the date for Lock-Your-Car Week. If possible, a representative of the National Automobile Theft Bureau should be asked to address the meeting. An alternative would be to have the PM review national and local trends in auto thefts stressing the importance of the forthcoming campaign. In conclusion, a representative of the sponsoring group might review the schedule of activities for the campaign.

PROGRAM ACTIVITIES

B-329. If the campaign is conducted in one day, a kick-off breakfast might be substituted for the orientation meeting. The program and the attendance might be similar if a weeklong campaign is planned and the orientation meeting were held. The kick-off breakfast might have a simple program with attendance limited to the sponsoring group and police representatives. An alternative to the breakfast would be a luncheon at which results to the moment are reported. Whether the campaign will last a week or a day, the general activities will be similar.

B-330. With the assistance of the MP officers, the installation should be zoned according to the established MP patrol areas where possible. The size and number of the zones will depend on the number of police and volunteer personnel that will be available. However, the zones should cover most of the installation's service areas, troop billet areas, and family-housing areas.

B-331. The inspection teams are each composed of a uniformed patrolman and three to five sponsoring group members, cover their assigned zones and place flyers under the windshield wipers of all cars found to have keys in the ignition or to be unlocked. Under no circumstances should the flyers be placed inside the cars, even through open windows.

B-332. It is suggested that flyers be ordered early enough to allow a local printer to inscribe an overleaf statement such as "This public service is provided as a courtesy of the MP force." The tally cards should be used to record the number of cars inspected, the number that were unlocked, and the number with keys in the ignition.

SCHOOL PROGRAM

B-333. The crime-prevention office might sponsor a poster contest for art students. It should be announced at least a month before the campaign to allow the entries to be placed in public in advance of the date. The PMO should present awards in an office ceremony. The ceremony should take place at noon on the campaign day or midweek if the campaign is longer.

PUBLIC ADDRESSES

B-334. Close to or during the campaign date, addresses by crime-prevention professionals should be scheduled for programs of service clubs, women's clubs, PTOs, and other civic groups. Well in advance of the campaign, the PAO should be visited by the PM and the other sponsors to develop comprehensive internal-information programs in support of the installation's Project Lock. Plans for the public-affairs program should include advance publicity for the campaign, coverage of events during the campaign, and wrap-up coverage following its completion. During the inspection days, findings on how many cars were unlocked and how many had keys in the ignition should be reported regularly to a headquarters (preferably to the MP unit).

CONCLUSION

B-335. Project Lock has been outlined to provide basic suggestions for a "Lock Your Car" day or week. No procedure can be designed to fit all needs or circumstances, and variations often will be desirable. However, if the suggestions and the materials contained in this section are used, Project Lock will not be difficult to organize and conduct. It provides better publicity opportunities than most public-service projects. It presents an opportunity for active participation by a number of crime-prevention groups. It has been field-tested and found to be an outstanding success.

SECTION VI — EVALUATION

B-336. Evaluations of installation crime-prevention programs (see Figure B-9, page B-80) do not rely on control groups, a tight control of variables, or elaborate statistical analyses to produce worthwhile results. Most often, the resources to conduct a formal evaluation that will stand up to rigorous academic scrutiny are not available.

CRIME-PREVENTION PROGRAMS

B-337. Several inherent difficulties in data collection on crime-prevention programs make it difficult to determine with 100 percent accuracy that a particular reduction in the crime rate was a result of a particular crime-prevention measure (unless a very elaborate analysis and control system are used). These difficulties include the control of variables, the displacement effect, and unreported crime.

CONTROL OF VARIABLES

B-338. In its simplest form, the type of evaluation most commonly used in academic or scientific settings seeks to determine the relationship between two variables. By varying the independent variable (for example, the dosage of a drug), the effect on the dependent variable (for example, a pulse rate) is determined while all other variables (such as food intake) are held constant.

B-339. Although more complex in form, the same model can be used to evaluate complex programs. Experimental and control groups can be selected. The "treatment" can be administered by researchers or those taught by researchers. The results can be analyzed for their statistical significance. However, since crime-prevention programs deal with human subjects, certain complications arise. The degree of success may have nothing to do with the efficacy of the program, but only with the way it was introduced or with the personal predilections of the groups involved. There is no standard population; human beings are not standardized as mice are for laboratory purposes. A program found successful at one installation may be a failure in another.

B-340. These considerations also apply in the evaluation of crime-control programs. This evaluation is further complicated by another problem—the people whose behavior is to be modified (the offenders) cannot be treated directly or separated into experimental and control groups; they will not stand up and be counted. Although public-health programs often encounter this problem, they often deal with physical cause-and-effect links between treatment and improvement. The same is not true for crime-control programs.

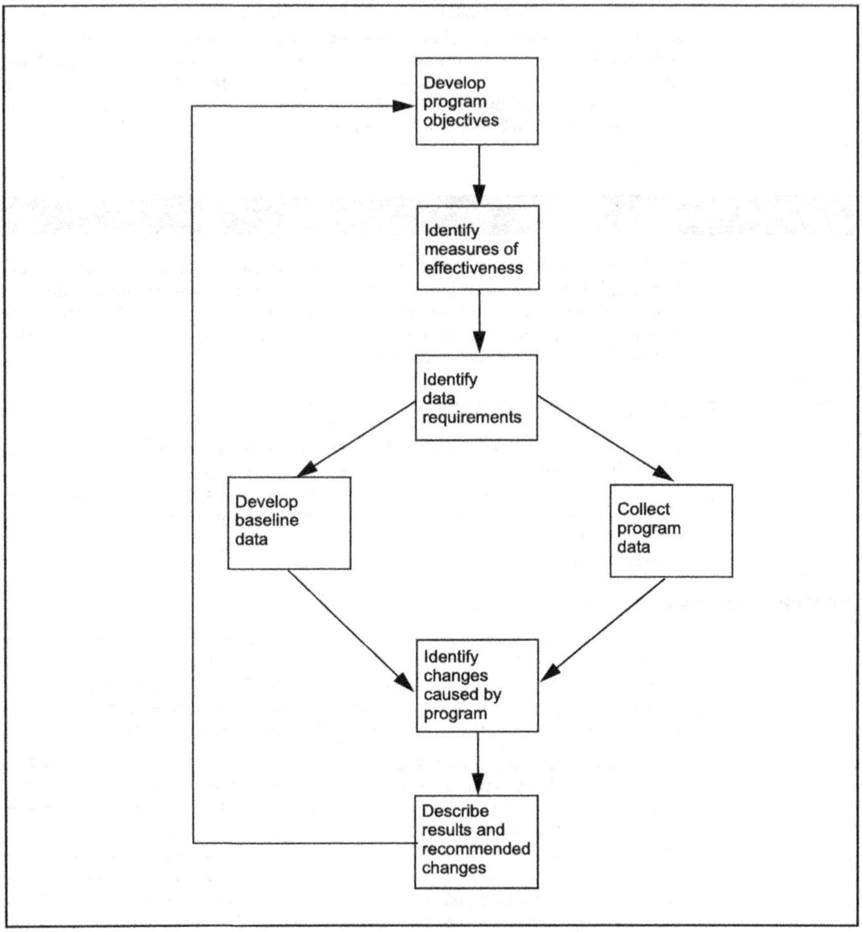

Figure B-9. Crime-Prevention Program Evaluation

The effectiveness of these programs is normally determined by looking at statistics of reported crime and arrests, which are more indirect indicators.

B-341. In a crime-control program, it may be impossible to classify variables as dependent and independent; they may all affect and be affected by each other. Furthermore, because of the difficulty in determining why people behave the way they do, a number of intervening and antecedent variables may go unnoticed. Police programs designed to reduce crime may have their most direct effect on the behavior of the general public toward the police, which in turn affects the crime rate.

B-342. Evaluations are not necessarily restricted to the analysis of objective crime data; they can also include subjective considerations and perceptions. These subjective evaluations can be of significant benefit in augmenting the statistical analyses of the program's results. They are especially helpful in assessing why and how a program worked and whether a statistical outcome is actually evidence that the program was successful. Interviews of participating agency personnel and residents of the program's area are usually used to supply this information. They can give the evaluator new insight into the actual program operation.

DISPLACEMENT EFFECT

B-343. In many cases where crime reductions have been measured and attributed to programs, it is unclear whether there has been an actual reduction in crime or whether the crime has been displaced. The amount of displacement depends to an extent on the offender's characteristics. An opportunistic offender can be pictured as having a relatively elastic demand—if the risk is too high, he will forgo the crime. An addict offender is typically pictured as having a relatively inelastic demand for the product because of his inelastic demand for drugs—despite the risks, he needs the product.

B-344. Deterrents may have little effect on perpetrators of expressive crimes. These are crimes in which the perpetrator is emotionally involved and is expressing these emotions. Most assaults and homicides fit this category. On the other hand, deterrents may have a strong effect on instrumental crimes, those that are seen by the offender only as a means to an end (usually money). If alternative avenues to the same end are made more attractive by comparison, the offender may well be deterred. Deterrence may produce a diversion to legal alternatives to crime; it also may cause displacement to illegal alternatives.

Displacement to Other Crimes

B-345. There is no immutable law that says a burglar cannot hold up a liquor store and a robber cannot burglarize a warehouse. If a specific crime or a set of crimes is the target of a crime-control program, offenders may decide to avoid the target crimes and ply their trade in other ways. Some offenders will be deterred from all crime if their crime specialty is the object of a crime-control program, but the extent of this deterrence should not be overestimated. The statutory categories of crime should not be confused with categories that serve to classify offenders.

B-346. In some cases, the result of displacing offenders to other crimes is beneficial. If the targeted crimes are more serious than the ones to which offenders are diverted, the net effect on the program may be the reduced danger to society. Of course, the converse may also be true; closing off the vulnerable and more easily protected targets of crime may cause an offender to commit more serious crimes with a net increase in the danger to society.

B-347. In some instances, the individual effect may be substantial but the overall effect may be negligible. Protecting a small fraction of premises against burglary will reduce the number of crimes committed against them, but the burglary rate against unprotected premises may go up.

Displacement to Other Tactics and Targets

B-348. Offenders can change their manner of committing a crime when a new program is established to counter their activity. One example of this took place in 1969 in a section of the Bronx that was having a rapid increase in outdoor crime. The crimes took place primarily in the evening hours when people were returning from work. The program instituted by the police consisted of intensive sweeps of randomly selected city blocks, coupled with plainclothes police officers patrolling the streets. It succeeded in reducing the number of offenses committed during the evening hours, but at the expense of increasing the number taking place in the late afternoon when patrolmen were taking their lunch hours or were occupied with school crossings or shift changes.

Displacement to Other Areas

B-349. The most frequently discussed type of crime displacement is from one area to another. For instance, it has been cynically suggested that the goal of the New York subway police is to chase crime into the streets where it belongs. More seriously, some recent police-helicopter-program evaluations have been questioned because they did not consider possible area displacements.

B-350. One type of boundary of interest to crime displacement is the jurisdictional boundary between the installation and surrounding cities. It has been conjectured that the crime reduction experienced in some central cities has been at the expense of the surrounding suburbs that have experienced increased crime rates.

B-351. An initial study of crime displacement was performed for the Washington, DC, area. It concluded that, although the decrease in Washington's crime rate was concurrent with an increase in the suburban crime rate, there is no evidence that the reduction in reported crime in Washington, DC, has resulted in a corresponding crime increase in the nearby suburbs.

B-352. The area-displacement effect can be measured with some degree of reliability. Three zones can be defined for the purposes of the measurement—the area containing the crime-control program (zone 1), a border around the area (zone 2), and the area chosen as the control area (zone 3). The width of the border may depend on the type of program implemented. If the program involves police helicopters, a quarter-mile-wide border may be necessary; for a patrol car, one or two blocks may suffice.

B-353. Crime rates before program initiation should be determined for all three zones. If zone 2 records a greater increase in crime than zone 3 while zone 1's crime rate decreases, then the increase in zone 2 can be attributed to two factors—

- The general increase in crime rate verified by any increase in zone 3.
- The increase caused by a displacement of crime from zone 1.

B-354. A displacement of this crime does not mean that the program is ineffective. It may suggest that the program should be expanded for all three zones.

UNREPORTED CRIME

B-355. Crime statistics are based on crimes reported to the police. It is well known that many crimes go unreported. Victimization studies can determine the extent of unreported crime and its change from year to year by area of the country or the reasons for failure to report them. These victimization studies are best suited to determining long-term effects. They are not well suited to most crime-control evaluations in which short-term changes must be assessed.

B-356. The amount of unreported crime is important but not for planning crime-control programs affecting police activities. The extent of unreported crimes is of little significance unless a program affects it. If a program encourages reported crime, the reported-crime rate may increase despite the program's effectiveness.

B-357. Ironically, a lowered reported-crime rate may be the direct result of an increase in the actual crime rate. Taking reports from victims of crimes occupies a substantial amount of a patrolman's time. Many of these crimes are minor and have no potential for solution. In an effort to increase the police department's time on patrol, some police chiefs have stopped the practice of sending a patrolman to get reports from the victim of a minor crime. This requires the victim to travel to the police station to report the crime. If the crime is minor or is seen by the victim to be unsolvable or the theft is not covered by insurance, the victim may decide not to inconvenience himself by going to the police station to report the crime. Therefore, the number of crimes reported to the police may drop. This may lead to a larger number of unreported crimes and prevent a complete picture to the crime-prevention counsel. Conversely, an actual decrease in crime due to the increased effectiveness of the police may produce an increase in the reported crime rate.

CRIME RATES

B-358. It has been pointed out that the crime rates, as presently calculated, do not reflect the true situation. For example, the rape rate should be calculated by dividing the annual number of rape cases by the number of women (since they are the population at risk). You would expect that the rate of commercial burglaries would be less in a residential area than a commercial area. When calculated on the basis of "per thousand people," this would be true; however, these rates should be obtained by dividing the number of cases by the number of commercial establishments (the population at risk) in each area.

B-359. The victim or the target is only one aspect of the crime. The offender can also be calculated into the rate. For example, the potential offenders in stranger-to-stranger crimes are usually considered to be males between 16 and 25 years of age. Therefore, one would expect fewer of these crimes in a city full of pensioners and retirees than in a city of the same population but with a higher proportion of young men. This fact is of minor importance in evaluating crime-control programs, since the age distribution of people in a city or a section of a city does not normally change greatly over the evaluation period. However, the former factor (the population at risk of becoming victimized) can be misleading if it is not taken into account. If possible, crime rates in experimental and control areas should be compared to the population that risks becoming victims of the target crimes.

MEASURES OF EFFECTIVENESS

B-360. The goals of the program determine the criteria that are used to measure its effectiveness. These goals and criteria should not be seen as confining; the evaluator should be amenable to broadening the criteria, especially if the program to be evaluated is a new one. For example, the program might be beneficial in some unforeseen way, wholly outside the original criteria. Conversely, the program may be an overall failure but a success according to the evaluation. It may be that the specified measures were the wrong ones to use for the program or should not have been used alone.

B-361. Programs aimed at controlling crime should not be evaluated solely for their effect on crime. Most programs cannot, by their very nature, focus on one specific objective alone. They normally are multifaceted in their effect and should be evaluated with respect to all of their facets. Similarly, the measures of effectiveness discussed in this section may not be adequate for every crime-control program, but they comprise some of the more useful measures that can be used.

B-362. This section concentrates on the two evaluation types—internal and external. Internal and external refer to whether the evaluation is conducted on the program's inner workings and logic or the external effect of the program (which depends on the program type). An internal evaluation of a crime-control program involving the use of new police patrol techniques would include the analysis of police response time and how it was effective in controlling crime or why it was successful in one area and not in another. The external evaluation would focus only on the effectiveness of the program in reducing crime rates or solving crimes, not on how or why or the conditions under which the results were achieved.

B-363. Evaluating how well a program achieved its goals is not the only purpose of an evaluation; how and why the results were achieved are of equal importance. External measures relate to the former evaluation; internal measures are concerned with the latter. The following examples will further serve to highlight the differences between these measures:

- Many crime-control programs depend on good community relations in order to achieve their goals. In these cases, a public-affairs campaign is often instituted concurrent with the crime-control program. The success of the public-affairs campaign should not be interpreted as program success. It may be a necessary part of the program, but it does not substitute for the results of the program in controlling crime. Testimonials from people involved in the program should also be considered only as a supplement to the evaluation based on external measures.

- A study undertaken for the President's Crime Commission showed that for certain types of incidents, the probability of arrest increased as the response time decreased. As a result of this finding, many police departments purchased new equipment or tried novel techniques to reduce response time without first determining whether their workloads included enough of the incidents for which quick response is useful. If this measure (response time) is to be used, it should be recognized as an internal measure and not substituted for the external evaluation.

INTERNAL MEASURES

B-364. Each program will have its own internal measures of effectiveness based on the logical elements of which it is constituted. This section covers only the internal measures of effectiveness that are common to most crime-control program evaluations. The measures covered include the crime rate, the clearance rate, the arrest rate, the crime-seriousness index, and the fear of crime.

CRIME RATE

B-365. The crime rate (the number of a specified type of crime committed per resident in a specified time period) is normally considered to be a measure of deterrence. If the crime rate decreases, it is presumed that potential offenders have modified their behavior to some extent and have committed fewer crimes. This is based on the assumption that the program has made the target crimes unattractive by increasing the actual or perceived risk of apprehension, by reducing the expected return from the crime, or by making alternative forms of behavior more attractive than the target group of offenses.

B-366. These deterrent effects use different means for their accomplishment. Most crime-control programs are police-oriented and concentrate on the risk-related aspects of deterrence. Victim-oriented programs focus on reducing the expected return. Many social and recreational programs deal with making alternatives more attractive. Regardless of the orientation of the programs, their deterrent effects are determined by measuring reported crime rates.

B-367. Reported crime rates can be changed by a number of factors, some of which are misleading. The public may feel that the police are becoming less effective in dealing with crimes and, therefore, report them less often. Conversely, if the public perceives that the police are becoming more effective, they may begin to report crimes that previously would have gone unreported. Another apparent crime-rate reduction may be due to the police not recording crimes that have been reported to them. Displacement effects that can produce misleading crime-rate reductions were discussed earlier.

B-368. There may also be an actual reduction in crime due to a program's deterrent effect. In some cases, the reduction in crime can be attributed to psychological deterrence. That is, the police department may have instituted some change (such as painting all police cars canary yellow) in preexisting patterns of operation that may cause a change in the behavior patterns of potential offenders. This type of deterrence is rarely long-lived.

B-369. On the other hand, there may have been a change instituted by the police that had the desired effect of increasing the actual risk of apprehension and, therefore, reducing the number of target offenses. An example of this is the police-operated burglar alarm of commercial establishments. In the experimental program, the number of alarms were increased almost tenfold compared to the (nonalarmed) control establishments. There was 1 capture in 36 control-group burglaries (2.8 percent), while there were 12 captures in 46 experimental-group burglaries (26 percent). Crime displacements to other crimes, tactics, targets, and areas reduced the actual effectiveness of the program, but this example shows that a significant change can be made in the actual risk of apprehension. Preliminary results indicate that the rate of increase of commercial burglaries has been decreased from about 15 percent per year to about 0 percent (at the expense of a greater increase in residential burglaries).

B-370. It is difficult but useful to distinguish between actual deterrence (due to an actual increase in risk) and deterrence that is purely psychological in nature (due to a perceived increase in risk). If it is suspected that part of the deterrent effect may be transient, a long-term study would be of benefit. In this way, the half life of the psychological deterrence can be gauged, which can give some indication of the extent to which resources should be committed to the program.

B-371. Some forms of psychological deterrence are almost entirely counterproductive. They may appear effective to those who would not commit a crime and ineffective to those who are "in the business" and study the presumed deterrent more closely. For example, a tear-gas pen may give a person a sense of security that is entirely without foundation. It may be dangerous to him if he actually attempts to use it when faced with an assailant.

B-372. One investigator has pointed out that for given criminal situations, nondelinquents perceive a higher risk of apprehension than do delinquents; in all probability, the delinquents have a more realistic assessment of the situation. A purely psychological deterrent may have the unfortunate effect of making only a cosmetic improvement. This gives the general population the impression that there has been a change for the better, while in reality the situation may not have changed or may have changed for the worse because of the division of resources to a nonexistent solution.

B-373. The crime rate can be used as a measure of effectiveness. However, the evaluator should delve into the determination of the crime rate to see if any change in the rate reflects a change in reporting procedures or the deterrent effect (with tangible evidence).

CLEARANCE RATE

B-374. The clearance rate is normally considered to be a measure of the ability of police to solve crimes. A cleared crime is one in which the police have identified the offender and have sufficient evidence to arrest him. The clearance rate is the percentage of total crimes that were cleared.

B-375. This measure of effectiveness should be used with care. A decreasing clearance rate may not mean that a police department is becoming less effective, and an increasing clearance rate may not mean that it is becoming more effective. This is due to a number of factors, primarily the public's conception of the role of the police with respect to crime and the present method of collecting crime data.

B-376. Often overlooked in discussion about crime is the role of the public in assisting the police. Police rely on community support to legitimize their authority as well as to help them carry out their work. If a segment of the community becomes alienated from the police (for whatever reason) and offers them little assistance in pursuing offenders, crime rates in these areas may rise. However, it is not only alienation of community groups that reduces the ability of the police to deal with crime; the profit motive is also to blame. Many store owners that have been robbed refuse to give their clerks time off (with pay) to help the police in their investigation. They absorb the loss of a robbery easily (it rarely comes close to the amount lost from shoplifting, employee theft, and damaged goods) and are unwilling to increase it by helping the police. They may feel that the chances of apprehending the offender are too slim, or they may be afraid of retribution if the offender discovers their assistance. They may also be afraid that their insurance will be cancelled.

B-377. If a police department begins a drive to increase its clearance rate, the increase may be forthcoming without any real change in police effectiveness. A survey of three police departments found that arrests for felonies were not made by the police in about 43 percent of the cases, in which there was probable cause, while the police were accompanied by witnesses. Making arrests in such instances would inflate the clearance rate quite easily. However, it should be noted that the police officer has a great deal of discretion in the exercise of his power of arrest. He may feel that the arrest charges will not hold up. One measure of the arrest quality is the percentage of arrests that leads to prosecutions.

B-378. In summary, clearance rate can be a measure for determining the effectiveness of crime-control programs. Its use can be increased by careful selection and specification of the crime categories that are studied, by determining the manner in which the crimes were cleared, and by determining if there has been a change in where the police draw the line in the exercise of their discretion.

ARREST RATE

B-379. Another measure of effectiveness that is often used as a determinant of crime-control effectiveness is the arrest rate, calculated either per police officer or per resident for a specified time period. Most of the considerations concerning the clearance rate (discussed above) also apply to the arrest rate. However, the arrest rate is distinguished from the clearance rate by an additional factor—it is not related to the total number of offenses. For example, the number of arrests for drug violations has risen considerably over the past few years. However, this increase is indicative of the extent of the problem, not of the effectiveness of the solution. It has been described how drug arrests may be traded off against arrests for other offenses and vice versa, especially when informal arrest quotas are established. Therefore, the use of the arrest rate by itself does not appear to be appropriate as a measure of the effectiveness for most crime-control programs.

CRIME-SERIOUSNESS INDEX

B-380. Among the many criticisms of crime statistics is the contention that, even if the data were reliable and complete, we would still have only a count of the number of incidents without an indication of their relative seriousness. The crime-seriousness index was proposed to include some of the major disutilities of crime typically committed by juveniles. Crimes are weighted according to the degree and nature of injury to the victims—whether they were intimidated and the nature of the intimidation or whether premises were forcibly entered, and the kind and value of property stolen. The weights were determined by requesting a sample of people to estimate the relative seriousness of various crimes.

B-381. All of the factors used to determine the weight are (or should be) included in offense reports. It would not be difficult to calculate an incident-seriousness score based on these reports, either for a specific evaluation or as a matter of course. Use of the seriousness index has also been proposed to measure the relative performance of law-enforcement agencies.

B-382. The crime-seriousness index is not the ultimate weighting scheme. The seriousness appears to be calculated more from the offender's viewpoint and the event than from the victim or society's viewpoint. For example, most people would consider the murder of a robbery victim by his assailant to be more serious than the murder of one spouse by the other. With regard to property loss, there is a difference between loss suffered by an individual who is insured and one who is not covered.

B-383. The loss relative to the individual's income is also an important factor; the theft of a $100 television from a low-income family has a much greater impact than the loss of $10,000 of jewels from a wealthy family. Perhaps a better index of the relative value of property loss to the victim would be the value of the loss in relation to the amount of the individual's discretionary income (that is, income not used for the basic necessities of life). Of course, such information is not available on police crime reports.

FEAR OF CRIME

B-384. It has been pointed out that the perceived risk of crime is greater than the actual risk of crime, and that this perceived risk does not seem to be correlated with the actual crime rate. Unless the public feels safer in proportion to its increased actual safety, the full potential of improvements will not have been reached. Therefore, the goal of a crime-control program can be broadened to include not only improved public safety (deterrence), effectiveness (clearance rate), and reduced crime impact (seriousness); but also improved, more accurate, public perceptions of safety as well.

B-385. Measurements of perceived safety can be both direct and indirect. Public-opinion surveys with regard to perceptions about crime and safety have been made frequently. It is also possible to gauge the effect of this fear using indirect measures by observing what people do rather than what they say. The number of patrons of movie theaters and restaurants at night (or other observations of this type of activity) could be used to gauge the fear of crime.

B-386. A reliable measure of the public's perception of public safety has not been developed. Additional research is being done and needs to be done before this type of measure of effectiveness can be used with confidence.

Appendix C

Intelligence, Counterintelligence, and Threat Analysis

Intelligence and counterintelligence make up the first line of defense in an antiterrorism program. A well-planned, systematic, all-source intelligence and counterintelligence program is essential. The role of intelligence and counterintelligence in antiterrorism is to identify the threat. Additionally, counterintelligence provides a warning of potential terrorist attacks and provides information for counterterrorism operations. This appendix provides the elements of the intelligence cycle that have particular importance in a viable antiterrorism program. Effective intelligence and counterintelligence support requires effort, planning and direction, collection and analysis, production, investigation, and dissemination. The entire process provides decision makers with information and timely warnings upon which to recommend antiterrorism actions.

INFORMATION SOURCES

C-1. The primary sources of intelligence and counterintelligence for the antiterrorism program are open-source information, criminal information, government intelligence and counterintelligence, and local information.

- **Open-source information.** This information is publicly available and can be collected, retained, and stored without special authorization. The news media is an excellent open source of information on terrorism. The media reports many major terrorist incidents and often includes in-depth reports on individuals, groups, or various government counterstrategies. Government sources include congressional hearings; publications by the Defense Intelligence Agency (DIA), the FBI, the Central Intelligence Agency (CIA), the Department of State (DOS), and the national criminal justice reference services. Additionally, there are private data services that offer timely information on terrorist activities worldwide. Terrorist groups and their affiliates may also have manuals, pamphlets, and newsletters that reveal their objectives, tactics, and possible targets.
- **Criminal information.** Both military and civil law-enforcement agencies collect criminal information. Because terrorist acts are criminal acts, criminal information is a major source for terrorist intelligence. Commanders must work through established law-enforcement liaison channels because the collection, retention, and dissemination of criminal information are regulated. Local military criminal investigative offices of the CID; the Naval Investigative Service Command (NISCOM); the Air Force Office of Special Investigations (AFOSI); and Headquarters, US Marine Corps,

Criminal Investigations Division, maintain current information that will assist in determining the local terrorist threat.

- **Government intelligence and counterintelligence.** The Community Counterterrorism Board (CCB) is responsible for coordinating with the national intelligence agencies concerning combating international terrorism. These agencies include the CIA (the lead agency), the DIA, the National Security Agency (NSA), the DOS, the Department of Justice (DOJ), the FBI, the Department of Energy (DOE), the Federal Aviation Administration (FAA), the Department of Transportation (DOT) (including the USCG), and the DOD. Service intelligence and counterintelligence production organizations include the US Army Counterintelligence Analysis Center; the Navy Antiterrorism Analysis Center (NAVATAC); Headquarters, US Marine Corps, Counterintelligence; and the US AFOSI Operations Center. These organizations compile comprehensive intelligence and counterintelligence for distribution on a need-to-know basis throughout the services. In combatant commands, the J2 is responsible for the intelligence fusion center. The Counterintelligence Support Officer (CISO) provides interface between the combatant command, the component commands, and the joint staff.
- **Local information.** Other valuable sources of information are the soldiers, civil servants, family members, and individuals with regional knowledge (such as college faculty or members of cultural organizations). Local crime or neighborhood watch programs can be valuable sources of information and can serve as a means to keep individuals informed in dispersed and remote areas. Intelligence exchanges with local government agencies through cooperative arrangements can also augment regional information.

RESPONSIBILITIES OF US GOVERNMENT LEAD AGENCIES

C-2. The FBI is responsible for collecting and processing domestic terrorist information. Overseas, terrorist intelligence is principally a CIA responsibility; but the DOS, the DIA, and the HN are also active players. The MI activities are conducted according to Presidential executive orders, federal law, status of forces agreements (SOFAs), memorandums of understanding (MOUs), and applicable service regulations. Responsibilities of intelligence activities include the following:

- The combatant commander (through the commander's J2 and the CISO in coordination with the DIA, the CIA, the embassy staff, the country team, and applicable HN authorities) obtains intelligence and counterintelligence specific to the AO. The commander issues intelligence and counterintelligence reports, advisories, and assessments to the units within his command or those operating within his command's AO. This network is the backbone for communicating intelligence and counterintelligence information and advisories and for warning of terrorist threats throughout the region.
- The Secretaries of the military departments were asked (in DOD Directive 2000.12) to ensure that a capability exists to receive and evaluate data from a service perspective and that the capability exists

to disseminate all relevant data on terrorist activities, trends, and indicators of imminent attack. To accomplish this task, each Secretary appoints an MI agency (INSCOM, NISCOM, or AFOSI) to conduct intelligence and counterintelligence activities directed against terrorists and to detect, neutralize, or deter terrorist acts. To accomplish this mission, the military department intelligence agency establishes counterintelligence offices on an area basis to collect and disseminate information to combatant commanders. Each military department's intelligence agency—

- Coordinates with appropriate US and HN agencies.
- Provides overall direction and coordination of the service counterintelligence effort.
- Operates a 24-hour operations center to receive and disseminate worldwide terrorist threat information to and from the combatant command's J2, applicable service staff elements, subordinate commands, and national agencies.
- Provides service commanders with information on terrorist threats concerning their personnel, facilities, and operations.
- Investigates terrorist incidents for intelligence, counterintelligence, and force-protection aspects (with the FBI or HN authorities).
- Provides terrorist threat information in threat briefings.
- Conducts liaison with representatives from federal, state, and local agencies (as well as HN agencies) to exchange information on terrorists.
- Provides international terrorism summaries and other threat information to supported commanders. On request, provides current intelligence and counterintelligence data on terrorist groups and disseminates time-sensitive and specific threat warnings to appropriate commands.

- Service criminal investigative services (such as the CID, the NISCOM, and the AFOSI) collect and evaluate criminal information and disseminate terrorist-related information to supported installation and activity commanders as well as to the service lead agency. As appropriate, criminal investigative elements also conduct liaison with local military or security police and civilian law-enforcement agencies.
- Intelligence staff elements of commanders at all echelons—
  - Report promptly all actual or suspected terrorist incidents, activities, and early warnings of terrorist attacks to supported and supporting activities, to the local counterintelligence office, and through the chain of command to the service lead agency.
  - Initiate and maintain liaison with the security police or PMO; local military criminal investigative offices; local counterintelligence offices; security offices; HN agencies and; as required, other organizations, elements, and individuals.
  - Develop and present terrorism threat-awareness briefings to all personnel within their commands (in cooperation with the local counterintelligence offices).
- Law-enforcement staff elements will—

  - Report all actual or suspected terrorist incidents or activities to their immediate commander, supported activities, and service lead agency through established reporting channels.
  - Initiate and maintain liaison with local counterintelligence offices and military criminal investigative offices.
  - Maintain liaison with federal, HN, and local law-enforcement agencies or other civil and military antiterrorism agencies as appropriate.
- Installation, base, unit, and port security officers—
  - Report all actual or suspected terrorist incidents or activities to their immediate commander, supporting military law-enforcement office, other supported activities, local counterintelligence office, and local military criminal investigation office.
  - Conduct regular liaison visits with the supporting military law-enforcement office, counterintelligence office, and local criminal investigation office.
  - Coordinate with the supporting military law-enforcement office and counterintelligence office on their preparation and continual updating of the threat assessments.
  - Assist in providing terrorism threat-awareness training and briefings to all personnel and family members as required by local situations.

INFORMATION REQUIREMENTS

C-3. To focus the threat analysis, intelligence and counterintelligence officers develop information requirements to identify targets using the following terrorist considerations—

- Organization, size, and composition of group.
- Motivation.
- Long- and short-range goals.
- Religious, political, and ethnic affiliations.
- International and national support (moral, physical, and financial).
- Recruiting methods, locations, and targets (students).
- Identity of group leaders, opportunists, and idealists.
- Group intelligence capabilities and connections with other terrorist groups.
- Sources of supply and support.
- Important dates, such as religious holidays.
- Planning ability.
- Internal discipline.
- Preferred tactics and operations.
- Willingness to kill.
- Willingness for self-sacrifice.
- Group skills, demonstrated or perceived (for example, sniping, demolitions, masquerade, industrial sabotage, airplane or boat

operations, tunneling, underwater, electronic surveillance, poisons, or contaminants).

- Equipment and weapons (on hand and required).
- Transportation (on hand and required).
- Medical-support availability.
- Means and methods of C^2.
- Means and method of communication.

THREAT ANALYSIS AND ASSESSMENT

C-4. The preparation of the terrorist threat analysis is a continual process of compiling and examining all available information to identify terrorist targeting of US interests. A vulnerability analysis is a continual process of compiling and examining information on a facility's security posture. The threat analysis is then paired with the facility's vulnerability analysis to create the threat and vulnerability assessment. Threat analysis is an essential step in identifying the probability of a terrorist attack. To enhance the capability to collect and analyze information from many sources, the DIA maintains a terrorism database. The combatant command's J2 and CISO (in coordination with the DIA) focus this database information and regional information toward the intelligence and counterintelligence needs specific to the security of the command. However, this terrorism database is limited to foreign terrorist groups because of limitations on US intelligence-collection operations. A country's threat assessments, information and biographies about terrorist organizations, and incidents in the database can be disseminated to commands. Commands at all echelons then augment or refine the DIA's threat analysis to focus on their area of interest. This process, operative across the full range of military operations, promotes coordination between all levels of the intelligence, counterintelligence, and law-enforcement communities; broadens acquisition channels; and enhances timely distribution of information to the supported commander.

C-5. Several factors complicate intelligence and counterintelligence collection and operations. The small size of terrorist groups, coupled with their mobility and cellular organization, make it difficult to identify the members. Unlike other criminals, terrorist cadres often receive training in counterintelligence and security measures from foreign intelligence agencies or other terrorists. Additionally, the traditional orientation of police organizations is toward individual criminals, while MI organizations focus on conventional forces. Terrorist activity, therefore, requires some degree of reorientation for police and MI and counterintelligence collection and operations.

C-6. An intelligence system's ability to provide critical and timely information to the user depends not only on efficient collection and processing, but also on the ability to organize, store, and retrieve this information rapidly. This capability, coupled with early warning, careful observation, and assessment of threat activity, enhances the probability of accurately predicting the types and timing of terrorist attacks.

C-7. Commanders must carefully exercise judgment in estimating both the existing terrorist threat and the need for changes in antiterrorism measures. Key questions are—

- What has changed (mission, political climate, installation and unit personnel or equipment, terrorist capabilities)?
- What affect will the changes have on the security posture?

C-8. Extraordinary security measures, unless part of a deliberate deception during critical or high-threat situations, draw attention and detract from mission accomplishment. Sound physical security, personnel who are aware, accurate threat and vulnerability assessments, and well-rehearsed response plans reduce the probability of a successful terrorist venture. The goal is to make an attack too difficult or the level of risk unacceptable to the terrorist.

DETERMINATION OF THE THREAT LEVEL

C-9. This threat-analysis methodology is used by the DIA, the joint staff, and the unified and specified commands for selecting the level of threat for an installation. It is applicable in an overseas setting, but fails to address issues unique to the sustaining base within CONUS. In CONUS there is a lack of intelligence and counterintelligence information from the CIA, the DIA, or the military services for the CONUS-based threat. That information must be extracted from law-enforcement channels at the local, regional, and national levels. In that context, the factors mentioned in this appendix are not as clear as they are within the intelligence process in place overseas. A modified version of this appendix should be considered for assessing threat levels in CONUS.

C-10. Threat levels within CONUS historically have been either low or negligible. This trend will most likely continue at least through the next decade. Domestic groups not covered by DOD intelligence reporting pose the greatest threat to the CONUS-based military. Due to the lack of reporting and information on these groups, the domestic terrorist groups are not currently factored into the current terrorist threat program. Therefore, the law-enforcement community must become a key player in establishing the threat levels in the context of the recommended model. A possible methodology would be the establishment of an additional threat level between low and medium—one that allows local commanders more flexibility in implementing additional security measures. Table C-1 shows the procedure for determining the threat level.

Table C-1. Threat Levels

Threat Level	
Critical	Factors 1, 2, and 5 are present. Factors 3 or 4 may be present.
High	Factors 1, 2, 3, and 4 are present.
Medium	Factors 1, 2, and 4 are present.
Low	Factors 1 and 2 are present. Factor 4 may be present.
Negligible	Factors 1 and/or 2 may be present.
Explanation of Factors	
Factor 1: Existence. A terrorist group is present, assessed to be present, or able to gain access to a given locale.	
Factor 2: Capability. The acquired, assessed, or demonstrated level of capability to conduct terrorist attacks.	
Factor 3: Intentions. Recent demonstrated anti-US terrorist activity or stated and/or assessed intent to conduct such activity.	
Factor 4: History. Demonstrated terrorist activity over time.	
Factor 5: Targeting. Current credible information on activity indicative of preparations for specific terrorist operations and/or specific intelligence that shows an attack is imminent.	

Appendix D

Crisis-Management Plan

The following pages highlight areas of concern in crisis-management planning. This plan is not meant to be all-inclusive or rigidly followed. Figure D-1 is a sample format only. It does not reflect a format developed and approved for use with OPLANs or contingency plans (CONPLANs) prepared by the CINCs to fulfill tasks assigned in the Joint Strategic Capabilities Plan (JSCP) or as otherwise directed by the Chairman of the Joint Chiefs of Staff. Figure D-2, page D-4, is a sample of the Crisis-Management-Plan Checklist, which is Annex A or Appendix H to the crisis-management plan. This checklist will help ensure that the plan is sound.

Copy No. ________
Issuing Headquarters
Place of Issue
Date of Issue

Crisis-Management Plan

Ref: Maps, charts, and other relevant documents.

Time Zone: X

Task Organization: List units organized to conduct antiterrorism operations. Include attachments, supporting roles, and the delegation of operational control as necessary.

1. Situation. Identify essential information to understand ongoing events.

a. Terrorist force. Identify the terrorist's composition, disposition, methods of operation, and estimated strengths and capabilities that could influence the crisis-management operation. Refer to an appropriate annex.

b. Response force. Explain the response force's abilities and responsibilities. The response force's abilities can influence the crisis-management mission.

c. Attachments and detachments. Address here or refer to an annex.

d. Assumptions. Provide assumptions used as a basis for this plan (for example, the strength of the response force to be supported and the support available from other agencies).

Figure D-1. Sample Crisis-Management Plan

(1) Tactical-situation possibilities. Obtained from the commander's planning guidance.

(2) Personnel situation. Provided by the personnel officer.

(3) Logistics situation. Provided by the logistics officer.

(4) Legal-situation possibilities. Provided by the SJA.

2. Mission. Identify the antiterrorism mission (for example, detect, deter, contain, and neutralize terrorist threats and actions aimed at the disruption of the installation).

3. Execution.

a. Concept of operations. State the commander's tactical plan. The purpose is to inform. It may address how the commander will conduct combat-terrorism operations. It provides enough detail to ensure proper action by subordinates in the absence of specific instructions. If the required details are extensive, address them in an annex. If an operation involves two or more distinct phases, designate each phase and use subparagraphs (for example, Phase I and Phase II).

b. Tasks. Identify specific tasks for each command element charged with executing a crisis-management mission. When giving multiple instructions, itemize and indicate the priority or sequence.

c. Coordinating instructions. Include coordination and control measures applicable to two or more command elements.

4. Service Support. Provide a statement of service-support instructions and arrangements supporting the crisis-management operation. Use the following subparagraphs as required:

a. General. Outline the general plan for service support.

b. Materiel and services. Address supply, transportation, labor, and services required.

c. Medical evacuation and hospitalization. Provide the plan for evacuating and hospitalizing sick, wounded, or injured personnel. Address evacuation responsibilities and the air-evacuation policy.

d. Personnel. Provide required information and instructions to supporting unit personnel.

(1) Maintenance of unit strength.

(a) Strength reports. Provide instructions for submitting status reports. Include requirements for routine and special reports.

(b) Replacements. Address validating existing personnel requisitions, instructions for submitting requisitions, and instructions for processing and removing replacements.

(2) Personnel management. Address military and civilian personnel and civilian detainee management procedures.

(3) Development and maintenance of morale.

(a) Morale and personnel services. Provide postal and financial services, religious activities, personal hygiene, and special services activity information.

(b) Graves registration. Include evacuation procedures and handling personal effects.

Figure D-1. Sample Crisis-Management Plan (continued)

(4) Maintenance of discipline, law, and order. Obtain this guidance from the PMO/security officer.

(5) Miscellaneous. Include personnel administrative matters not specifically assigned to another coordinating staff section or included in preceding subparagraphs.

e. Miscellaneous. Provide special instructions or special reports not covered in preceding paragraphs.

5. Command and Signal. Provide instructions for the command and operation of communications-electronics equipment. Communications-electronics instructions may refer to an annex but should list the index and issue number of the command, control, and communications (C^3) operation instructions in effect. If not already issued, give instructions for the control, coordination, and establishment of priorities in the use of electromagnetic emissions. Command instructions include subordinate and higher unit CP locations and designated alternate CPs.

6. Acknowledgement Instructions.

/s/

Commander

Annexes as applicable

Distribution:

Figure D-1. Sample Crisis-Management Plan (continued)

Annex A

Crisis -Management Plan Checklist

Yes	No	
		1. Intelligence and/or Counterintelligence.
____	____	Does the plan allow for the threat-analysis process (collection, analysis, production, and dissemination) to help identify the local threat?
____	____	Does the plan consider restrictions placed on the collection and storage of information?
____	____	Does the plan indicate an awareness of sources of information for the threat-analysis process (MI, counterintelligence, federal agencies, and state and local authorities)?
____	____	Does the plan allow for liaison and coordination of information (such as establishing a committee)?
		2. Threat Assessment.
____	____	Does the plan identify the local threat (immediate or long term)?
____	____	Does the plan identify other threats (such as national and international groups that have targeted or might target US installations)?
____	____	Does the installation incorporate factors for assessing the threat? Does it address —
____	____	Geography of the area concerned?
____	____	Law-enforcement resources?
____	____	Population cultural resources?
____	____	Communication capabilities?
____	____	Does the plan establish a priority of identified weaknesses and vulnerabilities?
____	____	Is the threat assessment periodically updated?
		3. Security Countermeasures.
____	____	Does the plan have specified THREATCONs and recommended actions?
____	____	Do security countermeasures include a combination of physical operations and sound-blanketing security measures?
____	____	Do the THREATCONs correspond to DOD 0-2000.12-H, Appendix BB?

Figure D-2. Sample Crisis-Management-Plan Checklist

Yes	No	
		4. OPSEC.
____	____	Have procedures been established that prevent terrorists from readily obtaining information about plans and operations (for example, not publishing the commanding general's itinerary and safeguarding classified material)?
____	____	Does the plan allow for in-depth coordination with the installation's OPSEC program?
____	____	Has an OPSEC annex been included in the CONPLAN?
		5. Personnel Security.
____	____	Has the threat analysis identified individuals vulnerable to terrorist attacks?
____	____	Has a training program been established to educate both military and civilian personnel in the proper techniques of personnel protection and security commensurate with the local threat and the type of position held?
		6. Physical Security.
____	____	Are special-threat plans and physical-security plans mutually supportive?
____	____	Do security measures establish obstacles to terrorist activity (such as guards, HN forces, lighting, and fencing)?
____	____	Does the special-threat plan include the threats identified in the threat statements of higher headquarters?
____	____	Does the physical-security officer assist in the threat analysis and corrective action?
____	____	Does the installation have and maintain detection systems and an appropriate assessment capability?
		7. Security Structure.
____	____	Does the plan indicate that the FBI has primary domestic investigative and operational responsibility in the US and US territories?
____	____	Has coordination with the SJA been established?
____	____	Does the plan allow for close cooperation between principal agents of the military, civilian, and HN communities and federal agencies?
____	____	Does the plan clearly indicate parameters for the use of force, including briefing any elements augmenting MP assets?
____	____	Is there a mutual understanding between all local agencies (military, local, FBI resident or senior agent-in-charge, HN forces, and local law enforcement) that might be involved in a terrorist incident on the installation regarding authority, jurisdiction, and possible interaction?

Figure D-2. Sample Crisis-Management-Plan Checklist (continued)

Yes	No	
____	____	Has the SJA considered the ramifications of closing the post (such as possible civilian union problems)?
____	____	Does the plan identify the DOS as having primary investigative and operational responsibilities overseas?
		8. Operations-Center Training.
____	____	Has the operational command and coordination center been established and exercised?
____	____	Is the operations center based on the needs of the installation while recognizing manpower limitations, resource availability, equipment, and command?
____	____	Does the plan include a location for the operations center?
____	____	Does the plan designate alternate locations for the operations center?
____	____	Does the plan allow for the use of visual aids (chalkboards, maps with overlays, and bulletin boards) to provide status reports and countermeasures?
____	____	Does the plan create and designate a location for a media center?
____	____	Have the operations and media centers been activated together within the last quarter?
____	____	Does the operations center have SOPs covering communications and reports to higher headquarters?
____	____	Does the operations center offer protection from a terrorist attack?
		9. Reaction-Force Training.
____	____	Has the force been trained and exercised under realistic conditions?
____	____	Has corrective action been applied to shortcomings and deficiencies?
____	____	Has the reaction force been formed and mission-specified trained (for example, building entry and search techniques, vehicle assault operations, countersniper techniques, and equipment)?
____	____	Has the reaction force been tested quarterly (alert procedures, response time, and overall preparedness)?
____	____	Has responsibility been fixed for the negotiation team? Has the negotiation team been trained and exercised under realistic conditions?
____	____	Does the negotiation team have the proper equipment?
		10. General Observations.
____	____	Was the plan developed as a coordinated staff effort?
____	____	Does the plan outline reporting requirements (logs, journals, and after-action reports)?

Figure D-2. Sample Crisis-Management-Plan Checklist (continued)

Yes	No	
____	____	Does the plan address the media's presence?
____	____	Does the plan include communication procedures and communication nets?
____	____	Does the plan consider the possible need for interpreters?
____	____	Does the plan consider the need for a list of personnel with various backgrounds to provide cultural profiles on foreign subjects and victims as well as to assist with any negotiation efforts?
____	____	Does the plan provide for and identify units that will augment MP assets?
____	____	Does the plan delineate specific taskings for each member of the operations center?
____	____	Does the plan provide for a response force for each phase of antiterrorism activity (initial response, negotiation, and assault)?
____	____	Does the plan designate service-support communications?
____	____	Does the plan make provisions for the notification of an accident-and-incident control officer?
____	____	Does the plan provide for EOD support?
____	____	Does the plan take into consideration the movement from various locations, including commercial airports, of civilian and military advisory personnel with military transportation assets?
____	____	Does the plan allow for the purchase or use of civilian vehicles, supplies, and food (if needed)? (This includes items used to satisfy a hostage demand.) Does the plan make provisions for paying civilian employees overtime if they are involved in a special-threat situation?
____	____	Does the plan take into consideration the messing, billeting, and transportation of civilian personnel?
____	____	Do appropriate personnel have the necessary language training?
____	____	Is WMD support available?

Figure D-2. Sample Crisis-Management-Plan Checklist (continued)

Appendix E
Office Security Measures

The office environment should afford executives the greatest degree of physical security. Executives usually work in facilities where attackers must pass by guards, security checkpoints, office workers, aides, or secretaries before reaching them. Unfortunately, the high media value of attacking executives in security strongholds where they are clearly associated with government activity increases the value of such attacks to terrorists. Hence, there may be a need to add security measures to offset the escalating capability of attack on more secure office areas by terrorist groups.

PHYSICAL-SECURITY SURVEY

E-1. A thorough physical-security survey of an office facility should be conducted. Offices of defense components attached to US embassies abroad should have these surveys performed by the DOS. Other DOD facilities should have surveys performed by the cognizant physical-security and facilities-engineering staffs. The best way to approach a physical-security site survey is to think like an intruder. Consider how approaches to the installation or facility could be made, how access to the building that houses executive offices could be gained, and how attacks on offices or other frequently used facilities could be mounted.

SECURITY-ENGINEERING ASSESSMENT

E-2. The next step in evaluating the need for supplemental physical-security measures is a thorough and detailed assessment of the weapons and tactics that terrorists might use to attack the structure in which DOD executives work. Security engineers and architects need technical threat data or assessments containing the following information—

- The mode of attack, such as—
  - Standoff weapons (man-portable AT/antiaircraft weapons, sniper rifles, rock grenades, and mortars).
  - Close combat weapons (submachine guns, pistols, knives, and garrotes).
  - Contact weapons (bombs, incendiary devices, and mines).
- Perimeter penetration aids (such as power tools, hand tools, or explosives), if used.
- The time of attack.
- The attacking force's size.
- The anticipated degree of outside support or autonomy.

E-3. Engineering design requirements are developed from the security engineering assessments. The data is used to—

- Assess the ability of building components to resist the effects of the threat.
- Identify appropriate security window-glazing materials and window treatments to determine what is required to achieve the desired penetration resistance times for anticipated threats.
- Calculate the total amount of delay time. This time is achieved by using camouflage, deception, barriers, and security devices to permit response forces to reach the scene of a terrorist attack in time to thwart the attack, capture or eliminate the terrorists, and rescue executives and their staffs or dependents.

TECHNICAL ASSESSMENT OF RESPONSES

E-4. After establishing a basic-design threat, engineers need data on the anticipated performance of response forces to be arrayed against the design threat and the expected or desired behavior of the protected executive. Some specific information needed includes—

- The response force's size, capability, supporting weapons, response time, and estimated effectiveness against the range of attacks.
- The desired options for the executive's protection—evacuate on warning, on detection, only if attacked, or only if forced to capitulate or do not evacuate.

E-5. Security planners need to know how long the structure that houses executives can withstand an attack before help arrives. Matching threat capabilities and anticipated operations by response forces establishes significant physical-security-system performance parameters. These can be quantified and used to develop detailed plans, drawings, and physical-security equipment-acquisition plans.

PHYSICAL-SECURITY ENHANCEMENT MEASURES

E-6. Several physical-security measures intended to provide additional protection for executives can be considered based on the requirements defined through the detailed analyses outlined above. The primary purpose of such measures should be to increase the time required by persons outside an installation to reach the executives housed at an installation. A secondary purpose of such measures should be to reduce or eliminate hazards to executives that might result from violence in the vicinity. Examples of physical-security measures to consider are—

- Increase the threat-detection time by installing sensors on perimeters and barriers. This includes—
  - Combining surveillance systems including seismic, acoustic, and IR sensors at or beyond the outer perimeter.
  - Supplementing surveillance systems with CCTV/imaging IR systems tied into the alert response-force staging area.

- Extending restricted areas or exclusion zones and relocating access-control points from the executive office area to a point closer to the installation's boundary.
- Enlarging and extending intrusion-detection sensors from within the installation to its perimeter, allowing the IDS to collect additional data necessary and sufficiently classify and identify an intrusion before the response force arrives.
- Enhancing both the number and the phenomenology of surveillance and detection systems within the executive office area as well as approaches leading to and from it in conjunction with measures listed below.

- Increase the threat's delay time between the perimeter and the executive office building. This includes—
  - Installing vehicle barriers and realigning roadways to eliminate straight, level stretches of road in excess of 50 meters in length.
  - Increasing concentric rings of fences, Jersey barricades, planters, bollards, and vehicle/personnel barriers.
  - Enhancing access-control areas supplemented by fire doors/security doors kept in a closed condition between the entrance to the building that houses executive offices and the executive office area.
- Confuse, camouflage, and deceive observers by hiding an executive's location. Accomplish this by—
  - Relocating executives to buildings not usually associated with office activities (barracks, motor pools, research and development [R&D] facilities, and so forth).
  - Constructing office areas in the barracks, motor pool, R&D facilities, and so forth.
  - Adding executive styles, decorative lighting, and window treatments to several different areas of office buildings to minimize the differences in external appearances between executive and nonexecutive offices.
- Increase the delay time between the entrance to the building that houses executives and the executive office area. Execute this by—
  - Adding fire doors, access-control points, dead-end corridors, and midcorridor physical barriers to complicate access to the executive office areas.
  - Adding security devices that, when activated, disrupt the intruder's ability to retain his thought processes (for example, flashing strobe lights, fog generators, noise generators, sirens, and fire-extinguishing systems).
- Increase the delay time by making access more difficult within the executive office structure. This may be accomplished by—
  - Substituting high-security doors and door frames for standard doors in areas leading to or from executive offices.
  - Installing high-security grating, wire mesh, or other materials to bar access to the executive office area through utility tunnels or conduits.

- Strengthening walls, floors, and ceilings by substituting steel-plate, concrete-filled, steel-reinforced cinder blocks or other ballistic-resistant materials for plaster/lath or wallboard room dividers, thereby protecting against explosive devices that are used as tools to breach a barrier.
- Increase the protection for building occupants against weapons and explosives effects. This includes—
  - Substituting blast- or bullet-resistant panels for glass windows or adding a fragment-retention film at least 4 millimeters thick to the interior of glass windows.
  - Adding exterior screens/plates to cover window areas and protect against gunfire and grenade/bomb fragments.
  - Installing blast curtains, metal blinds, metal shutters, or other window treatments in executive offices to protect interior space from glass shards and other small projectiles.
  - Strengthening walls to resist weapons and explosives effects by adding steel plates, reinforced concrete, or other retrofitting measures.
  - Adding steel plates or other ballistic materials in crawl spaces above dropped ceilings or extending walls separating the executive office area from other portions of an office building from floor to floor, thereby preventing unobserved and undetected access to the space between dropped ceilings.
- Increase the hold time to contain penetrators by—
  - Adding positive-action controls to a facility's doors and gates so that gates default to a closed and locked condition unless manually released.
  - Adding positive-action controls to access-control areas so that persons inside an access-control area can neither advance nor withdraw without affirmative action by a security officer posted outside the access-control area.

E-7. These measures are used to facilitate the apprehension of terrorists. There may be some instances when defeating terrorist attempts to gain access to the executive enhances the security of the executive and the response force. This is accomplished by channeling the terrorists out of the facility and installation along one route, leaving alternative routes available to evacuate executives and other key personnel.

E-8. Install emergency executive support facilities (including a safe haven and an emergency evacuation facility) by—

- Installing helicopter landing aids on a structure's roof or on an adjacent field far removed from parking areas.
- Installing a safe haven or other reinforced security structure adjacent to a helicopter landing facility to provide a secure waiting place for executives until a rescue helicopter with additional supporting air and ground units can extract the executives.

Appendix F

Physical-Security Plan

It is essential and in the best interest of security that each installation, unit, or activity maintains and uses a detailed physical-security plan. The plan should include at least special and general guard orders, access and material control, protective barriers/lighting systems, locks, and IDSs. All physical-security plans have the potential of being classified documents and must be treated accordingly. Figure F-1 depicts a sample physical-security plan.

Map Reference

Copy No. _______
Issuing Headquarters
Place of Issue
Date of Issue

Physical-Security Plan

1. Purpose. State the plan's purpose.

2. Area Security. Define the areas, buildings, and other structures considered critical and establish priorities for their protection.

3. Control Measures. Define and establish restrictions on access and movement into critical areas.

a. Categorize restrictions as to personnel, materials, and vehicles:

(1) Personnel access:

(a) Establishment of controls pertinent to each area or structure.

- Authority for access.
- Criteria for access.
  - Unit personnel.
  - Visitors.
  - Maintenance personnel.
  - Contractor personnel.
  - National guard.
  - Emergency response teams (police, fire, ambulance, and so forth).

Figure F-1. Sample Physical-Security Plan

(b) Identification and control.

- Description of the system to be used in each area. If a badge system is used, a complete description covering all aspects should be used in disseminating requirements for ID and control of personnel conducting business on the installation.
- Application of the system:
  - Unit personnel.
  - Visitors to restricted areas.
  - Visitors to administrative areas.
  - Vendors, tradesmen, and so forth.
  - Contractor personnel.
  - Maintenance or support personnel.
  - Fail-safe procedures during power outages.

(2) Material control.

(a) Incoming.

- Requirements for admission of material and supplies.
- Search and inspection of material for possible sabotage hazards.
- Special controls on delivery of supplies or personal shipments in restricted areas.

(b) Outgoing.

- Documentation required.
- Controls, as outlined in paragraph 3a(2a).
- Classified shipment not involving nuclear/chemical material.

(c) Nuclear/chemical material.

- Controls on movement of warheads/chemicals on the installation.
- Controls on shipments or movement of training warheads/chemicals.
- Controls on pickup or delivery of warheads/chemicals outside the installation.

(3) Vehicle control.

(a) Policy on search of military and privately owned vehicles.

(b) Parking regulations.

Figure F-1. Sample Physical-Security Plan (continued)

(c) Controls for entrance into restricted and administrative areas:

- Military vehicles.
- POVs.
- Emergency vehicles.
- Vehicle registration.

b. Indicate the manner in which the following security aids will be implemented on the installation:

(1) Protective barriers:

(a) Definition.

(b) Clear zones.

- Criteria.
- Maintenance.

(c) Signs.

- Types.
- Posting.

(d) Gates.

- Hours of operation.
- Security requirements.
- Lock security.
- Barrier plan.

(2) Protective lighting system:

(a) Use and control.

(b) Inspection.

(c) Action taken in case of commercial power failure.

(d) Action taken in case of failure of alternate power source.

(3) Emergency lighting system:

(a) Stationary.

(b) Portable.

Figure F-1. Sample Physical-Security Plan (continued)

(4) IDSs:

(a) Security classification.

(b) Inspection.

(c) Use and monitoring.

(d) Action taken in case of alarm conditions.

(e) Maintenance.

(f) Alarm logs or registers.

(g) Tamper-proof provisions.

(h) Monitor-panel locations.

(5) Communications:

(a) Locations.

(b) Use.

(c) Tests.

(d) Authentication.

(6) Security forces: General instructions that would apply to all security-force personnel (fixed and mobile). Detailed instructions such as special orders and SOP information should be attached as annexes. Security-force facets include—

(a) Composition and organization.

(b) Tour of duty.

(c) Essential posts and routes.

(d) Weapons and equipment.

(e) Training.

(f) Use of MWD teams.

(g) Method of challenging with signs and countersigns.

(h) Alert forces:

- Composition.
- Mission.
- Weapons and equipment.
- Location.
- Deployment concept.

Figure F-1. Sample Physical-Security Plan (continued)

(7) Contingency plans: Required actions in response to various emergency situations. Detailed plans for situations (counterterrorism, bomb threats, hostage negotiations, disaster, fire, and so forth) should be attached as annexes.

(a) Individual actions.

(b) Alert-force actions.

(c) Security-force actions.

(8) Use of air surveillance.

(9) Coordinating instructions. Matters that require coordination with other military and civil agencies such as—

(a) Adjacent installations or units.

(b) State and local agencies.

(c) Similar host-country agencies.

(d) Federal agencies.

The coordination/interaction allows for an exchange of intelligence information on security measures being used, contingency plans, and any other information to enhance local security.

On an installation, the host activity shall assume responsibility for coordinating physical-security efforts of all tenants, regardless of the components represented, as outlined in the support agreements and the host-activity security plan. Applicable provisions shall be included in, or be an appendix to, the support agreement. A formal agreement will contain definite assignment of physical-security responsibility for the items stored. The agreement should address—

- Maximum qualities to be stored.
- Physical safeguards to be used.
- Frequency of, and responsibility for, physical inventories or reconciliations.
- Reporting of losses for investigation.
- Lock and key control.
- The unit that has overall responsibility.

Procedures for authorization and ID of individuals to receipt for and physically take custody of Army property. The purpose of such coordination is protection in depth. Authority, jurisdiction, and responsibility must be set forth in a manner that ensures protection and avoids duplication of effort.

/s/ Commander

Figure F-1. Sample Physical-Security Plan (continued)

ANNEXES

F-1. Annexes to the plan should include, but are not limited to, the following. More information can be found in AR 190-13.

- **Annex A.** The installation threat statement (intelligence). This annex should contain the Terrorism Counteraction Plan (refer to AR 190-13).
- **Annex B.** A bomb-threat plan. As a minimum, the bomb-threat plan should provide guidance for—
  - Control of the operation.
  - Evacuation.
  - Search.
  - Finding the bomb or suspected bomb.
  - Disposal.
  - Detonation and damage control.
  - Control of publicity.
  - After-action report.
- **Annex C.** An installation closure plan.
- **Annex D.** A natural-disaster plan. This plan will be coordinated with natural-disaster plans of local jurisdictions. At a minimum, the natural-disaster plan should provide guidance for—
  - Control of the operation.
  - Evacuation.
  - Communication.
  - Control of publicity.
  - After-action report.
- **Annex E.** A civil-disturbance plan. It is the commander's responsibility to formulate a civil-disturbance plan based on local threats. (For example, commanders of chemical facilities should anticipate the need to develop crowd-control procedures to handle antichemical demonstrations.)
- **Annex F.** A resource plan to meet the minimum-essential physical-security needs for the installation or activity.
- **Annex G.** A communication plan. This plan is required to establish communications with other federal agencies and local law-enforcement agencies to share information about possible threats. The communications plan should address all communication needs for annexes B through F above.
- **Annex H.** A list of designated restricted areas.
- **Annex I.** A list of installation MEVAs.
- **Annex J.** A contingency plan. In most instances, it will be necessary to increase security for AA&E and other sensitive property, assets, and facilities during periods of natural disasters, natural emergencies, or increased threat from terrorists or criminal elements. Therefore, CONPLANs should include provisions for increasing the physical-security measures and procedures based on the local commander's assessment of the situation. Such contingencies may include hostage negotiations, protective services, and special-reaction teams. These

provisions should be designed for early detection of an attempted intrusion, theft, or interruption of normal security conditions.

- **Annex K.** Work-stoppage plan. This is a requirement for conducting a physical-security survey.

TACTICAL-ENVIRONMENT CONSIDERATIONS

F-2. In a tactical environment, the development of a physical-security plan is based on METT-TC (using the OPORD format and the higher headquarters' order). The order may be specific about the tasks the unit will perform. Time available may be limited and the scheme of maneuver may be dictated, but the leader must still evaluate the mission in terms of METT-TC to determine how MP elements can best carry out the commander's order.

F-3. Consider each of the following factors and compare courses of action to form a base for the physical-security plan. When the plan is firm, issue it as an order.

- Concepts for reconnaissance, coordination with adjacent and/or supporting units, and troop movement.
- Physical-security installation configurations and facilities. Areas to consider may include drop zones, landing zones, ranges, and training areas.

MISSION

F-4. The mission is usually the emplacement of defensive security rings to protect the populace against insurgents. The number of defensive security rings depends on the particular site and situation. The following questions must be evaluated:

- What is the mission?
- What specific and implied tasks are there to accomplish the mission?
- What is the commander's intent?

ENEMY

F-5. The commander identifies insurgent units operating in the area and tries to determine the type and size of the unit; the enemy's tactics, weapons, equipment, and probable collaborators; and the inhabitants' attitudes toward the insurgents. The following questions must be evaluated:

- What is known about the enemy?
- Where is the enemy and how strong is he?
- What weapons does the enemy have?
- What is the enemy doing?
- What can the enemy do in response to MP actions?
- How can we exploit the enemy's weaknesses?

TERRAIN AND WEATHER

F-6. The commander can use observation and fields of fire, cover and concealment, obstacles, key terrain, and avenues of approach (OCOKA) to plan for the physical-security defensive sites. The following questions must be evaluated:

- How will the terrain and weather affect the operation?
- How fast can movement be accomplished, and how much space does the terrain and unit formations take up?
- Will the weather affect the terrain or personnel?
- Has the weather already affected the terrain?

TROOPS

F-7. The commander must consider available equipment, the reaction time, reaction forces, communication assets, organization of troops, and medical support (if available). The following questions must be evaluated:

- What are the present conditions of vehicles and personnel?
- What is the status of ammunition and supplies?
- Who is best able to do a specific task?
- How much sleep have the soldiers had in the past 24 hours?
- What other assets are available to support the mission?
- How many teams/squads are available?
- What supplies and equipment are needed?
- What fire support is available and how can it be obtained?

TIME AVAILABLE

F-8. This factor is critical since the inhabitants must be ready to respond to an insurgent attack with little or no warning. The following questions must be evaluated:

- How much time is available to conduct planning?
- How long will it take to reach the objective?
- How long will it take to prepare the position?
- How much time do subordinates need?
- How long will it take the enemy to reposition forces?

CIVILIAN CONSIDERATIONS

F-9. The commander also must consider nonbelligerent third parties (such as dislocated civilians, personnel of international businesses and relief organizations, and the media). Every commander must prepare a site overlay that shows, as a minimum, the following:

- The attitude of the HN toward US forces.
- The population density near the objective.
- The condition of the local civilians.
- The possible effect of refugees and dislocated civilians on the mission.

Appendix G

Personal-Protection Measures

Terrorists frequently emulate military organizations as they develop, plan, train, and carry out terrorist attacks against DOD assets. Terrorists have a critical need for information regarding the whereabouts, habits, working environments, home environments, and other potential points of leverage against their targets. The three intelligence-collection methods used by terrorists against potential targets are human intelligence (HUMINT), photographic intelligence (PHOTINT), and signal intelligence (SIGINT).

PERSONAL PROTECTION

G-1. The measures that follow are useful in providing personal protection for US government employees and DOD civilian contractors in CONUS or OCONUS facilities.

OVERCOME ROUTINES

G-2. The reduced probability of success in kidnapping or killing a target makes the target far less desirable. Perform the following measures to prevent daily routines from being observed:

- Vary your route to and from work and your arrival and departure times.
- Vary your exercise schedule, using different routes and distances. It is best not to exercise alone.
- Do not divulge family or personal information to strangers.
- Enter and exit buildings through different doors, if possible.
- Avoid other routines.

MAINTAIN A LOW PROFILE

G-3. Americans are easy to identify in an overseas area. Perform the following measures to reduce easy ID:

- Dress and behave in public in a manner consistent with local customs. Items that are distinctively American should not be worn or displayed outside American compounds.
- Reduce visibility in the local community.
- Avoid flashing large sums of money, expensive jewelry, or luxury items.
- Avoid public disputes or confrontations, and report any trouble to the proper authorities.
- Ensure that personal information (home address, phone number, or family information) is not divulged.

PREPARE FOR UNEXPECTED EVENTS

G-4. All DOD personnel, contractors, and their family members should implement the following general measures:

- Get into the habit of checking in with friends and family.
- Know how to use the local phone system.
- Know the locations of civilian police, military police, government agencies, and the US embassy.
- Know certain key phrases in the local language.
- Set up simple signal systems that can alert family members or associates that danger is present.
- Carry ID showing your blood type and any special medical conditions.
- Keep personal affairs in good order.
- Avoid carrying sensitive or potentially embarrassing items.

WORKING ENVIRONMENT

G-5. The working environment is not immune from attempted acts by criminals or terrorists. DOD installations in CONUS and OCONUS usually provide a level of basic security comparable or superior to the basic level of security provided in the surrounding community. The following are general practices that can help reduce the likelihood of a terrorist attack:

- Establish and support an effective security program.
- Discourage the use of office facilities to store objects of significant intrinsic value unless it is mission essential.
- Train personnel to be alert for suspicious activities, persons, or objects.
- Arrange office interiors so that strange or foreign objects left in the room will be recognized immediately.
- Provide for security systems on exterior doors and windows.
- Ensure that access-control procedures are rigorously observed at all times for access to—
  - The installation.
  - Buildings within an installation.
  - Restricted or exclusion areas within buildings.
- Use an ID badge system containing a photograph.
- Identify offices by room number, color, or object name and not by rank, title, or the name of the incumbent.
- Avoid using nameplates on offices and parking places.

OFFICE PROCEDURES

G-6. In an office, the following steps can be taken to make intelligence collection and targeting more difficult for terrorists:

- Telephone and mail procedures:
  - When answering the telephone, avoid using ranks or titles.
  - When taking telephone messages, do not reveal the whereabouts or activities of the person being sought.

  - When leaving telephone messages, place them in unmarked folders; do not leave them exposed for observers to identify caller names and phone numbers, persons called, and messages left.
  - When opening mail, use a checklist to help identify letter bombs or packaged IEDs.
- Visitor-control procedures:
  - Place strict limitations on access to the executive office area.
  - Lock doors (from the inside) from the visitor-access area to executive offices or other restricted areas of a facility.
  - Ensure that receptionists clear all visitors before they enter inner offices.
  - Permit workmen or visitors access to restricted areas or exclusion areas under escort and only with proper ID. Confirm the work to be done before admitting workmen to restricted areas of the facility.
  - Limit publicity in public waiting areas to information that does not identify personnel by name, position, or office location.
  - Avoid posting unit rosters, manning boards, or photo boards where visitors or local contractors can view them.
  - Restrict the use of message boards, sign-in/-out boards, and other visual communications to general statements of availability.
- General working procedures:
  - Avoid carrying attaché cases, briefcases, or other courier bags unless necessary.
  - Avoid carrying items with markings that identify the owner by rank or title, even within the office environment.
  - Avoid working alone late at night and on days when the remainder of the staff is absent.
  - Ensure that office doors are locked when the office is vacant for any lengthy period, at night, and on weekends. If late-night work is necessary, work in conference rooms or internal offices where outside observation is not possible.
  - Ensure that the security office retains the office keys.
  - Ensure that papers, correspondence, communications materials, and other documents are not left unattended overnight.
  - Ensure that maintenance activity and janitorial services in key offices, production offices, or maintenance facilities are performed under the supervision of security personnel.
  - Prohibit the removal of property, material, or information stored on any media from the facility without proper written authorization.
  - Consider prohibiting the importation of property, material, or information stored on any media into the facility unless such items have been properly inspected.
  - Lock offices not in use to prohibit unauthorized access of stored material that could be used to hide IEDs or intelligence-collection devices.
  - Minimize the use of vehicles or vehicle markings that make it possible to readily identify the vehicle and its occupants as US-government or DOD-contractor personnel.

- Ensure that all personnel have access to some sort of duress alarm to annunciate and warn of a terrorist attack.
- Ensure that secretaries and guard posts are equipped with covert duress alarms that can be used to alert backup forces.
- Avoid placing office furnishings directly in front of exterior windows.

SPECIAL PROCEDURES FOR EXECUTIVE ASSISTANTS

G-7. The following suggestions are intended to be a guide for secretaries and executive assistants who may find themselves performing personnel-security duties as collateral duty. Executive assistants and security personnel should regularly train and exercise procedures used in case they must evacuate mission-critical personnel to safe havens.

- Request the installation of physical barriers (such as electromagnetically operated doors) to separate offices of senior executives from other offices.
- Request the installation of a silent trouble-alarm button with a signal terminating in the security department.
- Admit visitors into the executive area when they are positively screened in advance or are personally recognized.
- Do not inform unknown callers of an executive's whereabouts, home address, or telephone number.
- Store a fire extinguisher, a first aid kit, and an oxygen bottle in the office area.
- Remain calm and listen carefully when receiving a threatening call.
- Do not accept packages from strangers until satisfied with the individual's identity and the nature of the parcel.
- Keep travel itineraries for all personnel confidential.
- Distribute daily schedules for senior officers and civilian officials on a limited basis.

HOME ENVIRONMENT

G-8. The following discussion is intended to assist personnel in formulating plans to obtain housing outside US government compounds or DOD facilities. Personnel assigned to government housing may also find the antiterrorism and security tips presented below helpful in reducing the threat of violence and loss of property.

G-9. For general residential-security routines, discuss with family members the importance of—

- Varying routines in their daily activities.
- Blending in with the local environment.
- Avoiding unnecessary publicity and photographs that identify individual family members.
- Being alert to individuals, parked or abandoned vehicles, unusual utility work, or gatherings of people inconsistent with the residential environment.

SECURITY PRACTICES AT HOME

G-10. The following measures are specifically recommended for residential implementation. These measures are an extension of office antiterrorism-security practices.

- Do not use nameplates or uniquely American symbols on the exterior of residences occupied by DOD personnel overseas.
- Do not use nameplates on parking places, and avoid parking private or government vehicles in the same location day after day.
- Ensure that all family members answer the telephone politely but that they provide no information as to the name of the occupants until the caller's identity has been established.
- Treat all telephone conversations as though anyone who wanted to listen in was doing so.
- Examine carefully all mail delivered to the residence.

SOCIAL AND RECREATIONAL ACTIVITIES

G-11. DOD personnel are encouraged to participate in many social and recreational activities. The following precautions are recommended:

- Respond to formal social invitations in person (where possible) or by direct telephone contact.
- Be attentive to the security environment of social gatherings.
- Avoid the development of patterns with respect to time of arrival or departure at social events.
- Avoid prolonged presence at social functions where there is a high concentration of persons thought to be terrorist targets.
- Refrain from excessive use of alcohol at social functions; remain clearheaded and unimpaired.
- Vary routes to and from social events held at a central facility.
- Minimize appearances in uniform or formal attire.
- Decline invitations to appear in publicity photos.
- Participate in recreational activities within the American compound or at a DOD installation whenever possible.

NOTE: Refer to DOD 0-2000.12-H, Graphic Training Aid (GTA) 19-4-3, and Joint Services (JS) Guide 5260 for further guidance and explanation regarding protective measures.

Appendix H

Bombs

Terrorists have frequently used homemade devices or IEDs to carry out their attacks against DOD personnel, facilities, and assets. The IEDs are ideal terrorist weapons. They are relatively inexpensive to make, and the components of many IEDs are common items that can be obtained from many sources and are difficult to trace. The IEDs can be large or small and be designed so that they are transported to the attack site in components for last-minute assembly. Such design concepts make detection more difficult and provide an additional increment of personal safety to the terrorists.

GENERAL

H-1. The use of IEDs can enhance the violence that gives terrorist groups their ability to intimidate or coerce a target population. The detonation itself creates a highly visual, newsworthy scene, even hours after the detonation occurs. Bombs can detonate anywhere, without apparent reason and without warning. The use of bombs in a terror campaign emphasizes the authorities' inability to safeguard the public and maintain law and order. Bombs are ideal weapons because they can be designed to give terrorists opportunities to escape from the scene of their crimes.

CONCEALING BOMBS

H-2. Given the question, "Where have terrorists placed bombs in the past, and where should we look for them?" results in no easy answer. Table H-1, page H-2, lists a few obvious locations that should be examined. Terrorists who use bombs as their weapons of choice can be very creative in designing and placing their weapons.

H-3. Bombs can be found anywhere people can place them. Without becoming paranoid and seeing a bomb under every rock and behind every tree, the practical answer to the above questions is: "Where they can be easily placed without the bomber being caught."

DAMAGE AND CASUALTY MECHANISMS

H-4. The IEDs and other explosive devices inflict casualties in a variety of ways, including the following:

- Blast over pressure (a crushing action on vital components of the body; eardrums are the most vulnerable).
- Falling structural material.
- Flying debris (especially glass).
- Asphyxiation (lack of oxygen).

Table H-1. Potential IED Hiding Places

Outside Areas	
• Trash cans • Dumpsters • Mailboxes • Bushes	• Street drainage systems • Storage areas • Parked cars
Inside Buildings	
• Mail parcels or letters • Inside desks/storage containers • Ceilings with removable panels • Areas hidden by drapes or curtains • Recent repaired/patched segments of walls, floors, or ceilings	• Restrooms • Trash receptacles • Utility closets • Boiler rooms • Under stairwells
In Plain Sight	

- Sudden body translation against rigid barriers or objects (being picked up and thrown by a pressure wave).
- Bomb fragments.
- Burns from incendiary devices or fires resulting from blast damage.
- Inhalation of toxic fumes resulting from fires.

H-5. It is impossible to calculate a single minimum safe distance from an IED or other explosive device. The safe distance varies with each device and its placement. As a rule, the farther away from a bomb, the safer the intended or collateral targets are. Blast effects, fragmentation injuries, and injuries resulting from flying debris diminish greatly as the distance between a bomb and possible targets increase. The amount of material in the device, the type of explosive material, the manner in which the device is constructed, and the location or the container in which it is placed all have a bearing on the specific destructive potential for each IED.

H-6. The following are four general rules to follow to avoid injury from an IED:

- Move as far from a suspicious object as possible without being in further danger from other hazards such as traffic or secondary sources of explosion (such as POL storage).
- Stay out of the object's LOS, thereby reducing the hazard of injury because of direct fragmentation.
- Keep away from glass windows or other materials that could become flying debris.
- Remain alert for additional or secondary explosive devices in the immediate area, especially if the existence of a bomb-threat evacuation assembly area has been highly publicized.

H-7. Some terrorists have used two especially devious tactics in the past to intensify the magnitude of casualties inflicted by bombing attacks. In some instances, they have detonated a small device to lure media attention and curiosity seekers to the site; a larger, more deadly device has detonated some time after the first device, thereby inflicting a large number of casualties.

H-8. Other terrorists have used a real or simulated device to force the evacuation of a facility only to detonate a much more substantial device in identified bomb-threat evacuation assembly areas. These attacks are especially harmful because the evacuation assembly areas often concentrate government or commercial office workers more densely than they are when dispersed throughout their usual workplaces.

TELEPHONIC THREATS

H-9. When receiving a telephonic threat, treat the call seriously. Often, an anonymous telephone call is made regarding a bomb or an IED. See Figure H-1, page H-4, for information to record/obtain when receiving these calls

H-10. When an anonymous warning or threat is received, initiate the bomb-threat data card and notify the PMO, security police, security forces, or other law-enforcement/security offices immediately. Local SOPs will determine subsequent actions. Immediate action may include a search without evacuation, the movement of personnel within the establishment, a partial evacuation, or a total evacuation. The following criteria helps determine what immediate action to take:

- Factors favoring a search before the movement of personnel:—
  - There is a high incidence of hoax telephone threats.
  - Effective security arrangements have been established.
  - Information in the warning is imprecise or incorrect.
  - The caller sounded intoxicated, amused, or very young.
  - The prevailing threat of terrorist activity is low.
- Factors favoring movement of personnel before searching:
  - The area (post or base) is comparatively open.
  - Information in the warning is precise as to the matters of location, a description of the device, the timing, and the motive for the attack.
  - A prevailing threat of terrorist activity is high.

EVACUATION DRILLS

H-11. Evacuation and search drills should be performed periodically under the supervision of the installation's or unit's senior officer. The drills should be held in cooperation with local police if possible. Personnel in adjacent buildings should be informed of drills to avoid causing unnecessary alarm.

H-12. Evacuation procedures depend on the circumstances. Prepare, publicize, and rehearse evacuation plans in advance. Address alarm systems, assembly areas, routes to assembly areas, personnel-evacuation responses, building and area clearances, and evacuation drills.

PERSONNEL-EVACUATION RESPONSE

H-13. The bomb-threat alarm system should be easily distinguished from the fire alarm. When the alarm sounds, personnel should—

- Lock up or secure all classified materials.
- Conduct a quick visual search of their immediate working area.

Instructions: Be calm. Be courteous. Listen, do not interrupt the caller. Notify supervisor/security officer by prearranged signal while caller is on line.

Name of Operator____________________Time________Date________

Caller's Identity

Sex: ❑ Male ❑ Female ❑ Adult ❑ Juvenile Approximate age: Years________

Origin of Call

❑ Local ❑ Booth ❑ Internal (From within bldg)
❑ Long Distance If internal, leave line open for tracing the call.

Voice Characteristics		Speech		Language	
❑ Loud	❑ Soft	❑ Fast	❑ Slow	❑ Excellent	❑ Good
❑ High Pitch	❑ Deep	❑ Distinct	❑ Distorted	❑ Fair	❑ Poor
❑ Raspy	❑ Pleasant	❑ Stutter	❑ Nasal	❑ Foul	❑ Other ____
❑ Intoxicated	❑ Other____	❑ Slurred	❑ Lisp		
			❑ Other ____		

Accent	Manner		Background Noises	
❑ Local	❑ Calm	❑ Angry	❑ Factory Machines	❑ Trains
❑ Not Local	❑ Rational	❑ Irrational	❑ Bedlam	❑ Animals
Region____	❑ Coherent	❑ Incoherent	❑ Music	❑ Quiet
❑ Foreign	❑ Deliberate	❑ Emotional	❑ Office Machines	❑ Voices
Race____	❑ Righteous	❑ Laughing	❑ Mixed	❑ Airplanes
			❑ Street Traffic	❑ Party Atmosphere

Bomb Facts

Pretend difficulty with your hearing. Keep caller talking.

If caller seems agreeable to further conversation, ask questions like —
When will it go off? Certain Hour - Time Remaining - What kind of bomb? - Where are you now?
How do you know so much about the bomb? - What is your name and address?

If building is occupied, inform caller that detonation could cause injury or death.
Did caller appear familiar with plant or building by his description of the bomb location?

Write out the message in its entirety and any other comments on a separate sheet of paper and attach to this checklist.

Action To Take Immediately After Call

Notify your supervisor/security officer as instructed. Talk to no one other than as instructed by your supervisor/security officer.

Figure H-1. Sample Bomb-Threat Data Card

- Open windows (wherever possible).
- Leave the building, taking only valuable personal belongings.
- Leave doors open and immediately proceed to the assembly area.

H-14. Opening the building will reduce internal damage due to blast effects. It will also somewhat mitigate the extent of debris flying out of or falling from the building should a detonation occur.

ASSEMBLY AREAS

H-15. Choose the routes to the assembly area so that personnel do not approach the IED at any time. Preselect the routes to the assembly area, but devise a system to inform personnel of the location of the suspected IED and alternate routes. Routes prevent confusion and bunching and avoid potential hazards (such as plate glass, windows, and likely locations of additional IEDs).

H-16. Assembly areas should be preselected and well known to personnel. Establish a clearly defined procedure for controlling, marshaling, and checking personnel within the assembly area. If buildings or establishments are in a public area, coordinate the assembly areas with local police. Assembly areas are selected using the following criteria:

- Locate assembly areas at least 100 meters from the likely target or building (if possible).
- Locate assembly areas in areas where there is little chance of an IED being hidden. Open spaces are best. Avoid parking areas because IEDs can be easily hidden in vehicles.
- Select alternate assembly areas to reduce the likelihood of ambush with a second device or small-arms fire. If possible, search the assembly area before personnel occupy the space.
- Avoid locating assembly areas near expanses of plate glass or windows. Blast effects can cause windows to be sucked outward rather than blown inward.
- Select multiple assembly areas (if possible) to reduce the concentration of key personnel. Drill and exercise personnel to go to different assembly areas to avoid developing an evacuation and emergency pattern that can be used by terrorists to attack identifiable key personnel.

BUILDING AND AREA CLEARANCE

H-17. Establish procedures to ensure that threatened buildings and areas are cleared. Prevent personnel from reentering the building. Establish a cordon to prevent personnel from entering the danger area. Establish an initial control point (ICP) as the focal point for the PMO and for MP control.

H-18. Cordon suspicious objects to a distance of at least 100 meters, and cordon suspicious vehicles to a distance of at least 200 meters. Ensure that nobody enters the cordoned area. Establish an ICP on the cordon to control access; relinquish ICP responsibility to the PMO or local police upon their arrival. Maintain the cordon until the PMO, security police, security forces, or local police have completed their examination or stated that the cordon may stand down.

SEARCHING FOR A SUSPECTED IED

H-19. Searches are conducted in response to a telephonic threat or a report of an unidentified object on or near premises occupied by DOD personnel. The following types of searches may be used when searching for a suspected bomb or IED:

- An occupant search is used when the threat's credibility is low. Occupants search their own areas. The search is completed quickly because occupants know their area and are most likely to notice anything unusual.
- A team search is used when the threat's credibility is high. The search is very thorough and places the minimum number of personnel at risk. Evacuate the area completely, and ensure that it remains evacuated until the search is complete. Search teams will make a slow, thorough, systematic search of the area.

H-20. The following procedures should be followed if a search for explosive devices must be conducted before qualified EOD teams arrive:

- Make an audio check, listening for unusual sounds.
- Sweep the area visually up to the waist, then sweep up to the ceiling. Do not forget the tops of cabinets and cupboards.
- Perform a thorough and systematic search in and around containers and fixtures.
- Pass search results as quickly as possible to the leader responsible for controlling the search area. Do not use a radio; it may detonate the explosive.

H-21. Circumstances might arise in the case of a very short warning period. In other instances, a threat of a bomb against some facilities (if true) might necessitate the evacuation of a very large area. In these circumstances, searching for the presence of an explosive device to identify its location, appearance, and possible operating characteristics may be warranted.

H-22. Personnel who have not been trained in IED search and ID techniques should not search for explosive devices. Two types of errors are very common—the false ID of objects as IEDs and the incorrect ID of IEDs as benign objects. Depending on the devices used to arm and trigger an IED, the search process could actually result in an explosion.

SEARCH ORGANIZATION

H-23. The person controlling the search should have a method of tracking and recording the search results (such as a diagram of the area). Delegate areas of responsibility to the search-team leader, who should report to the person controlling the search when each area has been cleared. Pay particular attention to entrances, toilets, corridors, stairs, unlocked closets, storage spaces, rooms and areas not checked by usual occupants, external building areas, window ledges, ventilators, courtyards, and spaces shielded from normal view.

DISCOVERY OF A SUSPECTED IED

H-24. When a suspicious object has been found, report its location and general description immediately to the nearest law-enforcement or supervisory person. Do not touch or move a suspicious object. Instead, perform the following steps:

- If an object appears in an area associated with a specific individual or a clearly identified area—
  - Ask the individual/occupant to describe objects they have brought to work in the past few days.
  - Ask for an accounting of objects.
  - Ask for a verbal description/ID of objects.
- If an object's presence remains inexplicable—
  - Evacuate buildings and surrounding areas, including the search team.
  - Ensure that evacuated areas are at least 100 meters from the suspicious object.
  - Establish a cordon and an ICP.
  - Inform personnel at the ICP that an object has been found.
  - Keep the person who located the object at the ICP until questioned.
  - Avoid reentering the facility to identify an object that may or may not be an IED.

REACTING TO AN EXPLODED IED

H-25. The following procedures should be taken when an explosive/IED detonates at a DOD facility:

- For explosions without casualties—
  - Maintain the cordon. Allow only authorized personnel into the explosion area.
  - Fight any fires threatening undamaged buildings without risking personnel.
  - Report the explosion to the PMO, security police, security forces, or local police if they are not on the scene.
  - Report the explosion to the installation operations center even if an EOD team is on its way. Provide as much detail as possible, such as the time of the explosion, the number of explosions, the color of smoke, and the speed and spread of fire.
  - Ensure that a clear passage for emergency vehicles (fire trucks, ambulances, and so forth) and corresponding personnel is maintained.
  - Refer media inquiries to the PAO.
  - Establish a separate information center to handle inquiries from concerned friends and relatives.
- For explosions with casualties—
  - Select a small number of personnel to help search for casualties.
  - Assign additional personnel the responsibility for maintaining the cordon to keep additional volunteers searching for casualties.

Maintain the cordon until the EOD team verifies no further presence of bombs/IEDs at the site and the fire marshal determines that risk of additional injury to searchers from falling debris is acceptable.
  - Prepare a casualty list for notification of next of kin; delay publication of the list until its accuracy is determined.
  - Arrange for unaffected personnel to contact their next of kin immediately.

H-26. Civilian management officials and subordinate military commanders continue to have important personal roles to fulfill during a bomb/IED attack on DOD personnel, facilities, and assets. Perform the following procedures when reporting an attack:

- Pass available information to the operations center.
- Avoid delaying reports due to lack of information; report what is known. Do not take risks to obtain information.
- Include the following information in the report:
  - Any warning received and if so, how it was received.
  - The identity of the person who discovered the device.
  - How the device was discovered (casual discovery or organized search).
  - The location of the device (give as much detail as possible).
  - The time of discovery.
  - The estimated length of time the device has been in its location.
  - A description of the device (give as much detail as possible).
  - Safety measures taken.
  - Suggested routes to the scene.
  - Any other pertinent information.

H-27. Perform the following procedures when providing emergency assistance to authorities:

- Ensure that the PMO, security police, security forces, and other emergency-response units from local police, fire and rescue, and EOD teams are not impeded from reaching the ICP. Help maintain crowd control and emergency services' access to the site.
- Evacuate through the doors and windows of buildings.
- Assist the on-scene commander by obtaining a building diagram showing detailed plans of the public-service conduits (gas, electricity, central heating, and so forth), if possible. If unavailable, a sketch can be drawn by someone with detailed knowledge of the building.
- Locate, identify, and make witnesses available to investigative agency representatives when they arrive on the scene. Witnesses include the person who discovered the device, witnessed the explosion, or possesses detailed knowledge of the building or area.

H-28. Performing the above steps will provide substantial assistance to the crisis-management team and give other personnel constructive, supportive actions to take in resolving the crisis. Care must be exercised, however, that additional explosive devices are not concealed for detonation during the midst of rescue operations. These attacks add to the physical damage and emotional devastation of bomb/IED attacks.

H-29. The use of bombs and IEDs during terrorist attacks against DOD personnel, facilities, and assets is a common occurrence. The procedures outlined in this appendix are intended to help a DOD facility respond to an attack before an explosive device detonates. The procedures are also intended to help mitigate the consequences of an attack in case efforts to find an explosive device and render it inoperable are not successful. Incurring the costs to DOD facilities and installations of detecting an explosive device and terminating a terrorist incident before the device can detonate are almost always preferable rather than exercising plans and options to respond to a detonation. Several of the security measures discussed will help reduce the likelihood of a successful bomb/IED attack against DOD assets.

Appendix I

Executive Protection

DOD Directive 2000.12 recognizes a need to provide protection to military officers and DOD civilians who are assigned to high-risk billets, who are (by the nature of their work) high-risk personnel, or who are assigned to facilities identified as high-risk targets. The directive defines these terms as follows:

- High-risk billet. Authorized personnel billet (identified and recommended by appropriate authority) that because of grade, assignment, travel itinerary, or symbolic value may make personnel filling them an especially attractive or accessible terrorist target.
- High-risk personnel. US personnel and their family members whose grade, assignment, travel itinerary, or symbolic value may make them an especially attractive or accessible terrorist target.
- High-risk target. US facilities and material resources that, because of mission sensitivity, ease of access, isolation, or symbolic value may be an especially attractive or accessible terrorist target.

NOTE: For purposes of this appendix, the term executive will be applied to all persons requiring additional security protection who are assigned to high-risk billets, designated as high-risk personnel, or identified as high-risk targets.

SUPPLEMENTAL SECURITY MEASURES

I-1. The specific supplemental security measures that may be furnished to executives are subject to a wide range of legal and policy constraints. US law establishes stringent requirements that must be met before certain security measures may be implemented. DOD regulations, instructions, and legal opinions may further constrain the implementation of some protective measures described in this chapter. The SOFAs and MOUs between the US and a foreign government will also limit the use of some supplemental security measures. Leases and other conditions imposed by contract for purchase of land or buildings by the US for DOD use may also limit the application of certain security techniques. All of these constraints should be carefully considered when conducting security surveys, developing plans, and implementing additional security measures to protect high-risk personnel.

EXECUTIVE PROTECTION GOALS

I-2. In the discussion that follows, several measures are outlined that can afford senior military officers and DOD personnel additional protection against terrorist acts. The purpose underlying these measures is to—

- Increase the interval of time between detecting a threat and the onset of hostile action against executives and their dependents.
- Increase the amount of time required by terrorists to gain physical access to executives from the onset of hostile actions, whether the executives are at home, at the office, or in transit.

I-3. The implementation of supplemental security measures should strive to achieve the following prioritized goals:

- Enhancements should hold the terrorist threat at bay until a response force arrives (delay at a distance).
- Enhancements in physical security should enable executives to flee to safety (delay to permit flight).
- Enhancements should permit the executive to retreat into a safe haven of sufficient strength and survivability. This should enable a response force to wage an effective counterattack to liberate executives and others accompanying them to a safe haven, including family members at home and colleagues and visitors at work (delay, hold, and counterattack).

I-4. The following supplemental measures should be applied with care. There is a clear trade-off between increasing the level of physical security at the office and at home and preserving the anonymity of executives, thereby avoiding telltale signs of activity that point to prominence or criticality. These measures can be expensive. Expense can be measured not just in terms of dollars, but also in terms of changes to organizational routine. Therefore, three questions must be resolved before implementing bold, disruptive, and expensive supplemental security enhancements:

- What are the most cost-effective means of enhancing the security of executives at risk?
- How many changes in organizational routines and personal behaviors will have to be made for security measures to be effective in reducing the risk of terrorist attacks and the vulnerability of executives to such attacks?
- What are the anticipated costs of additional security measures in terms of dollars, organizational functionality, and mission capability?

I-5. Security enhancements can be made to improve the security of executives and can be even more effective if executives and their families take full advantage of and reinforce those measures. If executives do not change their behavior to accommodate additional security and protective measures, then the behaviors can effectively defeat the purpose of additional protection. Additional increments of security can be obtained to defeat virtually any threat. However, there is a point at which it is no longer cost-effective to add layer upon layer of protective measures to defeat a threat.

RESIDENTIAL SECURITY MEASURES

I-6. While terrorist groups conduct intelligence operations to identify targets, mistakes have been made in the past. DOD personnel should avoid leasing residences previously used by representatives of governments or organizations known to be targets of various terrorist groups. DOD personnel leasing residences formerly used by representatives of such governments may be placing themselves unnecessarily at risk of being attacked as a result of mistaken identity.

I-7. An executive's entire lifestyle should be included in security surveys used to assess the need for supplemental physical-security measures at the office. The executive's home and transportation from home to office and back should also be examined for risk and vulnerability. The same principles used to identify supplemental security improvements in an office environment apply to an executive's home environment as well. The purposes of physical-security enhancements are to—

- Increase the amount of time terrorists need to initiate and complete an attack on executives while at home, thereby giving response forces more time to rescue executives and their dependents.
- Reduce potential harm to executives and their families because of a terrorist assault mounted against the residence.

I-8. The goals of enhanced residential physical-security measures are to—

- Increase the amount of time between detection of a threat and the onset of hostile actions.
- Delay the terrorists as long as possible. Prevent terrorist access to executives and their family members and make it difficult to leave the scene to escape prosecution. These measures should not further jeopardize the lives of executives and their family members.
- Provide a safe haven where executives and their family members may flee for security pending the arrival of a response force on the scene.

I-9. The following measures can be implemented selectively to help security personnel achieve these objectives:

- Increase the time interval between threat detection and the onset of hostile terrorist acts by—
  - Ensuring that all door locks and window clasps are working.
  - Ensuring that all doors and windows are properly secured to their frames and that the frames are properly anchored to the residential structure.
  - Locking driveway gates with a security lock to prevent entry.
  - Installing a through-door viewing device or visitor intercom.
  - Installing security lights to aid in viewing entrances.
- Increase the number of physical barriers between the outer perimeter of the residence and the interior of the residence by—
  - Adding heavy, remotely operated gates to all fences, walls, and perimeter barriers consistent with the penetration resistance of the barrier between the residence, the street, and adjacent neighbors.
  - Creating a vestibule or air lock between living quarters and the exterior of a residence, ensuring that no one can enter the residence directly from the outside.
  - Adding fire doors or security doors or gates between the residence's bedrooms and living areas.

- Increase the time required to penetrate exterior structural walls by explosives, hand-held power tools, and hand tools by—
  - Adding additional armor covered by aesthetically pleasing materials to exterior walls.
  - Adding a separate reinforced masonry wall around the residence.
- Increase the surveillance of the residence and decrease response time by—
  - Installing CCTV systems to permit remote viewing of all doors and windows accessible from the ground, nearby structures, trees, or easily acquired platforms (such as a van parked next to a wall).
  - Installing area IDSs between the residence's perimeter and the residence itself, varying the number and types of sensors, and adding backup communication channels between the IDS and a surveillance assessment/response dispatch center.
- Increase the residence's durability and survivability to a terrorist attack by—
  - Fitting windows with either venetian blinds or thick curtains to reduce the observability of activities within the residence and to reduce hazards of flying glass in case of nearby explosions or gunfire.
  - Installing backup power systems for security devices (surveillance systems, communication systems, and access-control systems).
  - Ensuring that backup communication is available with the installation or embassy's security department via a secure landline or two-way radio.
  - Fitting a panic-alarm bell to the outside of the house with switches on all floor levels. Such an alarm should also annunciate at the local police and cognizant DOD or DOS security office.
  - Installing a safe haven in the home.

TRANSPORTATION MEASURES

I-10. High-risk personnel are most accessible to terrorists while in transit in official or privately owned vehicles. Specific steps can be taken to reduce the vulnerability of executives in transit.

SPECIAL TRANSPORTATION IN TRANSIT FROM DOMICILE TO DUTY

I-11. As a rule, Congress has strongly opposed the provision of domicile-to-duty transportation by the federal government to its officers and employees. Only 16 officials are entitled by statute to such assistance. Congress did, however, grant authority to the President and the heads of executive agencies and departments to provide domicile-to-duty transportation under certain circumstances. According to the statute, "a passenger carrier may be used to transport between residence and place of employment an officer or employee with regard to whom the head of a Federal agency makes a determination, [provided] that highly unusual circumstances present a clear and present danger, that an emergency exists, or that compelling operational considerations make such transportation essential to the conduct of official business."

I-12. The phrase "highly unusual circumstances which present a clear and present danger" is understood to mean that—

- The perceived danger is real, not imaginary.
- The perceived danger is immediate or imminent, not merely potential.
- Proof is provided that the use of a government vehicle would provide protection not otherwise available.

I-13. Such a danger would exist where there is an explicit threat of terrorist attacks or riot conditions and such transportation would be the only means of providing safe passage to and from work.

I-14. The phrase "emergency exists" means that there is an immediate, unforeseeable, temporary need to provide home-to-work transportation for an agency's essential employees. The phrase "similarly compelling operational considerations" means that there is an element of gravity or importance for the need of government-furnished transportation comparable to the gravity or importance associated with a clear and present danger or an emergency. Congress suggested further, "in such instances, [it is expected] that home-to-work transportation would be provided only for those employees who are essential to the operation of the government."

I-15. The Secretary of Defense has the statutory authority to allow a CINC to use government-owned or -leased vehicles to provide transportation in an area outside of the US for members of the uniformed services and other DOD personnel under certain circumstances. These circumstances include and are limited to a determination by the CINC that public or private transportation in the area is unsafe or is not available. Under these circumstances, DOD may provide transportation (usually in government buses or passenger vans) to personnel and their family members if it will help the CINC and his subordinate commanders maintain the capability to perform or undertake assigned missions. This transportation is not intended for transporting personnel from their residences to their places of work. The Secretary of Defense and the Service Secretaries also have the statutory authority to provide transportation from home to duty stations and back on a limited basis. This authority is usually implemented by providing a nontactical armored vehicle (NTAV) to protect personnel.

I-16. It is a DOD policy to make NTAVs available where necessary to enhance the security of DOD personnel consistent with the requirements and limitations found in the statute. DOD issuances, service regulations, and CINC guidance stipulate detailed procedures by which DOD manages NTAV programs. The statute also establishes a procedure for Presidential waiver of the "buy American" requirement; DOD and service regulations provide for the delegation of Presidential authority from the President to the Secretary of Defense; to the Director, Defense Security Assistance Agency; and to the Director, DIA. DOD Instruction 5210.84 authorizes DOS acquisition and installation of light vehicle armoring to DOS specifications in local defense-component vehicles on a reimbursable basis. The level of protection provided to the Defense Component Office will comply with approved overseas security policy group armored-vehicle standards.

I-17. The DOD recognizes two classes of NTAVs—heavy and light. Heavy NTAVs are fully armored vehicles intended to protect occupants from attack by bombs; IEDs; grenades; and high-velocity, small-arms projectiles. Light NTAVs are less than fully armored vehicles and are intended to protect occupants from attack by medium-velocity, small-arms projectiles and at least some types of IEDs.

I-18. The dividing lines between heavy and light NTAVs have become less distinct over time as armoring techniques and materials have given greater capability to NTAVs that are not classified as heavy. As a practical matter, add-on vehicle-armoring kits are now in production which (when properly installed in an appropriately powered and suspended vehicle) will provide a level of protection approaching that of the heavy NTAVs.

Heavy NTAVs

I-19. Heavy NTAVs may be assigned to US personnel upon certification by a Service Secretary only under the following conditions:

- Highly unusual circumstances present a clear and present danger to the health and safety of a nominated protectee.
- Compelling operational considerations make such transportation essential to conducting official business.

I-20. If the physical-security survey concludes that a heavy NTAV is warranted, the nominated protectee's Service Secretary shall, on the advice and recommendation of a combatant commander, determine whether the use of a heavy NTAV is warranted. If so, the Service Secretary shall authorize the use of a vehicle for a renewable 90- to 360-day period. At the end of the period, the requirement will be reexamined and a recertification for the protection shall be issued by the Service Secretary.

I-21. Each of the services manages a portion of the DOD's NTAV program. Each service has issued supplementary mandatory guidance for processing requests for, as well as allocation and use of, these scarce assets.

I-22. Heavy NTAVs are complex systems requiring specialized maintenance and operation. Normally, they will be assigned to DOD personnel with a driver who has been properly trained in the operation and maintenance of the vehicle. The operator is not a chauffeur; he is an integral part of a supplemental security package provided by DOD to meet its obligations in protecting key assets.

Light NTAVs

I-23. Light NTAVs may also be provided to US employees and officers where highly unusual circumstances present a clear and present danger to the health and safety of a nominated protectee or compelling operational considerations warrant their use. This category of NTAV features add-on armoring. While they are a less-complex armoring system than those used in heavy NTAVs, light NTAVs afford substantial protection to occupants against a variety of threats. New developments in after-manufacture armoring kits for vehicles are occurring at a rapid pace, increasing the number of vehicle manufacturers and models for which other NTAV modifications are suitable. Each service and the DIA have instructions for implementing DOD policy that authorizes the use of other NTAVs to enhance personnel protection of high-risk persons.

PRIVATELY OWNED VEHICLES

I-24. High-risk personnel may wish to forego the use of POVs during periods of extreme risk. Considerations include selecting measures that—

- Deter secret entry, making undetected placement of IEDs in or under the vehicle difficult for terrorists to accomplish.
- Enhance the vehicle's ability to increase distance between it and pursuers.
- Assist response forces in case of an incident.
- Make the vehicle appear little different than its standard models.

INDIVIDUAL PROTECTIVE MEASURES

I-25. Executives can enhance their personal security in the office environment by—

- Discouraging staff members who are taking telephone messages from disclosing their whereabouts.
- Ensuring that caution is used when opening mail and being especially careful with letters or packages that might contain IEDs.
- Ensuring that access is strictly limited to their office area.
- Limiting publicity and keeping official biographies short. This includes using outdated photographs if a publicity photograph is essential.
- Ensuring that they are not working alone late at night and on days when the remainder of the staff is absent.
- Working in conference rooms or internal offices where outside observation is not possible if late-night work is necessary. Security officers should be notified of the work so that they can periodically look in.
- Ensuring that office furnishings are not placed directly in front of exterior windows.

OFFICIAL BUSINESS AWAY FROM THE OFFICE

I-26. The following suggestions reinforce efforts by executives to maintain the high level of security provided in the home or office environment while on official business outside of these locations:

- Discuss security requirements with the person planning the function.
- Travel to and from the function with escorts.
- Choose the route carefully.
- Avoid publicizing planned attendance at official functions (unless required).
- Attempt to sit away from both public areas and windows.
- Encourage the function's sponsor to close the curtains to minimize the likelihood that anyone outside will be able to see inside and determine who is attending the function. This is extremely important for an evening function, when a well-lit interior can be easily viewed from a darkened exterior.
- Request that external floodlights be used to illuminate the area around the building where an evening function will occur.

LOCAL OFFICIAL AND UNOFFICIAL TRAVEL

I-27. Executives can greatly enhance their personal security when conducting official and unofficial travel by following these general practices:

- Vary daily patterns, such as leaving and returning at different times.
- Consider escorts to and from work or travel with a neighbor.
- Establish a simple oral or visual duress procedure between executives and drivers (for example, a phrase or movement used by the executive or driver only if something is amiss).
- Vary taxi companies. Ensure that the ID photo on the license matches the driver. If uneasy for any reason, take another taxi.
- Attend social functions with others, if possible.
- Examine the car before entering to see if there has been any interference. A small mirror on a rod is a cheap and effective method to inspect underneath cars. Do not touch the vehicle until it has been thoroughly checked (inside, around, and under).
- Avoid leaving personal items exposed in the car (uniform items, service-issued maps, official briefcases, and so forth).

SECURITY PRACTICES WHILE DRIVING

I-28. Executives can take the following measures to enhance security while driving:

- Keep car doors locked. Do not open windows more than a few inches.
- Avoid overloading a vehicle, and wear seat belts.
- Park vehicles in parking areas that are either locked or monitored. Never park overnight on the street. Before entering vehicles, check for signs of tampering.
- Keep the trunk locked.
- Drive in the inner lanes to keep from being forced to the curb.
- Use defensive and evasive driving techniques. Drill with your driver by watching for suspicious cars and taking evasive action.
- Avoid driving close behind other vehicles (especially service trucks), and be aware of activities and road conditions two to three blocks ahead.
- Beware of minor accidents that could block traffic in suspect areas such as crossroads. Crossroads are preferred areas for terrorist or criminal activities because they offer escape advantages.

I-29. If a terrorist roadblock is encountered, use the shoulder or curb (hit at a 30- to 45-degree angle) of the road to go around it or ram the terrorist's blocking vehicle. Blocking vehicles should be rammed in a nonengine area, at a 45-degree angle, in low gear, and at a constant moderate speed. The goal is to knock the blocking vehicle out of the way. In all cases, do not stop and never allow the executive's vehicle to be boxed in with a loss of maneuverability. Whenever a target vehicle veers away from the terrorist vehicle, it gives adverse maneuvering room and presents a better target to gunfire.

INTERURBAN, NATIONAL, AND INTERNATIONAL TRAVEL SECURITY PRACTICES AND PROCEDURES

I-30. To enhance security in interurban, national, and international circumstances, executives should—

- Book airline seats at the last moment. Consider using an alias.
- Restrict the use of ranks or titles.
- Avoid allowing unknown visitors in the hotel room or suite.
- Keep staff and family members advised of the itinerary and subsequent changes. Clearly and emphatically restrict this information to those having a need to know.

HOME SECURITY PRACTICES AND PROCEDURES

I-31. To enhance security at home, executives should—

- Check the ID of persons entering the premises (electricians, plumbers, telephone-maintenance personnel, and so forth). When in doubt, call their office to verify their identity before allowing them in your home.
- Avoid opening the door to a caller at night until he is visually identified through a window or a door viewer.
- Close curtains in a room before turning on lights.
- Consider placing the telephone where you will not be seen from doors or windows when answering.
- Investigate the household staff (especially temporary staff members).
- Stay alert and be on the lookout for the unusual. Ensure that the home is locked and secure whenever the residence is unattended. Be cautious upon return and look for the movement of furniture or the placement of unusual wires.
- Note and report suspicious persons.
- Control house keys strictly.
- Park the car in a locked garage.
- Consider installing a panic-alarm bell to the outside of the house with switches located on all floor levels.
- Clear the area around the house of dense foliage or shrubbery.
- Test duress alarms (if available). Make certain that family members understand how they work as well as the importance of their use.
- Cooperate with law-enforcement personnel, and abide by their security recommendations.

SECURITY AT SOCIAL AND RECREATIONAL ACTIVITIES

I-32. The risk of terrorist incidents is always present for high-risk personnel or personnel assigned to high-risk billets. The following measures are intended to permit executives to live a close-to-normal life while still remaining mindful of the risks to their security.

- Ensure that the host is aware of and takes appropriate measures for your security.
- Have your personal staff assist a civilian host, if required.
- Arrange for visitors to be subject to adequate security control.

- Screen the invitation list, if possible.
- Vary the times of sporting activities (golfing, jogging, and so forth).

COMBATING-TERRORISM TRAINING FOR EXECUTIVES

I-33. Combatant commanders annually compile a list of high-risk billets in their AO. These lists are forwarded through the appropriate service personnel channels, enabling each service to identify, plan, and provide resources to meet training requirements. All personnel and adult family members en route to high-risk billets must attend the Individual Terrorism Awareness Course (INTAC) conducted at the US Army John F. Kennedy Special Warfare Center at Fort Bragg, North Carolina. During this one-week course, personnel will receive instruction in defensive-driving techniques and survival shooting as well as individual protective measures and hostage survival. These individuals should also attend the appropriate regional orientation course (Middle East, Asia/Pacific, Latin America, or Africa) offered at the US Air Force Special Operations School at Hurlburt Air Force Base (AFB), Florida. The service member whose duties will require frequent vehicle operation should attend an appropriate evasive-driving course. Information on current offerings may be obtained by contacting the service representative to the DOD Antiterrorism Coordinating Committee or the Combating Terrorism Branch in the Office of the Assistant Secretary of Defense (OASD) Special Operations/ Low-Intensity Conflict (SO/LIC).

TRAVEL TO POTENTIAL PHYSICAL-THREAT RISK AREAS

I-34. Personnel en route to potential physical-threat risk areas (as identified by the OASDSO/LIC) should attend one of the following courses:

- The Dynamics of International Terrorism Course conducted at the US Air Force Special Operations School at Hurlburt AFB, Florida. During this one-week course, personnel will receive lectures on threats by region (Europe, Middle East, Latin America, Asia/Pacific, and Africa), the history and psychology of terrorism, personnel combating-terrorism measures (vehicle, personal, airline, and physical security), and hostage survival.
- A Regional Orientation Course (Middle East, Latin America, Africa, Asia/Pacific) at the US Air Force Special Operations School at Hurlburt AFB, Florida. This one-week course offers personnel instruction in cultural, political/military, and individual security factors associated with the region.

I-35. Installation security personnel may also receive the above training if they have completed the Antiterrorism Instructor Qualification Course (AIQC) at Fort Bragg, North Carolina.

PROTECTIVE SECURITY DETAILS

I-36. Each service can provide bodyguards for key senior military officers, DOD civilians, other US officials, or foreign dignitaries requiring personal protection. Each Service Secretary is responsible for assigning protective security details (PSDs) to service members based on the recommendation of their counterintelligence and/or law-enforcement investigation staffs. The PSDs are assigned to DOD personnel who meet requirements established by service regulations. In general, PSDs may be assigned only to those personnel whose position or assignment places them at risk and whose continued availability to the National Command Authorities and the CINCs is vital to DOD's mission execution.

I-37. A PSD provides high levels of security to an executive by establishing a series of protective cordons around him. The establishment of defense in depth often means that the innermost protective layer is in close contact with the executive at all hours of the day and night.

I-38. A PSD is trained to maintain a low profile. It is concerned about the executive's visibility and its ability to blend into his surroundings. There is nothing more damaging to the security of an executive than the obvious, detectable presence of a PSD when all other measures to have him blend into the local environment have been successful. A PSD will strive to keep travel routes and means of transportation from being publicized. If this cannot be accomplished, the PSD may suggest editorial changes to the itinerary scheduled for release to limit details of planned travel from public disclosure. For example, routes to and from announced appointments usually do not need to be revealed.

I-39. During the course of a PSD's mission, its members may be asked to perform several different security functions. They may, for example, perform direct or indirect protection or escort duty. Direct protection is open and obvious; indirect is generally a surveillance measure. The security-guard unit may operate as an interior guard and may consist of one or more men stationed at fixed posts. A PSD's members should know the identity of each individual in the party of a protected official; executives can help by introducing them to each member of the official party.

I-40. The protected person's attitude is critical to the success of the PSD's mission. Executives have a right and a responsibility to make their wishes known with respect to their personal security. They also have an obligation to listen carefully to the head of the PSD who is trained and highly qualified to help make reasonable judgments about manageable risks. A PSD's members understand that their function is inherently intrusive and that executives can easily resent the loss of privacy that accompanies the protection offered. On the other hand, PSDs have jobs to do, not merely to protect executives, but to help safeguard mission-critical assets—senior military and civilian leaders.

I-41. One of the most demanding functions placed on a PSD is to limit the ability of individuals to circulate and approach the executive. This is often very frustrating to executives who wish to shake hands, engage in close conversations with visitors, and move freely and without impediment in a social situation. The PSDs are trained to strictly enforce limitations on the circulation of individuals, carefully checking each person for ID and ascertaining that he is authorized to be present at the occasion.

I-42. DOD personnel who are provided with PSDs and must conduct official business or hold social engagements in large rooms can take steps to minimize the disruptions to such functions. These steps include—

- Providing advance attendee lists to the head of the PSD.
- Having one or more members of the staff who know the attendees stand with PSD members and identify the attendees as they arrive.
- Informing attendees that they will be admitted only at specified entrances.

I-43. The PSD's members are highly trained security specialists. While in the company of executives, they will be accommodating and helpful. Executives should remember, however, that the primary function of the PSD's members is to protect them, not perform errands or carry out personal services. A PSD's members who are performing valet or other chores cannot effectively protect the senior officers or civilian officials to whom they have been assigned.

EXECUTIVE-PROTECTION SYSTEM INTEGRATION

I-44. This appendix has focused on supplemental security measures used to address terrorist threats to senior high-risk personnel within the DOD. Various methods and measures have been discussed that provide increments of security over and above the base level of security provided to all DOD personnel assigned to an installation, facility, activity, or unit. In making decisions to allocate protective resources to enhance the security of senior officers and senior DOD officials, it is essential to remember that measures must be applied systematically. Additional security measures implemented to protect high-risk personnel in the office environment must be carried over to official functions conducted outside the office. The security measures must also be extended to protected persons' private lives and, depending on the nature of the threat, the lives of their family members.

I-45. The converse is equally true. It makes no sense to provide domicile-to-duty transportation for a high-risk person and make no provision for additional protection at home, at the office, and at official business and social functions. In view of the total costs of security measured in dollars, time, and inconvenience to protected persons, their staffs, colleagues, and families, it may be more prudent to radically alter living and working arrangements than to try to augment security in a piecemeal manner. For example, it might be prudent to house high-risk personnel within a DOD installation rather than to try to secure a detached, private residence at a substantial distance from the operations base of a response force. The key to successful executive protection is to ensure that the level of protection afforded by physical-security measures, operational procedures in the office and at home, and PSDs is constant. The level of protection must be matched to the threat and must be sustainable. Executives have a special responsibility to set a personal example of combating-terrorism awareness; of attention to personal, family, office, information, and OPSEC concerns; and of combating-terrorism security measures implementation. By doing so, they make their colleagues and subordinates more aware and more conscious of their security environment and less likely to be victimized by terrorist attacks.

Appendix J

Resource Management

Programs need annual funding to operate. This funding is obtained by devising and documenting a resource program that looks seven years out. Looking ahead one or two years is generally not a problem. Commanders must be able to describe future requirements in the out years so that money will be available when the program arrives in future years.

FUNDING PROGRAMS

J-1. Physical security is dependent on integrated systems with budgetary constraints. This appendix serves to inform commanders of three basic funding programs contributing to physical security—

- The RJC6, which resources physical-security equipment purchase and maintenance.
- The QLPR, which resources law-enforcement operations (to include security guards and special-reaction teams).
- The VTER, which provides resources for projects and temporary programs that enhance any type of security due to an increase in terrorist threat.

PROJECTED REQUIREMENTS

J-2. Recurring VTER requirements are usually shifted to QLPR as standard requirements after a period of a few years. Installations send seven-year projected requirements to the MACOMs for submission into the Army's Program Objective Memorandum (POM), which is an annual significant event for resource managers at all levels. Installation requirements should be included in the POM build; otherwise, the installation's program loses visibility from the start of the funding process. Input format is determined locally, but a generic example of the format and the type of information requested is provided at Figure J-1, page J-2.

OBLIGATION PLAN

J-3. Budget execution deals with the current fiscal year (October through September), with the exception of some types of dollars (appropriations) that are multiyear (such as procurement dollars that are executable for three years). Once the resource manager notifies an installation of its available annual funding, an obligation plan by month or quarter is developed to display how and when the funds will be spent. This obligation plan is also used to forecast when the program will run out of money, which in turn will justify the submission of an unfinanced requirement (UFR) to obtain the proper level of resources. Examples of an annual obligation plan and UFR are located at Figures J-2 and J-3, page J-3.

Program Objective Memorandum (POM)

Installation: Fort McClellan, AL
Activity: Military Police School
Appropriation: Operation & Maintenance, Army (OMA)
Army Management Structure (AMS): 321731
Management Decision Package (MDEP): VTER
Element of Resource (EOR): 2500
Issue: Force Protection/Antiterrorism Courses

Resources	FY00	FY01	FY02	FY03	FY04	FY05	FY06
($000)	500	510	520	530	540	550	560

Justification/Impact:
Funding is required to continue the eight functional contract training courses conducted by the US Army Military Police School in Force Protection/Antiterrorism for the Army and selected DOD services and agencies. Courses focus on the five pillars of force protection and provide students with a variety of force-protection expertise to prevent, react to, and
contain a terrorist event. If funds are not provided, the projected annual student load of 2,500 cannot be trained.

Figure J-1. Sample POM

TYPES OF APPROPRIATIONS

J-4. The most common types of appropriations that managers will be involved with are day-to-day operations or mission (OMA) funds and procurement (OPA) dollars that must be used for major projects or equipment buys over the $100,000 threshold. Army Materiel Command (AMC) uses research, development, and engineering (RD&E) dollars for operations at US Army Test and Evaluation Command (TECOM) facilities and for testing new equipment. Construction projects over $500,000 use military construction (MILCON) funds.

J-5. Resources needed for physical security are the result of planning. Commanders include physical-security programs and improvements as a part of all annual budgets. When the situation changes based on METT-TC, physical security programs are reviewed, updated and, when necessary, approved. Commanders allocate resources consistent with the threat. Force-protection requirements provide the fundamental reasons for resourcing physical security. Security managers, force-protection officers, and PMs help identify security requirements and prioritize expenditures. When physical-security improvements are not properly planned, integrated, and budgeted for, commanders accept risks for physical-security failures.

Annual Obligation Plan
FY 99

Installation: Fort McClellan, AL
Activity: Military Police School
Appropriation: Operation & Maintenance, Army (OMA)
Army Management Structure (AMS): 321731
Management Decision Package (MDEP): VTER
Element of Resource (EOR): 2500
Issue: Force Protection/Antiterrorism

Phased Monthly Obligation Plan (Cumulative):

Month	$000
Oct	40
Nov	80
Dec	120
Jan	160
Feb	200
Mar	240
Apr	280
May	320
Jun	360
Jul	360
Aug	360
Sep	360

Figure J-2. Sample Annual Obligation Plan

Unfinanced Requirement (UFR)
FY 99

Installation: Fort McClellan, AL
Activity: Military Police School
Appropriation: Operation & Maintenance, Army (OMA)
Army Management Structure (AMS): 321731
Management Decision Package (MDEP): VTER
Element of Resource (EOR): 2500
Issue: Force Protection/Antiterrorism

Justification/Impact:
Funds in the amount of $360K have been provided to conduct the eight Force-Protection/Antiterrorism contract courses in the US Army Military Police School. This amount will only support training through June 99. An additional $120K is required to conduct training during 4th quarter FY99. Unless additional funds are provided, 625 students will not be trained.

Figure J-3. Sample UFR

Appendix K

Vulnerability Assessment

After a commander has obtained a threat analysis, he proceeds to complete the analysis by conducting the vulnerability and criticality assessments. (This appendix will discuss only the VA.) This process considers a mission review and analysis of the installation, base, unit, or port in relation to the terrorist threat. The review should assess the cost of antiterrorism measures in terms of lost or reduced mission effectiveness. It should then assess the level of acceptable risk to facilities and personnel given the estimated erosion of mission effectiveness. Often the best operational method and routine may be the worst to counter potential terrorist activities. This review and analysis is performed routinely and particularly for deployment.

ASSESSMENT CONSIDERATIONS

K-1. The installation, base, unit, or port assessment is derived from the results of the vulnerability and criticality assessments. The assessment provides the staff with the overall vulnerability to terrorist attack. The staff then develops the crisis-management plan, which addresses all terrorist threat levels regardless of the present level. The THREATCONs are then applied according to the local threat. The considerations are—

- **Vulnerability.** The VA is a self-assessment tool used to evaluate its vulnerability to terrorist attack. The more vulnerable an installation, base, unit, or port is, the more attractive it becomes to terrorist attack.
- **Criticality.** The criticality assessment identifies key assets and infrastructures located on and adjacent to the installation, base, unit, or port. These assets are normally symbolic targets that traditionally appeal to a specific terrorist group (such as headquarters buildings and monuments). It addresses the impact of the temporary or permanent loss of key assets or infrastructures to the ability of the installation, base, unit, or port to perform its mission. The staff determines and prioritizes critical assets. The commander approves the prioritized list. The assessment—
  - Selects key assets.
  - Determines whether critical functions can be duplicated under various attack scenarios.
  - Determines the time required to duplicate key assets or infrastructure efforts if temporarily or permanently lost.
  - Determines the vulnerability of key assets or infrastructures to bombs, vehicle crashes, armed assault, and sabotage.
  - Determines the priority of response to key assets and infrastructures in case of fire, multiple bombings, or other terrorist acts.

- **Damage.** The damage assessment determines the ability of the installation, base, unit, or port to plan for and respond to a terrorist attack against key assets and infrastructures.
- **Recovery procedures.** The recovery-procedures assessment determines the capability to recover from the temporary or permanent loss of key assets and infrastructures. Based on this assessment, the staff establishes recovery procedures to ensure the continued ability to perform the mission.

THREATCON LEVELS

K-2. Specific security measures should be directly linked with THREATCON levels. These considerations are—

- **THREATCON Normal.** This THREATCON level exists when a general threat of possible terrorist activity exists but warrants only a routine security posture.
- **THREATCON Alpha.** This THREATCON applies when there is a general threat of possible terrorist activity against personnel and facilities (the nature and extent of which are unpredictable) and when circumstances do not justify full implementation of THREATCON Bravo measures. It may be necessary to implement measures from higher THREATCONs either resulting from intelligence or as a deterrent. The measures in this THREATCON must be capable of being maintained indefinitely.
- **THREATCON Bravo.** This THREATCON applies when an increased and more predictable threat of terrorist activity exists. The measures in this THREATCON must be capable of being maintained for weeks without causing undue hardship, affecting operational capability, or aggravating relations with local authorities. While in Bravo, the installation should bring manning levels and physical-protection levels to the point where the installation can instantly transition to THREATCON Charlie or Delta.
- **THREATCON Charlie.** The transition to THREATCON Charlie must be done on short notice. It is a result of an incident occurring or the receipt of intelligence indicating that some form of terrorist action against personnel and facilities is imminent. Charlie measures should primarily focus on manning adjustments and procedural changes. Security forces will usually enhance their security presence by acquiring additional manning or by adjusting the work-rest ratio (such as moving from a 3:1 to a 6:1 ratio). At Charlie, off-installation travel should be minimized.
- **THREATCON Delta.** Since the transition to THREATCON Delta is immediate, Delta measures should primarily focus on manning adjustments and procedural changes. THREATCON Delta applies in the immediate area where a terrorist attack has occurred or when intelligence has been received that terrorist action against a specific location or person is likely. The security force's manning level is usually peaked in Charlie; therefore, Delta's additional manning will usually come from an augmentation force. Once in Delta, nonessential operations will cease in order to enhance the security and response

posture. Normally, this THREATCON is declared as a localized condition.

K-3. With exception of THREATCON Normal, all THREATCON levels have certain measures associated with them. These measures are listed in JP 3-07.2, Appendix J, and AR 525-13.

ASSESSING VULNERABILITY

K-4. A VA addresses the consequences of terrorist attacks in terms of the ability of units, installations, commands, or activities to accomplish their assignments successfully, even if terrorists have inflicted casualties or destroyed or damaged DOD assets. The VA focuses on two broad areas—

- Preventing and, failing that, substantially mitigating the effects of a terrorist act.
- Maintaining emergency preparedness and crisis response.

K-5. The VA provides the commander with a tool to assess the potential vulnerability of an installation, base, unit, or port activity; but it is not a substitute for sound judgment. The VA must stand on its own and be supported by valid considerations. Typically, a small group of knowledgeable individuals develop the VA. The VA team consists of personnel with required areas of expertise. Some of these team members are the—

- Assessment-team chief.
- Physical-security specialist.
- Structural engineer.
- Infrastructure engineer.
- Operations-readiness specialist.
- Intelligence and/or counterintelligence specialist.

K-6. The functions and responsibilities of each team member are outlined in JP 3-07.2. The following paragraphs provide basic information regarding these areas:

- The assessment-team chief's key responsibilities include overseeing the management, training, and performance of the vulnerability-team members; finalizing the assessment-team out briefing; and preparing the population dynamics and risk assessment.
- The physical-security specialist is responsible for the security and safety of the installation, facility, and personnel.
- The structural engineer examines a variety of potential terrorist weapon effects and structural responses. This function serves to better protect personnel from shocks and blasts by reducing damage through the technically appropriate use of standoff measures, hardening, blast shielding, and shatter-resistant window film (such as Mylar®) as described in Chapter 3. The structural engineer's main responsibility is threat and damage assessment from terrorist weapons estimates and suggestions for threat protection or damage-mitigation measures.
- The infrastructure engineer examines protection against the effects of WMD, protection against terrorist-incident induced fires, and utility systems that can be used to minimize terrorist-incident casualties

(including elements of power, environmental control, and life-support systems). The primary responsibilities include infrastructure security and fire, safety, and damage control.

- The operations-readiness specialist examines plans, procedures, and capabilities for crisis response, consequence management, and recovery operations should a terrorist incident occur. The main responsibilities of this position include emergency-medical and individual-readiness assessments.
- The intelligence and/or counterintelligence specialist has the primary responsibility of performing logical analyses and preparing possible conclusions regarding terrorist targets and target vulnerabilities. These are based on processed intelligence information and knowledge of terrorist capabilities and methods in view of US installation, facility, and personnel safety and security practices.

CONDUCTING THE ASSESSMENT

K-7. Upon its arrival, the assessment team provides an in briefing for the commander, staff, and designated technical point of contact. Site personnel should conduct a site-familiarization briefing and tour. Administrative activities may include establishing the team support area, setting up equipment, scheduling team and technical points of contact meetings and discussions, ensuring classified-material control, establishing a personnel locator, and organizing materials for the out briefing and site folder. Each assessment-team member conducts the assessment based on the specific responsibilities for each function within his area.

POST-ASSESSMENT ACTIVITIES

K-8. Within 30 days of the visit, a summary narrative report and an annotated briefing should be delivered to the installation commander. Follow-on assistance for the commander may be applicable in areas of technical characteristics of improvement options, cost estimates, and generic sources of materials and equipment.

DRILLS AND EXERCISES

K-9. Multiechelon war gaming of possible terrorist attacks is the best test, short of an actual incident, to analyze the response of an installation, base, unit, or port. Drills and exercises test suspected vulnerabilities and antiterrorist measures. These exercises and drills also train the staff as well as reaction-force leadership and help maintain a valid threat assessment by identifying and adjusting to changing threat capabilities.

Glossary

AA&E arms, ammunition, and explosives

ABCS Army Battle Command System

AC alternating current

admin administration

ADP automated data processing

AF Air Force

AFB Air Force base

AFM Air Force manual

AFMAN Air Force manual

AFOSI Air Force Office of Special Investigations

AFR Air Force regulation

AIQC Antiterrorism Instructor Qualification Course

AIS automated information system

AL Alabama

AM amplitude modulated

AMC Army Materiel Command

AMS Army management structure

AO area of operations

AP armor piercing

Apr	April
AR	Army regulation
AR-PERSCOM	Army Reserve Personnel Command
ASP	ammunition supply point
AT	antitank
AT/FP	antiterrorism/force protection
attn	attention
Aug	August
AWG	American wire gauge
bldg	building
BMS	balanced magnetic switch
BTO	barbed-tape obstacle
BUPERS	Bureau of Naval Personnel
C^2	command and control
C^3	command, control, and communications
cav	cavalry
CB	citizen's band
CCB	Community Counterterrorism Board
CCIR	commander's critical information requirements
CCTV	closed-circuit television
CD-ROM	compact-disk, read-only memory

cdr	commander
CG	command guidance
chap	chapter
CIA	Central Intelligence Agency
CID	Criminal Investigation Division
CINC	commander in chief
CISO	counterintelligence support officer
CMU	concrete-masonry unit
CONEX	container express
CONPLAN	contingency plan
CONUS	continental United States
CP	command post
CPWG	crime-prevention working group
CQ	charge of quarters
CRIMP	Crime Reduction Involving Many People
CTA	common table of allowance
DA	Department of the Army
DARE	Drug Abuse Resistance and Education
DC	direct current
DC	District of Columbia
Dec	December

DIA	Defense Intelligence Agency
DOD	Department of Defense
DOE	Department of Energy
DOJ	Department of Justice
DOS	Department of State
DOT	Department of Transportation
DS	direct support
DTM	data-transmission media
DTOC	division tactical operations center
EDM	emergency-destruct measures
EECS	electronic entry-control system
EOD	explosive-ordnance disposal
EOR	element of resource
EPW	enemy prisoner of war
equip	equipment
ESS	electronic security system
FAA	Federal Aviation Administration
FBI	Federal Bureau of Investigation
FCC	Federal Communications Commission
Feb	February
FIS	foreign-intelligence services

FM	field manual
FM	frequency-modulated
ft	foot, feet
FY	fiscal year
G2	Assistant Chief of Staff, G2 (Intelligence)
GHz	gigahertz
GTA	graphic training aid
HN	host nation
HQ	headquarters
HUD	Housing and Urban Development Administration
HUMINT	human intelligence
Hz	hertz
IAW	in accordance with
ICP	initial control point
ID	identification
IDS	intrusion-detection system
IED	improvised explosive device
IG	inspector general
IID	improvised incendiary device
in	inch(es)
INSCOM	US Army Intelligence and Security Command

INTAC Individual Terrorism Awareness Course

IPB intelligence preparation of the battlefield

IR infrared

ISS information systems security

J2 Intelligence Directorate (Joint Command)

Jan January

JS Joint Service

JSAT Joint Security Assistance Training

JSCP Joint Strategic Capabilities Plan

JSIIDS Joint-Service Interior Intrusion-Detection System

Jul July

Jun June

K one thousand

kHz kilohertz

LED light-emitting diode

liq liquid

LOS line of sight

LOTS logistics over the shore

LP listening post

LRA local reproduction authorized

MACOM major Army command

maint	maintenance
Mar	March
MCO	Marine Corps order
MDEP	management decision package
MDMP	military decision-making process
METT-TC	mission, enemy, terrain, troops, time available, and civilian considerations
MEVA	mission-essential or vulnerable area
MI	military intelligence
MILCON	military construction
MILPO	military personnel office
MILVAN	military van
min	minimum
mm	millimeter(s)
MO	modus operandi
MOU	memorandum of understanding
MP	military police
MPACS	Military Police Automated Control System
MPMIS	Military Police Management Information System
MPR	military-police report
MS-DOS	Microsoft®-disk operating system
MWD	military working dog

N/A	not applicable
naut	nautical
NAVATAC	Navy Antiterrorism Analysis Center
NBC	nuclear, biological, and chemical
NCIC	National Crime Information Center
NCO	noncommissioned officer
NISCOM	Naval Investigative Service Command
No.	number
Nov	November
NSA	National Security Agency
NTAV	nontactical armored vehicle
NVD	night-vision device
O	official
OASD	Office of the Assistant Secretary of Defense
OCOKA	observation and fields of fire, cover and concealment, obstacles, key terrain, and avenues of approach
OCONUS	outside the continental United States
Oct	October
OMA	operations and maintenance, Army
OPA	operations and procurement, Army
OPLAN	operation plan
OPORD	operations order

OPSEC operations security

Pam pamphlet

PAO public affairs office(r)

PD probability of detection

PERSCOM Personnel Command

PHOTINT photographic intelligence

PI police intelligence

PIR passive infrared

PM provost marshal

PMO provost marshal office

POL petroleum, oil, and lubricants

POM Program Objective Memorandum

POV privately owned vehicle

PS physical security

PSD protective security detail

PSI physical-security inspector

PTO Parent-Teacher Organization

PX post exchange

R&D research and development

RD&E research, development, and engineering

ref reference(s)

RF radio frequency

RII relevant information and intelligence

ROI report of investigation

RORO roll on/roll off

RPG rocket-propelled grenade

/s/ signed

S2 Intelligence Officer (US Army)

SAW squad automatic weapon

SDNCO staff duty noncommissioned officer

SDO staff duty officer

Sep September

SIGINT signal intelligence

SJA staff judge advocate

SO/LIC Special Operations/Low-Intensity Conflict

SOFA status of forces agreement

SOP standing operating procedure

St. Saint

STANO surveillance, target acquisition, and night observation

stat statute

STC sound-transmission coefficient

STD standard

STU secure telephone unit

TAACOM	Theater Army Area Command
TB	technical bulletin
TDY	temporary duty
TECOM	US Army Test and Evaluation Command
TEMPEST	Terminal Electromagnetic-Pulse Emanation Standard
THREATCON	threat conditions
TM	technical manual
TMDE	test, measurement, and diagnostic equipment
TNT	trinitrotoluene
TRADOC	US Army Training and Doctrine Command
TSC	triple-standard concertina
UCMJ	Uniform Code of Military Justice
UFR	unfinanced requirement
US	United States
USACE	US Army Corps of Engineers
USACIDC	US Army Criminal Investigation Command
USAMPS	US Army Military Police School
USCG	US Coast Guard
VA	vulnerability assessment
VIP	very important person
vol	volume
WMD	weapons of mass destruction

Bibliography

AR 12-15. *Joint Security Assistance Training (JSAT) Regulation* (Navy Instructions 4950.4; AFR 50-29). 28 February 1990.

AR 50-5. *Nuclear and Chemical Weapons and Material—Nuclear Surety.* 3 October 1986.

AR 50-6. *Nuclear and Chemical Weapons and Materiel, Chemical Surety.* 1 February 1995.

AR 190-11. *Physical Security of Arms, Ammunitions, and Explosives.* 30 September 1993.

AR 190-12. *Military Police Working Dogs.* 30 September 1993.

AR 190-13. *The Army Physical Security Program.* 30 September 1993.

AR 190-14. *Carrying of Fire Arms and Use of Force for Law Enforcement and Security Duties.* 12 March 1993.

AR 190-22. *Searches, Seizures and Disposition of Property.* 1 January 1983.

AR 190-27. *Army Participation in National Crime Information Center.* 28 May 1993.

AR 190-51. *Security of Unclassified Army Property (Sensitive and Nonsensitive).* 30 September 1993.

AR 190-56. *The Army Civilian Police and Security Guard Program.* 21 June 1995.

AR 190-59. *Chemical Agent Security Program.* 27 June 1994.

AR 380-5. *Department of the Army Information Security Program.* 25 February 1988.

AR 380-10. *Technology Transfer, Disclosure of Information and Contacts with Foreign Representatives.* 30 December 1994.

AR 380-19. *Information Systems Security.* 27 February 1998.

AR 380-67. *The Department of the Army Personnel Security Program.* 9 September 1988.

AR 381-20. *The Army Counterintelligence Program.* 15 November 1993.

AR 405-20. *Federal Legislative Jurisdiction.* 1 August 1973.

AR 525-13. *Antiterrorism Force Protection (AT/FP): Security of Personnel, Information, and Critical Resources.* 10 September 1998.

AR 600-8-14. *Identification Cards for Members of the Uniformed Services, Their Family Members, and Other Eligible Personnel* (AF 36-3026[I]; BUPERS I 1750.10A; MCO P5512.1B; CG M5512.1; Manual 29.2, Instructions 1 and 2). 1 March 1998.

AR 604-5. *Personnel Security Clearance, Department of the Army Personnel Security Program Regulation* 1 February 1984.

DA Form 2028. *Recommended Changes to Publications and Blank Forms.* 1 February 1974.

DA Form 2806-R. *Physical Security Survey Report.* April 1985.

DA Form 2806-1-R. *Physical Security Inspection Report (LRA).* April 1985.

DA Form 2819. *Law Enforcement and Discipline Report.* May 1988.

DA Form 3975. *Military Police Report.* May 1988.

DA Pam 190-12. *Military Working Dog Program.* 30 September 1993.

DA Pam 190-51. *Risk Analysis for Army Property.* 30 September 1993.

DOD 0-2000.12-H. *Protection of DOD Personnel and Activities Against Acts of Terrorism and Political Turbulence.* February 1993.

DOD 4160.21-M. *Defense Reutilization and Marketing Manual.* August 1997.

DOD Directive 2000.12. *DOD Antiterrorism/Force Protection (AT/FP) Program.* 13 April 1999.

DOD Instruction 2000.16. *DOD Combating Terrorism Program Standards.* 10 May 1999.

DOD Instruction 5210.84. *Security of DOD Personnel at US Missions Abroad.* 22 January 1992.

Federal Specification FF-L-2740. *Locks, Combination.* 12 October 1989.

FM 5-34. *Engineer Field Data.* 30 August 1999.

FM 19-4. *Military Police Battlefield Circulation Control, Area Security, and Enemy Prisoner of War Operations.* 7 May 1993.

FM 55-20. *Army Rail Transport Units and Operations.* 1 June 00.

GTA 19-4-3. *Individual Protective Measures.* July 1997.

JP 3-07.2. *Joint Tactics, Techniques, and Procedures for Antiterrorism.* 17 March 1998.

JS Guide 5260. *Service Member's Personal Protection Guide: A Self-Help Handbook to Combating Terrorism.* July 1996.

TB 9-2300-422-20. *Security of Tactical Wheeled Vehicles.* 27 August 1988.

TM 5-805-8. *Builders' Hardware.* 20 January 1992.

TM 5-820-4. *Drainage for Areas Other Than Airfields* (AFM 88-5, Chap 4). 14 October 1983.

TM 5-853-1. *(O) Security Engineering Project Development* (AFMAN 32-1071, Vol 1). 12 May 1994.

TM 5-853-2. *(O) Security Engineering Concept Design* (AFMAN 32-1071, Vol 2). 12 May 1994.

TM 5-853-3. *(O) Security Engineering Final Design* (AFMAN 32-1071, Vol 3). 12 May 1994.

TM 5-853-4. *Security Engineering Electronic Security Systems.* 12 May 1994.

USACE STD 872-50-01. *Entry Points for US Army Installations.* To be published within the next six months.

USACE STD 872-90-02. *FE5 Chain-Link Security Fence Details.* May 1992.

USACE STD 872-90-03. *FE6 Chain-Link Security Fence Details.* May 1992.

USACE STD 872-90-04. *FE7 Chain-Link Security Fence Details for Non-Sensored Fence.* May 1992.

USACE STD 872-90-05. *FE7 Chain-Link Security Fence Details for Sensored Fence.* May 1992.

USACE STD 872-90-06. *FE8 Chain-Link Security Fence Details for Sensored Fence.* May 1992.

USACE STD 872-90-07. *FE5 Chain-Link Fence Gate Details.* May 1992.

USACE STD 872-90-08. *FE6 Chain-Link Fence Gate Details.* May 1992.

USACE STD 872-90-09. *FE8 Chain-Link Fence Gate Details.* May 1992.

Index

Field Manual
No. 3-19.15

Headquarters
Department of the Army
Washington, DC, 18 April 2005

Contents

Distribution Restriction: Approved for public release; distribution is unlimited.

*This publication supersedes FM 19-15, 25 November 1985.

Page

Preface

Field Manual (FM) 3-19.15 addresses continental United States (CONUS) and outside continental United States (OCONUS) civil disturbance operations. Today, United States (US) forces are deployed on peacekeeping, peace enforcement, and humanitarian assistance operations worldwide. During these operations, US forces are often faced with unruly and violent crowds intent on disrupting peace and the ability of US forces to maintain peace. Worldwide instability coupled with increasing US military participation in peacekeeping and related operations requires that US forces have access to the most current doctrine and tactics, techniques, and procedures (TTP) necessary to quell riots and restore public order.

In addition to covering civil unrest doctrine for OCONUS operations, FM 3-19.15 addresses domestic unrest and the military role in providing assistance to civil authorities requesting it for civil disturbance operations. It provides the commander and his staff guidance for preparing and planning for such operations. The principles of civil disturbance operations, planning and training for such operations, and the TTP employed to control civil disturbances and neutralize special threats are discussed in this manual. It also addresses special planning and preparation that are needed to quell riots in confinement facilities are also discussed.

In the past, commanders were limited to the type of force they could apply to quell a riot. Riot batons, riot control agents, or lethal force were often used. Today, there is a wide array of nonlethal weapons (NLW) available to the commander that extends his use of force along the force continuum. This manual addresses the use of nonlethal (NL) and lethal forces when quelling a riot.

Appendix A complies with current Army directives, which state that the metric system will be incorporated into all new publications.

The proponent for this publication is HQ, TRADOC. Send comments and recommendations on Department of the Army (DA) Form 2028 (*Recommended Changes to Publications and Blank Forms*) directly to Commandant, US Army Military Police School (USAMPS), ATTN: ATSJ-DD, 401 MANSCEN Loop, Fort Leonard Wood, MO 65473-8929.

Unless this publication states otherwise, masculine nouns or pronouns do not refer exclusively to men.

Chapter 1

Operational Threats of the Civil Disturbance Environment

As the Cold War ended with the fall of the Berlin Wall and the collapse of Soviet communism in 1989, the Army was no longer concerned with containing the aggressive Soviet Union. The US Army began to transform because the battlefield on which it would fight had changed. Operations Just Cause, Joint Endeavor, Desert Shield, and Desert Storm have all been recognized as containing nontraditional battlefields. Following these operations, the Army was called upon to conduct peacekeeping, peace enforcement, and humanitarian assistance operations. These operations required US forces to protect refugees and humanitarian workers and keep hostile factions separate to enforce peace.

GENERAL CAUSES FOR CIVIL UNREST

1-1. In these modern times, demonstrations, civil unrest, public disorder, and riots happen for a number of reasons. Some of these reasons are economic hardships, social injustices, ethnic differences (leading to oppression), objections to world organizations or certain governments, political grievances, and terrorist acts. An event can be triggered by a single cause or a combination of causes. For example, operations in the Balkans involving civil unrest and riots were the result of ethnic hatred, a lack of civil authority, food shortages, a revolution, and religious-based fighting factions.

1-2. Demonstrations may range from simple, nonviolent protests that address specific issues, to events that turn into full-scale riots. Gatherings in protest are recognized rights of any person or group, regardless of where US forces may be operating. This fundamental right is protected under the *Constitution of the US*. During peacekeeping or peace enforcement operations, US forces should never violate basic civil or human rights. Most protesters are law-abiding citizens who intend that their protests be nonviolent, but some protest planners insist that the event involve some kind of violence. Often, in the media, protesters can gain sympathy for their cause by prompting authorities to take physical action against them. Violence is often the result of demonstrators beginning to conduct unlawful or criminal acts and authorities (who are responsible for the safety and welfare of all) enforcing the laws of the municipality, state, or nation. The depth of violence is determined by the willingness of demonstrators to display and voice their opinions in support of their cause.

1-3. Commanders must be aware of the possibility that some individuals or groups within an organized demonstration may have the intent to cause disruption, incite violence, destroy property, and provoke the authorities. The tactical situation and actions of the crowd should dictate control and enforcement options. Agitators and criminal infiltrators within the crowd can lead to the eruption of violence. Inciting a crowd to violence or a greater intensity of violence through the use of brutish enforcement tactics should be avoided. Publicity can be detrimental to authorities and beneficial to crowds because it can further their causes.

1-4. Community unrest results in urban conflicts that arise from highly emotional social and economic issues. Economically deprived inner city residents may feel that they are treated unjustly or ignored by people in power and authority. Tensions can build quickly in a community over a variety of issues, such as hunger, poor employment opportunities, inadequate community services, poor housing, and labor issues. Tension in these areas creates the potential for violence. When tensions are high, it takes a small (seemingly minor) incident, rumor, or act of injustice to ignite groups within a crowd to riot and act violently. This is particularly true if community relations with authorities are part of the problem.

1-5. Significant ethnic differences in a community can create an atmosphere of distrust, even hatred. Unrest among ethnic groups competing for jobs, living areas, and sparse essentials can cause an eruption of civil disorder and/or riots. As emotions run high, violence becomes likely.

1-6. Terrorist organizations may infiltrate groups within a demonstrating crowd. These terrorist groups may intend to embarrass their government or other governments. Terrorist infiltrators can be used to provoke crowds as a diversion, as part of a demonstration, or as cover for terrorist acts.

CROWD DEVELOPMENT

1-7. Crowds are a gathering of a multitude of individuals and small groups that have temporarily assembled in the same place. These small groups are usually comprised of friends, family members, or acquaintances that represent a group belief or cause. Individuals assume a sense of anonymity—they are viewed as just another face in the crowd. People in small groups are known only to companions in their group and to others in the gathering that have come from the same neighborhood or community. Commanders must consider how the individuals assembled and how they are interacting during the gathering process.

GATHERINGS

1-8. The assembly process of a gathering refers to the movement of people from different locations to a common location within a given period. This largely determines who participates. Creating a gathering from a crowd is a process with a beginning, middle, and end. As shown in *Figure 1-1* a crowd has an assembly process that leads to the gathering of the crowd, which is always followed by a dispersal process.

1-9. Gatherings are often assisted by the activities of individuals or groups with a specific agenda, such as yelling catchy slogans and cheers that everyone can easily pick up and join in on. Some groups are so well organized that they can prestage leaders to infiltrate a gathering. This creates unity, even inciting newcomers to join their cause. The first phase of the gathering is the assembly process. It can occur in one of two ways—impromptu or organized.

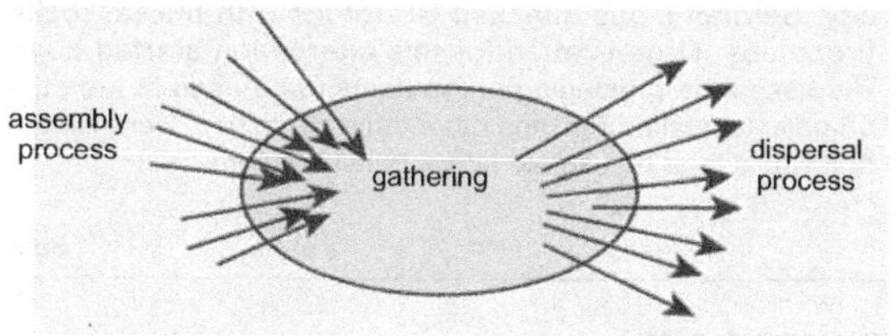

Figure 1-1. Crowd Building

IMPROMPTU GATHERINGS

1-10. Impromptu gatherings usually develop informally and are mostly done by word of mouth (one person to another or one group to another). Participants spread intelligence by telling one another when, where, and what is happening and inviting them to participate. An example of an impromptu gathering would be a gathering at a secured food distribution point after receiving information (by word of mouth) that a large truck carrying much-needed supplies and food is about to arrive. In this case, hunger would be the driving force causing the migration of people to the food distribution facility.

ORGANIZED GATHERINGS

1-11. Like an impromptu gathering, an organized gathering can also involve individuals and groups passing on information to one another. Passing intelligence of police activity and occurrences and on when and where events will take place helps organizers to prestage participants. Rarely is one represented group responsible for pulling together a gathering. Organized gatherings rely heavily on established groups that attract people to gather. Recent examples of these well-organized groups are anarchists, anti-globalization groups, and anti-free-enterprise groups. Groups representing extreme religious faiths and ethnic organizations have been common too. Some gatherings involve groups from rival, or even warring, factions.

1-12. Organized gathering processes rely more on centralized planning and organization. One or more of the groups offer organizers lists of individual names and groups, which they contact as potential participants. Modern technologies are available to these organizers through telephone banks, mailing lists, or e-mail addresses. In preparation for a long-term event, some group organizers have the means to provide logistical support of transportation, food, and water to participants.

> During the preelection period, Bosnian-Serbs (in a bid not to lose any of the territory they had won during the war) began to organize this fiercely contested city. Serbian mobs attacked US forces with bricks, rocks, metal bars, and firebombs. Organization for this altercation started days before, as many witnessed the prestaging of the items to be used in the attack. Wheelbarrows, 55-gallon metal drums, and other large containers were filled with the ammunition that would later be used to attack peacekeepers.
>
> Brcko, Bosnia
> September 1996

CROWD BUILDING

1-13. The second phase of the gathering process is the building of the crowd. Old concepts of crowds based on stereotypes are no longer true. Studies show that—

- Crowds are made of many entities, and all participants are not the same.
- Crowds are not made of isolated individuals but of a minority of individuals and a majority of small groups who may or may not be acquainted with each other.
- Groups and individuals in the crowd are not unanimous in their motivation.
- Groups and individuals in the crowd are usually not anonymous to each other, especially different groups gathered at the same place.
- Individuals within a crowd are not given to unique emotional displays; although known within a group, most individuals would not want to draw attention to themselves.
- Groups within a crowd will often act in unison in an attempt to achieve their agendas.
- Crowds are generally not unique or distinguished by violence or unlawful conduct.

DISPERSAL PROCESS OF A GATHERING

1-14. The last and final phase of any gathering of a crowd is the dispersal process. It involves the movement of people from the common location where they assembled to one or more alternate locations. The dispersal process ends the gathering of a crowd or at least begins its decline. The dispersal can occur on a routine, emergency, or coerced basis.

ROUTINE DISPERSAL

1-15. The routine dispersal may be specified in advance. It can also be included in the assembly instructions given by the organizers of an event.

EMERGENCY DISPERSAL

1-16. An emergency dispersal occurs when people evacuate an area in result of an unexpected crisis, such as a fire, explosion, bomb threat, or terrorist act. Individuals in such an emergency quickly recover from the initial shock. They keep their wits about them and improvise ways of extricating themselves and their companions from the dangerous situation. However, tragedies have occurred in some emergency dispersal situations. Therefore, it is important that forces on the scene are careful not to misinterpret what is occurring.

COERCION DISPERSAL

1-17. Coercion dispersal is caused by the use of force at some level. This is not necessarily the best way to force the dispersal of a crowd. The negotiated management of crowds is the preferred method and has proven to be highly successful in getting crowd organizers to police themselves, especially if the demonstration and/or protest leaders are available and willing to participate.

CROWD DYNAMICS

1-18. Understanding crowds and how individuals and groups form crowds through the gathering process is important because these issues apply to the dynamics of a crowd. Under most circumstances, gathered crowds are orderly and present little or no problems for authorities. Crowds consist of people who, although very motivated and passionate, are also subject to their own need for creature comforts. Rainy, cold, and nasty weather has a way of disheartening all but the few highly motivated and disciplined individuals. If problems exist, they usually fall into the following three categories:

- **Public disorder.** Public disorder is a basic breach of civic order. Individuals or small groups assembling have a tendency to disrupt the normal flow of things around them.
- **Public disturbance.** Public disturbance is designed to cause turmoil on top of the disruption. Individuals and groups assembling into a crowd begin chanting, yelling, singing, and voicing individual or collective opinions.
- **Riot.** A riot is a disturbance that turns violent. Assembled crowds become a mob that violently expresses itself by destroying property, assaulting others, and creating an extremely volatile environment.

1-19. Being part of a crowd of people has certain effects on different people. Each individual in a crowd is susceptible to behaving in a way that is contrary to their normal behavior. Some reasons for these behaviors are as follows:

- Crowds provide individuals with a sense of anonymity. With so many others, an individual realizes that he is just another face in the crowd, giving a sense of invulnerability.
- Crowd and individual behaviors are impersonal by nature. The "them-against-us" attitude affords those within the crowd the ability to freely (without hesitation or reservation) be verbally abusive, throw objects, or attack anyone who gets in the way.

- Crowds provide individuals with the idea that their moral responsibilities have shifted from themselves as a person to the crowd as a whole. Large numbers of people discourage individual behavior, and the urge to imitate others within the crowd is strong. Individuals look to others around them for cues of what to do next, disregarding their own background and training. Often, it is only the strong, well-disciplined person who can resist the prevailing behavior of a crowd.
- Crowd behavior influences the actions of both the disorderly individuals of the crowd and the authorities tasked to control them.

1-20. Individuals within a crowd are at times driven by deep-felt emotions. Emotional contagion is the most dramatic psychological factor of crowd dynamics. It provides the crowd with a temporary bond of psychological unity. Lasting long enough, this unity can push a simple organized crowd into a mob. Normal law and authority are rejected en masse under these conditions, increasing the potential for violence and panic to erupt.

1-21. Panic can erupt quickly, especially when crowds turn into mobs. Individuals within the mob can easily sense that their safety and well-being are at risk, putting them in a "fight-or-flight" mentality. Adding to the panic and confusion is the use of riot control agents (RCAs) by authorities in an attempt to gain control. Individuals in a mob, during the heat of confrontation, may attempt to leave and find that there are no escape routes and that roads are blocked. This can often lead to violent, physical attacks.

CROWD TYPES

1-22. Gaining an understanding of crowds and the dynamics that cause individuals or groups to join together in a crowd should be every leader's responsibility. Active, expressive, acquisitive, and hostile are not acceptable terms used to describe a crowd. They are the motives inferred from the actions the crowd takes. No gathering of a crowd is persistently or exclusively active, expressive, acquisitive, or hostile. For the sake of description, crowds can be identified as casual, sighting, agitated, or mob-like.

CASUAL CROWDS

1-23. Casual crowds are identified as individuals or small groups with nothing in common to bind them together. If they have an agenda, it is their own. They arrive separately and leave separately. Casual crowds are made up of individuals or small groups occupying the same common place, such as a shopping mall where these individuals or small groups meet.

SIGHTING CROWDS

1-24. Sighting crowds are similar to casual crowds with one additional element—an event. People migrate as a crowd to sporting events, are attracted to fires and accidents, and attend music concerts. Individuals or small groups gather at these events for the same purpose. It is the event and/ or one's curiosity that compels a crowd to come together.

AGITATED CROWDS

1-25. Agitated crowds add responses that are based on the elements (people, space, and event). Individuals with strong emotional feelings within a crowd can quickly spread and infect the rest of the crowd. As more people within the crowd become emotionally involved, a sense of unity can develop, causing changes in the overall demeanor of the crowd. Yelling, screaming, crying, and profane name-calling are all associated with an agitated crowd.

MOB-LIKE CROWDS

1-26. Mobs have all the elements found in the first three types of crowds, with the addition of aggressive, physical, and sometimes violent actions. Under these conditions, individuals within a crowd will often say and do things they usually would not. Extreme acts of violence and property damage are often a part of mob activities. Mobs consist of (or involve) the elements of people and groups being mixed together and becoming fluid.

CROWD TACTICS

1-27. During a public disorder or disturbance, individuals and small groups within a crowd use any number of tactics to resist authority and disrupt and add turmoil in order to achieve their goals. These tactics can be unplanned or planned and violent or nonviolent. The more organized and purposeful a crowd becomes, the more likely a tactic will be used.

NONVIOLENT BEHAVIOR

1-28. As explained earlier in this chapter, most gatherings of individuals and small groups into a crowd do not involves violent behavior. A public disorder or disturbance usually involves some harmless name-calling, demonstrations to express views, corporate yelling and chanting, and even singing and dancing. In some instances, there may be property damage and the erection of makeshift barricades. All four types of crowds can be nonviolent, but mobs and agitated crowds have the greatest tendency to turn violent. Nonviolent actions of a crowd are disruptive because they are in direct conflict with what authorities want them to do, such as refusing to leave when directed, locking arms, and sitting in front of or around areas and buildings that the authorities are attempting to clear.

VIOLENT BEHAVIOR

1-29. A crowd that becomes a mob can be very violent and destructive. Although some nonviolent activities occur, violent crowds strike out physically at bystanders or others in the crowd, destroying both private and government property, setting fires, and employing bombs (in extreme cases). The only limitations for violent crowds are their own imaginations, the training of their leaders, and the materials readily available.

1-30. Violent crowds may erect barricades and physical barriers to impede the movement of authorities and prevent them from entering certain areas and/or buildings and to hide their activities. Common materials used for barricades include vehicles, felled trees, furniture, felled light poles, fencing materials, or anything else on hand and readily available.

1-31. Setting fires and preventing firefighters access to the structure by blocking it is another violent crowd tactic. Often, fires are set to create confusion in authorities or as a diversion to another activity.

RIOTS

1-32. A riot is one or more groups or individuals who are part of a larger crowd that involves threats of violence against persons or property. In some cases, a crowd will continue to gather until it evolves into a riot.

1-33. Riots vary considerably in both targets and players. A communal riot, for instance, deals with deep-seated ethnic, religious, and language differences. Commodity riots involve an attack on property by acts of vandalism, looting, and arson. Protest riots, such as the riot around the World Trade Organization (WTO) Assembly in Seattle, Washington (30 November 1999), illustrated individuals and groups aggressively and sometimes violently acting out or voicing their opposition to the assembly. The Democratic Convention in Chicago, Illinois, in 1968 reflects a riot that directly targeted police and authority in general. Celebration riots occur across the US as a result of home team victories in sporting events, among other reasons. Celebrating crowds look to make the moment more memorable through raucous acts that demonstrate their joy or happiness, for example, the riots that took place in Chicago in 1992 as a result of the Chicago Bulls winning the National Basketball Association (NBA) Championship.

1-34. Commanders and leaders must be aware that highly organized groups of protesters and/or demonstrators have developed tactics to disrupt the control force. These crowd tactics were published in handbooks for communist organizers during the Cold War, and today the Internet contains sites that are devoted to sharing these tactics with the general public and various protest groups. Although these tactics are somewhat outdated, commanders and leaders must be aware of them and their potential consequences. Many of these groups, such as anarchists, often mock the tactics employed against large crowds; for example, they may identify the police as "Darth Vader Cops" when the control force response is in full riot control gear. Commanders must be aware that well-organized crowds may attempt to engage the control force, surround it, and overpower it by sheer numbers. To avoid this, commanders must do a detailed terrain analysis (intelligence preparation of the battlefield [IPB]) of the area to include all approach and exit routes in the area. The control force flanks have to be protected to avoid being enveloped by the mob. To avoid being enveloped, the on-site commander should keep the crowd at a comfortable distance from the control force formation, which is usually the lethal zone of his NL munitions. Establishing predetermined rally points for the control force is critical in the event that this type of tactic is used by the crowd.

TACTICS USED TO DEFEAT AUTHORITIES

1-35. Organized mobs will try to defeat the control force by employing several different types of tactics. These tactics include the following:

- Constructing barricades.
- Using Molotov cocktails, smoke grenades, rocks, and slingshots.
- Feinting and flanking actions.

BARRICADES

1-36. By constructing barricades, the rioters are trying to protect themselves from assault by the control force. If the control force does advance toward the barricade, rioters hope that the barricade will split the control force and leave it vulnerable to actions by the mob.

MOLOTOV COCKTAILS, SMOKE GRENADES, ROCKS, AND SLINGSHOTS

1-37. Angry mobs will often attempt to disrupt the control force by throwing rocks and other projectiles. As the situation deteriorates, the mob may escalate the violence by using a battery of slingshots that will pellet the control force with a barrage of projectiles ranging from small rocks to marbles. They may also use smoke grenades (homemade or store-bought) to mask their movements. Finally, Molotov cocktails are used against personnel, employed vehicles, and portions of the control force. The mob will attempt to gain a position above an armored vehicle to enable them to drop a Molotov cocktail into an open hatch.

FEINTING AND FLANKING ACTIONS

1-38. Organized mobs may attempt to disrupt the movement of the control force by feinting an assault. When the control force moves to blunt the assault, the mob will assault the exposed flanks in an attempt to split the control force and envelop a portion of it.

WEAPONS

1-39. Mobs will often use various types of weapons against authorities to achieve their agendas and goals. It is extremely important that leaders train soldiers to recognize possible threats of the various weapons used by crowds or mobs, enabling the soldiers to react with the appropriate minimum force.

VERBAL ABUSE

1-40. In almost every instance of a civil disturbance or riot, verbal abuse will be an aggressive tool. Obscene language, racial remarks, taunts, ridicules, and jeers should always be expected. It is apparent that the purpose for using verbal abuse is to anger, demoralize, and provoke a physical response. Undisciplined, untrained soldiers who face such an attack could cause the situation to escalate. Just one provoked action of a soldier could be interpreted as an act of brutality by the media.

PERCEIVED INNOCENT PEOPLE OR THE WEAK

1-41. Women, children, and the elderly have often been used as pawns in civil disturbance operations. On many occasions, the innocent and the weak are moved to the front of the crowd and used as a barrier. With the innocent and the weak directly facing authorities, aggressive and violence-prone individuals or groups behind them attempt to provoke authorities to react. Often, this tactic is to exploit the situation in the eyes of the media for propaganda value. Despite the perception of sometimes being weaker, women, children, and the elderly can be just as committed to the insurgency as those behind them.

BARRICADES

1-42. Barricades are made of anything large or heavy enough to impede or prevent the movement of authorities. Trees, vehicles, fires, and furniture are all examples of items used to erect a barricade.

THROWN OBJECTS

1-43. In many cases, a crowd will throw anything in their reach that can be picked or picked up. Rotten fruit and/or vegetables, rocks, bricks, bottles, sticks, and pieces of lumber are most often used because they are readily available. More sophisticated types of thrown objects are improvised bombs, such as petrol bombs or tennis balls with nails sticking out of them. Never discount that these individuals and groups could acquire hand-thrown explosives, such as sticks of dynamite or grenades.

VEHICLES AND OTHER OBJECTS

1-44. Vehicles and other objects can be placed on higher ground (atop a hill or in a building) to disrupt civil disturbance operations. For example, a vehicle or tire can be set on fire and rolled down a hill, carts or barrels filled with hazardous or flammable materials can be pushed down a hill, or burning furniture and petrol bombs can be thrown from two-story or higher buildings or toward a formation of soldiers.

FIRE

1-45. Rioters have set buildings, vehicles, and other structures on fire to block the advances of authorities and to create confusion and diversions. If property damage is the goal of a crowd, fire is an effective tool. A sniper, for example, may set a building on fire to divert attention or provide maximum target possibilities. Other tactics used, particularly in the Balkans, were to flood the lower floors or basements of structures with flammable liquids or gas and then ignite it in an opportune moment.

FIREARMS

1-46. Sniping or massing fire from within the crowd, in buildings, or other covered positions can (at times) be effective tools for terrorists using a crowd to cover their acts. Firearms are anything that fires a projectile toward a target of opportunity.

EXPLOSIVES

1-47. A bomber's imagination is the only limiting factor in the use of explosives. Explosives can be used as simple diversions to block the advances of formations or provide an escape for rioters. The worst use of explosives is to purposely kill, injure, or demoralize authorities that are protecting and assuring the safety of everyone.

Chapter 2

Control Force Operations

Information is the key to developing civil disturbance plans. One must know who the demonstrators are; when, where, and why they are demonstrating; what their capabilities are; and what their possible course of action (COA) is. A commander's need for current, valid information cannot be overemphasized. The commander must learn as much as he can about the participants (their motivations, strategies, tactics, targets, and dedication). The more knowledge a commander has about the participants, the better equipped he is to counter their actions. He needs sound information to decide how to best use his available resources.

CIVIL DISTURBANCE PLANNING CONSIDERATIONS

2-1. The side that possesses better information and uses that information to gain understanding has a major advantage over its opponent. Forces that have this advantage can use it to bring about changes in attitudes, decisions, and actions in assembled demonstrators. In planning for crowd control or civil disturbance operations, planners must decide what data is needed to develop threat assessments. This decision is based on a multitude of information sources.

STANDARDIZED POSTURES IN COMPANY TACTICAL STANDING OPERATING PROCEDURES

2-2. Within squads, platoons, and companies, equipment may be increased or decreased, as needed. Some examples are as follows:

- Choose the M9 pistol for extraction and apprehension teams. The use of a long weapon (for example, an M16 with an M203, an M4 with an M203, or a 12-gauge shotgun) with NL munitions capability is also recommended, especially for overwatch personnel.
- Add nonstandard weapons such as shotguns for greater NL capabilities.

NOTE: The shotgun is used to protect the M203 gunner as he reloads.

- Add NL munitions to existing organic weapons systems, such as the M203.
- Arm soldiers in the front line of the formation with their standard weapon. If the weapon is a long weapon, it should be carried across the back from left to right or vice versa, with the butt up and muzzle down. Ensure that the weapon is cleared and that the magazine is in the appropriate ammunition pouch.

- Balance the mix of weapons and munitions according to the mission, enemy, terrain, troops, time available, and civilian considerations (METT-TC).
- Establish command relations and determine who has or gives the authority to fire NL munitions.
- Maximize the distance and barriers between the crowd and control formations. Use NL munitions to create a standoff distance.
- Maintain a lethal overwatch of the control force at all times.
- Maintain a reserve force at all times to reinforce the control force, as needed.
- Create NLW range cards for static positions.
- Consider environmental conditions and their effect on the performance of NL munitions. Proper storage, periodic inspections, and the rotation of NL munitions are critical to maintaining the effectiveness and viability of NL munitions.

WEAPON AND AMMUNITION CONFIGURATIONS

2-3. Individuals designated as NL shooters must have the means to transition to lethal rounds, if required. Lethal rounds are carried separately from NL rounds so that the shooter will not confuse them in the heat of confrontation. However, in the heat of confrontation where a lethal option is necessary, NL rounds can be used lethally by adjusting the point of aim to a vulnerable part of the body and closing in on the distance to the target.

2-4. Squad leaders should designate NL shooters within their squad. Ideally, the squad should not change its organization to accommodate NLW.

2-5. Commanders should not dispatch NL patrols. They should plan a combat or security patrol with NL capabilities. Soldiers are never sent in harm's way without lethal protection. NLW are only considered additional tools for the mission and not a mission in itself. There is no such thing as a NL mission.

2-6. Soldiers manning a static position should have NL capabilities depending on METT-TC. Shotguns and the M203 work well at static positions.

2-7. Recovery, apprehension, and/or extraction teams should be established before deployment. Team members should be equipped with personal protective equipment (PPE), an M9, NL munitions and weapons, and Flex-Cufs®. They should have some training in open-hand control, pain compliance, and handcuffs and/or Flex-Cufs.

2-8. Crowd control formations (see *Chapter 6*) should be well trained and well rehearsed. Rapid, coordinated movements of a well-trained and well-rehearsed control force can often be a strong enough deterrent. A lethal overwatch marksman always covers the control force.

2-9. During a NL engagement, the use of designated marksmen (DM) provides confidence and safety to those facing a riot. The DM in an overwatch position scans the crowd to identify threats and designate personnel for recovery and lethal rounds firing (if a lethal threat is presented). They are ideally suited for flank security and countersniper operations.

NONLEHTAL PLANNING

2-10. The fundamental principles of mission planning are not changed by the incorporation of NLW into a plan. NLW are intended to be operational-enhancement tools. The first concern of a commander is the successful completion of the mission. NL munitions provide a commander with a wider range of response options, but these munitions are not a replacement for lethal capabilities. Commanders must always be capable of answering a lethal attack with a lethal response.

2-11. When a commander commits his soldiers and equipment to a crowd control situation, he commits his forces with the additional tools of NL capabilities. This is an addition to the force continuum that the force commander now has available to him. Commanders that properly employ NL munitions and weapons have a tactical advantage over those who rely completely on lethal means.

2-12. When developing a tactical standing operating procedure (TSOP) (see *Chapter 6*), units should consider the following:

- Crowd control formations.
- Extraction teams.
- Apprehension teams.
- Lethal overwatch marksmen and/or observer teams.
- Reserve and/or security forces.

MILITARY WORKING DOG TEAMS

2-13. Military working dog (MWD) teams may be employed with a control force formation as a method of increasing crowd apprehension about approaching or engaging the formation. The teams should be in the rear of the formation in plain sight of the crowd, but in front of the command element and the M33A1 squad RCA disperser. The MWD teams work back and forth behind the formation as an intimidation measure. The presence of the MWD may produce a profound psychological effect on the crowd. These teams may also be used to help control individuals who have been captured by the recovery and apprehension teams.

> **CAUTION**
> Do not unleash an MWD on a crowd.

COMBAT CAMERAMEN

2-14. Video and still cameramen should make a photographic record of the individuals in the crowd who are leaders and instigators. Events must be documented to hold personnel, factions, gangs, or groups accountable for acts that violate law, destroy property, or cause physical harm. Electronically recording events aids in the prosecution of such cases and eliminates the sense of anonymity that people in large crowds often feel.

INFORMATION OPERATIONS

2-15. Information superiority helps forces to anticipate problems and requirements. It allows commanders to control situations earlier and with less force, creating the conditions necessary to achieve the optimal end state. Public affairs, psychological operations (PSYOP), and civil military operations are activities that will allow the commander to control situations earlier and with less force.

2-16. Information is available from a multitude of sources. A diversity of sources is the best approach because it prevents biased perspectives. Primary sources are as follows:

- Open sources.
  - Libraries.
  - Newspapers and news periodicals.
  - Radio and television.
  - Internet.
- Law enforcement sources.
  - Local law enforcement agencies.
  - National law enforcement agencies.
- Military sources.
  - Department of Defense (DOD) intelligence community (most restrictive source).
  - Local military intelligence (MI) field offices.

2-17. Police intelligence operations (PIO) is one of the five military police functions. This function is a process of actively and passively collecting information that is of a police, criminal, or combat nature. As military police perform the other four functions of maneuver and mobility support (MS), area security (AS), law and order (L&O), and internment and resettlement (I/R), they are gathering information that supports, enhances, and contributes to the commander's protection program, situational awareness, and battlefield visualization by portraying relevant threat information that may affect operational and tactical environments.

2-18. IPB is a continuous process for analyzing the threat and the environment of a specific geographic area. During the IPB process, the Intelligence Officer (US Army) (S2) or Assistant Chief of Staff, G2 (Intelligence) (G2) uses all available databases, intelligence sources and products, and related MI discipline to analyze the threat and the environment. The PIO function supports this process by providing the S2 with collected police, criminal, and combat information that can directly and significantly contribute to the success of the MI effort. In addition to combat information, the PIO function provides additional information on possible criminal threats and COAs. This is intended to support the S2 IPB process and can be used by the commander to upgrade force protection. (See *FM 19-10.*)

THREAT ANALYSIS

2-19. Crowd control requires its own thought process. Emphasis should be on prevention rather than confrontation. In combat, military forces are taught to fight and eliminate threats. In crowd control, military forces must deal with noncombatants that have internationally recognized rights. These rights must be respected while maintaining public order. This is an issue that law enforcement agencies (LEAs) have been struggling with for years.

2-20. Dealing with crowd control incidents is a matter of using some basic guidelines. It also includes asking the right questions in a logical manner so that key issues are not omitted. The commander must use these guidelines in context with METT-TC and the location of the incident (CONUS or OCONUS).

PREINCIDENT PLANNING

2-21. This phase of planning begins before the incident and is initiated at the operational level with guidance from the strategic level. The planning includes guidance on crowd control and addresses responsibilities, training, organization, operating procedures, use of force (CONUS), and/or rules of engagement (OCONUS). The most difficult and productive decisions are those made in the preincident planning process.

Avoid Confrontation

2-22. Crowd situations are highly unpredictable, but one thing seems certain—confrontation will likely cause crowd resistance. When pushed, people tend to resist opposition to the realization of their purposes.

Focus on Prevention

2-23. Planning should key in on the prevention of unfavorable outcomes. Experience has shown some LEAs attempting to help crowds accomplish their goals within the law that have been beneficial and even led to conceding some violations for the purpose of avoiding confrontation. However, LEAs maintain a law enforcement presence, which signifies social restraint. LEAs also decide when and where they will not compromise and the amount of force to use.

Define Goals

2-24. When defining a goal, deciding what must be accomplished is the first step. Defining the goal is fundamental. However, actually working toward and accomplishing the goal is easier said than done, as the process can drive the situation. Commanders and leaders must be aware of this. The military force must focus on what they are trying to accomplish.

BACKGROUND INFORMATION AND INTELLIGENCE

2-25. Seek to know as much as possible about social protest groups (within the limits of the law, see *Appendix B*) before an incident. This will provide insight into the organization and its functions and provide a warning as to what to expect. When you know as much about these organizations as possible, then it is possible to anticipate their next move. See *Appendix C*.

2-26. Continually assess what is to be accomplished against what others are trying to accomplish. Some groups may not have goals that conflict with those of US forces, but their actions may. Other groups may have goals and activities that do conflict. In either case, the assessment of group goals compared to the goals of US forces helps to understand and avoid potential conflicts.

CIVIL DISTURBANCE OPERATIONS PREPARATION

2-27. Counterdemonstration workgroups are a mixture of traditional and nontraditional staff proponents brought together to coordinate the resources to be employed by task force (TF) units before a potential civil disturbance. Meetings of the counterdemonstration workgroup are held on a regular basis (generally weekly) and are chaired by the Assistant Chief of Staff G3 (Operations and Plans) (G3). The following staff proponents may be involved, and others may be involved as needed:

- Public affairs office (PAO).
- Joint military commission (JMC), if organized.
- Provost marshal office (PMO).
- PSYOP.
- G2, Assistant Chief of Staff G5 (Civil Affairs) (G5), Assistant Chief of Staff G6 (Signal) (G6).
- Civil affairs (CA).
- Army airspace command and control (A2C2).
- Chaplain.
- Engineer.
- Surgeon.
- Fire support element (FSE).

RELATIONSHIPS DEVELOPMENT

2-28. Working relationships between commanders and protest group leaders are increasingly seen as the best means for preventing bad outcomes in crowd situations. This is called the negotiated management model of crowd control. LEAs in large metropolitan cities in the US and Europe practice it.

2-29. Open dialog helps develop working relationships between commanders and protest group leaders, providing an opportunity to communicate clearly. Talking allows group leaders to tell authorities and tactical commanders what they want to accomplish. It also allows authorities and commanders to tell group leaders what they are prepared to do and how they might respond to certain crowd behaviors. Such communications can do much to resolve issues and prevent violence.

2-30. Commanders may not be able to talk to all leaders before a crowd assembles. Some groups do not have recognized leaders and are ad hoc organizations. Other groups may have several leaders, but only some of those leaders will negotiate. In such cases, meetings with some leaders may tell you who the other leaders are and provide critical information. Negotiations may also encourage more moderate leaders to do things that will support the commander.

2-31. Initiate communication with group leaders to work out issues before an event. Commanders should make a concentrated effort to maintain a working relationship with the leaders of protest groups.

2-32. Communication also means persuasion. Commanders should make a concerted effort to win over demonstration leaders. Commanders may tell group leaders that they want to help them complete their mission in a safe manner. They should encourage demonstrators to protest in an acceptable way and, if possible, offer favors to get them to do just that.

NOTE: Commanders must only offer what they are willing and able to deliver.

2-33. Several rules apply in negotiations, and credibility is key. Only communicate necessary information and those actions that authorities and commanders intend to do. Negotiations are made from a position of strength, and by negotiating an agreement may be reached. However, contingencies should be developed in the event that the agreement is violated.

2-34. Working relationships between commanders and protest group leaders often result in protest groups policing themselves. This is one of the basic premises of negotiated management. Protesters are more likely to listen to their own leaders, as opposed to listening to US forces. Protest organizers are likely to buy into a cooperative effort and agree to ground rules set during preprotest negotiations. Many protest leaders seek to stay on the right side of the law.

ACCEPTABILITY

2-35. Deployed US forces will find themselves engaged in crowd control operations under difficult circumstances. The host nation (HN) may have groups that do not accept the presence of US forces, which will make imposing order and the protection of citizens difficult at best. US forces will be under intense media and political scrutiny (an environment much like what LEAs operate in on a day-to-day basis).

2-36. Winning in this environment is not like winning in combat. US forces may appear to be invincible and formidable, but they risk being portrayed as oppressors. Thus, US forces can lose by appearing to win. Groups that perceive themselves as oppressed will readily seek victimhood in an effort to gain the support of public opinion. Winning in this environment is about seizing and holding the moral high ground. US forces must maintain the authority and legitimacy of what they are doing.

2-37. Projecting a favorable image will require outreach to local leaders and citizens. It will also require developing a relationship with the media. For example, the Los Angeles Sheriffs Department often invites the media to accompany their tactical commander during crowd control situations. This shows that they have nothing to hide. It also provides an opportunity for individuals to see the commander's side of an event.

2-38. In this environment, commanders must consider how actions will play among several audiences—local, allied, US, and international. The media effect is inescapable. In some cases, it will be difficult to accommodate all of these audiences. However, the most important audience will be the US public, which is key to the continued support of US forces.

CROWD ASSESSMENT QUESTIONS

2-39. Assessing crowds requires its own set of questions. These questions should be answered before a crowd assembles or as quickly as possible if a gathering occurs without notification. In some cases, all the questions may not be answered. These questions represent a logical way to think about crowds. The questions are as follows:

- **Who are they? What is the overarching identity of the crowd? Are they strikers, ethnic factions, or social protesters? Do they identify themselves as strikers, ethnic groups, religious factions, or protesters against some perceived social injustice?** Understanding who they are will indicate what they may do. It may be possible to determine identities and goals from advance assembling instructions, leaflets distributed to bystanders, placards and banners, and chants and songs.
- **What are their goals?** What the group wants to accomplish by assembling could determine the extent to which they can be accommodated once they have assembled. They may only seek recognition for their cause (being seen and heard). If so, this goal is usually easy to accommodate. However, some groups may have more demanding goals, for example the demonstrators in Seattle who sought to stop the WTO. Goals that cannot be accommodated make confrontation very likely.
- **What is the composition of the crowd and are there any known factions?** Seattle demonstrations against the WTO were comprised of groups that were protesting environmental issues, wages, and child labor laws. Differing goals and the resulting friction were evident between such organizations as the Ruckus Society and the American Federation of Labor-Congress of Industrial Organizations (AFL-CIO). Factions within a crowd represent threats and opportunities.
- **What are they capable of doing?** Protest groups often claim that they will assemble large numbers of people to produce some disruptive action. However, there is often a difference between the claim and reality. An organization may claim that it can mass a 100,000 people, but in reality can only get 40,000. Organizers exaggerate for the following reasons: they want to boost the morale of their own people and they want the media to report that they have strength in numbers. Studying the past activities of a group may provide indications of what they are capable of doing in the future.
- **What are their traditional behaviors or cultural repertoires?** What people do during protests is not universal. It varies with the group and the culture. Social protest organizations and striking unions will carry placards and banners. Other groups will protest in a more quiet way, like the 5,000 women in Sarajevo whose standard Sunday behavior was to sit and block traffic. They were protesting the loss of male relatives in the Bosnian war. Understanding the goals of

the protesters can be helpful in deciding how to respond to their behaviors.

- **When and where will they assemble?** Every protest organizer has a time and place for assembling and perhaps a destination for the crowd to move toward. If the organizer attempts to mobilize large numbers of participants, the time and place for assembling and dispersing must be made known in the instructions. This information may be stated in the mobilization instructions or disseminated by an informal network, such as word of mouth. Such impromptu networks in densely populated areas can enable rapid assembling.
- **Where will they go?** Many crowds have destinations. Organizations may march a specified distance to ensure that their cause gets sufficient attention. Commanders need to know the route to minimize disruptions to the rest of the community. In Los Angeles, LEAs attempt to reroute traffic and prevent congestion caused by protest marches. It is also necessary to provide security along the route to prevent counterdemonstrators from confronting the marchers. This will prevent an even greater problem for the community.
- **What are the possible targets of violence?** Riots, in particular, may focus on target facilities. In the 1992 Los Angeles riots, gun stores were major targets. In developing countries, targets may be more basic. In Somalia, throngs of people stormed food supply facilities. In Haiti, riots broke out over trash dumps where US forces had discarded the remnants of meals, ready to eat (MREs).
- **What is the worst-case scenario?** The worst-case scenario must be recognized for the sake of avoidance. This may be when a peaceful crowd degenerates into a violent riot.
- **When and where will they disperse?** Crowds have a life cycle that includes how they disperse (see *Chapter 1*). Commanders must consider this. It is essential that there be one or more avenues that individuals can use to disperse. These dispersal routes should be clearly marked, visible, and open-ended. There may be a need for US forces to assist the crowd with dispersal. For example, in Los Angeles a 70,000-man march went from east to central Los Angeles. Once at their destination, they engaged in peaceful activities and then dispersed. The LEAs provided transportation back to the assembly area (AA); otherwise, the protesters would have returned to the AA in a disorganized and unsupervised manner.
- **Are there plans for subsequent gatherings?** A crowd may disperse for a short time so participants can take care of personal needs. This could be days or only a few hours.

LEGAL CONSIDERATIONS

2-40. The legal authority in which a peace operation is conducted defines the parameters of the operation. All commanders, leaders, and soldiers must know the legal authority that regulates their operations.

MANDATES

2-41. There are two types of mandates—United Nations (UN) and non-UN. Mandates are the authority under which an operation is conducted. UN mandates are created from resolutions by the UN Security Council (UNSC) or the UN General Assembly (UNGA) that authorize and define an operation. Non-UN mandates are created from treaties, accords, resolutions, or agreements evolving from other international or regional organizations.

STATUS OF FORCES AGREEMENT AND STATUS OF MISSION AGREEMENT

2-42. A status of forces agreement (SOFA) and a status of mission agreement (SOMA) are key documents that define the legal authority and responsibilities of a force and force personnel participating in an operation. A SOFA or SOMA may be a treaty or memorandum of understanding. It is an agreement negotiated between the UN and the host countries, which details the rights, privileges, immunities, and nature of services to be provided to the force and its personnel and their responsibilities and obligations.

INTERNATIONAL LAW

2-43. Legally, international law affects most stability operations and support operations (SOSO). These laws consist mainly of international agreements, treaties, and customary international law, which include the agreements and laws known as the law of war. International agreements prescribe the rights, duties, powers, and privileges of nations relative to particular undertakings. International agreements will affect US participation in SOSO in the—

- Right of US forces to enter a foreign country.
- Status of US forces while in the foreign country.
- Construction and operation of US bases of operations.
- Aircraft overflights and landing rights.
- Processing of claims for any damage to persons and property of the foreign country.

2-44. These agreements or customary international laws govern all aspects of operations carried out. In the absence of a viable HN government or proper international agreement, many aspects of the operations are controlled by domestic (US) or customary international law.

UNITED STATES LAW

2-45. Operations in peace and conflict must comply with US law, whether as a statute, executive order, regulation, or other directive from a branch or agency of the federal government. The *Uniform Code of Military Justice (UCMJ)* applies to questions of military justice. Various statutes, *Executive Order 12333; Section 3, Code of Federal Regulations (CFR);* and DOD and service regulations govern all intelligence-gathering activities. The staff judge advocate (SJA) must actively advise and participate in all activities from initial planning to redeployment.

HOST NATION LAW

2-46. If a viable HN government exists, whether at the national or the local level, all laws of the HN will apply to US forces in that country unless an international agreement provides otherwise. The types of laws that may inhibit US operations are in the fields of immigration, labor, currency exchange, the procurement of goods and services, customs and taxes, and criminal and civil liabilities. A SOFA usually covers these areas before an operation begins. Therefore, planners must understand the law to assess it and determine if it will adversely affect the operation. Assistance may be available from the local US embassy or the command judge advocate; if not, the command may have to rely on other sources for guidance. If local law hinders the operation, the commander must inform the US embassy and request that it negotiate a solution.

CIVIL AUTHORITIES

2-47. There will be times that US forces will deploy to nations where the infrastructure of civil authorities has survived the conflict and is functional. Local authorities and police are responsible for ensuring the safety of their citizens, including the UN High Commissioner for Refugees (UNHCR). US forces assist in this by providing general military security and by facilitating negotiations. In countries where the civil authority infrastructure has collapsed, the UN mandate may require the coalition of forces to help ensure the safety of citizens until their own civil authorities can be rebuilt.

RULES OF ENGAGEMENT

2-48. The rules of engagement (ROE) are the commander's rules for the use of force (RUF). Commanders are responsible for interpreting, drafting, disseminating, and training on the ROE. Ensure that the SJA assists throughout the ROE development process. This will ensure that the ROE will not improperly constrain actions, but are still consistent with domestic and international law and policies and orders of the chain of command.

2-49. Clearly stated ROE are published before Army forces are committed. The highest military authority, with input from subordinate commanders, will continually evaluate the ROE and modify them as appropriate. When the use of NLW has been authorized in the ROE, the decision to use NLW should be delegated to the lowest possible level, preferably the platoon or squad level. This requires that all personnel, not just leaders, have a clear understanding of the ROE and the commander's intent (see *FM 3-22.40*).

2-50. In peace operations, the use of force is restrained while the diplomatic solution is sought through negotiation or mediation. Accordingly, ROE are more restrictive in peace operations, making them more politically sensitive. In today's world, peace operations are usually conducted by a coalition of forces operating under the purview of the UN charter and customary international law. Therefore, the UN may mandate certain restraints. Using overwhelming force could compromise diplomatic efforts to reach a peaceful settlement. Even a single, relatively small engagement could jeopardize the legitimacy of the peacekeeping force and upset negotiations or mediations. Commanders must beware that in any confrontation ROE decisions made by soldiers can have strategic political implications on current and future operations. Therefore, it is vital that leaders and soldiers hold a common understanding of when, how, and to what degree force is to be used. This requires training.

MULTINATIONAL OPERATIONS

2-51. Multinational operations include military forces from different nations. These operations are usually interagency operations. Conducting operations with foreign military partners, like operations with civilian partners, is uncommon to many soldiers, so a clear understanding of this different environment is necessary.

2-52. The majority of US peace operations will be part of a UN peace operation. Their multinational character merits particular attention because national interests and organizational influence may compete with doctrine and efficiency. Commanders can expect contributing nations to adhere to national policies and priorities, which at times complicate the multinational effort.

2-53. When working within a multinational force and with allies, do the following:

- Train coalition partners and allies on tactics and techniques.
- Remember that teamwork and trust are essential.
- Achieve unity of effort.
- Consider equipment and communications capability and interface.
- Strive to eliminate sources of confusion and misunderstanding.
- Remember that military doctrine varies from country to country.
- Remember that national culture influences the way units operate.
- Remember that language barriers represent a significant challenge.

SCALABLE EFFECTS

2-54. Scalable effects concepts are a measured approach in response to a crowd gathering. By recognizing a use-of-force policy, soldiers must be taught and understand that they use the minimum force necessary. Without the appearance of a graduated response, the gathering crowd may consider actions as excessive, causing a possible escalation of hostilities or violence. Do the following to aid in the scalable effects process:

- Try to persuade the crowd to quietly disperse by talking with leaders
- Use translators as necessary.
- Let the first approach be by local authorities (the mayor or police).
- Pass out handbills requesting that the crowd return home.
- Use video and still cameras to photograph individuals and events for later use at trials.
- Give warnings before moving to the next level of force.

2-55. Ideally, the force should be positioned out of sight of the crowd. Have the troops move into position with few shouted commands. For maximum effect on the crowd, have them form their formation decisively and professionally. Do the following to aid in the formation process:

- Display force in a graduated manner (for example, a UH 60, hardened vehicles, and soldiers with charged weapons).
- Exploit the psychological effects of show of force.
- Demonstrate force (do not use on unarmed civilians).
- Escalate the PSYOP message.
- Highlight the target pointer.
- Demonstrate sniper precision strike capability.
- Employ riot control measures.
- Use NL munitions.
- Use RCAs (if approved).
- Move through the crowd using riot control formations and movement techniques.
- Use the graduated response matrix (GRM)

2-56. With the increased participation of US forces in contingency operations around the world, it accentuates the need to establish procedures for applying graduated military responses to situations that threaten these missions. Numerous GRMs and similar products exist throughout the military. These products graphically portray available responses in a graduated manner. The intent of these products is to give the on-scene commanders a list of options with which to control or defuse a situation before it gets out of hand. Most threats can be eliminated without loss of life or collateral damage by effectively applying the resources available.

PLAN A GRADUATED RESPONSE MATRIX

2-57. The planning for and development of a GRM begins with the mission analysis portion of the military decision-making process (MDMP). Some missions require US forces to enforce treaties or accords to protect the lives of civilians in uncertain or hostile environments (such as noncombatant evacuation operations [NEO]) or to provide large-scale humanitarian assistance. These missions require some sort of graduated response criteria to maintain order and prevent uncertain environments from becoming hostile. The development of a GRM requires the following seven steps:

- Identify the need for a GRM.
- Establish a team to develop the GRM.
- Develop targets.
- Coordinate staff functions.
- War-game.
- Receive command approval.
- Operate deliberately.

Identify the Need for a Graduated Response Matrix

2-58. Not all missions require a GRM. The decision to use a GRM requires careful consideration. Once it is agreed that a GRM is necessary, it requires guidance from the commander regarding the response options available.

Determining the appropriate responses is based on the facts, assumptions, and limitations identified during mission analysis.

2-59. Planners (staff) must agree on the intent of the GRM. The GRM can be used as a training and rehearsal tool. It provides leaders with the most likely vignettes that can be incorporated into a COA analysis, predeployment training, and rehearsals. The GRM can also be used as a handy reference during situations that require graduated responses.

Establish a Team to Develop the Graduated Response Matrix

2-60. Establish a GRM development team with a broad range of skills, knowledge, and professionalism. The following list of individuals makes up a typical GRM team:

- Provost marshal or senior military police advisor.
- FSE.
- Brigade legal officer.
- PSYOP representative.
- Land information warfare or information operations officer.

2-61. Since the GRM is designed to give commanders graduated options for dealing with both hostile and nonhostile threats to the mission, this team composition allows for target selection, application of the ROE, and attack methods using both NL and lethal means.

Develop Targets

2-62. The FSE, in conjunction with the S2 section, develops targets for both lethal and NL attacks. In the case of stability operations, these targets are usually not the conventional specific point or piece of equipment on the ground. They are more situational than specific. The GRM identifies situations or acts that subordinate elements could face during the mission. The sample GRM in *Figure 2-1* shows three possible situations or acts that on-scene commanders could expect to encounter. From the targeting standpoint, these are groups of more specific targets.

2-63. During mission analysis, the fire support officer (FSO) identifies both lethal and NL assets available to the unit. A tactical PSYOP team (TPT) attached to the unit is an example of a NL attack asset that should not be overlooked. The following are examples of what the FSO should look for:

- RCAs.
- TPTs.
- Electronic warfare assets.
- CA teams.
- Information operations (IO) teams.
- Artillery smoke projectiles.
- Aircraft (AH-64s, OH-58Ds, and AC-130s).
- Mortars.

2-64. The lethal assets described could be used in a NL show of force or demonstration to defuse a situation before it requires lethal force. The critical element of this mission analysis is not to focus solely on lethal assets. In

Crowd Action	Nonlethal Response			
	Command Presence/PSYOP	**Show of Force**	**Demonstration of Force**	**Riot Control Means (if approved)**
Unarmed Civilians	- Ensure that the on-site commander of the TPT directs the broadcast of the dispersal proclamation and/or passes handbills. - Escalate the tone of the dispersal proclamation from information to a warning of force.	- Display force along with escalating the dispersal proclamation. - Display force in a graduated manner, such as a helicopter hovering over a crowd or soldiers with charged weapons. - Exploit the psychological effect of a show of force.	- Do not demonstrate force toward unarmed civilians.	- Employ RCAs at the point of penetration. - Use PSYOP to exploit the psychological effect. - Move through the crowd using riot control formations and movement techniques.
Armed Civilians (knives, clubs)			- Display force along with escalating the dispersal proclamation. - Highlight the target pointer. - Demonstrate sniper precision strike capabilities.	
Armed Crowds/ Military (firearms)				**- Do not use RCAs—they may escalate the situation.**
Hostile intent/hostile act occurs by armed threat.				
Crowd Action	**Lethal Response**			
	Sniper Response	**Small Arms Direct Fire**	**Aerial Fires**	**Indirect Fires**
Unarmed Civilians	- Ensure that target leaders or troublemakers are targeted. - Use the minimum response necessary. - Exploit the psychological effect of an attack.	- Determine that sniper attack is insufficient. - Consider demonstrating capabilities. - Exploit the psychological effect of a lethal response. - Escalate gradually, starting with a small caliber, single round and work up to a large caliber, automatic.	- Determine that small arms direct fire is ineffective. - Use the minimum response necessary. - Use a minimal precision strike initially; use subsequent fires based on the situation. - Exploit the psychological impact of each strike.	- Determine that air assets are unavailable or ineffective. - Use the minimum response necessary. - Ensure that the response is directed by the on-site commander.
Armed Civilians (knives, clubs)			**- CAS/indirect fires must be authorized by the MACOM commander.** - Consider requesting permission for use when— • All lesser means have been ineffective. • There are physical eyes on target. • Proximity to civilians has been considered. • Risk to friendly forces/evacuees outweighs the risk of collateral damage.	
Armed Civilians/ Military (firearms)				

Figure 2-1. Sample GRM Card

stability operations, the Army wants to prevent acts of hostility first and then be prepared, if necessary, to apply lethal force.

2-65. Graduated responses can range from command presence through the show of force, a demonstration, the use of RCAs, and other techniques (such as the application of lethal force using snipers, small arms, AC-130s, and indirect fires).

Coordinate Staff Functions

2-66. To coordinate staff functions the rest of the GRM team assembles to complete the escalation sequence for each response. PYSOP and legal representatives are critical attendees during the escalation sequencing process. In the area of psychological operations, the TPT must exploit the effects of all responses.

2-67. The legal officer evaluates each escalation of force option and graduated response to ensure that it is consistent with the ROE. The GRM is designed to recommend applications of force consistent with the ROE, yet not limit the leader or individual soldier's right of self-defense. A GRM should clearly show that if a hostile act occurs, lethal options would be first and foremost.

2-68. In the case of lethal responses, the commander's guidance must again be applied. For example, lethal responses are allowed only in self-defense.

War-Game

2-69. Once the types of escalations for each potential graduated response are determined and annotated, the GRM must then be war-gamed. The staff must walk through each act or situation from the on-scene commander's standpoint.

Receive Command Approval

2-70. Once the GRM has been war-gamed, it must be submitted to the commander for approval. This is the final check to ensure that the GRM team has applied the commander's guidance correctly and met his intent.

Operate Deliberately

2-71. Operating deliberately affords commanders ample opportunity to plan and prepare for possible civil disorder situations. Through the effective gathering of information and a working cooperation with local government and police officials, commanders can often be made aware of dates, times, locations, and what groups may assemble before the operation. The purpose is to gather as much information as possible who is involved, where they are assembling, what incident promoted the activity, and what seems to be the prevailing attitude of the assembling crowd.

2-72. Civil disturbance operations are dynamic, ever-changing environments requiring effective communications both up and down the chains of command. They require advance preparation and planning using the established troop-leading procedures (TLP) outlined in *FM 3-19.4*. The eight-step process of TLP aids commanders and their subordinate leaders in planning and preparing for an operation. Most steps can be accomplished concurrently, but rarely is there enough time to go through each step in detail. Despite this, leaders must follow these steps to ensure that nothing is left out of the planning process. As outlined in *FM 3-19.4*, the following are the eight steps of TLP:

- Receive and analyze the mission.
- Issue a warning order.
- Make a tentative plan.
- Initiate movement.
- Conduct a reconnaissance.
- Complete the plan.
- Issue the order.
- Supervise, refine, and rehearse.

CONDUCT HASTY OPERATIONS

2-73. Conducting hasty operations seems to be the norm rather than the exception. Unlike deliberate operations, hasty civil disorder operations are reactionary in nature with little or no time for planning. In these situations, the event of a crowd gathering is already underway. There is very little, if any, advanced warning of the unfolding situation, and commanders are usually put in the position of sending their soldiers into an already volatile, and perhaps hostile, environment.

2-74. Commanders should be in immediate and constant communication with local civil and police authorities. The purpose is to gather as much information as possible about who is involved, where they are assembling, what incident promoted the activity, and what seems to be the prevailing attitude of the assembling crowd.

2-75. It is essential that commanders operating in environments where civil disturbance operations are likely ensure that their soldiers receive training at all levels. Commanders and subordinate leaders need to instinctively rely on the same TLP that they would in a deliberate operation. See *FM 3-19.4*.

CONDUCT TACTICAL OPTIONS

2-76. In the peacekeeping and peace enforcement environment, the use of conventional firearms or the threat of their use may not be the solution to a situation where US forces must separate two belligerent, hostile ethnic groups or prevent a similar group from entering an area that is off limits to them.

2-77. Crowd control options are often combined. Commanders choose their options based on an evaluation of the particular crowd. Commanders select any combination of control techniques and force options they think will influence the particular situation (METT-TC). Commanders must always try to choose the response that can be expected to reduce the intensity of the situation.

Monitoring

2-78. Monitoring an assembled crowd consists of gathering necessary intelligence and watching them to determine progress and development. It is this gathered information that helps commanders select the appropriate actions. Gathered information can often be used to help defuse a situation through persuasion.

2-79. Monitoring is continuous. Without timely information, responses could cause the situation to escalate beyond what it should have or cause a response to be inadequate based on the situation.

2-80. Commanders may task teams with specific missions to monitor crowd activity and note any new developments. Monitoring done by these observation teams gives the commander up-to-the-minute information so he can gauge crowd activity and intent in relation to the overall situation.

2-81. Observation teams monitor crowd activities to gather information. They observe and report on crowd size, location, and mood and on the developing situation. An observation team may consist of a marksman, a radio operator, and an observer equipped with binoculars. They may be posted strategically on rooftops and other high terrain that overlooks the crowd. Sometimes observers use helicopter-mounted observation devices. This also affords security for the operational forces. Commanders must know where their observation teams are located so they are not mistaken for teams of snipers.

2-82. The timely flow of information may allow the commander to influence the outcome of the situation with simple negotiations. Monitoring is appropriate when more decisive action is not feasible because of crowd size or when the intensity of a situation might escalate. It is particularly useful in large, nonviolent demonstrations. Monitoring can serve as an interim measure until more control forces arrive. It includes establishing communication with crowd leaders to convey official interest and intent to the crowd. Monitoring also includes efforts to gain the cooperation of crowd leaders.

2-83. Communication with crowd leaders and participants can help a commander control a situation without more severe measures. If communications exist with crowd leaders, the authorities may be able to divert the leaders or the crowd from their stated or apparent goal. Pressure can be put on leaders to channel the crowd into an area that minimizes disruption to the community and aids crowd control operations. March routes and demonstration areas can be limited to those that will help contain the crowd and reduce their potential for disrupting the community. Pressure can be positive (offering concessions) or negative (with deterrents).

2-84. If the commander can gain the cooperation of crowd leaders, it can decrease the potential for disorder, and if crowd leaders seek cooperation from authorities, officials should try to be accommodating. Crowd leaders can be placed in liaison positions between the crowd and the control force. Crowd leaders can be made responsible for managing the crowd by policing their own activities.

2-85. Taking still pictures or videotaping the faces of individuals within a crowd may prevent or reduce unlawful and violent acts. This could lessen their sense of anonymity. If needed, photographs or videotapes can be used as evidence for prosecution. To be effective, crowd members must see their presence being recorded. The photographer or cameraman should be in uniform to let the crowd know who he or she is. Recorders must be close enough to the crowd to be seen easily, but not close enough to be endangered.

Blocking

2-86. It is not uncommon for protests and demonstrations against the US or its forces to occur outside the main entrances to military installations or US Embassies. Because of the sensitive nature of equipment and information that can be found at these locations, blocking unauthorized entry is critical.

2-87. Blocking is the physical denial of advance upon a facility or an area by a crowd. Commanders may have to task their forces to block a crowd. Blocking physically denies crowd advance. Crowd control formations, particularly line formations, along with barricades can be used to block advancing crowds. Barricades of vehicles, concertina wire, and water-filled barrels are used to block or channel the movement of the crowd.

Dispersing

2-88. Dispersing is taking deliberate actions to fragment an assembled crowd in order to prevent the destruction of property or prevent injury. It is extremely effective against smaller crowds in congested urban environments. However, this may increase and spread lawless activity rather than reduce it. Thus, the commander's forces must control dispersal routes and the areas in which the dispersal will occur. Forces must protect the facilities that could be considered likely targets of opportunity for dispersing individuals or groups. Dispersal of the crowd may require apprehension of small groups still active in the area. Proclamations, shows of force, crowd control formations, and RCAs can disperse crowds.

2-89. Avoid (if possible) the dispersion of crowds into wide-open areas because it gives the crowd the opportunity to grow rapidly in size. The direction of crowd movement is difficult to anticipate and channel.

2-90. Once the crowd has started to disperse, it may occasionally be necessary for the forces to advance, keeping the crowd on the move and in the right direction. A crowd on the move should not be hurried to avoid panic. At no time should the crowd be cornered in a position where there is the perception of no escape. This invokes the "fight-or-flight" syndrome, possibly escalating violent activity.

2-91. Issuing a proclamation (see *Figure 2-2,* page 2-20) can help disperse a crowd. Proclamations officially establish the illegal nature of crowd actions and put the populace on notice that the situation demands special measures. Proclamations prepare the people for the presence of military authority and it tends to inspire respect. Proclamations support law-abiding elements and psychologically bolster military forces trying to restore order. They also demonstrate the gravity of the situation to all concerned and are excellent ways to make a commander's intentions known to a crowd. It is also a good way to reduce crowd size before direct action is taken. Commanders can make a verbal proclamation similar to the following statement: "*In the name of the President of the United States, I command that you disperse and retire peaceably to your homes.*" Such a proclamation may even make direct action unnecessary.

I. DISPERSAL

Unlawful Gathering

"Attention! Attention! This area must be cleared at once! Further unlawful behavior will not be tolerated. Clear this area at once or the necessary force to do so will be used."*

Demonstration

"Disperse and retire peaceably! Disperse and retire peaceably! Attention all demonstrators!"

"The demonstration in which you are participating ends at ______. The permit that was agreed to by the leaders of the demonstration expires at that time. All demonstrators must depart from the ______ by _____. All persons who wish to leave voluntarily may board the buses. These buses will go to the _____. Those who wish to take buses should move to ______. Those demonstrators who do not leave voluntarily by ______ will be arrested and taken to a federal detention center. All demonstrators are urged to abide by the permit."*

Warning of Looting

"Return to your homes! Someone may be looting them at this moment! During a disturbance, criminal activity is at its peak. Your family or your property may be in danger."*

II. EMPLOYMENT OF TROOPS

Employment

"Attention! Attention! Troops are present in this area. They are preparing to advance. Order must and will be maintained. Disperse peaceably and leave this area. To avoid possible injury, leave at once."

"Disperse now and avoid possible injury! Disperse now and avoid possible injury!"*

(Repeat until troops are committed.)

When Troops Are Committed

"Troops are advancing now. They will not stop until this crowd is dispersed and order is restored. To avoid injuries, leave the area at once. Return to your homes as peaceful citizens. Troops have their orders, and they will not stop until the crowd is dispersed. Do not get hurt. Leave this area."*

III. PRESENCE OF CHILDREN

(Used in conjunction with other announcements.)

"Attention! Attention! Do not attempt to cause further disorder. Disperse now in an orderly manner and avoid possible injury to children. Return at once to your homes."*

IV. ADDITIONAL INSTRUCTIONS

*Indicate the method, streets, and direction that the crowd should use when dispersing.

Figure 2-2. Sample Proclamation

2-92. In making a proclamation, a commander may consider imposing a time limit. However, the situation may change, and not imposing a time limit leaves the commander free to choose other actions when he wants. A proclamation must be specific in its instruction.

2-93. If a time limit is stated, it must be a reasonable length of time for the crowd to comply with the instructions. When drafting a proclamation, the commander must consult closely with the SJA. He must use the simplest language possible to maximize the effects of the proclamation. If proclamations must be translated to a local language, the translation must be made with great care.

2-94. The commander at the scene may direct that a proclamation be issued over public address (PA) systems. The force of the words used in the proclamation must be gauged to the composition of the crowd. If the crowd consists of normally law-abiding citizens who are presently assembled to show disagreement with an existing situation, the proclamation requires less force. On the other hand, if the crowd consists of militant rioters, the proclamation requires more force. The text may take a number of forms, depending on the situation.

2-95. A show of force is often a useful measure for dispersing a crowd. When troops arrive, the psychological impact of their arrival can be used. Soldiers can dismount from the helicopters, buses, or trucks in plain sight of the crowd, but they must be far enough away to prevent a provoked attack of thrown objects. The first echelon to dismount from the vehicle secures the surrounding area.

2-96. When small groups are scattered throughout a large disturbance area, a show of force can be made by marching troops conducting motor marches, conducting saturation patrolling, and setting up static posts. Sometimes marching well-equipped, highly-disciplined soldiers in view of a crowd may be all that is needed to convince them to disperse and retire peaceably. On the other hand, a show of force may attract people to an event. It may also provoke a nonviolent crowd into a violent confrontation. Intelligence can help best in preparing a show-of-force response.

Containing

2-97. Containment is the process of limiting a crowd to the area they are presently occupying. It is a suitable option when the crowd must be prevented from spreading to surrounding areas and communities. Additionally, it prevents those outside the already assembled crowd from joining the gathering. Containment is useful when apprehensions become necessary for preventing those in the crowd from escaping. Crowd control formations, perimeter patrols, and barriers are effective methods to accomplish containment. In all instances, caution must be used to avoid the "fight-or-flight" syndrome common to people feeling trapped with no escape.

2-98. Armored vehicles, armored security vehicles (ASVs), and up-armored high-mobility multipurpose wheeled vehicles (HMMWVs) are adaptable to roadblock operations as they can serve as barriers. They also provide added protection for the soldiers inside and those outside manning the roadblocks because they provide an easily accessible barrier for them to crouch behind.

2-99. Military vehicles traveling at close intervals in a column formation next to a crowd are largely a psychological barrier (see *Chapter 6*). They can be used to contain a large, fast-moving crowd. The moving cordon creates a temporary obstacle between the crowd and the line beyond which they will not be allowed to cross. A well-trained driver in a mobile cordon can do a better job than dismounted soldiers. In order to execute safe, quick reversals of direction for mobile cordons on narrow roads, the soldiers executing this formation must be highly-trained vehicle operators.

2-100. By-the-number commands issued over vehicle radios are the most common method for coordinating a cordon movement. Vehicles equipped with PA systems can also prove effective to issue both commands to troops and directives to the crowd. The cordon speed is no more than 5 miles per hour with an interval of at least 20 feet. Blue and red lights, sirens, blinkers, and horns may or may not be used. Armored vehicles can also serve as mobile command posts. When used properly, vehicles provide security, communications, and mobility.

DECIDE TO CONTAIN OR DISPERSE

2-101. Current crowd control doctrine places emphasis on crowd dispersal. Forced dispersal may result in a crowd breaking up into multiple groups that scatter over a large area. This may pose even greater public order problems and may pose a continued threat to control forces. A crowd is often better controlled by means of containment (confining its activities to a given area). A crowd has a limited duration, and its numbers are likely to diminish as individual needs take precedence over those of the crowd.

2-102. During the G8 summit protest in Ottawa, Canada, in June 2002, several thousand people showed up to demonstrate against the evils of world corporate leaders. Canadian Police determined that their best response to this event was a "soft response." Throughout the two-day event, hundreds of officers in standard patrol uniforms operated in four-officer teams, prestaged throughout the city. As the demonstrators began to assemble and eventually took to marching, officers provided escorts both in the front and rear of the enormous crowd along with officers marching to the flanks of the demonstration. This "soft approach" frustrated demonstrators, especially those who had it within their agenda to cause property damage and personal injury. Even more frustrating to the demonstrators was the heavy downpour of rain they endured on both days. Their organizations quickly disbanded out of concern for their own creature comforts. In the end, 10,000 people participated in the protest, with only three arrests made in two days.

USE PATROLS

2-103. Alert, aggressive patrolling of the disturbance area deters the gathering of crowds. The use of saturation patrolling is most effective for this purpose. It allows information to be collected, and it creates the psychological impression of the control force being everywhere at once.

2-104. Standard military transport vehicles provide mobility and communications for area coverage. Soldiers must be deployed with enough vehicles to provide the flexibility to handle all situations in the disturbance area. The vehicles can be modified with sandbags, wire screens, or similar materials to protect against sniper fire and thrown objects. If sniper fire is expected or has been encountered, then up-armored HMMWVs or light armored vehicles should be used for patrolling.

2-105. Armored vehicles patrolling an area of violence provide an added psychological effect and allow troops to maneuver close to snipers to make an apprehension. They may also be used as rescue vehicles to extract hostages or people surrounded by a hostile crowd.

2-106. Varying patrol routes and times for mounted and dismounted patrols keep lawbreakers from being able to pick a safe place or time to act. Patrols are particularly useful in preventing overt arson and violence. Patrol members can also spot and promptly report fires. Whenever possible, military patrols are integrated with civil police patrols. Joint patrols conserve military forces and aid civilian military communications. They also help troops become familiar with an area quickly. In addition, the civil police are available to make apprehensions, if needed. Patrol members must practice proper standards of conduct and fair treatment of civilians at all times. They are performing an important community relation and control function.

2-107. Motorized patrols, because of their speed and mobility, provide timely reconnaissance and broad area coverage. Motorized patrols are in radio contact with TF headquarters (HQ). They can make periodic contact with foot patrols and stationary posts, while patrolling their own areas. Moreover, motorized patrols can respond quickly to calls for help from other patrols and guard posts. Motorized patrols should have at least three vehicles with three men in each. Strength in numbers is necessary for protection. When planning patrol routes, avoid areas where the patrols can be isolated or placed in jeopardy. In addition, motorized patrols are equipped with fire extinguishers to put out small fires, thus reducing the burden on the fire department.

2-108. Besides foot and motorized patrols, air patrols provide a third dimension in directing the overall control effort in the disturbance area. They can perform reconnaissance and surveillance and provide near real time formation over the disturbance area. They are an excellent means of providing timely information to the commander concerning the demonstration area and its surrounding perimeter. They can monitor the actions of rioters, the extent of damage, the status of access routes, locations and conditions of road barriers, and other important conditions.

USE DELAYING TACTICS

2-109. AOs that units are assigned to patrol are often larger than the unit has resources to manage effectively. In addition, units must manage taskings, contingency plans, and other commitments, which will quickly drain available manpower. This problem will manifest itself when factions threaten to riot or gather in large numbers to demonstrate. In order for the commander to stall for time while he shifts unit assets, he may want to delay the arrival of buses and vehicles full of demonstrators. This technique is based on the assumption that demonstrators are moving to the demonstration site by vehicle. Key and secondary routes into the demonstration site must be analyzed by the S2 when he does the IPB.

NOTE: There should be an overwatch when this technique is attempted.

2-110. As demonstrator buses and/or vehicles move toward the demonstration site, soldiers in two HMMWVs block the road just ahead of the approaching vehicles. Once the demonstrators dismount their vehicles and begin to close in on the soldiers, the soldiers leapfrog (see *Figure 2-3*) 1 kilometer farther down the road to repeat the process again. This is repeated as often as necessary. This process, if repeated enough times, should discourage the demonstrators enough that they give up and go home. It should also provide enough time to shift assets to the demonstration site.

2-111. Demonstrators are just as capable of blocking secondary and key routes leading into a demonstration or riot site to prevent relief forces from assisting with riot control as US forces are capable of preventing demonstrators access to the site. One must remember that demonstrators are often well organized and have handheld radios and other items. A great example of this was the 1999 demonstration against the WTO in Seattle, Washington. Police had erected temporary portable fences in the downtown areas to keep demonstrators from disrupting the delegates. Determined to be disruptive, the demonstrators relocated the temporary fences by not allowing police to enter the area. Police were forced to back patrol cars over their own fences to gain access.

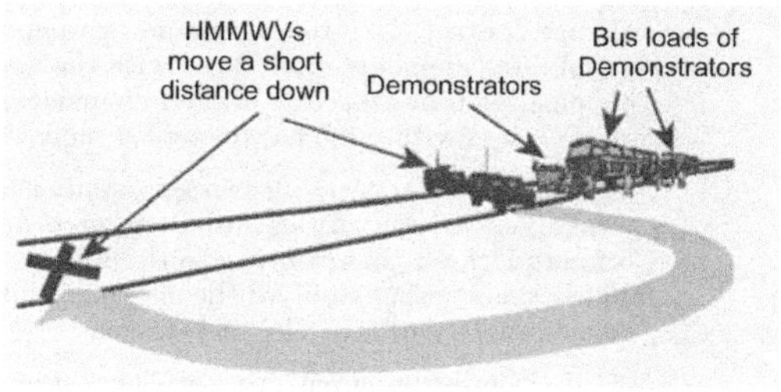

Figure 2-3. Delaying Tactics

Chapter 3

Apprehension, Search, and Detention

During the course of a civil disturbance, some members of the crowd may take part in unlawful activities, such as looting and assault. This will require the civil authorities or attached elements to apprehend, search, and detain people who are participating in the demonstration. Circumstances of the unfolding situation may require US military forces to be called upon to search, take custody of, and detain people who are participating in some form of protest and violating the law. Search, apprehension, and detention operations are conducted to halt these violations and to deter future violations.

LEGAL CONSIDERATIONS

3-1. All apprehensions are made by the civil police force, unless it is not possible for them to do so. Individuals must be known to have committed an illegal act, or there must be probable cause to believe that an individual committed such an act to be apprehended. If it becomes necessary for a military control force to apprehend or temporarily detain such violators, control force members often do so with the approval of the civil authorities. This ensures that all searches, apprehensions, and detentions (if necessary) are conducted within the parameters of the law.

3-2. Commanders operating overseas, similar to Bosnia and Kosovo, may find themselves without an established government or LEAs to work with. Commanders on the ground are then thrust into positions of authority, making them responsible for the actions, conduct, searches, apprehensions, and detentions performed by their troops.

3-3. If US military forces are called upon by the local authorities, certain policies have to be observed during search, apprehension, and detention operations. First, treat all people firmly but with reasonable courtesy and dignity. Remember that participating in a legal demonstration to express views is a right of all people, not just Americans in the US. The action, attitude, and behavior of US soldiers performing these operations are very important.

3-4. Our forces should never be seen as a military policing force on our own soil or as an occupying force in another country. Treating people with contempt, hostility, or excessive force increases the likelihood of resistance and violence. Searching people, placing them under apprehension, and detaining them without probable and just cause or without concern for their constitutional rights create problems and hinder due process. Numerous cases exist where authorities violated an individual's rights, resulting in the prosecution of police authorities or civil suits being waged against the officers and their leaders. This includes the military personnel assisting those officials.

3-5. It is very important that civil law enforcement (if available) be present and supervise all activities. Military personnel conducting a search or making an apprehension must carry out all the procedures carefully within the parameters of their training and the authority given to them. For each search and apprehension performed, the apprehension element must document the specific conduct of the person that violated the law.

3-6. Apprehending officers and supporting US servicemen must provide careful descriptive data for each subject. This description must be sufficient so that, at a later time, the suspected person can be clearly identified as the subject in question. Names and addresses of witnesses to the violation must be obtained. This information can be recorded on *Department of Defense (DD) Form 2708 (Receipt for Inmate or Detained Person)*. Apprehension teams must retain and tag each item of physical evidence, such as weapons or stolen goods, that supports the apprehension on *DA Form 4137 (Evidence/Property Custody Document)*. The team must give the owner a receipt for all property collected as evidence and retained.

TYPES OF SUBJECTS

3-7. There are two types of subjects (cooperative and uncooperative); they are as follows:

- **Cooperative.** Cooperative subjects actively and willingly comply with instructions. Cooperative subjects are usually compliant and offer no resistance to verbal directions or physical actions. However, if they provide any resistance to authority, it is usually passive. Some examples of this resistance are as follows:
  - Yelling.
  - Screaming.
  - Tightening the body.
  - Pulling away.
  - Letting the body go limp.
- **Uncooperative.** Uncooperative subjects will display preattack cues. Uncooperative subjects will actively resist being apprehended. Often, this resistance comes in the form of physical actions, such as—
  - Assaulting verbally.
  - Refusing to be handcuffed.
  - Refusing to be searched.
  - Attempting to push away, forcing a pursuit.
  - Wrestling with the apprehension team.
  - Striking out at the apprehension team with fists, hands and elbows, and feet.
  - Striking out at the apprehension team with weapons, such as knives, clubs, bats, bricks, rocks, or firearms.

APPREHENSION

3-8. US military forces will be in a supporting role to the local civilian authorities or acting as the control force in civil disturbance operations. When making apprehensions, it is best to use apprehension teams. These teams provide quick, organized responses to developing situations. Teams can be organized at the team, squad, or platoon level depending on the number of apprehensions expected. The intelligence information gathered helps to prepare for events where apprehensions may be necessary.

3-9. Each team (if possible) consists of a civilian police or apprehension officer, security element, and recorder. Team apprehension officers make the actual apprehension with the help of the security elements. If there is not a civilian police officer available to act as the apprehension officer, then it may be an officer or noncommissioned officer (NCO).

3-10. Ensuring that the apprehension process is properly conducted and documented, determine who is to be apprehended. Then inform the subject of why he is under apprehension and supervise his handling.

3-11. Security elements help the apprehension officer handle subjects. Security elements move, restrain, and search subjects under the supervision of the apprehension officer. The primary responsibility of the security elements is to watch the crowd and act as a blocking force, keeping the crowd from interfering with the apprehension. Recorders document the event, preferably by video, and complete *DD Form 2708*. If possible, recorders also photograph the apprehension officer with each subject. This aids in the identification process because it links the apprehension officer to the subject and provides the apprehension officer with documentation of the apprehension for use in court.

3-12. Electronically capturing the events as they unfold is essential in all civil disturbance operations. If personnel and equipment are available, videotape the events as they happen or, at a minimum, take still photographs to provide pictorial documentation that may be used later in court. Recording the scene before, during, and after the apprehension provides necessary evidence and can also help eliminate additional hostile and illegal activities. Having control force members take videos or pictures can help control crowd reaction. It reduces the feelings of anonymity that a crowd often enjoys and makes crowd members less prone to unruly or illegal acts.

3-13. Verbal commands given to the subject should be in a voice and manner appropriate for the purpose and the situation. Commands should be simple, concise, and in an understood language. Verbal commands should be convincing and convey an expectation that the instructions will be obeyed. With desperate, frightened people, sympathy combined with an attitude of firm assurance on the soldier's part may avoid additional violent behavior or trouble. An attitude of forceful authority may be needed with an uncooperative person. Commands given by soldiers must be done in a manner and context so that the subject has no confusion as to what the soldier is ordering them to do.

3-14. A subject will need to be transported from the area. In CONUS operations, civil authorities will usually be responsible for providing the transportation of a subject. Exceptions to this may be if the demonstration occurs on an installation. OCONUS operations may require US forces to provide vehicles, such as trucks, buses, or sedans, that will be used to transport a subject. Whenever possible, vehicles used to transport a subject should be modified with barriers to separate them from the drivers. If a barrier cannot be provided, a guard is to be placed in the vehicle. Passenger compartments are checked for items that could be used as weapons. When guards are required, the guard always sits with his weapon away from the prisoner. Prisoners must be placed where they can be best controlled, but they are not fastened to the vehicle with hand irons or Flex-Cufs. When seat belts are available they must be fastened. See *Soldier Training Publications (STPs) 19-95C1-SM and 19-95C24-SM-TG.*

3-15. Female escorts should always be used to transport a female subject. If female escorts are not available, measures must be taken to avoid false charges of molestation or sexual misconduct. Net control stations should be notified of the departure and arrival times and vehicle mileage readings (before and after). At least two male escorts must be present at all times when a female escort is not available. These procedures also apply when the male and female roles are reversed.

FLEX-CUFS

3-16. The two main positions used to place Flex-Cufs on a subject are the standing-supported and the prone.

3-17. When using the standing-supported position, the search man—

- Escorts the subject to a wall or other sturdy vertical surface (see *Figure 3-1*).

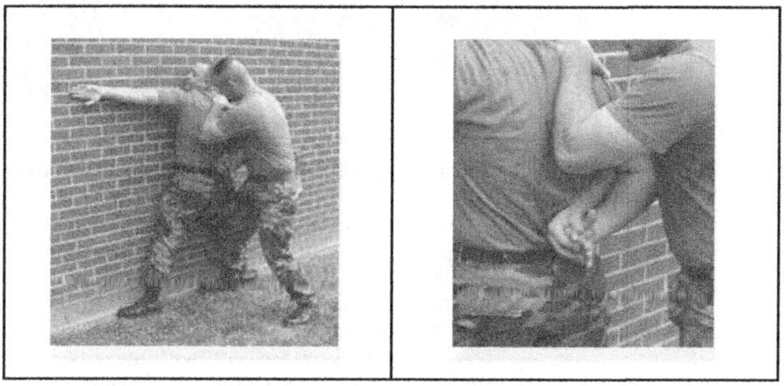

Figure 3-1. Standing-Supported Position

- Instructs the subject (verbally) to do the following:
  - Turn his head away.
  - Place his free arm (palm out, thumb down, and elbow locked) against the surface of the wall.

- Spread his feet until instructed to stop.
- Place his heels against the wall.

- Uses the leg closest to the controlled arm to pin the subject's leg (above the knee) to the wall.
- Reinforces verbal commands with pain compliance techniques if the subject fails to follow instructions.
- Maintains a double 90° hold on the controlled arm with his inside hand. With his free hand, the search man grasps the middle portion of the Flex-Cuf, keeping one cuff up and one cuff down.
- Places the first cuff (bottom cuff) on the subject's controlled arm, maintaining wrist manipulation (see *Figure 3-2*).
- Reaches through the free cuff with his outside hand and instructs the subject to place his free hand in the center of his back. With his outside hand, the search man takes the subject's free hand and completes the restraint by securing the cuff. He ensures that there is at least one-finger distance between the cuff and the subject's wrist (see *Figure 3-3,* page 3-6).

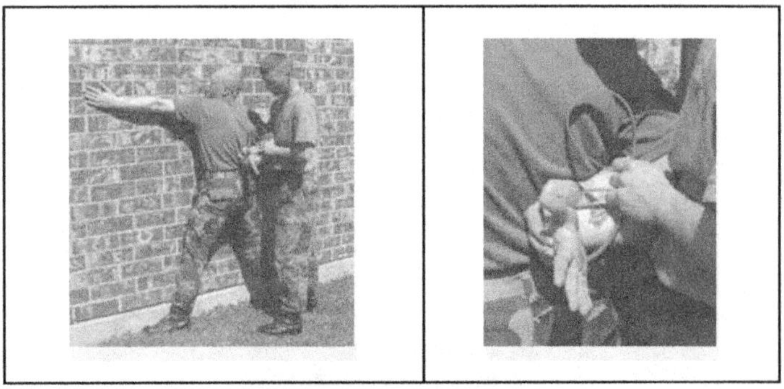

Figure 3-2. Cuffing the First Hand From the Standing Supported Position

3-18. When using the prone position, the search man—

- Grasps the Flex-Cuf with his outside hand, maintains tension on the subject's arm, and places the first cuff on the subject's controlled wrist (see *Figure 3-4,* page 3-6). Next, the search man secures the cuff, keeping at least a one-finger distance between the cuff and the wrist.
- Reaches through the free cuff with his outside hand and instructs the subject to place his free hand in the center of his back (see *Figure 3-5,* page 3-7).
- Reinforces verbal commands with pain compliance techniques if the subject fails to follow instructions. The search man bends the subject's controlled arm (using the natural break of the elbow as a hinge) toward the hand in the middle of the subject's back. With the outside hand, the search man takes the subject's free hand and completes the restraint by securing the cuff.

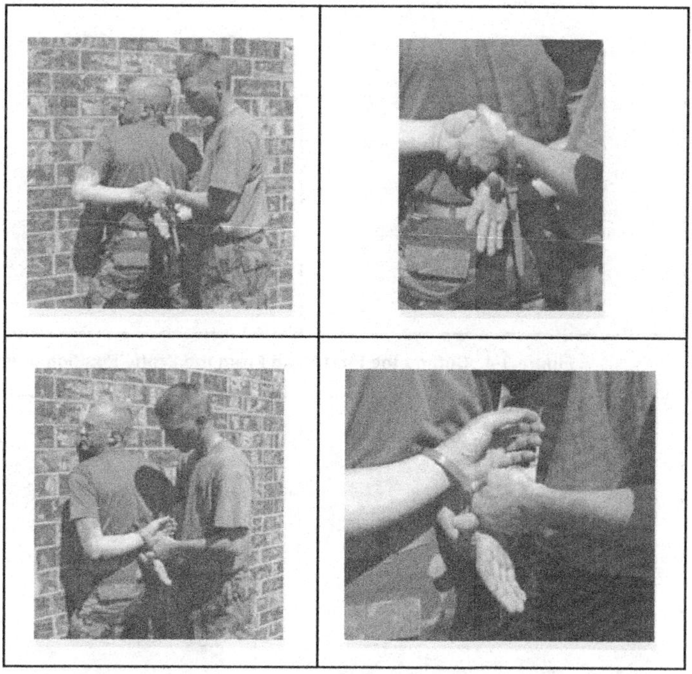

Figure 3-3. Cuffing the Second Hand From the Standing Supported Position

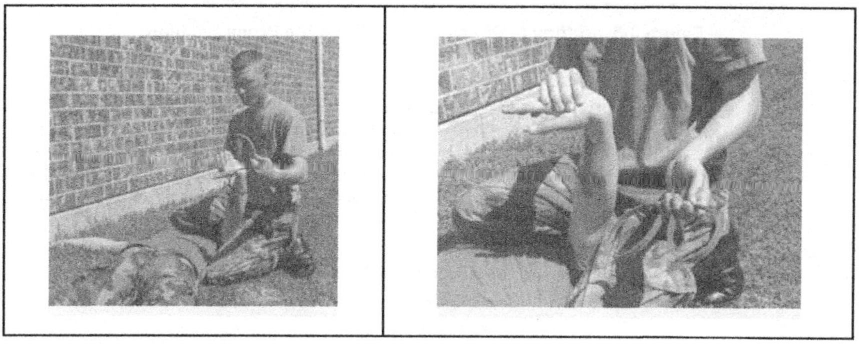

Figure 3-4. Cuffing the First Hand From the Prone Position

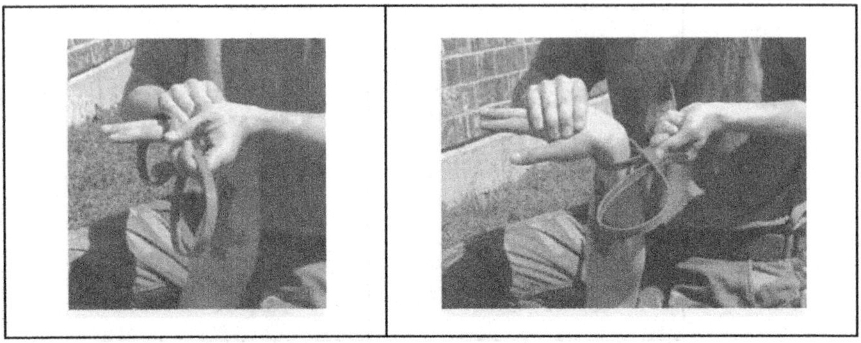

Figure 3-4. Cuffing the First Hand From the Prone Position (continued)

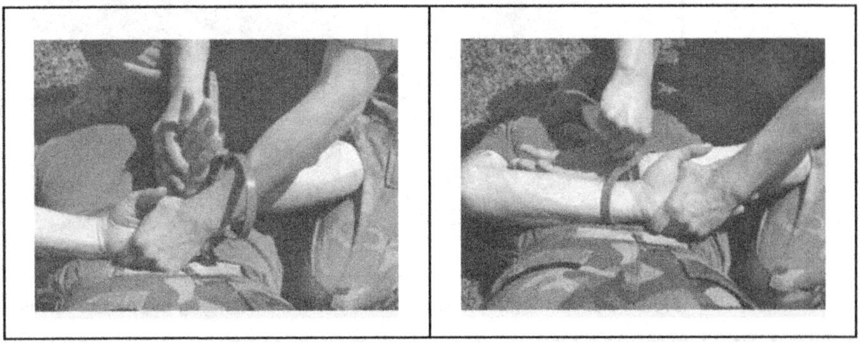

Figure 3-5. Cuffing the Second Hand From the Prone Position

SEARCHES

3-19. Initial contact with the subject is the most dangerous time during any apprehension; therefore, initial contact must be a swift and decisive team effort. Do not give the subject time to think or react. Distract his attention when possible. Apprehension teams are comprised of a search man, threat cover man, and security man (if necessary). Their responsibilities are as follows:

- The search man—
  - Issues verbal commands.
  - Conducts searches.
  - Handles radio communication.
  - Holsters his weapon before approaching a subject and informs the threat cover man that he is holstering his weapon.
  - Performs control procedures and handcuffs the subject.

  - Moves the subject to a safe area.
  - Removes weapons (if any) and sensitive evidence from the subject.
  - Disengages after the subject is restrained.
  - Controls and escorts the subject to a detention area.
- The threat cover man—
  - Provides cover for the search man.
  - Discourages escape and resistance.
  - Discourages hostile interference.
  - Alerts the search man of threats.
  - Resists distractions and maintains target acquisition or scanning.
  - Assesses threats constantly and prepares for the unexpected.
- The security man (if necessary)—
  - Provides cover for the search man.
  - Discourages hostile interference.
  - Alerts the search man of threats.
  - Resists distractions and maintains target acquisition or scanning.
  - Assesses threats constantly and prepares for the unexpected.

Search Principles

3-20. A subject should always be put at a disadvantage and searched immediately, but circumstances may dictate postponing the search until a safer place can be found. At a minimum, a quick pat down can be done en route to a safer location. The longer the apprehension takes, the greater the risk of drawing a hostile crowd and violent acts toward the control force.

3-21. Crowds often sympathize with a subject and try to help him or escalate the activities of the crowd toward violence. A subject's actions may also incite the crowd to violence. The longer a subject is the focal point of an apprehension, the easier it is to stir a crowd to hostile or violent acts. Apprehension teams should choose to move the subject out of the crowd's view. For the safety of the team, always apply Flex-Cufs or handcuffs and escort the subject to an area where a search can be done safely.

3-22. Search team members never operate alone; searches are conducted by a minimum of two team members. Initial steps must be taken to put the subject at a physical disadvantage. Before conducting a search (regardless of rank, sex, position, age, or physical condition), a team member applies the handcuffs or Flex-Cufs. Verbal commands must be short, clear, and distinct.

3-23. A search is conducted on everyone that has been apprehended to ensure that the team and those being apprehended are safe. The search man must abide by the following procedures when conducting a search:

- Search in a clear, secure area (when possible).
- Be ready to counter any resistance or an assault by the subject. Consider the subject armed until proven (by search) otherwise.
- Employ the appropriate level of force immediately, based on the actions of the subject.

- Do not step over the subject when he is at a disadvantage on the ground; go around him and remain out of sight.
- Do a 360° search of the subject's waistline for weapons. Search obvious and inconspicuous areas. Inconspicuous areas such as the groin, armpits, chest, and small of the back must be searched on both men and women; however, do not linger in these areas.

NOTE: Conduct same-gender searches when possible. If mixed gender searches are necessary for speed or security, conduct them in a respectful manner and avoid any action that could be interpreted as sexual in nature. To prevent allegations of sexual misconduct, the apprehension officer shows careful control of the soldiers who perform mixed-gender searches.

- Maintain security.
- Brief subjects quickly on what to do and who to follow.
- Remember that when conducting a search, a subject's body can be divided for a more efficient and thorough search. For example, search from head to toe or from side to side.

NOTE: The search man announces the items found and secures weapons and contraband.

- Military forces (when supporting civil authorities) may perform searches of people and property that are not usually subject to military law during a civil disturbance.
- Remember that it is always better for civilian police to search civilians and civilian property (when possible).
- Remember that if military forces are called upon to support police it is usually a last resort and there are not enough civilian officers available to accomplish the mission.
- Military personnel may search people incident to an apprehension.

3-24. The prescribed methods of searching someone are the frisk, the standing, and the prone. The method of search used largely depends on the situation and the mission. Searches are made not only of the subject and his personal effects such as purses, bags, and wallets, but also of the immediate surrounding area. This prevents the subject being searched from grabbing a weapon or destroying evidence. It may also be necessary for military forces to search private property, including vehicles, if there is—

- Reasonable belief that a person has committed or is committing a violent crime and is hiding in a building or a vehicle.
- Reasonable belief that a vehicle contains weapons or instruments of violence.
- Probable cause for searching a building or a vehicle.
- Probable cause to believe that (unless immediate action is taken) evidence of a crime will be destroyed before a warrant can be obtained.

NOTE: Search teams should always consist of two personnel—one to conduct the search, and the other to provide security and overwatch for the soldier conducting the search. Before being searched, a subject should always be restrained and put at a disadvantage by

securing his hands with handcuffs or Flex-Cufs. Ask the subject if there is anything in his pockets that could hurt the searcher, such as knives or needles.

Frisk Search

3-25. The frisk search (also known as the stop and frisk, pat down, and terry stop) is a search incident to a lawful stop. Subjects are not under apprehension at this point, but if it were an apprehension situation, handcuffs or Flex-Cufs would be applied first. Frisk searches are used when there is reasonable belief that a subject may be armed with anything that could be considered a safety threat.

3-26. Legal and regulatory circumstances for conducting a frisk search are covered in the *Manual for Courts-Martial (MCM)*. When a lawful stop is performed, the person stopped may be frisked for weapons when that person is reasonably believed to be armed and dangerous.

3-27. The primary purpose of a frisk search is safety. Frisk-search techniques are to be used only on a compliant subject. Frisk searches only cover the outer garments. Do not search the pockets of the individual unless a weapon is suspected. A frisk search begins when the search man informs the subject of the intent to search.

- During the preparation phase, the search man—
  - Tells the subject that he must perform a search for weapons for the subject's safety and his own.
  - Tells the subject to turn and face away from him and instructs the subject to keep his hands out to his sides where they are visible.

NOTE: The cover man providing security moves around the subject and positions himself out of reach and in front of the subject.

  - Instructs the subject to spread his feet until instructed to stop (more than shoulder width) and point his toes outward compromising his balance (see *Figure 3-6*).

NOTE: Do not kick at the subject's feet to spread them apart; it is unnecessary and unprofessional.

  - Tells the subject to bring both of his hands back to the small of his back with palms out, thumbs up, and the back of his hands together.
  - Instructs the subject not to move (see *Figure 3-7*).
- During the performance phase the search man—
  - Approaches the subject from behind (if possible) at a 45° angle, off center. The search man approaches with caution to ensure that his feet are to the outside of the subject's stance. See *Figure 3-8,* page 3-12.
  - Stays just inside arms reach and grips the subject's fingers with his support hand. The search man firmly grips at least one finger on each of the subject's hands (both thumbs, both index fingers, or an index and middle finger). An overhand or an underhand grip can be used effectively.

Figure 3-6. Approaching the Subject to Perform a Frisk Search

Figure 3-7. Giving Instructions to Subject Before Frisk Search

NOTE: The primary purpose of gripping the fingers is to control the subject's hands while performing the frisk search (see *Figure 3-9,* page 3-12). If the subject is going to resist or become aggressive, he will attempt to free his hands first. If the subject attempts to pull his fingers from the search man's grip, the search man must respond instantly by increasing his grip. If the grip on the fingers is lost, the search man should immediately push away from the subject and create a safe distance. This will allow the search team to respond with the necessary force to regain control of the situation.

Figure 3-8. Performing a Frisk Search

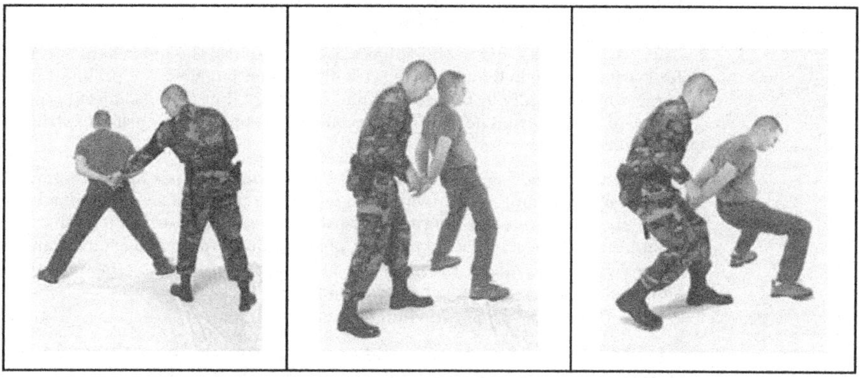

Figure 3-9. Gripping the Subject's Fingers During a Frisk Search

- Ensures that the subject always looks away from him.
- Places his strong hand on the outside of the subject's elbow (similar to the handcuffed escort position).

NOTE: The search man searches one side (the front, back, or side) at a time from head to foot.

- Divides the subject (visually) straight down his middle and does not cross the line while conducting the search.

NOTE: The search man is just feeling or searching for possible weapons, so an open hand should be used.

- Works at arm's reach. It is the sense of touch not the sense of sight that is being used. While performing the search, search men should stay behind and to the side of the subject. This provides the search man with a tactical advantage and prevention from being struck by the subject's elbows or knees if he should begin to resist.

3-28. When searching below the subject's knees, the search man takes a half step back and to the outside. He squats down, but does not drop his knees to the ground.

3-29. If no weapons are found and the subject is to be detained, the hands are secured with Flex-Cufs and the subject is escorted to the holding area. If the subject is to be released, collect information (name and address) and escort him away from the crowd.

Standing Search

3-30. Safety is the primary concern when dealing with a subject under apprehension. Therefore, the search man must protect himself by restraining subjects under apprehension before conducting the standing search. All subjects under apprehension have the propensity to turn violent or hostile. Putting a subject at a physical disadvantage provides safety for the subject and the search man. It limits the subject's ability to hide, destroy, or remove any evidence he may have on him.

3-31. Apply Flex-Cufs to the subject quickly during the apprehension. It is impossible to predict what a subject will do when put under apprehension. A subject may initially be cooperative and compliant, but as the subject becomes aware of his situation he may be overcome with emotion and begin to struggle or pull away.

3-32. There is no legal or regulatory prohibition against opposite-gender searches. Common sense prevails and whenever possible, searches should be done by soldiers of the same gender to avoid obvious questions and complaints. Once a subject is put at a physical disadvantage by cuffing and is more easily controlled, it affords the necessary time to get someone of the same gender there to conduct the search.

- During the preparation phase, the search man—

NOTE: Standing searches always begin with the subject in the handcuffed escort position (see *Figure 3-10,* page 3-14).

  - Takes his outside hand (farthest from subject) and places it on the outside of the subject's elbow while his inside hand (closest to subject) has a reverse-handshake grip of the subject's hand. At this point, both hands are on the same arm (see *Figure 3-10,* page 3-14).
  - Ensures that his position is on the same side of the subject's body being controlled (behind and to the side). When changing sides, change hand position.
  - Instructs the subject to spread his feet until instructed to stop (usually more than shoulder width) and point his toes outward, compromising the subject's balance.

Figure 3-10. Handcuffed Escort Position

NOTE: Do not kick at a subject's feet to spread them; it is unnecessary and unprofessional. If the subject becomes noncompliant or aggressive, pain compliance is usually the best tool to use, but only use enough force to gain compliance.

  - Uses his outside hand to conduct the search. His inside hand is always used to maintain control during the search.
- During the performance phase, the search man—

NOTE: Only one side (the front, back, and side) of the subject is searched at a time from head to toe (see *Figure 3-11*).

  - Divides the subject (visually) (straight down the middle using the spine as the centerline). He does not cross this line while conducting the search. Searches must be systematic, from head to foot. If there is an interruption, it must be started all over again.
  - Searches with an open hand, using a rub down or squeezing method. Do not only pat down like in the frisk search.
  - Works at arms reach. It is the sense of touch, not the sense of sight that is being used. To prevent the possibility of being struck by an elbow or knee, the search man stays behind and to the side of the subject.
  - Takes a half step back and to the outside when searching below the subject's knees. The search man squats down, but does not drop his knees to the ground. Maintaining positive control is essential throughout the search, so never the reverse handshake is never released.
  - Change sides when satisfied that all items have been removed from the side in which the search started. Without losing positive control, the search man conducts the search on the other side of the subject.

Figure 3-11. Performing the Standing Search

NOTE: Changing sides is accomplished by moving the outside hand (the searching hand) to the subject's uncontrolled handcuffed hand located on the outside. The outside hand and forearm will cross over the top of the reverse-handshake grip of the inside hand (see *Figure 3-12,* page 16).

- Crosses over the reverse-handshake grip with the outside hand and grips the subject's uncontrolled handcuffed hand in a reverse-handshake grip.
- Releases (quickly) the first reverse-handshake grip moving that hand to the subject's elbow as he sidesteps to the side yet to be searched. The search man repeats the standing search technique used on the first side of the subject.
- Gives strong verbal commands and uses only the force necessary to control the subject if he resists or becomes aggressive at any time during the search.
- Removes all items found during the search to be evaluated as potential evidence or a weapon. Items not considered a weapon or of evidentiary value will be returned to the subject at a later time.

Prone Search

3-33. Safety is the primary concern when dealing with a person under apprehension. Therefore, the search man must protect himself by restraining subjects under apprehension before conducting this search. Remember that all subjects under apprehension have the propensity to turn violent or hostile. That is the reason a prone search would be necessary. After performing an arm bar takedown on a noncompliant subject, it is especially important for the search man to put him at a physical disadvantage to provide for the subject's safety and his own. By doing so, the subject's ability to hide, destroy, or remove any evidence is limited.

Figure 3-12. Performing the Standing Search (Changing Sides)

3-34. The safest way to handcuff or Flex-Cuf a noncompliant subject is in the prone position, when the search man has greater control of the subject. It is usually the safest position in which to conduct a search

3-35. Handcuff the subject quickly during the apprehension. It is impossible to predict what a subject will do when put under apprehension. Often a subject may initially be cooperative and compliant; however, as the he becomes aware of his situation, he may become overcome with emotion and begin to struggle or pull away.

3-36. There is no legal or regulatory prohibition against opposite-gender searches. Common sense prevails and whenever possible, searches should be done by the same gender to avoid obvious questions and complaints. Once a subject is at a physical disadvantage by cuffing and is more easily controlled, it affords the necessary time to get someone of the same gender there to conduct the search.

NOTE: The subject is put into a physical disadvantage when he is taken facedown to the ground using an arm bar takedown and cuffed.

- During the preparation phase, the search man—
  - Ensures that the Flex-Cufs are not too tight.
  - Rotates his entire body around the subject's shoulder and center at the outside of the subject's hip with his shin over the subject's shoulder and one knee on the ground. At least one knee should remain over the subject's back (near side), with the other on the ground.
  - Pins the subject down by pressing his knee on the side of the subject's back or down on his closest elbow if resistance or aggression is encountered. He simultaneously grips the subject's closest handcuffed hand, grabs it with a reverse handshake, and applies pain for compliance.

    - Uses strong, forceful verbal commands, such as "Stop resisting, stop moving around."
    - Ensures that the subject's feet and ankles are crossed.

CAUTION

Stay alert. If the subject uncrosses his feet and ankles, it is usually a sign of preattack.

    - Removes all items found during the search so they can be evaluated as potential evidence. Items not considered a weapon or of evidentiary value will be returned to the subject at a later time.
- During the performance phase, the search man—

NOTE: The search man must be systematic from head to foot while conducting the prone search. If interrupted during the search, it is important to start from the beginning to ensure that nothing is missed. It is better to go over the same area two or more times than to miss something that can be used as a weapon.

NOTE: The search man searches only one side (the front, back, or side) of the subject at a time until that side is complete.

    - Divides the subject (visually) straight down the middle, using the spine as the centerline.

NOTE: He does not cross this line while conducting the search.

    - Searches with an open hand, using a rub down or squeezing method for best results. He does not just pat down the subject like in the frisk search.
    - Works at an arm's reach, searching only the area of the subject's body that is not in contact with the ground. The sense of touch, not the sense of sight, is being used.
    - Stops the search as the subject's knees are reached. He taps one of the subject's legs and firmly commands him to raise his foot up and back, allowing for an inspection of the bottom of his foot. He then searches from the knee to the ankle area. This is performed on both the subject's legs and feet.

NOTE: At this point, the entire back side of has been searched. What remains to be searched is the front side of the subject (the chest and front of the hips, waist, groin, and thighs).

    - Pivots back and faces the subject. Using the hand closest to the subject's head, the search man reaches across the subject's back and firmly grips the shoulder, rolling the subject up on the his side toward him to expose the areas not yet searched.
    - Rolls the subject back down on his chest once all areas have been searched.

3-37. The search man must be completely satisfied that he has found everything. He must be especially sensitive to areas where weapons are often concealed like the waistband, the small of the back, ankles, pockets, shoes, the tongue of the shoes, and the groin area.

NOTE: Before removing subjects from the area, a brief summary must accompany them. The summary must include the subject's name, what he is suspected of doing, the time and date of the incident, and who saw him do the action.

DETENTION

3-38. Civil authorities must provide adequate detention facilities for all subjects. Authorities must be prepared to detain large numbers of people. They may choose to expand existing detention facilities or set up temporary facilities to accommodate the extra load. If possible, large-scale arrests are delayed until sufficient detention facilities have been set up.

3-39. If US military forces are committed to support local authorities, commanders should coordinate with them to ensure that adequate detention facilities are available and to learn their locations and capacities. If there are more detainees than civil detention facilities can handle, civil authorities may ask the military to provide support by setting up and operating temporary facilities. Army correctional facilities cannot be used to detain civilians. A temporary military detention facility can be set up if—

- Federal troops have been employed according to provisions.
- The TF commander has verified that available civilian detention facilities can no longer accommodate the number of prisoners awaiting arraignment and trial by civil courts.
- The Army Chief of Staff has granted prior approval.

3-40. Use of the temporary facility ends as soon as civil authorities can take custody of the detainees. Military forces are responsible for the custody, health, comfort, and sustenance of all detainees in their facilities until custody is transferred to civil authorities. Temporary facilities cannot be used to confine people arraigned or convicted in civil courts.

3-41. Women are detained in their own facility. Temporary facilities are supervised and controlled by officers and NCOs trained and experienced in military correctional operations. Females guard females. If two female guards are unavailable, use two male guards or one male and one female guard (see *FM 3-19.40*).

3-42. The same operational procedures that apply to the management of an installation confinement facility (see *FM 3-19.40*) apply to the management of temporary detention facilities, except for training, employment, and administrative discipline. Guards and support personnel under direct supervision and control of military officers and NCOs need not be trained or experienced in military correction operations. They must be specifically instructed and closely supervised in the proper use of force, custodial procedures, and the completion of military and civilian forms and reports that may be used. Soldiers that may be tasked to operate a temporary detention facility should be familiar with the forms and reports used for civilian apprehensions too.

3-43. Temporary facilities are set up on the nearest military installation or on suitable property under federal control. Ideally, the facility should be close enough to the disturbance area to minimize transportation and escort needs. However, it should be far enough away not to endanger those being detained. Whenever possible, existing structures are adapted for this use, but construction may be needed to provide segregation for ensuring effective control and administration.

3-44. The basic structure must include the following:

- Search areas.
- Holding areas for incoming men and women.
- A processing area.
- Holding areas for men and women who have committed misdemeanors.
- Holding areas for men and women who have committed felonies or are violent.
- A holding area for detainees' property and evidence obtained during a search.
- A holding area for administrative support and records.
- A medical station.
- Latrines.

For more information on these structures see *FM 3-19.40.*

3-45. Facility personnel must ensure that proper sanitation is maintained. When large numbers of people are detained or processed through a facility, sanitation becomes a problem. Medical personnel must conduct regular health inspections to detect unsanitary practices and conditions.

3-46. Facilities must be organized for a smooth flow of traffic. Processing stations must be set up so there is a linear or circular sequence of movement. These stations are set up out of sight of the holding areas. If possible, they are separated by a door to reduce noise. To reduce distractions, each station may be partitioned. Detainees may be more cooperative if they are out of sight of each other.

3-47. Injured people are given prompt medical treatment and transportation to medical facilities when necessary. A medical aid station for screening detainees and treating minor injuries is set up inside or next to the detention facility. Treatment areas, however, must be out of sight of the processing and holding areas. If possible, access to the medical facility bypasses the holding and processing areas. Facility personnel may also consider setting up a separate holding area for injured detainees.

3-48. Detention facility operation plans must contain emergency procedures. As a minimum, the following areas must be addressed:

- Fire evacuation.
- Disturbance control.
- Facility security.

NOTE: Depending on the situation, other emergency procedures may be needed.

3-49. Upon arrival at the detention facility, the detainee is logged in and searched, regardless of whether a complete search has already been conducted by the apprehending team in the field. Separate search areas are set up for men and women. Weapons, contraband, flame-producing devices, suspected evidence, money, and high-value items are confiscated. Medications are also confiscated. Medical personnel screen detainees that have had medications confiscated. Receipts are provided for any property or evidence that is confiscated. Confiscated items are tagged, and the items are stored in a controlled-property area.

3-50. As a detainee is brought to the facility, a file is initiated. Assigned detainee case numbers are used on all paperwork, such as logs, evidence tags, reports, and visual documentation. All paperwork (including photographs) that began outside the detention facility is marked with the case number. Facility personnel may also use hospital identification tags. Using indelible ink, the case number is written and attached to the tag on the detainee's wrist. Different colors may be used to identify different subject classifications, such as misdemeanors, felonies, or violent offenders. If opposing factions are involved, procedures are established to ensure that members of opposing factions are not processed together or detained in the same holding area.

3-51. After a detainee has been searched and classified, he is taken to a processing station where his paperwork is processed. If subjects passively resist by going limp, they may be moved by wheelchair. This reduces the number of escorts needed and the fatigue to facility personnel. Paperwork is reviewed to ensure that all information is complete, including charges, witnesses, and reasons for the apprehension.

3-52. Equipment such as height charts, scales, fingerprinting equipment, and cameras must be available for completing the police report and identifying the subject. Detainees may use aliases and not carry identification, so pay special attention to obtaining as much information as possible about the detainee's physical characteristics.

3-53. After processing is complete, the detainee is placed in the appropriate holding area. Paperwork for the detainee is forwarded to the administrative section. Files are reviewed for completeness and to determine the disposition of the detainee. Information from the file may be placed in a computer to find repeat subjects. Computers may be used to file criminal information only.

3-54. Custody transfers and release procedures must be coordinated with civil authorities and the appropriate legal counsel. Every effort must be made to arraign subjects quickly. The purpose of a detention facility is not to keep people off the streets, but to aid in processing subjects through the legal system. To speed up the release process, planners should consider issuing citations or subpoenas for minor offenses.

3-55. Military force members may have to respond to writs of habeas corpus. Writs are court orders addressed to a prisoner's custodian. Such a writ directs the custodian to bring the prisoner to court to determine the legality of the prisoner's apprehension and detention. Custodians must ensure that all documentation concerning the case is brought to court with the detainee.

3-56. Military personnel must obey writs issued by federal courts. For writs issued by a state court, the custodian or his legal advisor should respectfully reply that US authority is holding the prisoner. The Office of the SJA can answer any questions and explain the correct procedures.

Chapter 4

Riot Shield and Riot Baton Techniques

A soldier may be required to be part of a riot control formation where his primary weapon is a riot baton and his only protection a riot shield. A riot baton in the hands of a well-trained soldier is an invaluable weapon in a crowd control situation and may also be the most appropriate weapon. A riot baton is a very versatile weapon that can be used as an offensive and/or defensive weapon. The riot shield offers protection from the top of the head to just below the knees.

OVERVIEW

4-1. Use of the riot baton and riot shield is based on METT-TC and the commander's choice-of-force option. Formations committed to a riot with riot batons must have lethal overwatch, lethal weapons, and NLW employed within the formation and reserve forces with additional NL and lethal capabilities. These forces are positioned to rapidly reinforce the control force formation.

4-2. During the termination phase of the disturbance (when the violence has subsided and is expected to remain that way), soldiers are placed in a defensive posture. They may be equipped with a riot shield and a riot baton to perform their tasks. The riot shield is primarily a defensive weapon, which may be used in an offensive mode. It is primarily used as an offensive weapon when the formation is in close contact with rioters. The riot baton can also be used when formations are on the offensive and in contact with the crowd. However, crowd control elements employed with riot shields and riot batons must have lethal weapons and NLW available.

WARNING

Riot control formations are never employed with riot shields and riot batons only. There must always be lethal overwatch elements in position and lethal and NL capabilities within the riot control formation with a ready reserve force armed with lethal and NL capabilities.

RIOT SHIELD

4-3. In the hands of a properly trained soldier, the riot shield has proven to be the first line of defense for a formation. The riot shield may be used as an offensive weapon when in contact with an aggressive crowd.

4-4. The riot shield may be held with one arm. It is secured to the weak side arm with a self-sticking strap system and a handle. This system is designed to make the riot shield an extension of the soldier's arm. The riot shield is held at a slightly inward angle. This allows for debris thrown at the soldier to fall harmlessly to the ground (see *Figure 4-1*).

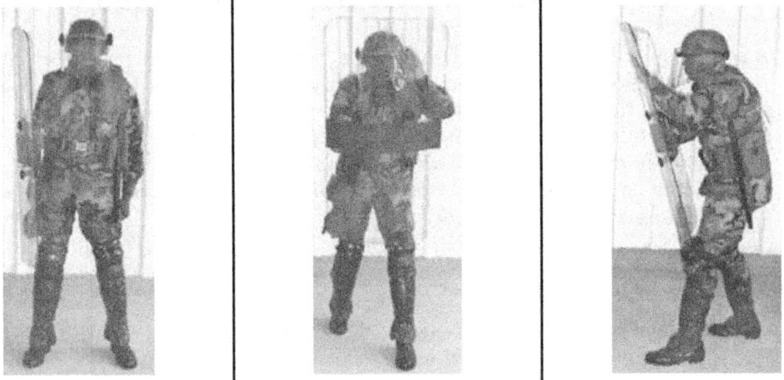

At the Ready | **On Guard** | **On-Guard Profile**

Figure 4-1. Riot Shield Positions

RIOT SHIELD CARRIES

4-5. The two basic riot shield carries are as follows:

- **At the ready.** The at-the-ready position is used while in a column formation or on the move toward hostilities. This position is designed for rapid movement. It should not be used when in contact with a crowd.
- **On guard.** The on-guard position is used when the riot control formation is in contact with the crowd or contact is eminent. It is used primarily in the following formations:
  - Line.
  - Wedge.
  - Echelon.
  - Diamond.
  - Circle.

DEFENSIVE AND RETENTION TECHNIQUES

4-6. The primary purpose of the riot shield is for defense. However, each riot shield holder must be proficient in riot shield retention techniques.

DEFENSE

4-7. The first line of defense for the formation is the riot shield. It is designed to withstand strikes from objects thrown by the crowd. In addition to protecting the riot shield bearer, it also protects other members of the formation from thrown objects.

RETENTION

4-8. When the control force is in contact with the crowd, the crowd will attempt to strip the riot shield from the bearer. When a subject grabs the top of the riot shield (see *Figure 4-2*), the riot shield holder slaps the riot shield with his strong hand near the subject's hands and gives the following commands (while continuing to slap the riot shield until the subject releases it):

- "Get back."
- "Get away."
- "Stop."

Figure 4-2. Riot Shield Retention Technique (Top Attempt)

4-9. If the riot shield is grabbed from the bottom (see *Figure 4-3,* page 4-4), then the riot shield holder forcefully drops to one knee, pinning the subject's fingers and/or hands to the ground. This must be done quickly and with force.

RIOT BATON

4-10. A riot baton in the hands of a properly trained soldier is a formidable weapon. However, it must be used in conjunction with other measures to be most effective. Soldiers must be trained with the riot baton to the point that its various techniques are automatic to them. This training must also include

Figure 4-3. Riot Shield Retention Technique (Bottom Attempt)

learning the vulnerable points of the human body so they can avoid areas that may cause permanent injury or death when struck.

> **WARNING**
> **The riot baton is never raised above the head to strike a subject in a club fashion. Not only is it likely to cause permanent injury; but it also gives an unfavorable image of the control force. The soldier is also vulnerable to an attack on his rib cage when his arm is raised.**

TARGET AREAS OF THE HUMAN BODY

4-11. The target areas of the human body are divided into three areas that represent the primary-, secondary-, and final-target areas. These areas are color coded as green (shown as gray in *Figure 4-4*), yellow (shown as white in *Figure 4-4*), and red (shown as patterned gray in *Figure 4-4*).

Primary-Target Areas

4-12. The primary-target areas are those areas which, when struck or restrained, cause the least amount of trauma to the body. These include the meaty or muscular areas of the body, such as the—

- Foot, shin, and instep.
- Inside and outside of the thigh.
- Lower abdominal region.
- Fore- and upper-arms.

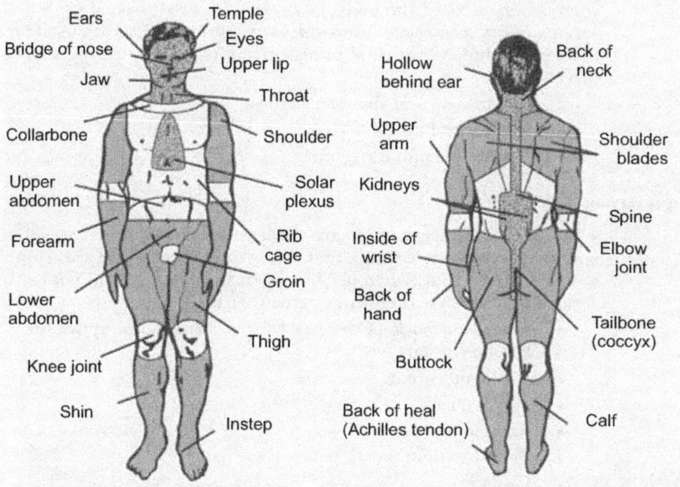

Green (lowest level of trauma) Primary-target areas:

Reasoning: Minimum level of resultant trauma. Injury tends to be temporary rather than permanent; however, exceptions can occur when applying as a restraint technique.

NOTE: When performing a block with a riot baton, the whole body is a green area, except for the head, neck, and spine.

Yellow (moderate level of trauma) Secondary-target areas:

Reasoning: Moderate level of resultant trauma. Injury tends to be more permanent, but may also be temporary when applying as a striking technique.

Red (deadly force) Final-target areas:

Reasoning: Highest level of resultant trauma. Injury tends to be permanent rather than temporary and may include unconsciousness, serious bodily injury, shock, or death.

Figure 4-4. Escalation of Trauma Chart

Secondary-Target Areas

4-13. The secondary-target areas are those areas of the body which, when struck or restrained, cause a moderate level of trauma to the body. This type of trauma tends to be longer lasting because the time to heal is longer than with other parts of the body. If these areas are struck with a riot baton, serious injury may occur. However, some of these areas are used for control or restraining holds that tend to lessen the threat of injury. Some secondary targets are the—

- Collar bone and shoulder blades.
- Elbows and knees.
- Upper abdominal region.

Final-Target Areas

4-14. The final-target areas are those areas which, when struck with a riot baton, cause serious trauma that tends to be fatal or critical. However, there are some areas that (when used for restraining only), will not cause death or serious injury. Some final-target areas are the—

- Front and back of the head (eyes, ears, nose, upper lip, and hollow behind the ear).
- Neck and throat.
- Upper chest.
- Xiphoid process.
- Spinal column.
- Groin area.

JOINTS OF THE HUMAN BODY

4-15. The joints of the human body are often described as hinges and, when manipulated in a certain fashion, allow for better control of the subject being restrained, especially if he is aggressive. When applied correctly, minimal pressure is needed to control the subject. When a subject attempts to resist, he creates additional pain for himself. By anatomically locking out these hinges, the subject will be forced to comply. When restraining a subject, emphasis must be placed on locking out two of the three joints of the upper body or two of the three joints of the lower body. The joints are as follows:

- **Shoulder.** Apply pressure to the shoulder to prevent the subject from freely moving the upper portion of his body.
- **Elbow.** Apply pressure on the elbow or just above the elbow on the ulner nerve.
- **Wrist.** Apply pressure by pancaking the hand to twist or lock out the wrist. Apply pressure at the break of the wrist and push the subject's thumb toward a stationary object (the wall or the ground) to lock the hand out.
- **Hip.** Apply pressure to the hip to knock the subject off balance or alter his center of gravity.

- **Knee.** Apply pressure to the knee by pushing outward on it to cause the subject to be placed off balance, or use the knee as a hinge for pain compliance.
- **Ankle.** Apply pressure to the ankle by clasping and turning it in a counterclockwise manner to lock the hinge out.

TYPES OF RIOT BATONS

4-16. The Army is currently fielding two types of riot batons, wooden and expandable. The most common riot baton is the 36-inch hickory riot baton with thong. There is also the 24- to 36-inch expandable riot baton, which has been added to the nonlethal capabilities set (NLCS). Regardless of which riot baton the soldiers are equipped with, all techniques for blocking and striking are identical. Similar to the riot shield, the riot baton offers soldiers both offensive and defensive capabilities.

4-17. Soldiers must be properly trained in all blocking and striking techniques. Improper use of the riot baton by an untrained soldier has the potential for creating a greater problem than what already exists.

NOMENCLATURE OF THE RIOT BATON

4-18. The riot baton is divided into four basic parts: the long end, long portion, grip portion, and grip end (see *Figure 4-5*). Although the expandable and wooden riot batons differ, their nomenclature is quite similar.

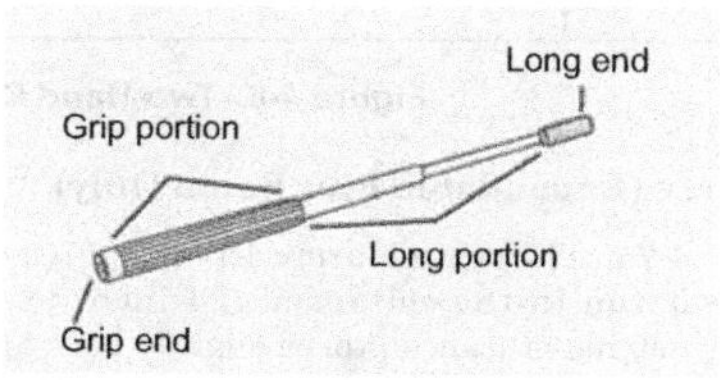

Figure 4-5. The Riot Baton

TYPES OF RIOT BATON CARRIES AND POSITIONS

4-19. As with any other weapon that the soldier will use, there are certain carries and positions from which the riot baton may be used offensively or defensively.

Two-Hand Carry

4-20. The two-hand carry is the most common carry for the riot baton. It is intended for executing all blocks and returning with strikes. When the expandable riot baton is in the collapsed position (at 24 inches), the long end of the riot baton is angled downward approximately 45°. When it is extended to 36 inches the long end is angled upward approximately 45°.

4-21. The wooden riot baton is secured with the soldier's hand by means of a leather thong. He hooks the thumb of his strong hand (palm down) in the thong, pulls the thong across the back of his hand, and grips the riot baton. This provides him with a very secure grip (see *Figure 4-6*).

4-22. The two-hand carry is intended for executing all blocks and returning with strikes. The soldier grips the riot baton with both hands. His strong hand grips the short end of the riot baton, and his weak hand grips the long portion of the riot baton, 2 to 4 inches from the end. His palm is down with his strong hand pulled in close against his strong side hip.

4-23. Wherever the long portion of the riot baton is pointed is where the soldier will strike or jab. The riot baton should be kept at a 45° angle to the ground. All blocks and two-hand jabs and strikes are executed from the two-hand carry.

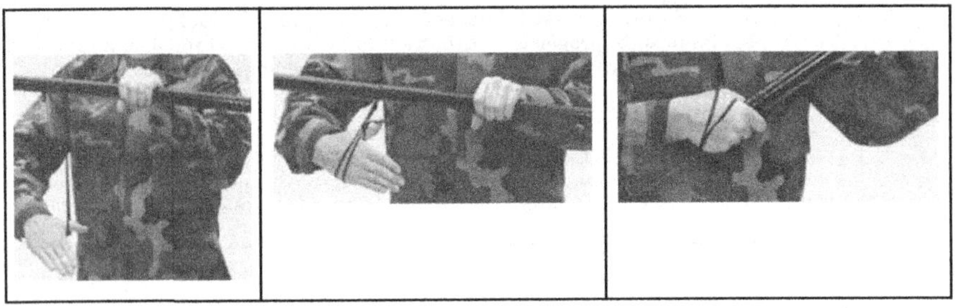

Figure 4-6. Two-Hand Carry

Outside-Arm Carry (Expandable Riot Baton Only)

4-24. The outside-arm carry (see *Figure 4-7*) is used with the expandable riot baton in the collapsed (24-inch) position only. Assuming a wide-based defensive stance (knees slightly bent and feet shoulder width apart), a soldier draws the grip portion of the riot baton with his strong hand. Without striking out or widely swinging the riot baton, he brings the long portion up against the outer part of his strong arm (the tricep area). The soldier's weak hand should be raised in a position to protect and block strikes from an aggressor. It is from this carrying position that a soldier can effectively use both the front and rear strikes. Front and rear strikes should never be performed with an expandable riot baton in the expanded position or with a 36-inch wooden riot baton.

> **CAUTION**
> The expandable and 36-inch wooden riot batons provide a greater risk of injury to the soldier's wrist.

Figure 4-7. Outside-Arm Carry

Vertical High-Profile Carry

4-25. Assuming a wide-based defensive stance, the soldier draws the riot baton with his strong hand by the grip portion. Once the riot baton has been drawn from the carrier, the soldier swings the riot baton in a downward motion, fully extending the arm and holding the riot baton down and parallel to the strong side leg. The soldier's weak hand is raised in a position to protect and block strikes from an aggressor (see *Figure 4-8*, page 4-10). Although this carry can be used with both riot baton systems, it is best suited for the expandable riot baton collapsed to 24 inches. The vertical high-profile carry clearly shows the soldier with a drawn riot baton in an effective defensive posture and ready for attack. A drawn riot baton in the hands of a trained soldier may offer just the kind of deterrence needed to stop the activity of an aggressor.

Vertical Low-Profile Carry

4-26. Assuming a wide-based defensive stance, the soldier draws the riot baton with his strong hand by the grip portion. Once the riot baton has been drawn from the carrier, the soldier swings the riot baton in a downward motion, fully extending his arm and holding the riot baton down and parallel to his strong side leg. His weak side hand is raised in a position to protect and block strikes from an aggressor (see *Figure 4-9*, page 4-10). Although this carry can be used with both types of riot batons, it is best suited for the expandable riot baton collapsed to 24 inches. This position portrays the soldier in an effective defensive position ready for an attack, but with a possible hidden weapon. The major difference between the vertical low-profile and high-profile carry is that in the low-profile carry the soldier hides the riot baton behind his strong side leg, allowing for some element of surprise to a potential aggressor. A drawn riot baton in the hands of a trained soldier may offer the deterrence needed to stop the activity of an aggressor.

Figure 4-8. Vertical High-Profile Carry

Figure 4-9. Vertical Low-Profile Carry

RIOT BATON BLOCKING TECHNIQUES

4-27. There are five basic types of riot baton blocking techniques that are executed from the two-hand carry position.

High Block

4-28. The high block is very effective in blocking a downward, vertical strike directed at the top of the head and shoulders. This type of downward, vertical strike may be from a club, pipe, or similar object. The high block is a two-step movement. To perform a high block the soldier performs the following steps:

Step 1. Raises the riot baton straight across his body until it is parallel to the ground (see *Figure 4-10*).

Step 2. Brings the riot baton straight-up in front of his body until the long portion of the riot baton is in front of his forehead (approximately 2 inches from his head). The riot baton should be in a horizontal position above his head and slightly angled down toward his weak hand, as it absorbs the shock from the strike (see *Figure 4-10*). The fingers of his weak hand should be open and behind the long portion of the riot baton when blocking the object to protect his fingers from being smashed by the strike.

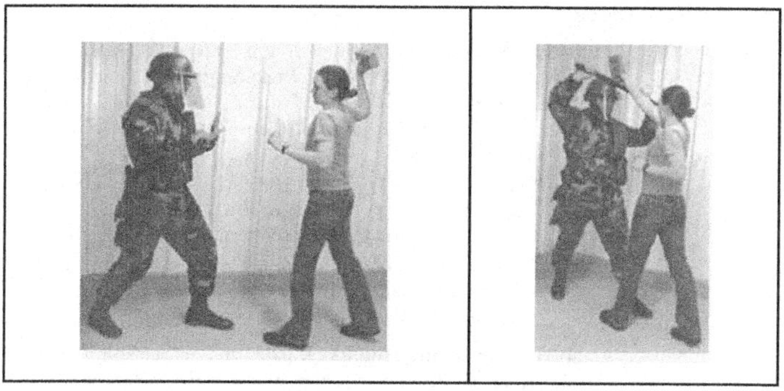

Figure 4-10. High Block (Steps 1 and 2)

Low Block

4-29. A low block is very effective in blocking an upward vertical strike directed at the groin, lower abdomen, chest, or chin. This upward vertical strike may be from an individual's foot, knee, or fist (for example, an upper cut to the chin). To perform a low block, the soldier performs the following steps:

Step 1. Raises the riot baton straight across his body until it is parallel to the ground (see *Figure 4-11*, page 4-12).

Step 2. Brings the riot baton straight down in front of his body, bending at his knees and keeping his body upright until the long portion of the riot baton is just below his knees (see *Figure 4-11*). The riot baton should be horizontal and parallel to the ground as it absorbs the shock from an upward strike. His weak hand fingers should be open and behind the long portion of the riot baton when blocking the object to protect his fingers from being smashed by the strike.

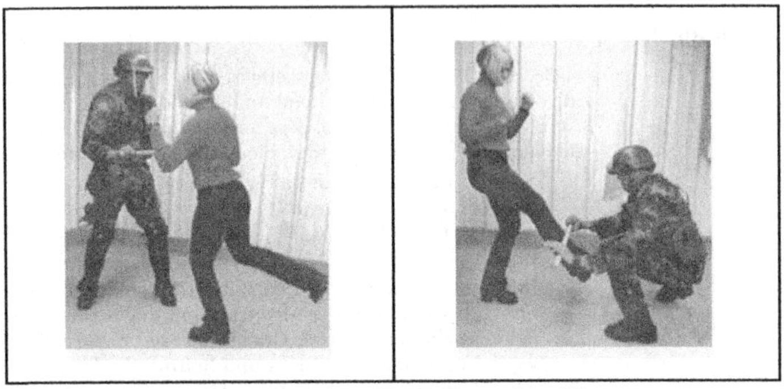

Figure 4-11. Low Block (Steps 1 and 2)

Strong Side Block

4-30. A strong side block is very effective in blocking a horizontal strike directed at the strong side of the head, neck, chest (or flank), or hip area. The horizontal strike may be from an individual's foot, fist, elbow, knee, or weapon. To perform a strong side block, the soldier performs the following steps:

Step 1. Raises the riot baton straight across his body in a vertical position where the grip portion is in his strong hand and the long portion is in his weak hand with the fingers straight up (see *Figure 4-12*).

Step 2. Turns his body laterally by rotating his hips toward his strong side and moving the riot baton across his body, keeping the riot baton vertical (see *Figure 4-12*).

4-31. The strong side block provides protection for a soldier from the front to the rear of his body. A riot baton held in the vertical position absorbs the shock of a strike coming from the strong side. The fingers of his weak hand should be open and behind the long portion of the riot baton, when blocking the object to protect his fingers from being smashed by the strike.

Weak Side Block

4-32. A weak side block is very effective in blocking a horizontal strike directed at the weak side of the head, neck, chest, or hip area. The horizontal strike may be from an individual's foot, fist, elbow, or knee. To perform a weak side block, the soldier performs the following steps:

Step 1. Moves the riot baton in a vertical position toward his weak side, ensuring that the grip portion is pointing straight down and the long portion is pointing straight up.

Figure 4-12. Strong Side Block (Steps 1 and 2)

Step 2. Keeps the fingers of his weak hand open and behind the long portion of the riot baton when blocking the object to protect his fingers from being smashed by the strike. At the moment of contact with the opposing strike, the blocking surface of the long portion should be at a right angle to the opposing object.

Step 3. Assess the situation following a weak side block, and take appropriate follow-up action, as needed.

Middle Block

4-33. A middle block is very effective in blocking a front horizontal strike directed at the face, throat, chest, abdomen, or hip. The horizontal strike may be from an individual trying to tackle near the waist or shove or punch at the face. To perform a middle block, the soldier performs the following steps:

Step 1. Moves the riot baton toward the middle of his body at a 45° angle, keeping the long portion pointed up and slightly forward of the grip portion (see *Figure 4-13*, page 4-14).

Step 2. Keeps the fingers of his weak hand open and behind the long portion of the riot baton when blocking the object to protect his fingers from being smashed by the strike. However, his weak hand may have to grip the long portion when pushing an individual away or forcing an individual down on the ground if the individual is trying to tackle.

RIOT BATON STRIKING TECHNIQUES

4-34. Seven strikes can be executed using the riot baton. Two of these strikes come from the outside-arm or vertical carry. The other five strikes come from the two-hand carry.

Figure 4-13. Middle Block

One-Hand Forward Strike

4-35. The one-hand forward strike is a very quick and effective offensive strike. It is usually employed as a countermeasure to an attack from the front that has been blocked effectively.

4-36. Target areas can be as high as the outside upper arm, down to the outer thigh region. An effective one-hand forward strike must be aggressively quick, with the purpose of distracting, disabling, or altering behavior. Therefore, those employing this technique must be sensitive to what part of the riot baton they use to strike the individual. Strike the target with the last 2 to 4 inches of the long end for optimal results.

4-37. A one-hand forward strike may be employed in one of the following three ways. When selecting available targets, use the principles in *Figure 4-4,* page 4-5.

- A horizontal manner.
- A downward diagonal manner to destabilize an individual.
- A vertical manner in front of the body to clear an aggressor's hands.

NOTE: When the one-hand forward strike is used in a vertical manner, keep the long portion straight up, not angled forward. By keeping the long portion up rather than angled forward, the chance of striking the individual in the face or head is significantly reduced. For the counterstrike to be effective, follow through with the technique when the baton makes contact with the aggressor and/or his object to destabilize the him. A pattern of movement such as a

forward shuffle, forward pivot, strong side step, or rear pivot may enhance this technique.

WARNING

Do not execute the one-hand forward strike with the wooden or expandable riot baton in the extended position. It has been shown to cause injury to the wrist of the user.

4-38. When using the vertical or outside-arm carry position (see *Figures 4-7,* page 4-9 and *4-8,* page 4-10), the soldier quickly moves the riot baton across his body using the strength and power of his hips, moving from his strong side to his weak side (see *Figure 4-14,* page 4-16). Ending this strike leaves the riot baton positioned under the weak side armpit in preparation for the one-hand reverse strike (see *Figure 4-14*). When delivering this strong side strike technique (in a horizontal or downward diagonal manner), ensure that the palm is facing up. In a vertical delivery, the palm will be toward the chest rather than up. Pause following the one-hand forward strike and assess the situation. Take appropriate follow-up action, as needed. If the situation does not call for additional strikes, avoid striking again.

One-Hand Reverse Strike

4-39. The starting position for the one-hand reverse strike is under the weak side armpit (see *Figure 4-14*). The one-hand reverse strike is used as a follow-up strike and is quick and effective.

4-40. Target areas can be as high as the outside upper arm, down to the outer thigh region. An effective one-hand reverse strike must be aggressively quick, with the purpose of distracting, disabling, or altering behavior. Therefore, those employing this technique must be sensitive to what part of the riot baton they use to strike the individual. Strike the target with the last 2 to 4 inches of the riot baton for optimal results.

4-41. A soldier may employee a reverse strike in one of the following three ways:

- A horizontal manner.
- A downward diagonal manner to destabilize an individual.
- A vertical manner.

4-42. In front of the body to clear an individual's hands. When the one-hand reverse strike is used in a vertical manner, keep the long portion straight up, not angled forward. By keeping the long portion up rather than angled forward, the chance of striking the individual in the face or head is significantly reduced. When selecting available targets, use the principles in *Figure 4-4,* page 4-5.

4-43. The one-hand reverse strike is considered a follow-up strike to the one-hand forward strike. As a follow-up strike, it is important that soldiers are taught the discipline of assessing the target. An aggressor may not require another strike, based on his condition and actions after enduring the one-hand forward strike.

Forward Strike | Target Assessment | Reverse Strike

Figure 4-14. One-Hand Forward and Reverse Strikes

4-44. When using the one-hand reverse strike, the solder quickly moves the straight baton across his body, using the strength and power of his hips (from weak side to strong side). When the strike is completed, the soldier's arm will be back in the outside-arm carry position (see *Figure 4-7,* page 4-9). The soldier ensures that his palm is facing down when delivering the counterstrike in a horizontal or diagonal manner. This will prevent injury to the user's wrist. In a vertical delivery, the palm will be toward the chest rather than up. Following the one-hand reverse strike, assess the situation and take appropriate follow-up action, as needed. If the situation does not call for additional strikes, the soldier should avoid striking again.

4-45. When the one-hand reverse strike is used in a vertical manner, keep the long portion straight up, not angled forward. By keeping the long portion up rather than angled forward, the chance of striking an individual in the face or head is significantly reduced. For the counterstrike to be effective, follow through with the technique when the baton makes contact with the aggressor and/or his object to destabilize him. A pattern of movement such as the forward shuffle, forward pivot, strong side step, or rear pivot may enhance this technique.

WARNING

This strike will not be executed with a 36-inch wooden or expandable riot baton in the extended position. It has been shown to cause injury to the wrist of the user.

Two-Hand Strong Side Horizontal Strike

4-46. Starting from the two-hand carry position, the two-hand strong side horizontal strike can be aggressively quick. The purpose of this strike is to create distance, distract, disable, or alter behavior. Strong side horizontal strikes are quick and effective offensive strikes, usually employed as a countermeasure to close frontal attacks that have been effectively blocked. When selecting available targets, use the principles in *Figure 4-4,* page 4-5.

4-47. The two-hand carry position enables soldiers to use riot batons at the 24- or 36-inch length. To effectively employ this strike, use the strength and power of the hips to thrust the riot baton in a horizontal manner. Simultaneously, pull back with the weak hand as the strong hand drives the grip end toward the target, striking a rib or the abdominal region of the aggressor (see *Figure 4-15*, page 4-18). Following the two-hand strong side horizontal strike, assess the target before following up with another strike. An aggressor may not require another strike, based on his condition and actions after enduring the one-hand forward strike.

NOTE: A pattern of movement such as a forward shuffle or forward pivot may enhance this technique.

Two-Hand Weak Side Horizontal Strike

4-48. An effective two-hand weak side horizontal strike must be aggressively quick to create distance, distract, disable, or alter behavior. The weak side horizontal strike is a quick and effective offensive strike usually employed as a follow-up strike to the strong side horizontal strike. It is a countermeasure designed to close frontal attacks that have been effectively blocked. When selecting available targets, use the principles in *Figure 4-4*. Maintaining a strong defensive stance after employing a strong side horizontal strike puts the soldier in position to execute a two-hand weak side horizontal strike.

4-49. Use of the two-hand carry position enables soldiers to use riot batons at 24 or 36 inches. To effectively employ this strike, use the strength and power of the hips to thrust the riot baton in a horizontal manner. Simultaneously, pull back with the strong hand as the weak hand drives the grip forward toward the aggressor, striking a rib or the abdominal region of the target (see *Figure 4-15*). Following the two-hand weak side horizontal strike, it is important to assess the aggressor. An aggressor may not require another strike, based on his condition and actions after enduring the one-hand forward strike.

NOTE: A pattern of movement such as a forward shuffle or forward pivot may enhance this technique.

Two-Hand Front Jab

4-50. The two-hand front jab starts from the two-hand carry position (see *Figure 4-16,* page 4-18). It is aggressively quick and creates distance, distracts, disables, and/or alters behavior. Front jabs are effective offensive strikes, usually employed as a countermeasure to charging and overpowering frontal attacks. When selecting available targets, use the principles in *Figure 4-4*.

Figure 4-15. Two-Hand, Strong Side, and Weak Side Horizontal Strikes

4-51. Use of the two-hand carry position enables soldiers to use riot batons at 24 or 36 inches. Use the strength and power of the hips to effectively employ this strike in a slightly downward or horizontal manner. Simultaneously, use both arms to shoot the long end of the riot baton straight out from the body to the target area. Immediately, pull back the riot baton to the two-hand carry position (see *Figure 4-16*).

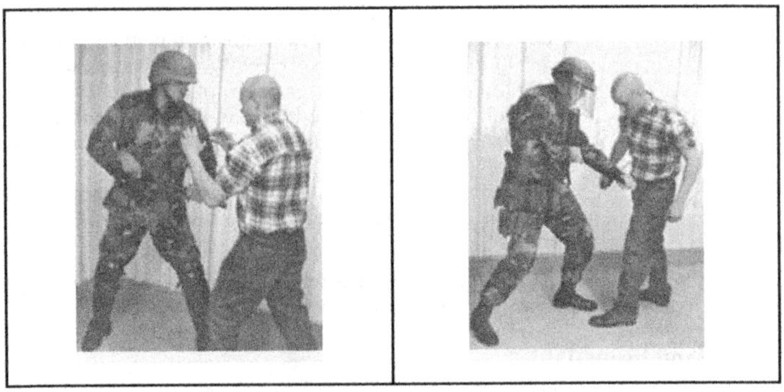

Figure 4-16. Front Jab

4-52. Forward jabs drive the long end of the riot baton toward the aggressor, striking a rib, a hip flexor, or the abdominal region (see *Figure 4-4*, page 4-5). Following the two-hand front jab, it is important that soldiers assess the aggressor before following through with another strike, as the aggressor may not require another strike.

NOTE: A pattern of movement such as a forward shuffle will enhance the power of this technique.

Two-Hand Rear Jab

4-53. Starting from the two-hand carry position, the two-hand rear jab is aggressively quick and creates distance, distracts, disables, or alters behavior. Rear jabs are quick and effective offensive strikes usually employed as a countermeasure to charging and overpowering attacks from the rear. When selecting available targets, use the principles in *Figure 4-4,* page 4-5.

4-54. Use of the two-hand carry position enables soldiers to employ riot batons at 24 or 36 inches. To effectively employ this strike, use the strength and power of the hips to thrust the riot baton in a slightly downward or horizontal manner. Simultaneously, turn the head to the rear to see and assess the target. Use both arms to shoot the riot baton straight back from the body to the targeted area and immediately pull back the riot baton to the two-hand carry position (see *Figure 4-17*).

NOTE: A pattern of movement such as a rear shuffle will enhance the power of this technique.

Figure 4-17. Rear Jab

Two-Hand Middle Strike

4-55. The two-hand middle strike starts from the two-hand carry position. It is aggressively quick and creates distance, distracts, disables, or alters behavior. It is proven to be an effective follow-up technique to a two-hand weak or strong side block or to destabilize an individual by pushing him back and away. The two-hand middle strike is used as a two-count movement. When selecting available targets, use the principles in *Figure 4-4*.

4-56. To perform a two-hand middle strike (see *Figure 4-18*, page 4-20) the soldier does the following:

Step 1. Brings the riot baton up to chest level and slightly off parallel to the ground.

Step 2. Steps forward aggressively, simultaneously thrusting the long portion of the riot baton forward in a horizontal manner. Both arms are fully extended at the end of the movement with a quick snappy return to the step 1 position. Use the entire body to generate the power.

NOTE: It is important that soldiers assess the aggressor before applying a follow-up strike. If the soldier steps forward as he employs the middle strike, it will afford greater impact and power.

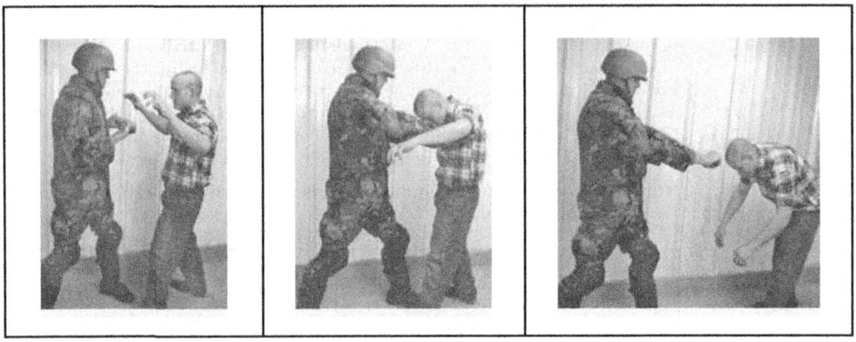

Figure 4-18. Middle Strike

Riot Baton Retention

4-57. A common technique an aggressor may use is grabbing for a soldier's riot baton or, in some cases, even taking the riot baton and using it against the soldier. This causes a disruption or distraction.

4-58. Maintaining physical control of the riot baton is essential for the safety of the soldier and the control force as a whole. Aggressors will generally attempt to grab the riot baton where they can get the strongest hold. Soldiers must resist the instinct of getting into a tug-of-war match over the riot baton, which often proves to be ineffective.

4-59. Using the momentum of the aggressor to pull the riot baton away, the soldier simply steps into or closer to the aggressor with the weak side foot. With the momentum created, the aggressor's hold on the riot baton is lessened. The soldier then rapidly uses the tracing-C technique to recover the riot baton.

4-60. A soldier using the tracing-C technique does the following:

Step 1. Pushes up on the riot baton.

Step 2. Pulls the riot baton around in an upward swing as if tracing a C in the air from bottom to top with the long end.

NOTE: This movement seems to wrap the aggressor's arms around themselves (see *Figure 4-19*).

Step 3. Drives (immediately with a quick and forceful movement) the long end of the riot baton straight down, as if slicing the C in half (see *Figure 4-19*).

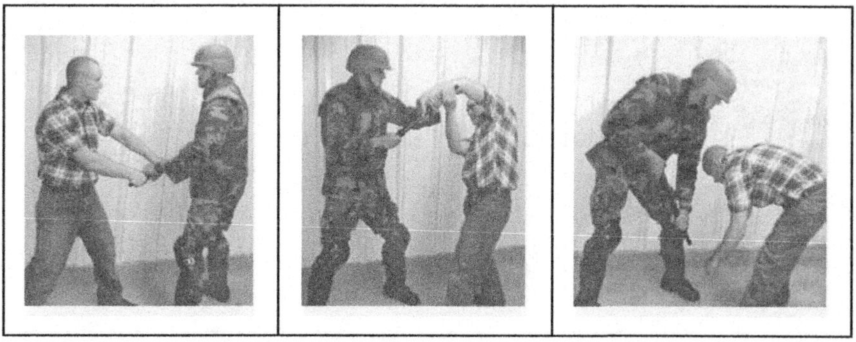

Figure 4-19. Riot Baton Retention

4-61. Once the soldier reaches the top of the C and the aggressor is tied up, it is impossible for the aggressor to maintain hold of the riot baton. Further actions such as a strike may or may not be necessary. Soldiers must be able to assess the situation and react accordingly.

WORKING AS A TEAM

4-62. Soldiers never have to rely solely on their own skills or abilities in riot control operations. Although individual skills were discussed in this chapter, it is important to remember that well-trained soldiers employing offensive, defensive, riot shield, and riot baton techniques will perform the collective task of crowd control well.

4-63. Riot control operations are dynamic and often chaotic. Soldiers armed with riot shields provide a first line of defense while those armed with riot batons and firearms provide overwatch as the formation moves toward engaging a hostile crowd. To work effectively and in concert, soldiers must be properly trained, drilled, and exercised. If an aggressor approaching a line of skirmishers violently grabs a riot shield, help is required to maintain control (see *Figure 4-20*, page 4-22). A soldier armed with a 36-inch wooden riot baton or an expandable riot baton (extended to 36 inches) can be of assistance. The soldier simply moves forward into a gap between the riot shields where he can quickly strike the top or side of the riot shield, slicing the long end across or straight down onto the hands and fingers of the aggressor (see *Figure 4-18*).

4-64. Riot batons are not the only way to effectively knock an aggressor's hands from a riot shield. Soldiers armed with 12-gauge shotguns or M16/M203s can just as effectively smash the hands or fingers of a persistent aggressor with the barrel of their weapons using the technique described above.

Figure 4-20. Clearing a Riot Shield

NONLETHAL MUNITIONS

4-65. With the inclusion of NL munitions, commanders now have additional tools that afford excellent standoff capabilities. This is critical because commanders now have intermediate options for dealing with a hostile crowd. No longer will a commander issue a proclamation, have it ignored by the crowd, and then be forced to use lethal munitions to backup the proclamation. Now he can use NL munitions to gain the same standoff distance but without the catastrophic results of lethal munitions.

4-66. NL munitions include (but are not limited to) a 12-gauge shotgun and 40-millimeter rounds. These blunt munitions have the capability of effectively engaging targets as close as 5 meters and as far as 50 meters. Range is dependent upon each round. These munitions are used from a distance to dissuade, discourage, and disperse the crowd. If the crowd disperses from 50 meters and soldiers are in close contact with the crowd, then the mission has been successful.

TAP-DOWN TECHNIQUE

4-67. Commanders must be familiar with the characteristics and capabilities of all rounds to use them effectively. When 12-gauge and M203 weapons are used within a line of skirmishers (usually located behind the riot shields), it is imperative that a method is developed, trained, and practiced to effectively operate as a team. Operators of these weapons may encounter problems in riot control situations when trying to identify and effectively engage legitimate targets. The tap-down technique is used to overcome target identification and engagement with NL munitions. If ordered to engage targets or the situation authorizes an engagement, soldiers armed with NL munitions may step forward, level their weapons, and fire upon legitimate targets (see *Figure 4-21*).

4-68. A soldier that is armed with a NLW and is using the tap-down technique must first tap on the shield holder's shoulder. Tapping on the shield holder's shoulder alerts him that the gunner is preparing to fire. In response to the tap, the shield holder drops to one knee while keeping the riot shield firmly affixed to their front for protection. The gunner takes a step forward and fills the gap. He then leans into the riot shield bearer's back with his knee and fires. This technique ensures that the long end of the weapon is extended beyond the riot shield.

WARNING

The long end of the weapon must be projected beyond the riot shield before firing. This is a critical safety step, which will prevent NL projectiles from hitting the riot shield and injuring soldiers.

4-69. Once soldiers firing munitions have fired, they raise their weapon back to a high port arms. The soldiers clear their weapons and tap the riot shield bearer a second time, signaling them to rise to a standing position.

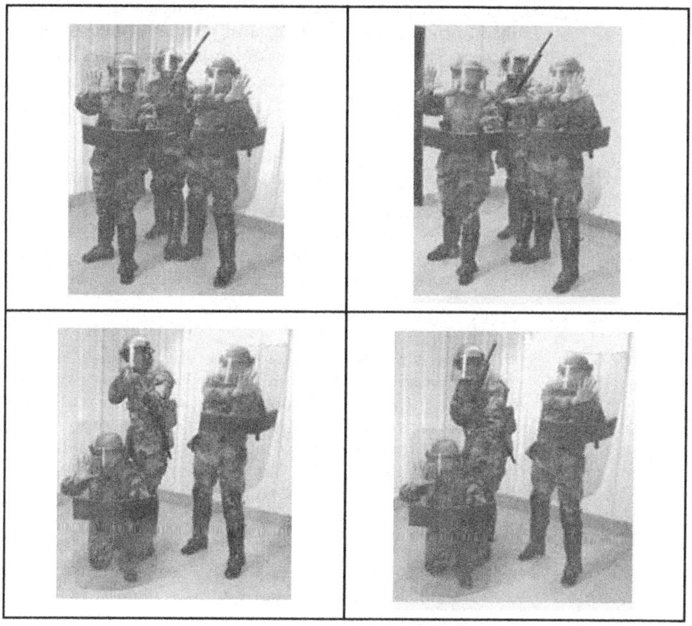

Figure 4-21. Tap-Down Technique

Chapter 5

Nonlethal Capabilities Set and Employment Considerations

During the past several decades, US forces have regularly performed peacekeeping and peace enforcement missions at a greater tempo than in the past. In the peacekeeping and peace enforcement environment, the use of conventional firearms or the threat of their use may not be the solution to a situation where US forces must separate two belligerent, hostile ethnic groups or prevent a similar group from entering an area that is off-limits to them. However, a show of force using NLW and NL munitions will cause crowds to disperse, separate, or leave the area with minimal casualities. The NLCS is a well-rounded, versatile package of both commercial off-the-shelf (COTS) and government off-the-shelf (GOTS) NL equipment and munitions. NLCS contents are divided into four distinct categories: personnel protectors, personnel effectors, mission enhancers, and training devices.

PERSONNEL PROTECTORS

5-1. Personnel protectors provide the individual soldier with added protection to the sensitive and vital areas of his body. They provide excellent protection for the individual soldier from trauma often inflicted by thrown objects (such as rocks, bricks, sticks, and bottles).

NONBALLISTIC RIOT FACE SHIELD

5-2. A nonballistic riot face shield (see *Figure 5-1,* page 5-2) has two adjustable positions and is constructed of hard durable plastic that securely attaches to the Kevlar® helmet. It may be worn with an M17 or M40 protective mask and is designed to protect the soldier's head, face, and neck from thrown objects. Nonballistic riot face shields will not protect from attacks with ballistic weapons, such as firearms. All soldiers operating in a civil disturbance environment should be outfitted with riot face shields.

NONBALLISTIC FULL-LENGTH SHIELD

5-3. A nonballistic full-length riot shield (see *Figure 5-2,* page 5-2) is a body shield constructed of hard durable plastic. Left- and right-handed personnel can easily use the nonballistic full-length riot shield. It is designed to protect the soldier's face, torso, and upper legs from thrown objects. Nonballistic full-length riot shields will not provide protection from ballistic weapons. Only the front line of skirmishers and those providing lateral protection to both flanks of the formation will be outfitted with nonballistic full-length riot shields.

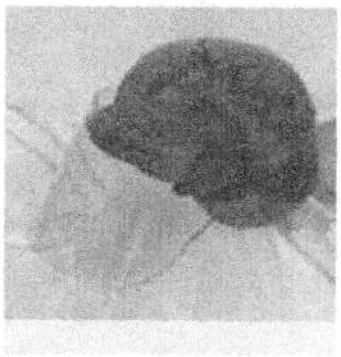

Figure 5-1. Nonballistic Riot Face Shield

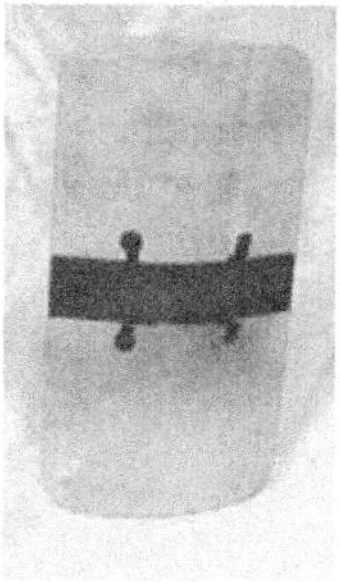

Figure 5-2. Nonballistic Full-Length Riot Shield

NONBALLISTIC SHIN GUARD

5-4. The nonballistic shin guards (see *Figure 5-3*) are constructed of hard durable plastic and are easily fastened to a soldier's legs. They are designed to protect the soldiers, feet, shins, and knees from thrown objects. They will not provide protection from ballistic weapons, such as firearms. All soldiers operating in a civil disturbance environment should be outfitted with shin guards.

BALLISTIC RIOT FACE SHIELD

5-5. The ballistic riot face shield (see *Figure 5-4*) has two adjustable positions constructed of acrylic and bullet-resistant materials that provide the wearer with Level IIIA protection. The ballistic riot face shield easily attaches to the Kevlar helmet and is capable of being worn with either an M17 or M40 protective mask. It will protect the soldier from hand-thrown objects. Because of its excessive weight (3.4 pounds), the ballistic riot face shield should not be worn for standard riot control operations. Ballistic riot face shields are best suited for use within specialized teams associated with riot control operations; for example, teams responsible for the search and apprehension of identified criminals or building- and alley-clearing operations.

Figure 5-3. Nonballistic Shin Guards

Figure 5-4. Ballistic Riot Face Shield

BALLISTIC FULL-LENGTH RIOT BODY SHIELD WITH LIGHT KIT

5-6. The ballistic full-length riot body shield (see *Figure 5-5,* page 5-4) is constructed from Spectra Shield® bullet-resistant material that provides a soldier with Level IIIA protection. It will protect the soldier from thrown objects. Because of its excessive size (18 pounds and 20 by 36 inches), the ballistic full-length riot body shield should not be used as part of a standard riot control operation. Ballistic full length riot body shields are best suited for use within specialized teams associated with riot control operations; for example, teams responsible for the search and apprehension of identified criminals or building- and alley-clearing operations.

Figure 5-5. Ballistic Full-Length Riot Body Shield

BALLISTIC SHIN GUARDS

5-7. Ballistic shin guards (see *Figure 5-6*) are constructed of Kevlar KM2 materials and provide a soldier with Level IIIA protection to the feet, shins, and upper legs. They protect the soldier from hand-thrown objects. There are three different sizes, and because of their excessive weight (7 to 10 pounds), ballistic shin guards should not be deployed as part of a standard riot control operation. Ballistic shin guards are best suited for use within specialized teams associated with riot control operations; for example, teams responsible for the search and apprehension of identified criminals or building- and alley-clearing operations.

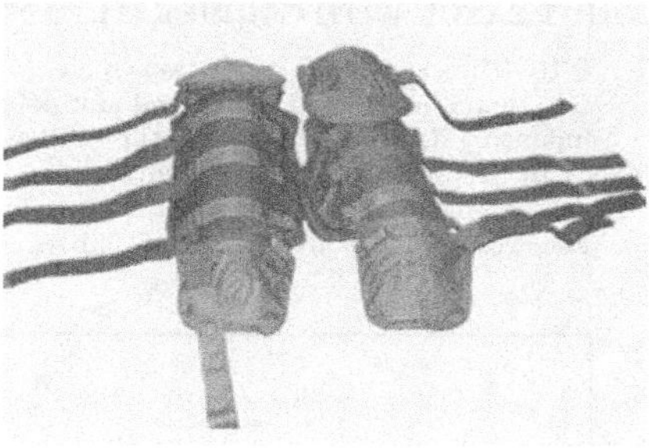

Figure 5-6. Ballistic Shin Guards

PERSONNEL EFFECTORS

5-8. Personnel effectors are those items that provide soldiers with an NL response without the necessity of always coming into direct physical contact with aggressors. Personnel effectors are 12-gauge and 40-millimeter ammunition that inflict blunt trauma. These munitions are compatible with the shotgun and the M203 grenade launcher already in the Army inventory. There are also diversionary and rubber ball grenades, which distract and inflict blunt trauma.

WOOD RIOT BATON WITH BELT RING (36 INCHES)

5-9. The 36-inch wood riot baton (see *Figure 5-7*) is made from solid oak. When held properly, the riot baton thong provides greater retention capabilities. Intended primarily for self-defense, the 36-inch wood riot baton can be used effectively as an offensive tool if it becomes necessary to keep rioters out of arm's reach of soldiers conducting crowd control operations.

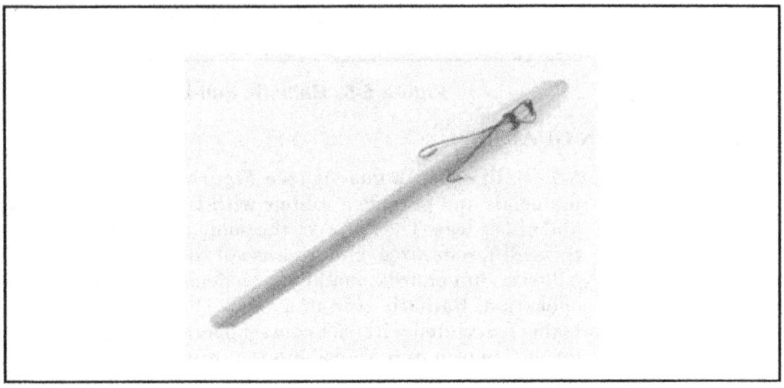

Figure 5-7. Wood Riot Baton

EXPANDABLE RIOT BATON WITH CARRIER (24 TO 36 INCHES)

5-10. The expandable riot baton (see *Figure 5-8*) provides the soldier with an NL means of crowd control and self-defense. The riot baton comes with a mounting device that attaches the riot baton to a belt. The riot baton is 24 to 36 inches in length and is primarily intended for self-defense. The expandable riot baton can be used effectively as an offensive tool if it becomes necessary to keep rioters out of arm's reach of soldiers conducting crowd control operations.

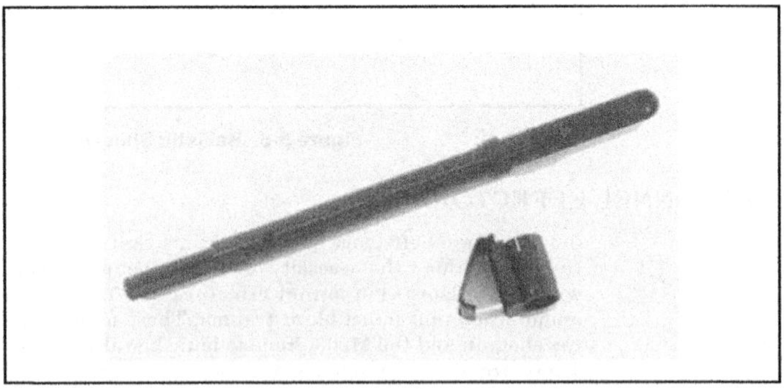

Figure 5-8. Expandable Riot Baton and Carrier

500 (3 INCH) 12-GAUGE SHOTGUN

5-11. The 12-gauge shotgun (see *Figure 5-9*) is a pump action shotgun currently in NLCS inventory. The pump action shotgun is chambered to take up to 3-inch shells. The 3-inch chamber allows for the use of M1012 and M1013 NL munitions. This shotgun also provides a visually distinct alternative to standard military weapons that may be desired based on mission considerations (force signature and acceptability).

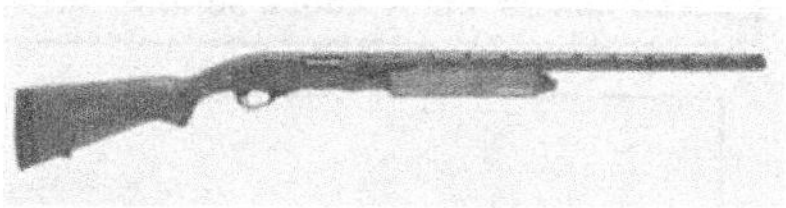

Figure 5-9. 12-Gauge Shotgun

M7 66-MILLIMETER LAUNCHER WITH M315 INSTALLATION KIT

5-12. The M7 (see *Figure 5-10*) is a 66-millimeter vehicle-mounted, NL, grenade-launching device that is mounted on a HMMWV. It is an indirect fire support system that can deliver the M98 distraction grenade that creates a flash-bang effect, L96A1 antiriot grenade, or M99 blunt trauma grenade that creates a sting-ball effect. The M315 installation kit is used to install an M7 discharger on the turret ring of appropriate HMMWV variants. An adjustable bracket allows the launch angle to be depressed for engaging targets at ranges of 50, 75, and 100 meters. The system enforces standoff distances and deters potential threats.

Figure 5-10. M7 Launcher

MODIFIED CROWD CONTROL MUNITION-GROUND EMPLACEMENT

5-13. Similar in nature to the claymore mine, the modified crowd control munition-ground emplacement (MCCM-GE) (see *Figure 5-11*) is an NL munition that can deliver six hundred .32-caliber rubber pellets and a flash-bang effect. The MCCM-GE will stop, confuse, disorient, and/or temporarily incapacitate area targets and/or personnel at ranges out to 30 meters.

WARNING

Avoid employing the MCCM-GE within 10 meters of personnel as it may cause serious bodily harm or death.

Figure 5-11. MCCM-GE

12-GAUGE, FIN-STABILIZED POINT, NONLETHAL CARTRIDGE (M1012)

5-14. The M1012 (see *Figure 5-12*) is a single projectile round made of hard rubber that is shaped like a bomblet and designed to be fired a at a single target. With a muzzle velocity of 500 feet per second, the M1012 has the effective range of no closer than 5 meters and no further than 30 meters. Engagement inside 5 meters could cause serious injury or death. Beyond 30 meters the kinetic energy dissipates to the point where the round becomes ineffective.

Figure 5-12. M1012

12-GAUGE, CROWD DISPERSAL, NONLETHAL CARTRIDGE (M1013)

5-15. The M1013 (see *Figure 5-13*) is a multiple projectile round with .23 caliber hard rubber pellets that is designed to be fired at and employed with the purpose of affecting multiple targets. With a muzzle velocity of 900 feet per second, the M1013 has an effective range of no closer than 5 meters and no further than 30 meters. Engagement inside 5 meters could cause serious injury or death. Beyond 30 meters, the kinetic energy dissipates to the point where the rubber pellets become ineffective.

Figure 5-13. M1013

40-MILLIMETER SPONGE GRENADE (M1006)

5-16. The M1006 (see *Figure 5-14*) is a single projectile round made of pliable foam rubber with a hard plastic back. The M1006 is designed to be aimed and fired at a single target. With a muzzle velocity of 265 feet per second, the M1006 has an effective range of no closer than 10 meters and no further than 50 meters. Engagements inside 10 meters could cause serious injury or death. Beyond 50 meters, the kinetic energy dissipates to the point where the round becomes ineffective.

Figure 5-14. M1006

40-MILLIMETER CROWD DISPERSAL CARTRIDGE

5-17. The crowd dispersal cartridge (CDC) (see *Figure 5-15*) is a multiple projectile round with .48 caliber hard rubber balls that is designed to be fired and employed with the purpose of affecting multiple targets. With a muzzle velocity of 450 feet per second, the CDC has an effective range of no closer than 10 meters and no further than 30 meters. Engagement inside 10 meters could cause serious injury or death and beyond 30 meters the kinetic energy dissipates until the rubber balls become ineffective.

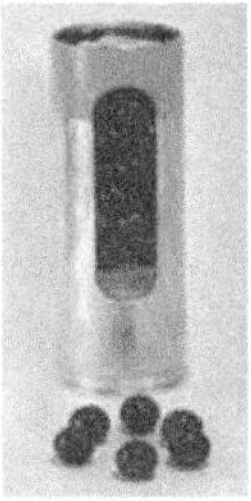

Figure 5-15. CDC

RUBBER BALL GRENADE (GG04)

5-18. The GG04 (see *Figure 5-16,* page 5-10) is a multiple projectile, flash-bang grenade with 100 .25-caliber hard rubber pellets. Each grenade has a fuse delay of 2.8 to 3 seconds with a flash measuring approximately 1 million CP and 180 decibels at 3.5 feet. At detonation, rubber pellets are dispatched at 360° with an effective range of 2 to 3 meters and a maximum engagement range of 15 to 20 meters. The rubber ball grenade is designed to be hand-thrown or muzzle-launched from a 12-gauge shotgun.

RUBBER BALL GRENADE LAUNCHING CUP AND AA30 12-GAUGE LAUNCH ROUND

5-19. The launch cup (see *Figure 5-17,* page 5-10) firmly attaches to the 12-gauge shotgun without any additional tools. It works in concert with the AA30 12-gauge launch round to propel the rubber ball grenade approximately 100 meters. The shooter simply inserts the full body and the safety lever of the grenade into the launching cup, pulls the pin, loads a 12-gauge round, and fires (approximately) at a 30° angle. The desired effect is to have the grenade explode approximately 7 to 10 feet above the target.

RELOADABLE FLASH-BANG GRENADE (M84)

5-20. The M84 (see *Figure 5-18,* page 5-10) is a reloadable flash-bang thrown grenade. Each grenade has a 1- to 2-second fuse delay, with a flash capability of 1.5 to 2.5 million CP and a bang of 168 to 175 decibels. Although it is part of the NLCS, use of the M84 is not recommended for use in crowd control situations. Specialty teams, such as security response teams, and search teams may use this device effectively.

Figure 5-16. GG04

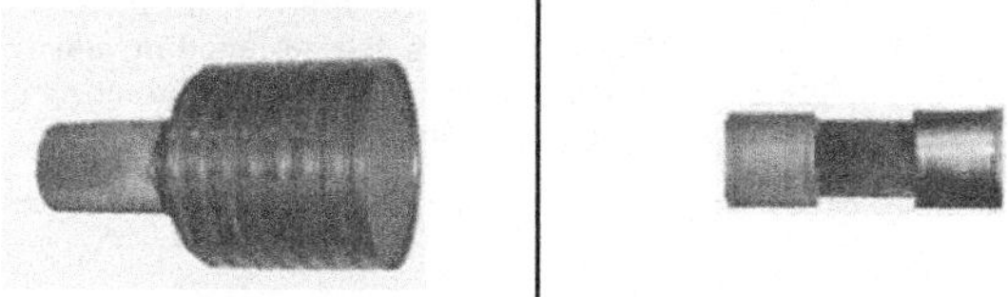

Figure 5-17. Launching Cup and AA30

Figure 5-18. M-84

FLEX-CUFS

5-21. Flex-Cufs consist of a tough pliable plastic band with a self-locking mechanism in the center for each end. When the Flex-Cuf (see *Figure 5-19*) is threaded the restraint bands circle around the wrists or ankles, impeding movement and securing the individual. They are lightweight and disposable and require a cutting tool or other instrument to remove.

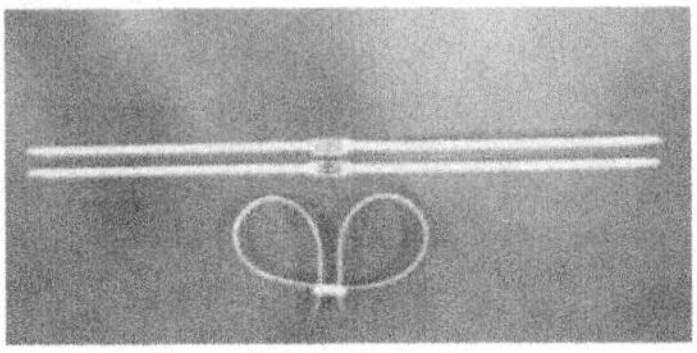

Figure 5-19. Flex-Cufs

INDIVIDUAL RIOT CONTROL AGENT DISPERSER (M36)

5-22. The M36 (see *Figure 5-20*) contains a Dibenz (F, -1 4-oxazepine [CR] solution. It can deliver 25 one-second bursts out to 12 feet. Individual disposable RCA dispersers are intended primarily for self-defense or to keep rioters out of arm's reach of soldiers conducting crowd control formations or soldiers engaged in missions where a noncombatant exists.

Figure 5-20. M36

MIDSIZE RIOT CONTROL AGENT DISPERSER (M37)

5-23. The M37 (see *Figure 5-21*, page 5-12) is the size of a standard fire extinguisher that uses compressed air to force the RCA out to a range of 30 feet. It has the capacity to employ 18 bursts for 3 seconds per charge. It is excellent for providing a wide coverage of RCAs onto a hostile crowd while maintaining excellent standoff capability. The M37 can be refilled and is rechargeable. It can be filled with CR solution (liquid agent) or CS1 (dry agent). For training purposes, CR can be substituted with water and CS1 can be substituted with talcum powder.

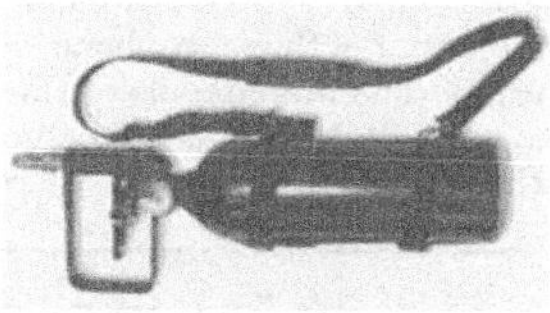

Figure 5-21. M37

SQUAD RIOT CONTROL AGENT DISPERSER (M33A1)

5-24. The M33A1 (see *Figure 5-22*) is designed to provide crowd control and protection at the squad level. It is capable of projecting a ballistic stream of RCAs beyond 25 feet in up to 25 half-second bursts. It consists of a frame and harness assembly, compressed-gas cylinder (agent container assembly), air pressure assembly, gun and hose assembly, multijet spray unit, and check valve assembly. The M33A1 can be refilled and is rechargeable. For training purposes, CR can be substituted with water and CS1 can be substituted with talcum powder.

Figure 5-22. M33A1

MISSION ENHANCERS

5-25. Mission enhancers aid the commander in various tactical situations. Mission enhancers include various types of devices used for illumination, vehicle denial, traffic control missions, and force protection enhancement.

PORTABLE BULLHORN

5-26. The portable bullhorn (see *Figure 5-23*) provides squad leaders with the capability to project their voices above the noise and commotion created by crowds and mobs. The megaphone has a built-in siren for alarm purposes that allows the operator to use the device as an ordinary megaphone and siren alarm during emergencies. The bullhorn is portable and easy to carry. The acoustic range of the megaphone for a clear voice output is ¾ to 1 mile. The power output of the megaphone is rated at 15 watts with a maximum rating of 20 watts. The dimensions of the megaphone are 14.5 inches with a 9-inch diameter bell. It weighs 3.5 pounds without batteries. Battery requirements for the megaphone are eight dry-cell size C batteries.

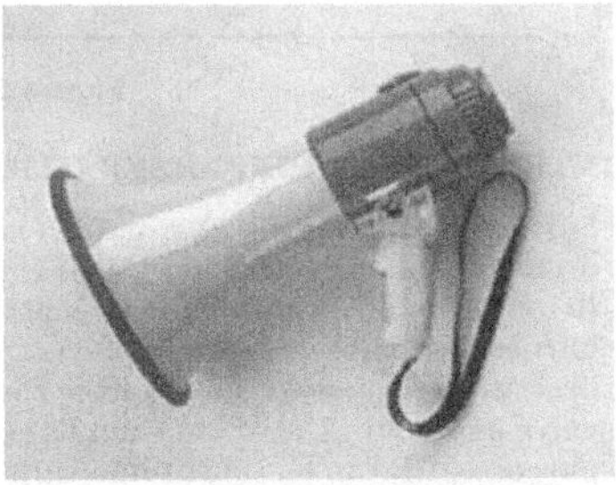

Figure 5-23. Portable Bullhorn

GROUND-MOUNTED BULLHORNS

5-27. Ground-mounted bullhorns (see *Figure 5-24,* page 5-14) are a critical communication enhancement device for conducting crowd control tactics. The bullhorn can facilitate communication with the crowd in conjunction with PSYOP support and assistance with the communication of commands to troops engaged in the crowd control process. The power output of the ground-mounted bullhorn is rated at 22 watts. The dimensions of the bullhorn are 13.75 by 9 by 14.75 inches, and it weighs 15 pounds.

INDIVIDUAL VOICE AMPLIFICATION SYSTEM (M7)

5-28. The M7 (see *Figure 5-25,* page 5-14) is a critical communication enhancement device for conducting crowd control tactics using RCAs while wearing the M40 protective mask. The M7 is fitted to the M40 protective mask to facilitate oral communication and increase the user's ability to communicate using radios and other devices.

Figure 5-24. Ground-Mounted Bullhorns

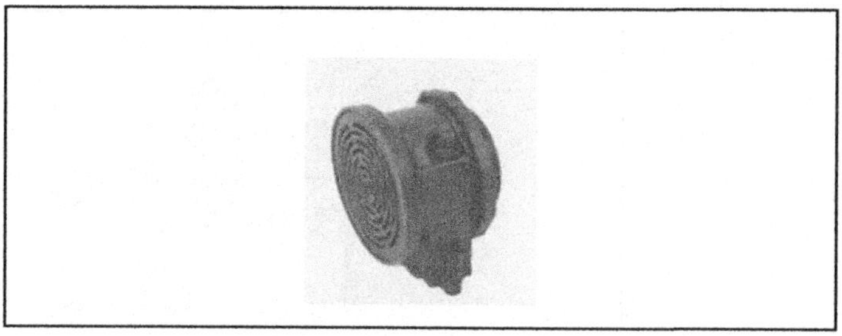

Figure 5-25. Individual Voice Amplification System

INDIVIDUAL HIGH-INTENSITY SEARCHLIGHT WITH HOLSTER

5-29. The individual high-intensity searchlight (see *Figure 5-26*) is used primarily for illumination in crowd control operations at night. An individual can use it for general illumination of the operational area to pinpoint agitators and threats, reduce the ability of the rioters to see troop formations and actions, enhance tactical deception and actions, and enhance tactical deception techniques for units conducting crowd control operations.

HIGH-INTENSITY LIGHT

5-30. A high-intensity light (see *Figure 5-27*) is intended for use in low light or night conditions. A high-intensity light can project a beam that will enable the user to identify an individual person up to 1,900 yards away. It has an adjustable beam spread of 1° to 15° and will run continuously at maximum power for up to 45 minutes. The internal power supply requires 4 hours for a full charge.

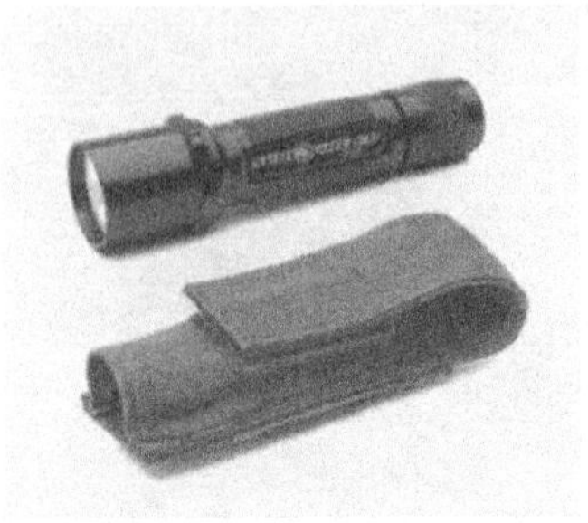

Figure 5-26. Individual High-Intensity Searchlight with Holster

Figure 5-27. High-Intensity Light

HIGH-INTENSITY LIGHT SYSTEM WITH 12-GAUGE SHOTGUN ATTACHMENT KIT

5-31. This 12-guage high-intensity light system (see *Figure 5-28,* page 5-16) is used to illuminate possible targets in crowd control situations where there is little or no light. It serves to blur or blind the aggressor's ability to see. It mounts easily to the end of the 12-gauge shotgun barrel.

40-MILLIMETER CARRYING POUCH

5-32. The 40-millimeter carrying pouch is made from durable canvas material, (see *Figure 5-29,* page 5-16) and can be easily affixed to load-bearing equipment (LBE) and/or a load-bearing vest (LBV) using Velcro straps. It is capable of carrying up to six 40-millimeter rounds.

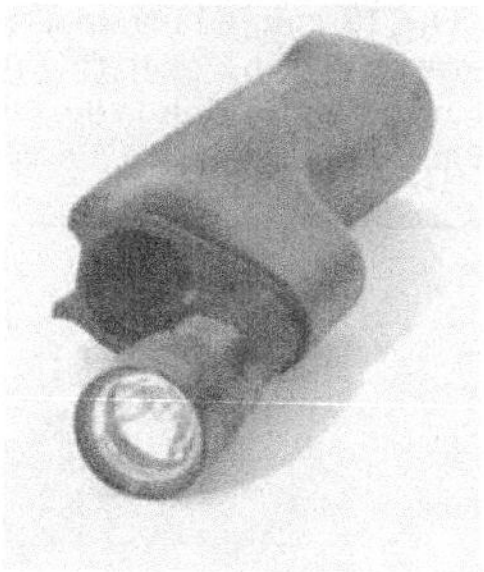

Figure 5-28. 12-Gauge High-Intensity Light System With 12-Gauge Shotgun Attachment Kit

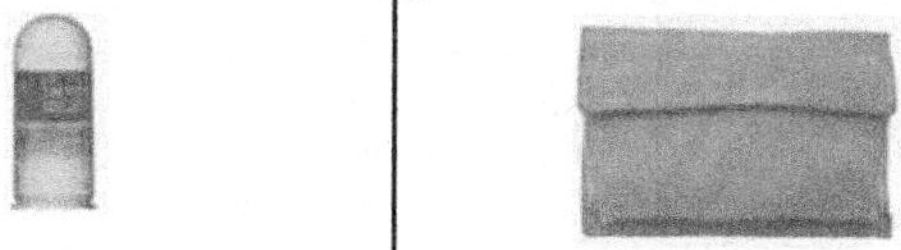

Figure 5-29. 40-Millimeter Carrying Pouch

12-GAUGE UTILITY POUCH

5-33. The 12-gauge utility pouch is made from durable canvas material (see *Figure 5-30*). It can be easily affixed to an LBE and/or LBV using metal clasps. The 12-gauge utility pouch is capable of carrying up to twenty-five 12-gauge shotgun rounds.

Figure 5-30. 12-Gauge Utility Pouch

12-GAUGE BUTTSTOCK CUFF

5-34. The 12-gauge buttstock cuff (see *Figure 5-31)* firmly attaches to the shotgun by sliding over the stock. Elastic straps aid in preventing the buttstock cuff from sliding or rolling. It holds up to seven 12-gauge cartridges, which give the soldier quick access to additional rounds.

Figure 5-31. 12-Gauge Buttstock Cuff

DIVERSIONARY AND/OR RUBBER BALL GRENADE POUCHES

5-35. The diversionary and/or rubber ball grenade pouch (see *Figure 5-32*) is an ammunition-carrying pouch made of durable canvas material. It is partitioned in a manner to provide the soldier with a means of holding up to six rubber ball grenades or six flash-bang grenades in separate compartments.

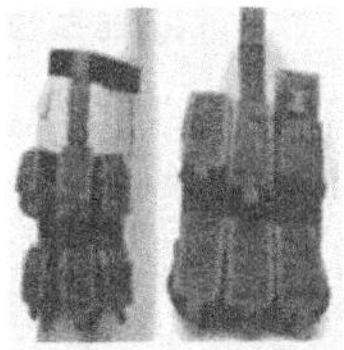

Figure 5-32. Diversionary and/or Rubber Ball Grenade Pouch

CALTROPS

5-36. The caltrop (see *Figure 5-33,* page 5-18) is a field-expedient area denial system. Securing approximately 50 caltrops to 550-type military cord allows a soldier to quickly toss and recover the system. Caltrops are used to deny vehicle and pedestrian movement in designated areas. The four prongs are approximately 2 inches tall and will puncture tires or boots. A soldier can easily employ the caltrop by scattering it in a designated area hidden by camouflage.

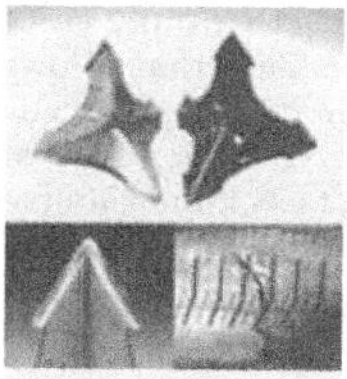

Figure 5-33. Caltrop

PORTABLE VEHICLE ARREST BARRIER

5-37. A portable vehicle arrest barrier (PVAB) (see *Figure 5-34*) is a lightweight, portable, manually-emplaced and recoverable-barrier entrapment device that is used (on command) to create a barrier that will stop vehicles being driven by known or suspected hostile forces with minimal damage to the vehicle and its occupants. Two individuals can easily emplace it in approximately 4 to 8 hours. It has the capability of stopping a light truck (up to 7,500 pounds) at speeds of up to 45 miles per hour with a stopping distance of less than 200 feet. It is not to be used as a stand-alone system. The PVAB may be used at dismount points, checkpoints, and roadblocks to prevent unauthorized access by wheeled vehicles into or out of areas.

Figure 5-34. PVAB

TRAINING DEVICES

5-38. Training devices that commanders can use to effectively prepare their soldiers are included in the NLCS. Training devices allow soldiers to learn and build confidence in the development of personal defensive skills. These are critical pieces of equipment because they allow soldiers to train at almost full speed, which provides reinforcement of realistic motor skills. Training in this manner has proven to be more effective than just striking out at the air.

IMPACT TRAINING SUIT

5-39. Impact training suits (see *Figure 5-35*) are made of closed-cell shock-absorbent foam that allows soldiers to hone their riot baton skills through realistic hand-to-hand training. Impact training suits absorb strikes inflicted by the current expandable and wooden riot batons. They provide protection for the head, face, hands, and legs without significant degradation of the wearer's mobility.

Figure 5-35. Impact Training Suits

TRAINING STRIKE BAGS

5-40. Training strike bags (see *Figure 5-36*) give the soldier a durable, yet absorbing surface to strike or kick for training. It also helps develops proficiency in open-hand control and riot baton techniques.

Figure 5-36. Training Strike Bag

PRACTICE GRENADE BODY (GG05)

5-41. The GG05 is blue in color and is easily identified as a practice grenade (see *Figure 5-37*). It is designed to replicate the live GG04 rubber ball grenade. It is made of durable rubber and can be hand-thrown or muzzle-launched from the 12-gauge shotgun. It can be reloaded.

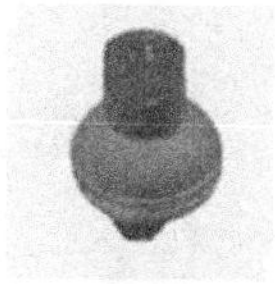

Figure 5-37. GG05

PRACTICE GRENADE FUSE

5-42. The practice grenade fuse (see *Figure 5-38*) is identified by its blue spoon and brown stripe. It easily screws into the body of the GG05. Once used, unscrew the practice grenade fuse from the body and reload it with a new fuse.

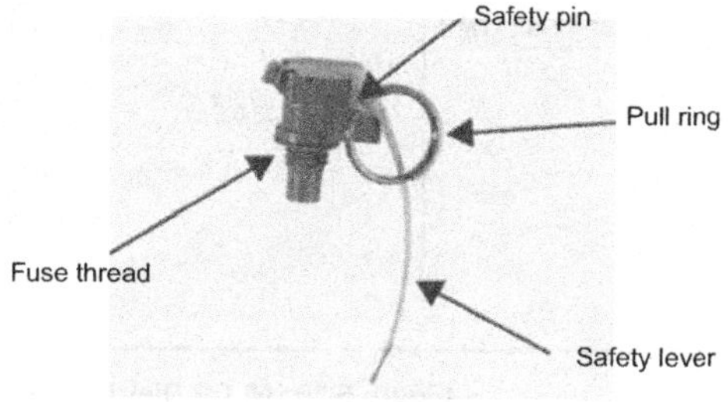

Figure 5-38. Practice Grenade Fuse

CORE CAPABILITIES

5-43. Core capabilities are those fundamental competencies that enable the commander to achieve the desired operational outcome. In the case of NLW, this includes providing a flexible means of response in order to protect friendly forces and/or influence the actions of potential adversaries and noncombatants. With the availability of the NLCS (see *Figure 5-39*) these goals can be achieved without resorting to lethal force and in a manner that will minimize collateral damage. The core capabilities associated with NL effects fall into the following two major categories:

NOTE: There are nine platoon equivalents within a light infantry battalion.

PLT Organization	Personnel and Weapons	Nonlethal Weapons & Equipment
PLT	7 personnel (5 M16/M4 and 2 M60/ M240)	3 riot face shields 3 sets of shin guards 3 sets of full-body restraints 3 individual voice amplification systems 3 individual RCA dispensers with carriers 1 portable bullhorn 1 ground-mounted bullhorn 1 high-intensity searchlight 1 set of caltrops 3 ballistic riot face shields 3 sets of ballistic shin guards 3 ballistic riot body shields 1 riot baton training suit 1 riot baton strike pad
SQD SQD SQD	3 personnel (3 M16/M4)	3 riot face shields 3 sets of shin guards 3 riot body shields 3 expandable riot batons with carriers 9 sets of cinch-strap restraints 3 individual voice amplification systems 3 individual RCA dispensers with carriers 3 individual high-intensity search lights with carriers (see *Figure 5-26*)
TM TM TM TM TM TM	24 personnel (12 M16/M4 and 6 M16/ M4 with M203 6 SAW)	24 riot face shields 24 sets of shin guards 100 sets of disposable flex-cuffs 24 individual voice amplification systems 24 individual RCA dispensers with carriers 6 expandable riot batons with carriers 11 grenade pouches 12 forty-millimeter grenade pouches 6 twelve-gauge ammunition pouches 6 twelve-gauge bitstock ammunition carriers

Figure 5-39. NLCS Distribution

- **Counterpersonnel.** NL counterpersonnel capabilities enable the application of military force with a reduced risk of fatalities or serious casualties among noncombatants or, in some instances, among enemy forces. There are several specific NL counterpersonnel capabilities to be explored. These include the means to influence the behavior and activities of a potentially hostile crowd, and the capability to bring a mob engaged in a riot under control. While there are many similarities in these two scenarios, each involves unique challenges, which may require radically different solutions. For more information, see *FM 3-22.40.*
- **Countermaterial.** NL countermateriel capabilities enhance operations by rendering equipment and facilities unusable without complete destruction. An NL countermateriel capabilities enable the

application of military force to defuse potentially volatile situations under circumstances in which more destructive conventional military means might prove counterproductive. For example, pre-emptive strikes against troublesome and aggressive nations may be politically unacceptable when only conventional weapons with the high risk of casualties are involved. With NL countermateriel capabilities, the enemy's ability to threaten its neighbors could be curtailed with less political risk by attacking only their weapons of war and their supporting infrastructure. For more information, see *FM 3-22.40*.

ADDITIONAL CAPABILITIES

5-44. NLW are only NL when fired within their parameters.

DANGER
If NLW are fired outside their parameters, they may cause serious injury or death.

5-45. If a situation turns from NL to lethal, an NL gunner only needs to change his point of aim and distance to inflict serious bodily harm and/or death to the threat. The closer the soldier is to the target, the greater the kinetic energy and the more severe the effects are to the threat.

EMPLOYMENT CONSIDERATIONS

5-46. NL munitions and equipment are not completely NL; they are NL by intent. The term NL does not guarantee zero mortality or nonpermanent damage. NLW can add flexibility to combat operations and enhance force protection by providing an environment in which friendly troops can engage threatening targets with a reduced risk of noncombatant casualties and collateral damage.

5-47. If the tactical situation dictates a NL response, a soldier who levels a weapon loaded with NL munitions must be trained in distance to the target and appropriate points of aim. Based on the soldier's understanding and training on the characteristics and capabilities of the round, they can easily discern and engage targets to the desired effect. If the tactical situation dictates a lethal response, soldiers can easily adjust the point of aim to a more vulnerable target location on the body and move in closer to the target. NL munitions can be used as a lethal response if the situation dictates and the intent of the soldier is to inflict death.

NONLETHAL WEAPONS CAPABILITIES IN FORMATIONS

5-48. The mix of NLW within the formation is based on METT-TC. For example, the first rank of the formation will have riot shields, riot batons, and their personal weapon (slung across their back [left to right] with the butt up and muzzle down or holstered). The second row will have a mix of shotguns and M203s. Shotguns and M203s may be used as area denial or point target weapons depending on the specific munitions used. The shotgun provides cover for the slower loading M203. The commander may move these weapons within the formation to meet the needs of the mission and to create distance between the formation and the crowd (see *Appendix D*).

Chapter 6

Civil Disturbance Formations

Control force formations, when properly employed and executed against a crowd, are some of the most practical methods of crowd control. The riot baton has been the weapon of choice for the control force. With the development of NLW, the commander now has more options at his disposal. Before NLW, the only other choices were chemical irritants, MWDs, and lethal force. Now commanders can create a standoff distance with NLW that reduces the risk of close fighting with the mob. During any crowd control operation, the commander must protect the control force with lethal overwatch. This overwatch may be in the form of marksmen and observer teams (sniper with observer) and a response force with lethal capabilities.

OVERVIEW

6-1. With advances in NLW, it is possible to create an effective standoff distance of 15 to 100 meters between the crowd and the control force formation. It is now possible to advance and disperse the crowd without coming into direct contact with it. Some NLW may also be used to assist in the identification and apprehension of instigators.

6-2. Control force formations are used to disperse, contain, and/or block a crowd. These formations are more effective in urban areas than in open areas, such as parks or fields. When control force operations are employed in urban areas, it is easier to split a crowd into smaller segments, isolate instigators, or funnel the crowd into the desired location by using buildings and other man-made structures. Commanders must be mindful of the fact that well-organized crowds may attempt to move to the flanks of a formation and/or get behind the formation to gain a tactical advantage.

6-3. Commanders must be aware of the limitations of control force formations. They are not the answer to all civil disturbance situations. Do not expose the formation to sniper fire and/or unnecessary violence simply for a show of force. Control force formations should be used when it is decided that a crowd poses a threat and must be dispersed or moved to a specified area. Before dividing a large crowd, the commander must consider that this might not solve the problem. It may worsen the problem by creating smaller elements that may engage the control force in small-mob tactics, such as sniping, looting, burning, and attempting to envelop the control force. Commanders must anticipate that breakaway crowds may engage in small-mob tactics. These tactics can be defeated by area control measures, such as building searches, saturation patrolling, and other tactics. After dispersing the crowd, the control force must not assume that there will be an immediate return to peaceful activities. The use of formations is only part of the total dispersal effort.

6-4. Commanders must assess and secure the areas of operations (AOs). This may be accomplished with the use of helicopter overflights and by visually securing rooftops, high elevations, and advantage points. Designated marksmen teams may be used to secure the rooftops and provide lethal overwatch for the control formation. The control force commander must be aware that designated marksmen teams are positioned on rooftops.

6-5. Crowd control formations and their support teams have more capabilities than just crowd dispersal. They also have the capability to apprehend and detain certain members of the crowd that the commander feels may instigate the crowd to further violence. Commanders must analyze the threat, determine the mission of the control force, and decide which control force formations will accomplish the mission according to METT-TC.

6-6. During CONUS operations, there are usually no enemy forces. However, commanders must analyze and prepare for any potential opposition. The formations discussed in this chapter are guidelines and may be adapted to fit any mission or situation. Whatever the modification, soldiers must stay in position and on line. Only through training and rehearsals will soldiers become proficient in basic formation movements and be able to adjust to changing situations. Soldiers must also be prepared to employ NLW. *Figure 6-1* is used as a key to identify soldiers within crowd control formations.

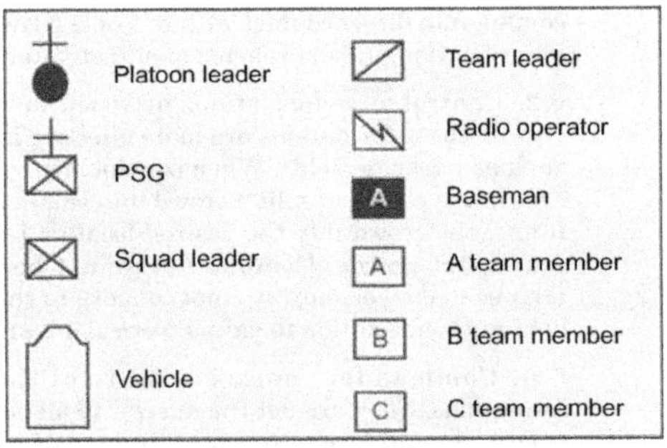

Figure 6-1. Symbols for a Formation Element

ACTIONS BEFORE MOVEMENT

6-7. Before moving the control force to the civil disturbance area and into a potentially hostile urban environment, the commander and his staff must review the preincident plans and expand them (as needed) based on current information. Because of the employment of forces against ambiguous threats,

IPB becomes critical. Information is the key to developing an appropriate response to civil disturbances.

6-8. Civil disturbance training must be part of the annual unit training plan. If a soldier is deployed to an area where civil disturbances are a recurring event, sustainment training must occur more frequently. Before moving the control force to the rally point, an intense rehearsal must be conducted by the leadership and the control force. A top-down review of the rules of engagement (ROE/rules for the use of force (RUF) must be accomplished. The ROE/RUF must be part of the unit training plan and each soldier must be trained to standard (see *Chapter 8*). The use of NLW (type specific) must be granted to the on-site commander from the approval authority.

6-9. Information is the key to developing plans for appropriate responses to a civil disturbance, and IPB is crucial to the success of the mission. During the planning phase (see *Chapter 2*), the information provided by the IPB will aid the staff in selecting various routes and rally points. The selection of ingress and egress routes must include a variety of ways in and out of the area. Direct and indirect routes and those with cover and concealment are necessary. The rally point must be cleared and secured by an advanced party capable of controlling the area.

RALLY POINT ACTIONS

6-10. Once at the rally point, the commander makes contact with the local police, civil authorities, or military officer. He then assesses the situation, decides the next COA, and issues orders, as appropriate. The platoon sergeant (PSG) forms the platoon into a column formation and places team leaders directly in front of their teams (see *Figure 6-2,* page 6-4). Team and squad leaders make last minute checks of their soldiers and await orders.

6-11. The commander places his lethal overwatch (designated marksmen [DM] teams) in areas that overlook the control force and the mob. They may be to the flanks and slightly to the front of the formation. It is their responsibility to protect the formation from lethal fire by constantly scanning the crowd.

6-12. When the commander receives word that DM teams are in place, he forms the control force into a crowd control formation and moves them quickly from the rally point to where the crowd is assembled. Depending on the situation, movement should be in a column formation or one of the three basic control force formations. Reserve forces are left at the rally point until ordered forward as additional lethal overwatch or to reinforce the formation.

CONTROL FORCE FORMATIONS

6-13. Control force formations, when properly employed and executed against a crowd of limited size, are one of the most practical methods of crowd control. Experience has indicated that the most frequently used control force

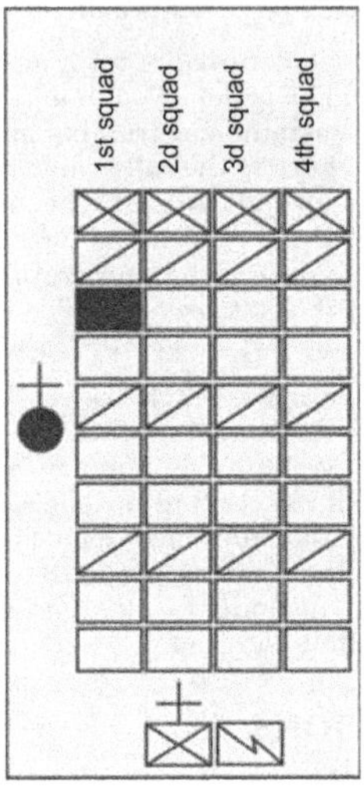

Figure 6-2. Column Formation

formations are the line, echelon, and wedge. The two lesser-used formations are the diamond and circle. Descriptions of all the formations are as follows:

- **Line formation.** The line is the basic formation and it is used more often because of its offensive and defensive applications. As an offensive formation, the line is used to push or drive crowds straight back, across an open area, or up a city street. As a defensive formation, the line is used to hold the crowd or to deny access to areas.
- **Echelon formation.** The echelon is an offensive formation used to turn or divert groups in open or built-up areas and to move crowds away from buildings, fences, and walls.
- **Wedge formation.** The wedge is an offensive formation that is used to penetrate and split crowds into smaller groups.
- **Diamond formation.** The diamond, when used as an offensive formation, is used to enter a crowd and is the formation of choice for extraction teams. As a defensive formation, the diamond is used when all-around security is required in open areas. The decision to use this formation is based on the conformation of the crowd.
- **Circular formation.** The circular formation is used for the same purpose as the diamond formation. The decision to use this formation is based on the conformation of the crowd.

6-14. There are many suitable variations of control force formations that may be employed, but appropriate commands and signals must be created to execute the formations. Because of their somewhat complicated nature and the coordination required for these formations, new variations must be trained, practiced, and rehearsed before they are used in a civil disturbance.

VEHICLES AND FORMATIONS

6-15. Vehicles may be employed with troops in control force formations, especially when a large riotous crowd is on hand. Although vehicles add strength to formations, certain precautions must be taken. Covering the windshield with sturdy, close mesh fencing and the standard safety glass will offer some level of protection to the occupants. Shields and/or mobile barriers may be built by mounting a wooden or metal frame strung with barbed wire across the front of a vehicle. Members of the formation should walk as near to the front corners of each vehicle as possible to keep rioters from attacking the sides and rear of the vehicles.

6-16. Up-armored HMMWVs, ASVs, or armored vehicles may be employed with the control force formation when determined by METT-TC. They present a strong psychological effect and offer protection for the occupants. Wheeled armored vehicles are best because they do not deface the pavement. Whenever vehicles are used in a crowd control formation, the commander must ensure that he is still able to see and control the formation. This may require him to occupy a position in a similar vehicle behind the formation or command subordinate leaders to move away from the front line and into line-of-sight (LOS) control. This will require additional hand-and-arm signals to allow the commander and subordinate leaders to communicate. Commanders can effectively use vehicles as primary communication tools by using horns and lights as signals to organize movement or actions. However vehicles are used, it takes a good measure of training and practice to be effective.

ELEMENTS OF A CONTROL FORCE FORMATION

6-17. Four elements make up the basic crowd control formation. They are as follows:

- **Base element.** The base element is made up of two ranks. The first rank is shield holders, while the second rank contains the NLW. This is the front line of the formation.
- **Support element.** The support element forms up in a column formation behind the base element and has a variety of uses. It may be used to replace base element members as needed, provide lateral or direct support, or perform extraction team operations. The support element helps the base element by performing the following three essential tasks:
  - **General support.** The general-support element is formed from an uncommitted squad in the platoon (generally the 2d squad). When a company is tasked as the control force element, one of the platoons becomes the general-support element. The element is in a column formation centered on and behind the main formation. From this formation, the general-support element can move to lateral or direct support, as needed.
  - **Lateral support.** The lateral-support element is used to protect the flanks of the formation. This is done by moving a set number of teams forward from the general-support element or by using the

end teams in the formation; however, this will make your formation smaller. Once in position, these teams become part of the formation, with their riot shields facing the flanks of the formation.

- **Direct support.** When direct support is ordered, the general-support element moves forward as they form their own formation. While moving forward, the squad and team leaders from the control force formation step back and allow the riot shield holders from the support element to step behind riot shield holders of the original formation. This allows the support riot shield holders to form up between and behind the riot shield holders of the original formation. Now there are two lines of riot shield holders between the crowd and NLW firers. This formation is the strongest of the formations and requires more planning and practice to master.

- **Command element.** The command element contains several different members. A general configuration for the command element is the platoon leader, the PSG, a radio operator, a video recorder operator (if required), and an interpreter (if required). This element does not have a fixed location within the formation and moves about as needed.
- **Lethal overwatch element.** The lethal overwatch element is a team formed from reserve security forces.

6-18. The reserve support element is not part of the control force formation until it is brought forward from the rally point to join the formation. It remains until released by the formation commander. Lethal overwatch teams are formed from the reserve force. Once lethal overwatch teams are deployed and in position, they are under the control of the formation commander. Communication between the lethal overwatch team and the formation commander is a priority because he approves target selection and engagement.

WEAPON POSITIONS

6-19. At the commander's discretion, members of the control force formation may carry their weapons in any of the positions prescribed in *FM 3-21.5*. However, the primary method of carrying weapons in the control force formation is at the safe-port position (see *Figure 6-3*). It allows the soldier to control both ends of the weapon while moving in and out of the formation and advancing on the crowd.

6-20. High-port position (see *Figure 6-3*) is a position of complete readiness. It is used whenever troops are in contact with a crowd that is showing resistance or not withdrawing. This is a tiring position and is hard to maintain for extended periods. Commanders must rest their troops at every opportunity by using less tiring positions.

Safe-port arms | High-port arms

Figure 6-3. Weapons Positions

COMMANDS

6-21. Commands are given to the control force formation orally or with hand-and-arm signals. Oral commands for the on-guard position are given in one count. All other commands are given in two counts, a preparatory command followed by a command of execution. However, oral commands cannot be relied on completely, so commanders must plan to use hand-and-arm signals too. When vehicles are part of the formation, radios become an alternate method for relaying commands. To improve communication, commanders may use hand-and-arm signals (see *Figure 6-4,* page 6-8) with oral commands. The basic commands used in control force formations are outlined in *Table 6-1,* page 6-9. These commands can be added to or combined to form more complex commands, therefore forming more complex formations.

MOVEMENTS

6-22. Control force formations move in the same manner as regular formations. This allows the commander to more precisely maneuver the formation by commanding the number of steps it should move and in what direction, such as "Five Steps, Forward March." The normal rate of march for entering and leaving a control force formation is double-timing. The half-step march is used when the formation is in direct contact with the crowd. This slows the formation down and allows for better command and control. The squad and team leaders echo preparatory commands and provide cadence while in control force formations.

INTERVAL AND DISTANCE

6-23. Interval is the lateral space between elements. Distance is the space between elements in a column. The usual interval and distance between soldiers in control force formations is 30 inches. The interval and distance can be adjusted based on METT-TC. If the crowd has to be physically pushed back, a close interval is preferred. This allows the riot shields to be overlapped, creating a stronger wall. During peaceful demonstrations, the interval could increase to double arm width, allowing the formation to cover more area and still control the crowd. The interval and distance could change several times during the course of operations before the crowd is dispersed.

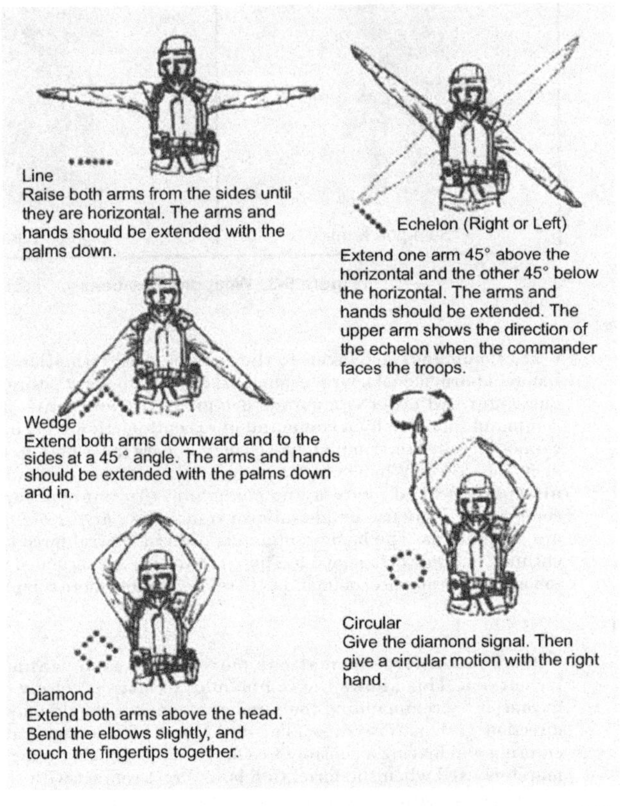

Figure 6-4. Hand-and-Arm Signals

Table 6-1. Basic Commands

Commands for	Command	Given by	Purpose
	Platoon On Line	Platoon leader and/or commander	To form a line formation.
	Platoon Wedge	Platoon leader and/or commander	To form a wedge formation.
	Platoon Echelon	Platoon leader and/or commander	To form an echelon formation.
	Platoon Diamond	Platoon leader and/or commander	To form a diamond formation.
	Platoon Circle	Platoon leader and/or commander	To form a circular formation.
Formations	Move	Platoon leader and/or commander	As a command of execution. **NOTE: The platoon leader and/or commander also identifies the location for the formation by pointing his arm to the desired location for the formation.**
	General Support	Platoon leader and/or commander	To place a specified unit in the rear of the base element.
	Lateral Support	Platoon leader and/or commander	To place support elements on the left or right flank of the base element.
Support	Direct Support	Platoon leader and/or commander	To move support elements forward to strengthen the base element.
	Open	Extraction team leader	To open a space in the formation.
	Up	Extraction team leader	To inform the extraction team to start moving back to the main formation.
	Weapon System Gunners, Number of Rounds, Type of Round, and Prepare to Fire	Platoon leader and/or commander	To ready NL gunners to fire weapons; for example, M203 gunners, three rounds, area target, prepare to fire.
Fire	Shield Down	NL gunners	To tell riot shield holders to get down on one knee and lower the riot shield to the ground.
	Shield Up	NL gunners	To tell riot shield holders to stand and raise the riot shield.
	Port Arms	Platoon leader and/or commander	To use when not in contact with the crowd.
Weapons	High Port	Platoon leader and/or commander	To tell NL gunners what position to hold their weapons in while the formation is in physical contact with the crowd. This allows for rapid targeting and firing of NL munitions.

EXTRACTION TEAM OPERATIONS

6-24. The extraction team is a squad (see *Figure 6-5*) that is generally formed from the support element, but could also be formed from reserve security forces. Once the team is formed, it is OPCON to the formation commander until the team's mission is complete and it returns to the rear of the formation. This team provides the riot control formation with the means to employ NL and lethal cover forward of the formation. It also provides the formation with the ability to remove and detain rioters downed by munitions.

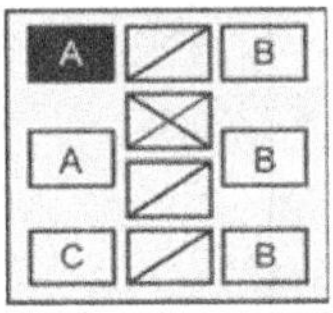

Figure 6-5. Extraction Team Formation

The extraction team—

- Extracts vehicles or personnel that are in immediate danger from the crowd.
- Detains and escorts downed rioters to the rear of the formation.
- Restrains and searches subjects.
- Covers confined or congested areas where a full riot control formation cannot be inserted.

6-25. The extraction team is usually a squad. While the recommended distance for deploying the extraction team from the control force formation is no farther than 10 meters, this may change based on METT-TC. The control formation protects the extraction team and provides immediate support, if necessary.

6-26. The extraction team may deploy from anywhere within the formation. When given orders to deploy, the squad leader of the extraction team gives the formation commands, identifies at least two personnel in the squad to conduct the search and apprehension of the subject, and leads the squad to the base line of the formation. He then extends his arm between the two riot shield holders and commands, "Open," while tapping the riot shield holders on the side of the shoulder. The two riot shield holders take one step backward and one step to the right or left. This clears a path for the extraction team. The extraction team leader then states the number of personnel leaving the formation, such as "Ten security personnel leaving." As the last extraction team member exits the formation he states, "Last Man." The two riot shield holders then return to their original position in the formation. The adjacent base element squad and/or team leaders pass the number of personnel in the team through the formation.

6-27. As the extraction team approaches an instigator or incapacitated demonstrator, the riot shield holders envelop the target and face in the direction of the nearest threat. NL gunners on the left and right cover their respective areas. The search and apprehension team immediately controls and restrains the target. To avoid confusion, one team member concentrates on controlling the subject while another team member applies a restraining system. When the search and apprehension team has the subject under control, the extraction squad leader sounds off with, "Up." Upon hearing the "Up" command, it is repeated by all extraction team members. Team members then grab a shoulder of the team member in front of them to backtrack into the formation. As the extraction team approaches the formation, the nearest base element leader will extend his arm and command, "Open," while tapping the riot shield holders on the side of the shoulder. The riot shield holders move one step backward and to the right or left to open the formation. As he enters the formation, the extraction team leader sounds off with the number of personnel entering the formation, such as "Ten security personnel and one demonstrator." The adjacent base element leader counts the number of personnel as they enter the formation. When the last man enters the formation, he states, "Last Man." The riot shield holders then return to their original places in the formation. The adjacent squad and team leaders in the base element pass information through formation that the extraction team has returned.

LETHAL OVERWATCH TEAMS

6-28. During a NL engagement, the use of a DM team provides confidence and safety to those facing the mob. If a lethal threat is presented, the DM team in an overwatch position (armed with a standard infantry rifle that is mounted with a high-powered scope) can scan the crowd, identify agitators and riot leaders for apprehension, or fire lethal fire if so ordered or warranted. It is also ideally suited for flank and countersniper operations.

6-29. The DM team sets the security overwatch and provides real-time intelligence and reporting that is vital to mission success. It provides coverage during the entire approach to the crowd, and its ability to select positions and provide cover fire (if and when warranted) is critical to the safety of the control force. Successful execution requires training, practice, and rehearsal with the control force formation.

6-30. The DM team is organized from a standard three- or four-man team. Each individual on the team is equipped with a standard-issue weapon. One individual has an infantry rifle with a scope, another has a set of binoculars, and two other individuals provide security for the team. Each team is equipped with a radio for communication with the HQ element of the control force formation.

6-31. The teams must have a visual advantage over the crowd to provide lethal protection to the formation. This is best accomplished by placing them on nearby rooftops, in the upper floors of buildings, and on hilltops. They must have the optical equipment to identify mob leaders, instigators, and/or individuals with weapons. Team integrity must be maintained. In organizations with three-man teams, one individual becomes the marksman, another the observer, and another the security. When teams have a 4th member, that individual becomes the video camera operator. This will help identify individuals for later apprehension and provides a record of events.

6-32. Two teams should be deployed to cover the flanks of the formation. Each team should be deployed to a position that is in front of the formation and in an overwatch position. As the formation moves forward, the teams leapfrog forward from the formation.

6-33. The reserve forces, located at the rally point, can provide another lethal force of some size that can be brought forward, as needed. Each soldier in the control force formation should have a sidearm (9-millimeter pistol) or an M16 or M4 rifle. The rifle is slung diagonally across his back with the rifle butt over his left shoulder and the muzzle below his right hip. Lethal munitions are in a separate ammunition pouch that is isolated from NL munitions. This is to prevent the mixing of NL and lethal ammunition.

6-34. Once the lethal overwatch teams are deployed, they are under the OPCON of the formation commander. The lethal overwatch element must have direct and secure communications with the formation commander. The number of teams deployed is based on METT-TC. Teams will leapfrog to keep ahead of the advancing control force formation with one team in the overwatch position at all times.

RESERVE FORCES

6-35. A large reserve of soldiers should be maintained during civil disturbance operations. Knowing that a large reserve force is available provides confidence and safety within the control force and helps prevent them from overreacting to provocative acts with disorderly and criminal elements in the crowd. When determining the number of reserve forces required, it should be resolved in favor of a large number. Lethal overwatch teams, forces to augment the control force, and apprehension teams come from the reserve forces.

6-36. Apprehension teams are a must when it is likely that a large number of people will be apprehended. These teams provide an organized response to what could be a chaotic situation. The teams can be organized at the team or squad level, depending on the number of apprehensions expected. These teams usually operate behind the base element. During peaceful demonstrations, these teams can be sent forward to remove demonstrators who refuse to disperse.

6-37. Each team consists of an apprehension, movement, and security element (added if the control force cannot provide security); a recorder; and a video or still photographer. The apprehension element is an officer or NCO and a civilian police officer (CONUS or OCONUS) or US marshal. It decides who will be apprehended, informs the subject of the reasons, supervises the handling of the subject, and ensures that the apprehension is properly conducted and documented. The apprehension team is responsible for the overall apprehension process, limiting the number of individuals appearing in court to document the apprehension. The movement element helps the apprehension element handle the subject. They move, restrain, and search the individuals being apprehended under the supervision of the apprehension element. The security element prevents the crowd from interfering with the mission. The recorder documents the apprehension on *DD Form 2708*. When possible, the video and/or still photographer recorder photographs each subject with the apprehension element. This provides the apprehension element with the necessary documentation for use in court. One individual may need to conduct many of the duties of an apprehension team based on METT-TC.

6-38. The video and/or still photographer record the situation before, during, and after the apprehensions. These images may also be used as pictorial documentation for use in court. It is possible that the presence of video and still cameras actively recording the situation can cause crowd members to be less prone to unruly or criminal acts.

6-39. The behavior of the selected individual to be apprehended is observed to decide if the subject is violent, cooperative, or passive. Unless there is good reason to believe otherwise, the subject must be considered dangerous. The mood of the crowd is observed to estimate what it might do. Is it only curious onlookers or does it support the individual targeted for apprehension? Will it interfere?

6-40. When apprehending an individual, the actions must be kept as simple as the situation allows. The element of surprise can give the apprehension team an advantage over the apprehended individual. It is best not to use surprise if the individual might panic and injure someone. Select an area for the apprehension that is to the advantage of the team and not the individual. Avoid apprehensions in crowded areas as these areas provide many avenues of escape and contain supporters of the apprehended individual.

6-41. All verbal commands given to the apprehended individual must convincing authority with short, clear, and distinct commands. Voice, attitude, and deportment must convey an expectation that the apprehended individual will comply.

6-42. The individual should be searched immediately; however, circumstances (violence or sympathizers for the individual being apprehended) may dictate temporarily postponing the search until the individual can be moved to a safer area. Flex-Cufs or hand irons may be applied.

SQUAD FORMATIONS

6-43. Civil disturbance training is a collective task. Using team and squad concepts, team leaders instruct their teams and squad leaders instruct their squads. Safety is key when using force on force training for the control force formation. Soldiers must understand that serious injuries may occur if they do not follow guidelines set by the commander.

6-44. The smallest formation is a squad formation, which is used in back of the main formation to cover side streets. Squad members must know their positions regardless of which control force formation they are in, squad to company.

6-45. Training must enforce soldiers' understanding of the need to stay informed for individual protection. Individual soldiers are much more vulnerable to attack when they break ranks and chase after crowd members. When an individual soldier breaks ranks he not only puts himself in danger, but the entire formation.

6-46. When moving a squad into a crowd control formation from a column formation, the squad leader takes one or more steps to the right of the squad and faces it. He gives the command and hand-and-arm signal for the formation he wants. Then he indicates where he wants the formation located by pointing in that general direction. The A team leader commands, "Follow Me." He places the baseman at the position indicated by the squad leader. The squad forms on the baseman position.

LINE FORMATION

6-47. The command for forming a squad in a line formation is "Squad Line Formation, Move." On the command of execution, the A team leader leads the baseman to the place indicated by the squad leader. The squad members align themselves in sequence with the baseman at normal intervals (see *Figure 6-6*). Based on the situation, the squad leader may designate a specific interval. He does this when he gives the preparatory command. If no interval is specified, the squad automatically forms using the normal interval.

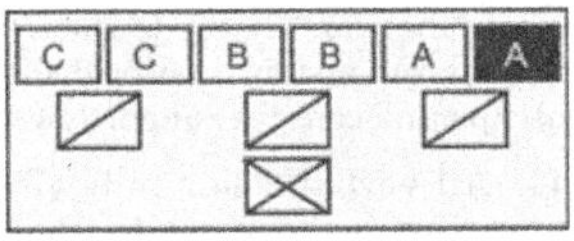

Figure 6-6. Squad Line Formation

ECHELON FORMATION

6-48. The command for having a squad form in an echelon formation is "Squad Echelon Right (or Left), Move" (see *Figure 6-7*). On the command of execution, the A team leader places the baseman at the location indicated by the squad leader. The squad members align themselves in sequence with the baseman, one pace to the side and one pace to the rear.

WEDGE FORMATION

6-49. The command for moving a squad formation into a wedge formation is "Squad Wedge, Move." On the command of execution, the A team leader moves the baseman to the place indicated by the squad leader. The A team lines up to the left of the baseman, one pace to the left and one pace to the rear of each preceding man. B team members align themselves with the baseman, one pace to the right and one pace to the rear of each preceding man. C team members align themselves in the same way to the left of the baseman (see *Figure 6-8*).

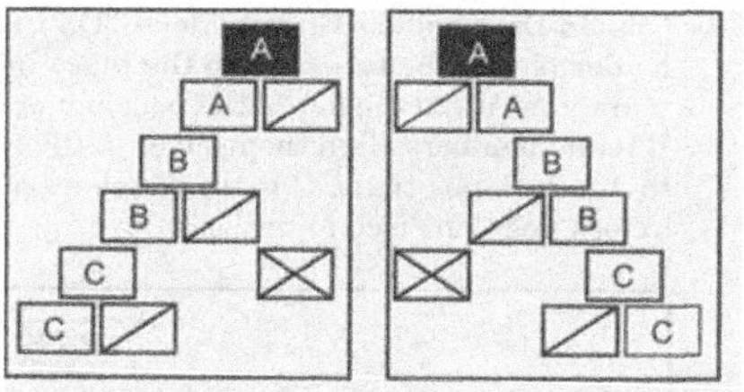

Figure 6-7. Squad Echelons Left and Right Formations

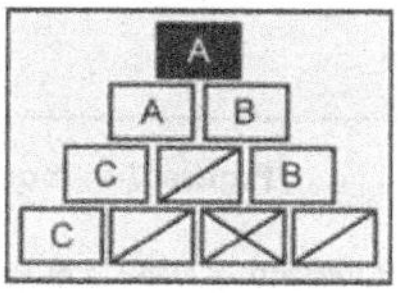

Figure 6-8. Squad Wedge Formation

ASSEMBLY

6-50. When assembling a squad from the line formation, the squad leader takes a position a sufficient distance to the rear of the squad and commands, "Squad Assemble." At the same time, he raises his right hand in the air and makes a circular motion. Squad members, who hear the command, automatically come to the port arms position. The A team leader and the baseman do an about-face movement.

6-51. The other squad members face toward the baseman. On the command of execution "Move," the squad leader points to the place where he wants the squad to assemble. The A team leader double-times to the designated spot, and the other members of the squad follow him. Team leaders fall into the file in their designated positions as they move to the AA. To assemble from a squad echelon formation, the steps are the same as a line formation.

6-52. To assemble from a squad wedge formation, the squad leader takes the same steps as for the line and echelon formations. B team members do a half left face, and the A team and C team members do a half right face. The C team pauses at the baseman position and allows the B team to clear the formation before moving to the AA.

DIAMOND AND CIRCLE FORMATIONS

6-53. The diamond and circle formations are used during extraction team operations. They are small formations used to penetrate the crowd or cover small areas. The decision on which formation to use is based on METT-TC and the ability of the squad to perform.

6-54. Commands for forming a squad in a diamond or circle formation is "Squad Diamond (or Circle), Move." On the command of execution, the A team leader moves the baseman to the place indicated by the squad leader. The A team lines up to the left of the baseman, covering the 8 to 12 o'clock positions. B team members align themselves to the right of the baseman, covering the 12 to 4 o'clock positions. C team members align themselves covering the 4 to 8 o'clock positions (see *Figure 6-9*).

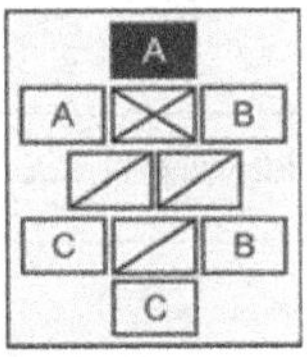

Figure 6-9. Squad Diamond or Circle Formation

6-55. To assemble from a squad diamond or circle formation, the squad leader takes the same steps as he would for a line formation. The A team leads the squad to the AA followed by the B team and then the C team. Teams fall into a column formation as they file back to the AA.

THREE-SQUAD PLATOON FORMATIONS

6-56. The platoon HQ consists of the following individuals: the platoon leader, the PSG, and a messenger or radio operator. When directed by the platoon leader, other individuals such as the selected marksman, interpreter, RCA disperser operators, and firefighters may augment the platoon HQ. In forming all control force formations from the column, the platoon leader moves to the right of the platoon and faces them when giving his commands. He gives the hand-and-arm signal and the preparatory command. On the command of execution "Move," he points to the approximate location where he wants the platoon to form. The platoon leader pauses between the preparatory command and the command of execution to allow each squad leader to issue a preparatory command to his squad.

Platoon Line Formation

6-57. The command for forming a platoon line is "Platoon on Line, Move." Immediately following the platoon leader's preparatory command, the squad leaders of the 1st and 3d squads give the command, "Follow Me." The leader of the 2d squad gives the command, "Stand Fast." On the platoon leader's command of execution, the 1st and 3d squads move forward to the point designated by the platoon leader. The 1st team member of the 1st squad is the baseman for the platoon formation. The 1st squad forms a squad line to the left of the baseman while the 3d squad forms a squad line to the right of the baseman and dresses with the 1st squad. After the 1st and 3d squads have cleared the platoon, the 2d squad leader gives the command, "Follow Me." The 2d squad splits and the A and B teams form a line to the right of the 3d squad.

The C team (with the squad leader) forms a line to the left of the 1st squad (see *Figure 6-10*).

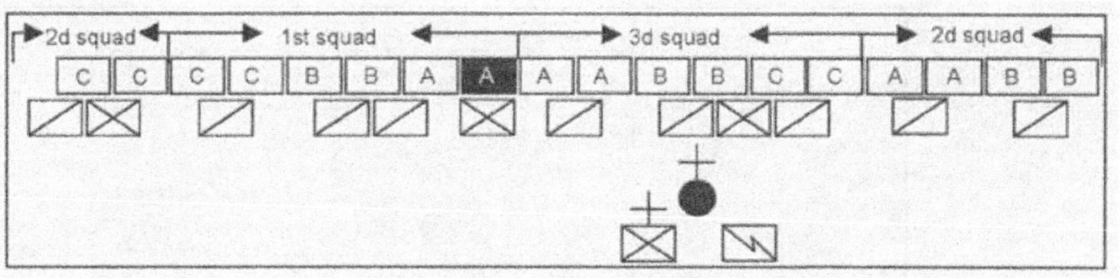

Figure 6-10. Platoon Line Formation

Assembling From a Platoon Line Formation

6-58. When assembling the platoon from the platoon line formation, the platoon leader and his HQ personnel take a position to the rear of the platoon. He then gives the command, "Platoon Assemble," while raising his right arm and making a circular motion above his head. Platoon members automatically assume the port arms position. The 1st squad leader and the baseman do an about face. The 3d squad leader moves to the head of his squad. All other members do an about-face movement toward the baseman. On the platoon leader's command of execution, the platoon leader points to where he wants the platoon to assemble. The 1st squad leader gives the command, "Follow Me," and moves his squad toward the AA. The squad moves to make a single file line to the baseman position followed by the C team of the 2d squad. As the 1st squad clears the baseman position, the 2d squad leader comes to a halt at the baseman position. The 3d squad leader gives the command, "Follow Me," and leads his squad to the AA followed by the A and B teams of the 2d squad. The A team leader of the 2d squad comes to a halt at the baseman position. Once the 3d squad has cleared the baseman position, the 2d squad leader gives the command, "Follow Me," and leads his squad to the AA. The 2d squad reassembles into their original column formation as they move to the AA. As the 1st squad comes to a halt (facing the crowd) in the AA, the 3d squad dresses with the 1st squad and leaves space for the 2d squad to return to its place in the formation. Once the platoon is formed into a column formation, HQ personnel take their position in the formation. The PSG then gives the command, "Countercolumn March," and moves the platoon to the rally point.

Platoon Line Formation With General Support

6-59. Whenever the command for establishing a crowd control formation contains the phrase "In Support" (without modification), it means that the support element is to remain in general support.

6-60. The command for forming a platoon line with the 2d squad in general support is "Platoon Line, 2d Squads in Support, Move." The 1st and 3d squads of the platoon line with general support execute the line, while the 2d squad remains in the column (see *Figure 6-11*). To assemble a platoon with a squad in general support, the procedure is the same as the platoon line, except that the 1st and 3d squads dress with the 2d squad.

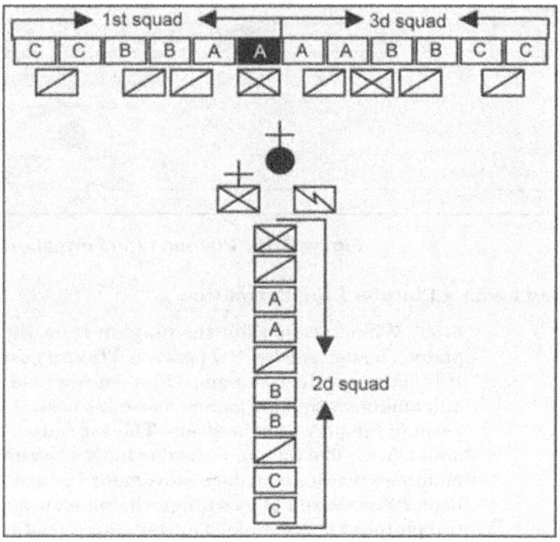

Figure 6-11. Platoon Line Formation With General Support

Platoon Line Formation With Lateral Support

6-61. The command for forming a platoon line with one squad in lateral support is "Platoon Line, 2d Squad in Lateral Support, Move." The 1st and 3d squads execute the line as before, and the 2d squad stands fast. After the line has been formed, the 2d squad leader gives the command, "Follow Me." The A and B teams move to the right flank while the squad leader and the C team move to the left flank of the formation. The 2d squad forms columns behind the last men on the line formation. If no direction is given, the 2d squad splits to cover both flanks. A squad may be used to cover one flank by giving the command, "2d Squad Right (or Left) Lateral Support" (see *Figure 6-12*).

6-62. The 2d squad may be moved from general to lateral support at any time by the platoon leader. The platoon leader commands, "2d Squad, Lateral Support, Move." To have the 2d squad join the line from either support position, the platoon leader commands, "2d Squad, Extend the Line, Move." The 2d squad leader commands, "Follow Me." On the command of execution, the squads set up individual lines, closing and dressing on the existing line. To assemble the platoon from a platoon line with lateral support, follow the same steps as for the platoon line.

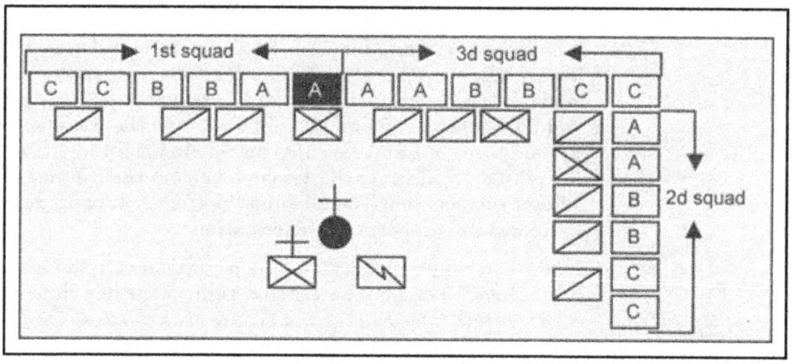

Figure 6-12. Platoon Line Formation With Lateral Support (Right)

Platoon Line Formation With Direct Support

6-63. The command for forming a platoon line with one squad in direct support is "Platoon Line, 2d Squad in Direct Support, Move." The 1st and 3d squads execute a line as before. The 2d squad executes a squad line directly to the rear and centered on the 1st line. The men in the supporting line cover the intervals between the men in the 1st line. To do this, the squad and team leaders of the 1st and 3d squads must take two steps to the rear to allow space for the 2d squad to move into place. After seeing the squad and team leaders move back from the line, the 2d squad leader commands, "Follow Me," and leads his squad into position (see *Figure 6-13*).

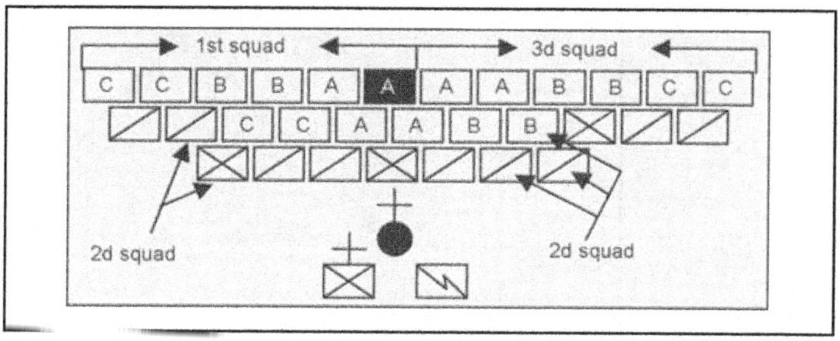

Figure 6-13. Platoon Line Formation With Direct Support

6-64. To assemble a platoon with direct support, the procedure is the same as the platoon line, except that the 2d squad moves to the AA first. To assemble the support squads from any position to general support, the platoon leader commands, "2d Squad in Support, Move." The 2d squad then returns to a column behind the line formed by the other two squads.

Platoon Echelon Right and Left Formations

6-65. The command for forming a platoon echelon right is "Platoon Echelon Right, Move." After the platoon leader's preparatory command, the squad leader of the 1st squad commands, "Follow Me." The squad leaders of the 2d and 3d squads command, "Stand Fast." On the command of execution, the 1st squad moves out and executes an echelon right at the location designated by the platoon leader. As the 1st squad clears the column, the 3d and 2d squads extend the echelon. The 2d squad performs support to this formation in the same ways it supports a line formation.

6-66. The command for forming a platoon echelon left is "Platoon Echelon Left, Move." The platoon echelon right is formed in the same manner as the echelon left (see *Figure 6-14*). *Figures 6-15; 6-16,* page 6-22; and *6-17,* page 6-23 show support formations in a left echelon.

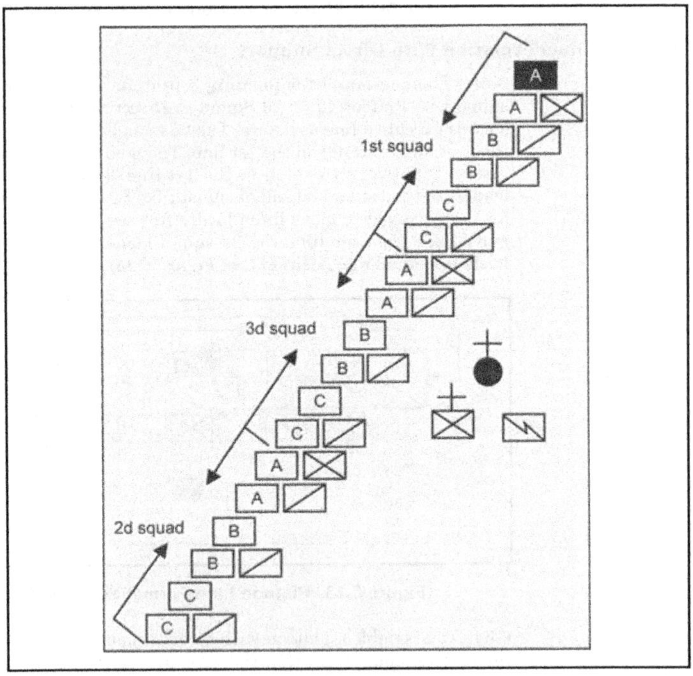

Figure 6-14. Platoon Echelon Left Formation

6-67. To assemble the platoon, the commander takes his position to the rear of the formation with the platoon HQ. On the preparatory command, "Platoon Assemble," the 1st squad leader does an about-face movement while all other

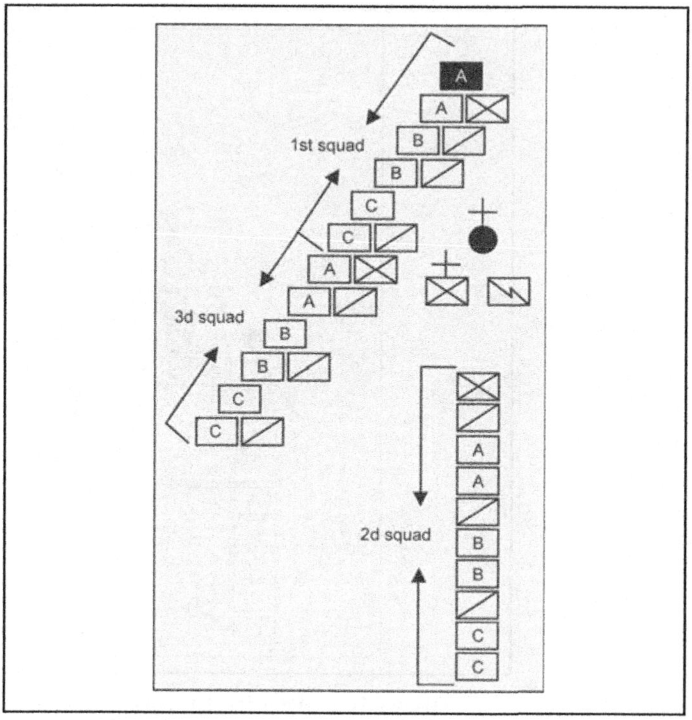

Figure 6-15. Platoon Echelon Left Formation With 2d Squad in General Support

members assume the port arms position and face the baseman. On the command, "Move," the 1st squad moves first, followed by the 3d and 2d squads. Each squad dresses with the 1st squad as it reaches the AA. The assembly of the echelon right formation is the same as the echelon left.

Platoon Wedge Formation

6-68. The command for forming the platoon wedge formation is "Platoon Wedge, Move." After the platoon leader's preparatory command, the squad leaders of the 1st and 3d squads command, "Follow Me." At the same time, the 2d squad leader commands, "Stand Fast." On the platoon leader's command of execution, the 1st and 4th squads move directly to the front. The 1st squad executes an echelon left on the baseman while the 3d squad executes an echelon right. When the last man of the 1st and 4th squads clear the 2d squad, the 2d squad leader commands, "Follow Me," and moves out to the left and right. The 2d squad splits and the A and B teams go to the right and the C

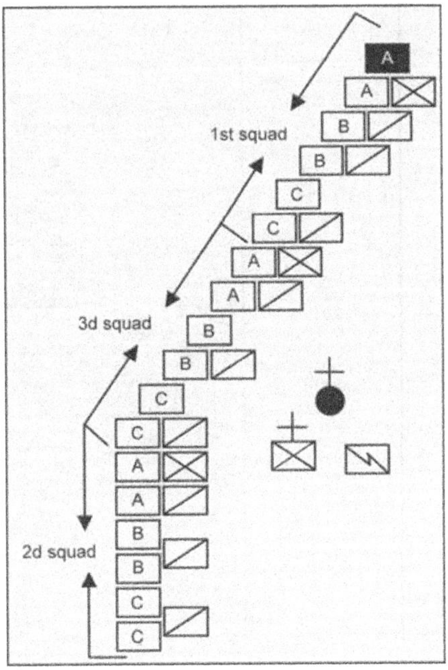

Figure 6-16. Platoon Echelon Left Formation With 2d Squad in Lateral Support

team goes to the left to extend the formation (see *Figure 6-18*). Assembly of the wedge formation is the same as the platoon line.

Platoon Wedge Formation With General Support

6-69. The command for forming a platoon wedge formation with support is "Platoon Wedge, 2d Squad in Support, Move." The 1st and 3d squads execute the wedge while the 2d squad remains in a column formation (see *Figure 6-19,* page 6-24).

Platoon Wedge Formation With Lateral Support

6-70. The command for forming a platoon wedge formation with lateral support is "Platoon Wedge, 2d Squad in Lateral Support, Move." The 1st and 3d squads execute the wedge while the 2d squad stands fast. After the 1st and 3d squads have cleared the column, the 2d squad leader commands, "Follow Me." If no direction is given, the 2d squad splits to cover both flanks. A squad

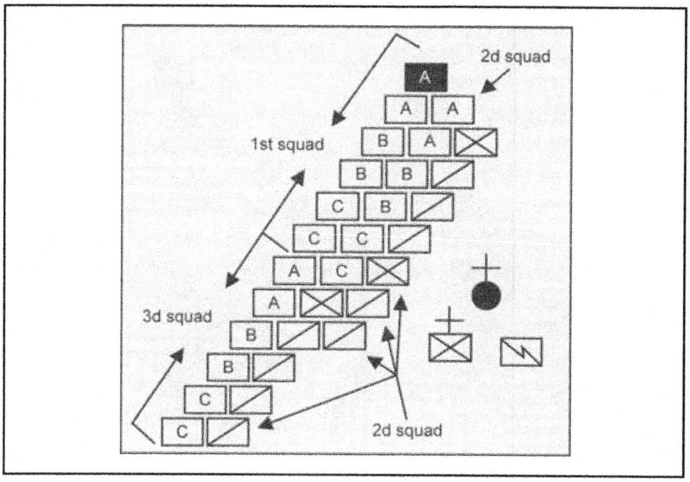

Figure 6-17. Platoon Echelon Left With 2d Squad in Direct Support

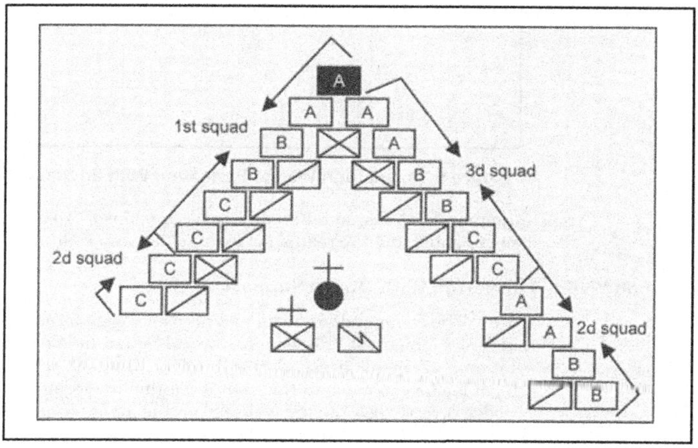

Figure 6-18. Platoon Wedge Formation

may be used to cover one flank by commanding, "2d Squad Right (or Left) Lateral Support" (see *Figures 6-20* and *6-21*, page 6-25).

6-71. To move the 2d squad from general support to lateral support, the platoon leader commands, "2d Squad, Lateral Support, Move." To have the 2d squad join the wedge from general or lateral support, the platoon leader

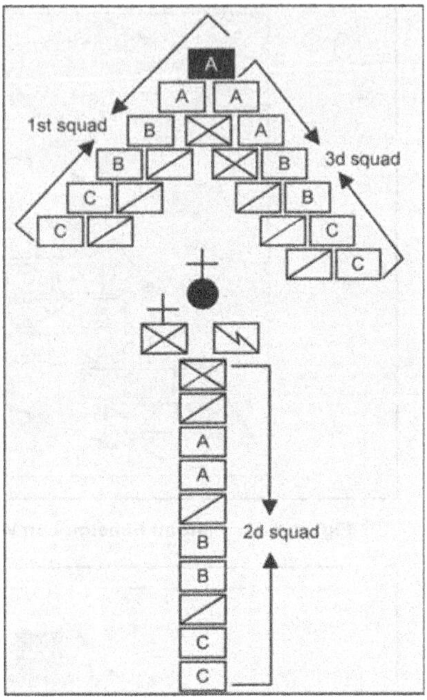

Figure 6-19. Platoon Wedge Formation With 2d Squad in General Support

commands, "2d Squad, Extend the Wedge, Move." To extend the wedge, the 2d squad splits and the teams move to the left and right, respectively.

Platoon Wedge Formation With Direct Support

6-72. The command for forming a platoon wedge formation with the 2d squad in direct support is "Platoon Wedge, 2d Squad in Direct Support, Move." The 1st and 3d squads execute a wedge. The 2d squad waits for the squad and/or team leaders to move to the rear, as in the line formation, before executing a wedge directly behind and centered on the main formation. The individuals in the supporting wedge cover the intervals between individuals in the leading wedge (see *Figure 6-22,* page 6-26).

Platoon Diamond and Circle Formations

6-73. These formations are used in open areas or where the crowd may be able to envelop the formation. The diamond and circle formations are primarily defensive in nature and difficult to maneuver. The decision about which formation to use is based on the ability of the platoon to perform them.

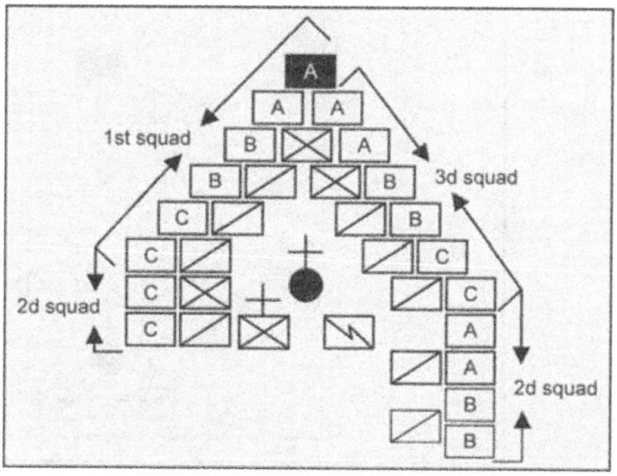

Figure 6-20. Platoon Wedge Formation With 2d Squad in Lateral Support

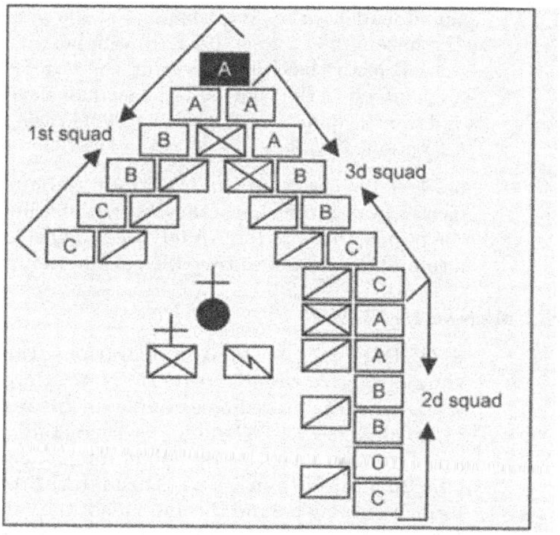

Figure 6-21. Platoon Wedge Formation With 2d Squad in Right Lateral Support

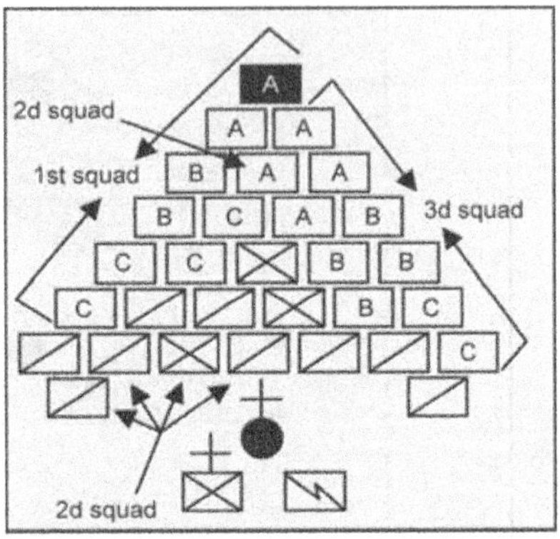

Figure 6-22. Platoon Wedge Formation With 2d Squad in Direct Support

The command for having a platoon form in a diamond or circle formation is "Platoon Diamond (or Circle), Move." On the command of execution, the 1st squad leader moves the baseman to the place indicated by the squad leader. The baseman becomes the 12 o'clock position while the 1st squad lines up to the left of the baseman covering the 8 to 12 o'clock positions. The 3d squad aligns itself to the right of the baseman, covering the 12 to 4 o'clock positions and the 2d squad aligns itself, covering the 4 to 8 o'clock positions (see *Figures 6-23* and *6-24*).

6-74. To assemble from a platoon diamond or circle formation, the HQ element takes the same steps as that of a line formation. The 1st squad leads the platoon back to the AA followed by the 2d and 3d squads. Squads fall into a column formation as they file back to the AA.

RELIEF ELEMENTS IN PLACE

6-75. During crowd control operations, the platoon leader may rotate his squads to give them a rest. He is not limited to using only those squads specified in the preceding examples as the lead elements of the formations. He can replace the base elements by commanding direct support to the 1st squad and then directing the 1st squad into general support. The support element (the 2d squad) then passes through the 1st squad. Once the 2d squad is formed directly behind the 1st squad, the 1st squad leader commands, "Right Face." The members of the 1st squad do a right-face movement and the 2d squad steps quickly though the 1st squad and forms the base formation. The 1st squad leader then commands, "Follow Me," and leads his squad to the general support position. This procedure is often necessary when protective masks are used and chemical irritants employed. The 1st squad then replaces the 3d squad in the same manner. Platoons participating in company operations are relieved in the same manner.

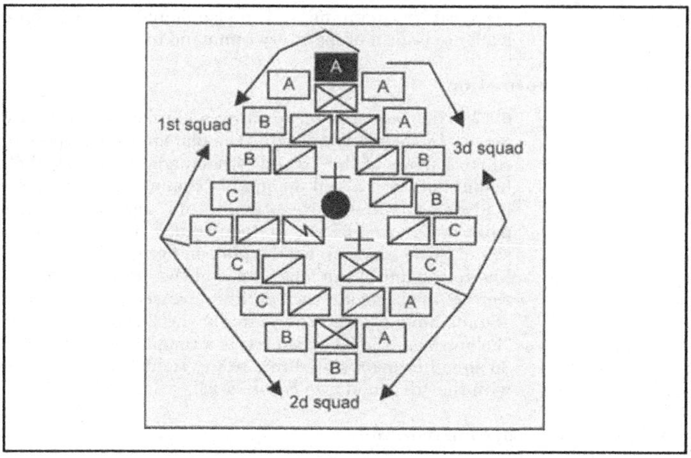

Figure 6-23. Platoon Diamond Formation

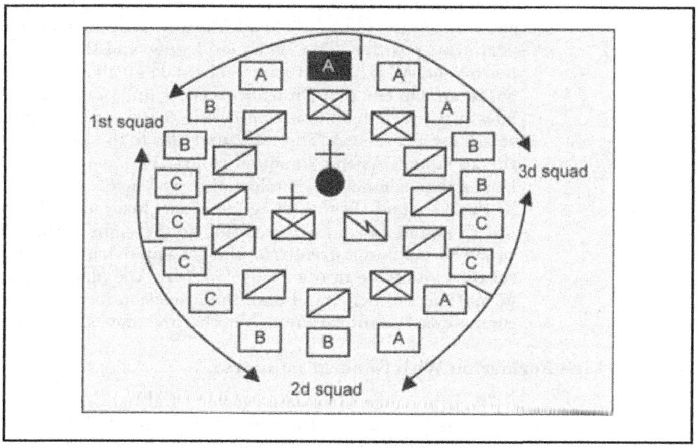

Figure 6-24. Platoon Circle Formation

FOUR-SQUAD PLATOON FORMATIONS

6-76. When forming control force formations from the column, the platoon leader moves to the right of the platoon and faces it when giving his commands. He gives the hand-and-arm signal and the preparatory command. On the command of execution, "Move," he points to the approximate location where he wants the platoon to form. The platoon leader pauses between the preparatory command and the command of execution to allow each squad leader to issue a preparatory command to his squad.

Platoon Line Formation

6-77. The command for forming a platoon line formation is "Platoon Line, Move." Immediately following the platoon leader's preparatory command, the squad leaders of the 1st and 4th squads command, "Follow Me." The squad leaders of the 2d and 3d squads command, "Stand Fast." On the platoon leader's command of execution, the 1st and 4th squads move forward to the point designated by the platoon leader. The 1st A team member of the 1st squad is the baseman for the platoon formation. The 1st squad forms a squad line to the left of the baseman and the 4th squad forms a squad line to the right of the baseman with all lines dressed to the right. After the 1st and 4th squads have cleared the platoon, the 2d and 3d squad leaders command, "Follow Me." The 2d squad forms a squad line to the left of the 1st squad. The 3d squad forms a squad line to the right of the 4th squad. All squads dress with the 4th squad (see *Figure 6-25*).

Platoon Line Formation Assembly

6-78. When assembling from the platoon line formation, the platoon leader and his HQ personnel take a position to the rear of the platoon. He then gives the command, "Platoon Assemble," while raising his right arm and making a circular motion above his head. Platoon members automatically assume the port arms position. The 1st squad leader and the baseman do an about-face movement. All other platoon members do an about-face movement toward the baseman. On the platoon leader's command of execution, he points to where he wants the platoon to assemble. He commands, "Follow Me," and moves his squad toward the AA. The 1st squad files to the baseman position, followed by the 2d squad. As the 2d squad clears the baseman position, the 4th squad leader then commands, "Follow Me," and leads his squad to the AA, followed by the 3d squad. As the 1st squad comes to a halt facing the crowd in the AA, the 2d squad assumes its position to the right of the 1st squad. As the 4th squad assembles, it dresses on the 1st squad, leaving space for the 3d squad to return to its place in the formation. Once the platoon is formed into a column formation, HQ personnel take their position in the formation. The PSG then commands, "Countercolumn, March," and moves the platoon to the rally point.

Platoon Line Formation With General Support

6-79. Whenever the command for establishing a crowd control formation contains the phrase "In Support Without Modification," it means that the support element is to remain in general support. The command for forming a platoon line with two squads in general support is "Platoon Line, 2d and 3d Squads in Support, Move." The 1st and 4th squads execute the line, while the 2d and 3d squads remain in the column (see *Figure 6-26,* page 6-30). To assemble a platoon with two squads in general support, the procedure is the same as that of the platoon line, except that the 1st and 4th squads dress on with the 2d and 3d squads.

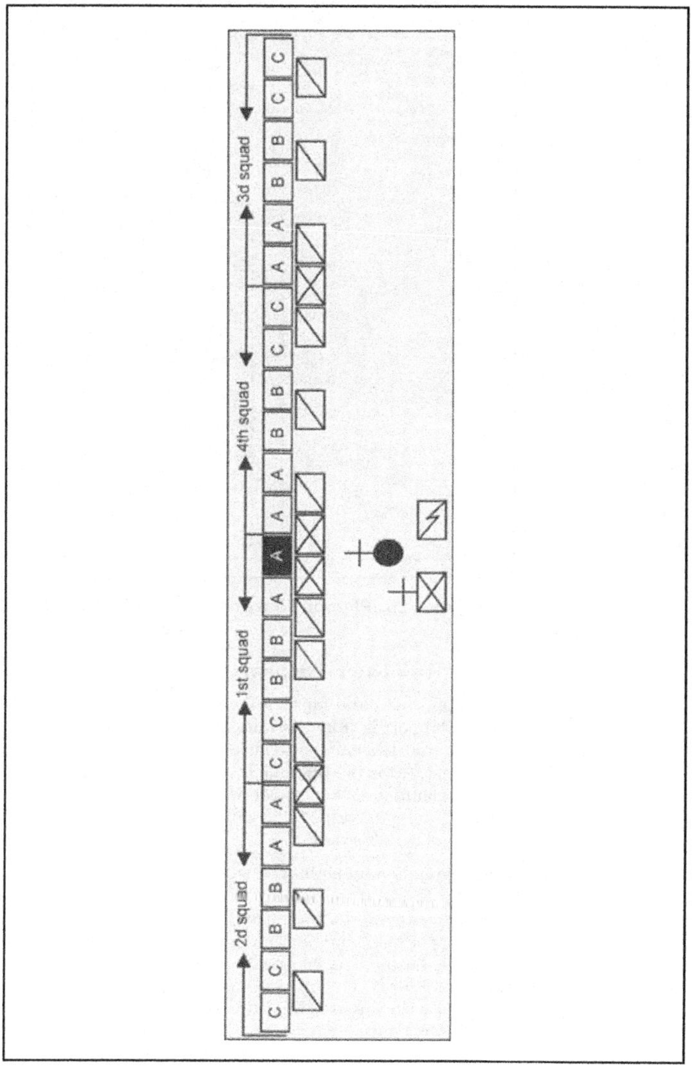

Figure 6-25. Platoon Line Formation With Four Squads

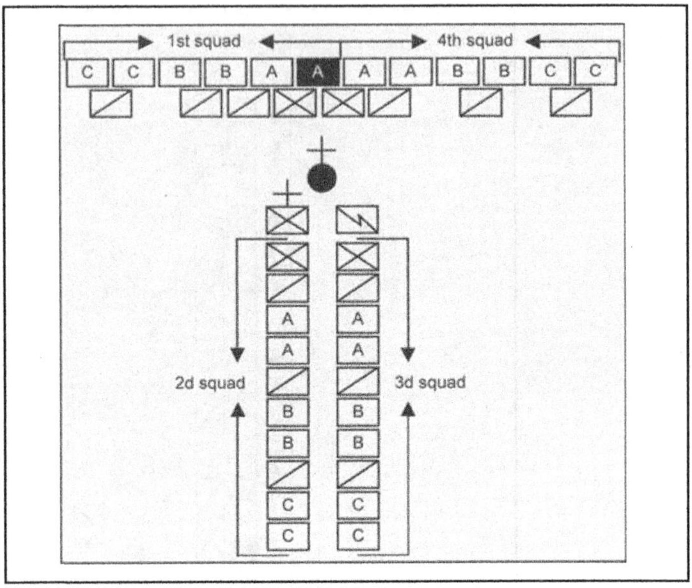

Figure 6-26. Platoon Line Formation With d and 3d Squads in General Support

Platoon Line Formation With Lateral Support

6-80. The command for forming a platoon line formation with two squads in lateral support is "Platoon Line, 2d and 3d Squads in Lateral Support, Move." The 4th and 1st squads execute the line as before, and the 2d and 3d squads stand fast. After the line has been formed, the squad leaders of the 2d and 3d squads command, "Follow Me." The 2d squad forms a column behind the last man of the 1st squad. The 3d squad forms a column behind the last man of the 4th squad (see *Figure 6-27*).

6-81. Squads may be moved from general support to lateral support at any time by the platoon leader. He commands, "2d and 3d Squads, Lateral Support, Move." To have the 2d and 3d squads join the line from either support position, the platoon leader commands, "2d and 3d Squads, Extend the Line, Move." The 2d and 3d squad leaders command, "Follow Me." On the command of execution, the squads set up individual lines, closing and dressing with the existing line. To assemble the platoon from a platoon line with lateral support, the procedure is the same as that of the platoon line, except lateral support squads face forward until they move to the AA.

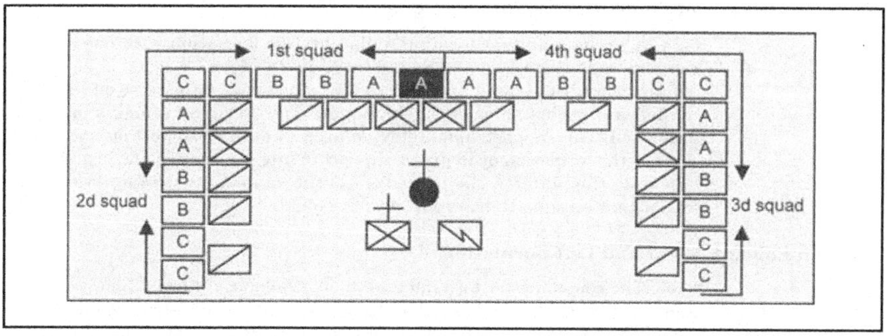

Figure 6-27. Platoon Line Formation With 2d and 3d Squads in Lateral Support

Platoon Line Formation With Direct Support

6-82. The command for forming a platoon line with two squads in direct support is "Platoon Line, 2d and 3d Squads in Direct Support, Move." The 1st and 4th squads execute a line as before. The 2d and 3d squads execute a 2d line directly to the rear and centered on the 1st line. The members in the direct support line cover the intervals between members in the baseline. To do this, the squad and team leaders of the 1st and 4th squads must take two steps to the rear to allow space for the 2d and 3d squads to move into place. After seeing the squad and team leaders move back from the line, 2d and 3d squad leaders command, "Follow Me," and lead their squads into position (see *Figure 6-28*).

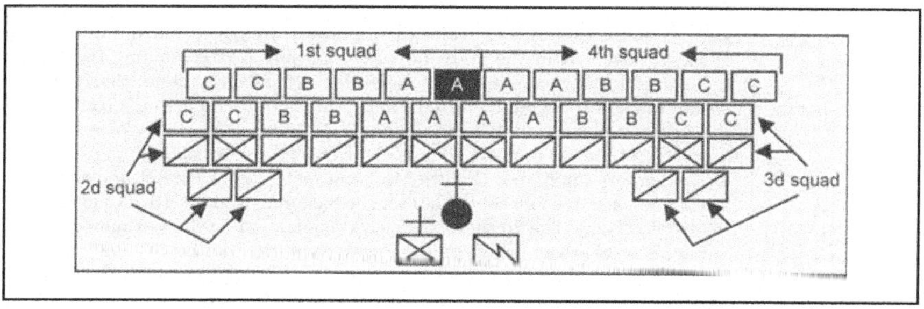

Figure 6-28. Platoon Line Formation With 2d and 3d Squads in Direct Support

6-83. To assemble the support squads from any position to general support, the platoon leader commands, "2d and 3d Squads in Support, Move." The 2d and 3d squads then return to a column behind the line formed by the 1st and 4th squads. To assemble the entire platoon, the procedure is the same as that of a platoon line, except that the 2d squad moves to the AA first, followed by the 3d, 1st, and 4th squads.

Platoon Line Formation With One Support Squad

6-84. The command for forming a platoon line formation with one squad in support is "Platoon Line, 3d Squad in Support, Move." The 1st squad moves out and executes a squad line. The 2d and 4th squads form a line on the side of the left and rights sides of the 1st squad. The 3d squad remains in column formation. The support squad may be used in lateral support on one or both sides of the formation or in direct support of any segment of the formation. To assemble the platoon, the procedure is the same as that of a platoon line, except that all squads dress with the 3d squad.

Platoon Echelon Right and Left Formations

6-85. The command for forming a platoon echelon right formation is "Platoon Echelon Right, Move." After the platoon leader's preparatory command, the squad leader of the 1st squad commands, "Follow Me." The squad leaders of the 2d, 3d, and 4th squads command, "Stand Fast." On the command of execution, the 1st squad moves out and executes an echelon right at the location designated by the platoon leader. As each squad clears the column, the next successive squad moves out and extends the echelon that has been formed. To assemble the platoon, the procedure is the same as that of a platoon line, except that all squads file back in order.

6-86. The command for forming a platoon echelon left formation is "Platoon Echelon Left, Move." The platoon echelon left is formed in the same manner as the echelon right, but in reverse order (see *Figure 6-29*). To assemble the platoon, the procedure is the same as that of an echelon right. See *Figures 6-30,* page 6-34*; 6-31,* page 6-35*;* and *6-32,* page 6-36, for examples of echelon formations with support.

Platoon Wedge Formation

6-87. The command for forming the platoon wedge formation is "Platoon Wedge, Move." After the platoon leader's preparatory command, the 1st and 4th squad leaders command, "Follow Me." At the same time, the 2d and 3d squad leaders command, "Stand Fast." On the platoon leader's command of execution, the 1st and 4th squads move directly to the front. When the last members of the 1st and 4th squads clear the 2d and 3d squads, the 2d and 3d squad leaders command, "Follow Me," and move out to the left and right. The 1st squad executes an echelon left on the baseman and the 4th squad executes an echelon right. The 2d squad forms an echelon left on the last man of the 1st squad. The 3d squad forms an echelon right on the last man of the 4th squad. This completes the wedge formation (see *Figure 6-33,* page 6-36). To assemble the platoon, the procedure is the same as that of a platoon line.

Platoon Wedge Formation With General Support

6-88. The command for forming a platoon wedge formation with two squads in support is "Platoon Wedge, 2d and 3d Squads in Support, Move." The 1st and 4th squads execute the wedge while the 2d and 3d squads remain in column formation (see *Figure 6-34,* page 6-37).

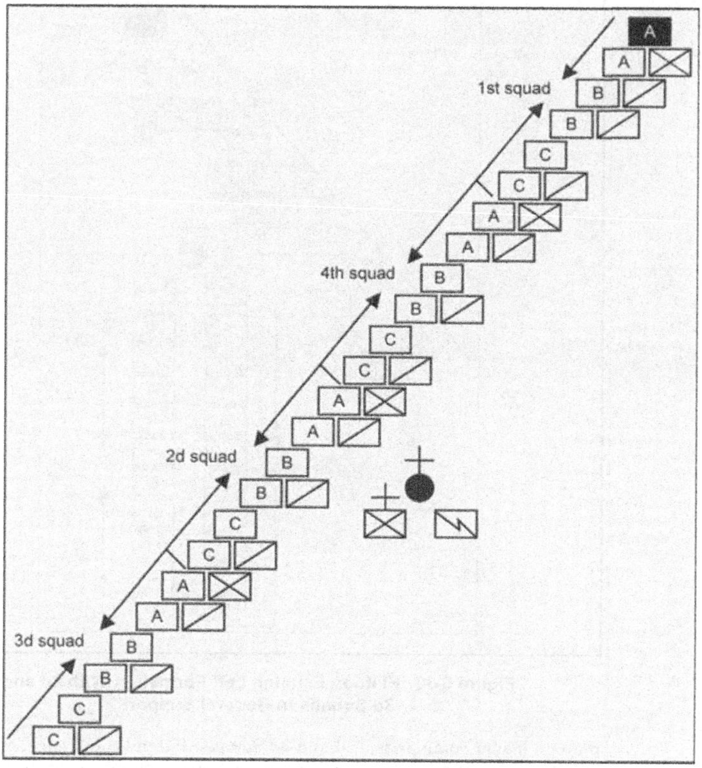

Figure 6-29. Platoon Echelon Left Formation With Four Squads

Platoon Wedge Formation With Lateral Support

6-89. The command for forming a platoon wedge formation with lateral support is "Platoon Wedge, 2d and 3d Squads in Lateral Support, Move." The 1st and 4th squads execute the wedge while the 2d and 3d squads stand fast. After the 1st and 4th squads have cleared the column, the 2d and 3d squad leaders command, "Follow Me," and move out to the left and right. The 2d squad forms a column formation behind the last man of the 1st squad. The 3d squad forms a column formation behind the last man of the 4th squad (see *Figure 6-35,* page 6-38).

6-90. To move the 2d and 3d squads from general to lateral support, the platoon leader commands, "2d and 3d Squads, Lateral Support, Move." To have the 2d and 3d squads join the wedge from general or lateral support, the

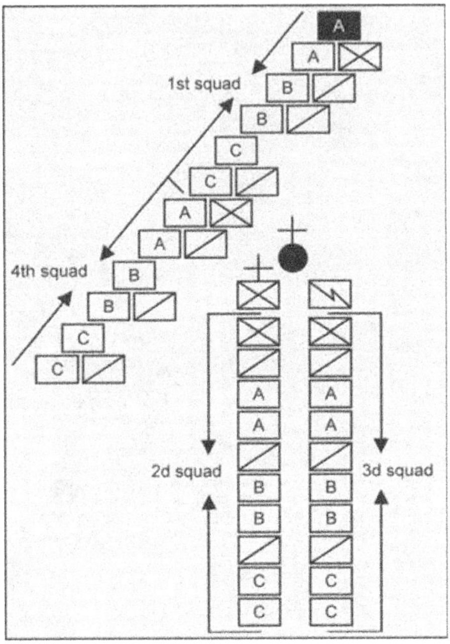

Figure 6-30. Platoon Echelon Left Formation With 2d and 3d Squads in General Support

platoon leader commands, "2d and 3d Squads, Extend the Wedge, Move." The 2d and 3d squad leaders command, "Follow Me," and move out to the left and right to extend the wedge on the 1st and 4th squads.

Platoon Wedge Formation With Direct Support

6-91. The command for forming a platoon wedge with two squads in direct support is "Platoon Wedge, 2d and 3d Squads in Direct Support, Move." The 1st and 4th squads execute a wedge. The 2d and 3d squads execute a wedge and close in on the leading wedge. The members in the supporting wedge cover the intervals between members in the leading wedge. To do this, the squad and team leaders of the 1st and 4th squads must take two steps to the rear to allow space for the 2d and 3d squads to move into place. After seeing the squad and team leaders move back from the line, the 2d and 3d squad leaders command, "Follow Me," and lead their squads into position (See *Figure 6-36,* page 6-38). To move the 2d and 3d squads back to general support, the platoon leader commands, "2d and 3d Squads Support, Move." Then the 2d and 3d squads move to their general support positions. To

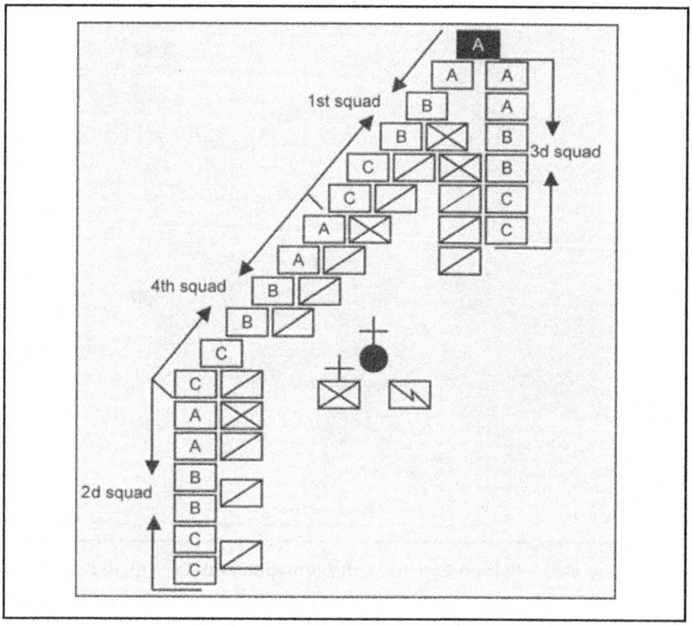

Figure 6-31. Platoon Echelon Left Formation With 2d Squad in Lateral Support

assemble a platoon from a platoon wedge with direct support, the procedure is the same as that of a platoon wedge, except that the 2d squad assembles first, followed by the 3d, 1st, and 4th squads.

Platoon Wedge Formation With One Support Squad

6-92. The command for forming a platoon wedge formation with one squad in support is "Platoon Wedge, 3d Squad in Support, Move." The 1st squad moves out and executes a squad wedge. The 2d and 4th squads form echelons left and right on the 1st squad. The 3d squad remains in the column formation.

Platoon Diamond and Circle Formations

6-93. The platoon diamond and circle formations are used in open areas or where the crowd may be able to envelop the formation. They are primarily defensive in nature and hard to maneuver. The decision concerning which formation to use is based on the ability of the platoon to perform them. The command for having a platoon form in a diamond or circle formation is "Platoon Diamond (or Circle), Move." On the command of execution, the 1st squad leader moves the baseman to the location indicated by the squad leader. The baseman becomes the 12 o'clock position. The 1st squad lines up to the

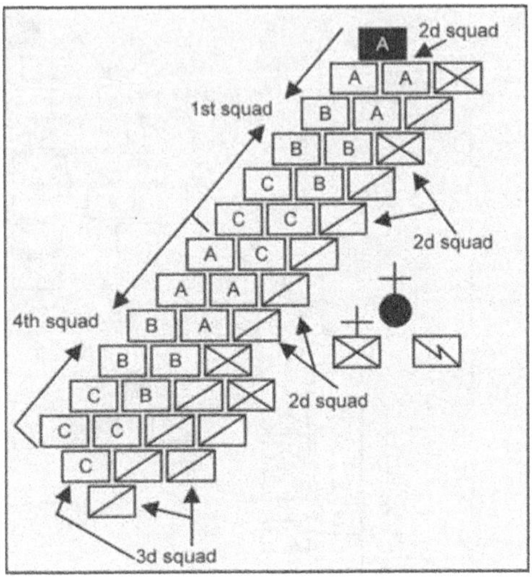

Figure 6-32. Platoon Echelon Left Formation With 2d and 3d Squads in Direct Support

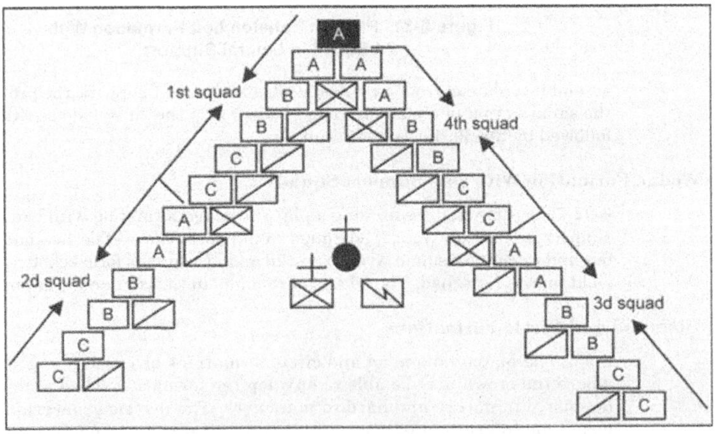

Figure 6-33. Platoon Wedge Formation With Four Squads

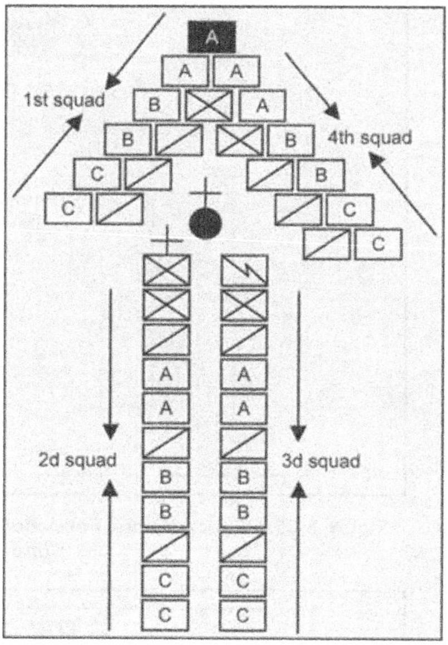

Figure 6-34. Platoon Wedge Formation With 2d and 3d Squads in General Support

left of the baseman covering the 9 to 12 o'clock positions. The 4th squad aligns themselves to the right of the baseman covering the 12 to 3 o'clock positions. The 3d squad aligns itself covering the 3 to 6 o'clock positions. The 2d squad aligns itself covering the 6 to 9 o'clock positions.

6-94. To assemble a platoon from a platoon diamond or circle formation, the HQ element uses the same procedures as that of a line formation. The 1st squad leads the platoon back to the AA followed by the 2d, 3d, and 4th squads. The squads fall into a column formation as they file back to the AA.

COMPANY FORMATIONS

6-95. When a company commander orders his company into control force formations from a column, he moves to the left near the head of the column so he can be seen by the platoon leaders. He then faces the company. As he gives his preparatory command, he points to where he wants the formation. If he does not indicate a location, the company forms immediately in front of the lead platoon.

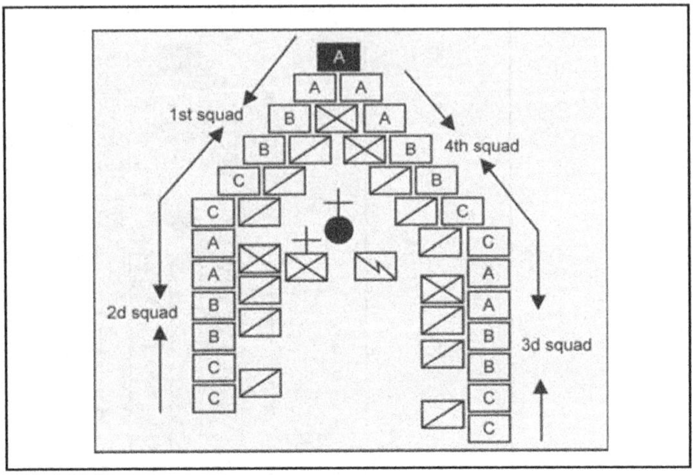

Figure 6-35. Platoon Wedge Formation With 2d and 3d Squads in Lateral Support

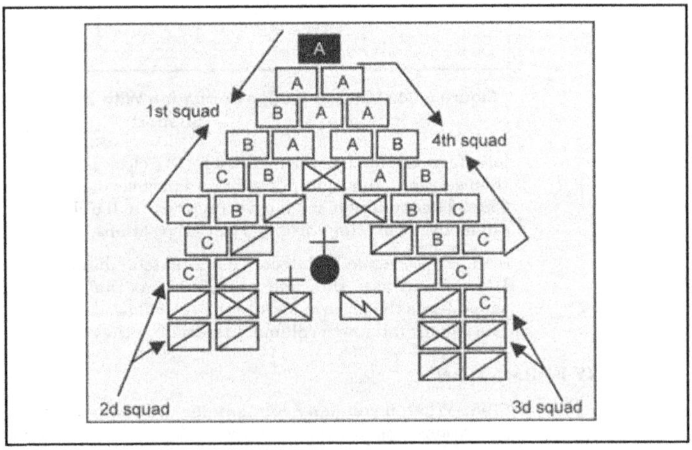

Figure 6-36. Platoon Wedge Formation With 2d and 3d Squads in Direct Support

Relieving Elements in Place

6-96. During crowd control operations, the platoon leader may rotate his squads to give them a rest. See *paragraph 6-76,* page 6-27.

Company Line in Depth Formation

6-97. The command for forming a company line in depth formation (see *Figure 6-37*) is "Company Line in Depth, Move." After the company commander's preparatory command, each platoon leader gives the command, "Platoon Line." The squad leaders follow with their commands to their squads to form the platoon line. On the command of execution, each platoon establishes a platoon line immediately to its front.

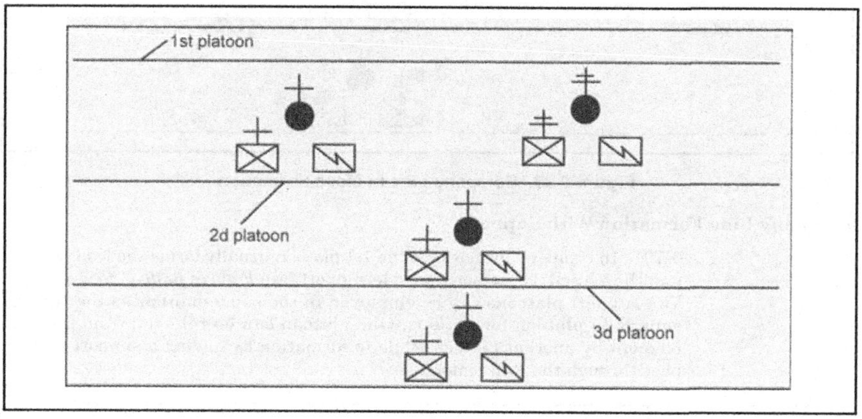

Figure 6-37. Company Line in Depth Formation

6-98. If the commander wants a more formidable formation, he commands, "2d Platoon, Direct Support, Move." The squad and team leaders from the 1st platoon take two steps backward. The 2d platoon moves forward and covers the intervals between the members of the lead platoon. The platoon leader of the 3d platoon then moves his platoon forward to occupy the position formerly held by the 2d platoon.

6-99. If the 3d platoon is called on to support the company line, it moves forward and takes up a position directly behind the 2d platoon in the same manner. Platoon leaders and PSGs help each other in controlling the company.

Company Line in Mass Formation

6-100. The command for forming a company line in mass formation is "Company Line in Mass, Move" (see *Figure 6-38,* page 6-40). The company line in mass is the same as a company line with both platoons in direct support. Instead of forming it in stages, the company commander indicates his desire for this formation in his initial command. The platoons each form a line, and the 2d and 3d platoons close on the 1st platoon without further command.

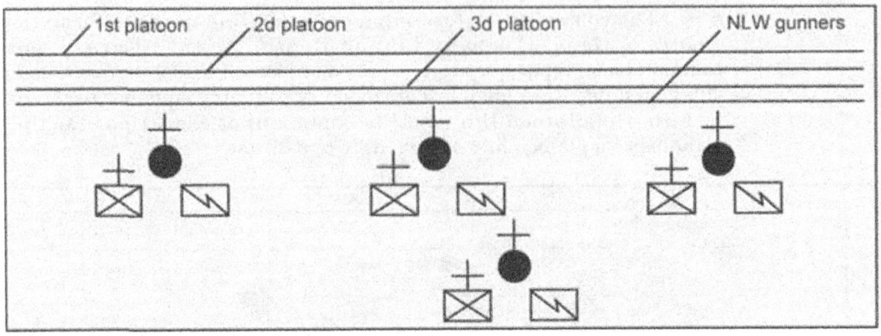

Figure 6-38. Company Line in Mass Formation

Company Line Formation With Support

6-101. In company formations, the 1st platoon usually forms the lead element and the 2d and 3d platoons provide support (see *Figures 6-39; 6-40;* and *6-41*). The support platoons can be employed in the same manner as the support squads in platoon formations. One platoon can be relieved from the lead element by another platoon while in formation by having a support platoon pass through the lead platoon.

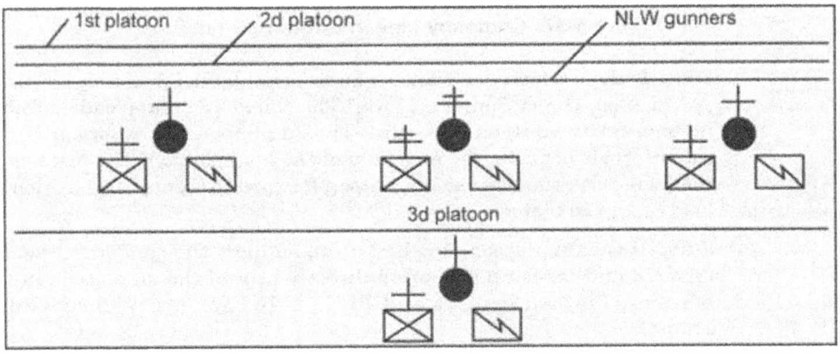

Figure 6-39. Company Line in Mass Formation With One Platoon in Depth

Company Echelons and Wedge Formations

6-102. Company echelons (see *Figures 6-42,* page 6-42; *6-43,* page 6-43; *6-44,* page 6-43; and *6-45,* page 6-44) and company wedge formations (see *Figures 6-46,* page 6-45; *6-47,* page 6-46; *6-48,* page 6-46; and *6-49,* page 6-47) are formed in the same manner and with the same variations as the company line formation. A company assembles from a crowd control formation in the same manner as a platoon or a squad does. However, due to the size of company

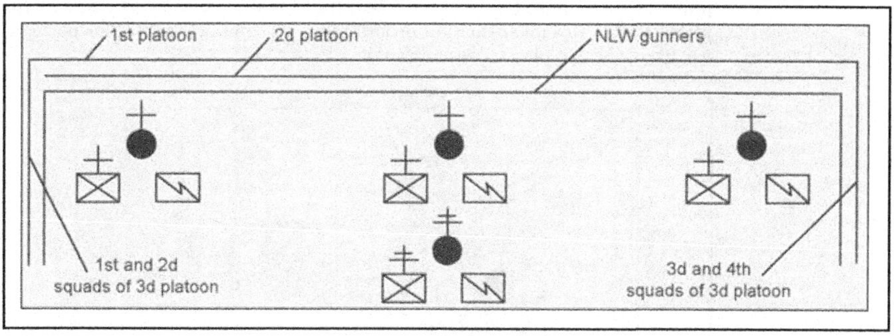

Figure 6-40. Company Line in Mass Formation With One Platoon in Lateral Support

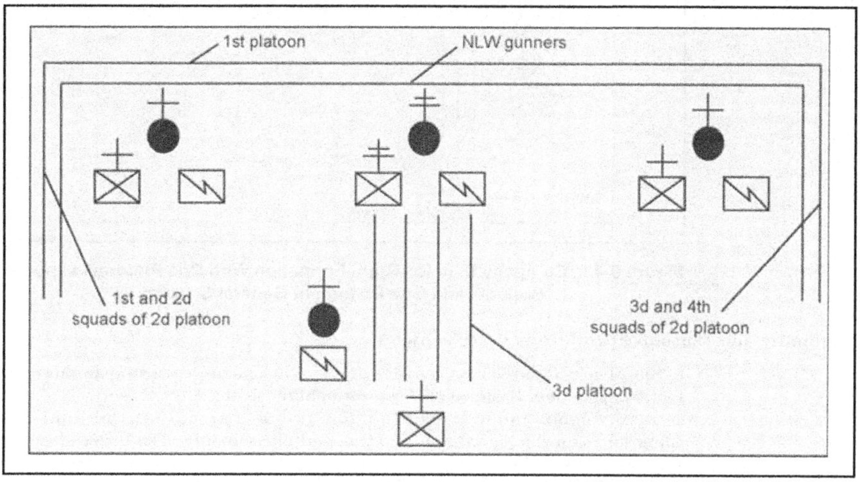

Figure 6-41. Company Line, One Platoon in Lateral Support and One Platoon in General Support

formations, the commander must consider the area that the company is operating in when he assembles the company. If he follows the guidelines of platoon assembly, he will have no difficulty. When assembling the company, the commander takes a position to the rear of the formation and commands, "Company Assemble." The platoon leaders take charge of their respective platoons. Usually, the 3d platoon assembles first, just behind the formation. After the 3d platoon is assembled, the 2d platoon assembles in front of the 3d platoon and the 1st platoon assembles in front of the 2d. All platoons assemble facing the commander. The 2d and 3d platoons then dress with the 1st platoon. As each platoon is assembled, they are ordered to port arms by the platoon leader to await further orders from the commander. If four platoons are used, the 4th platoon assembles first.

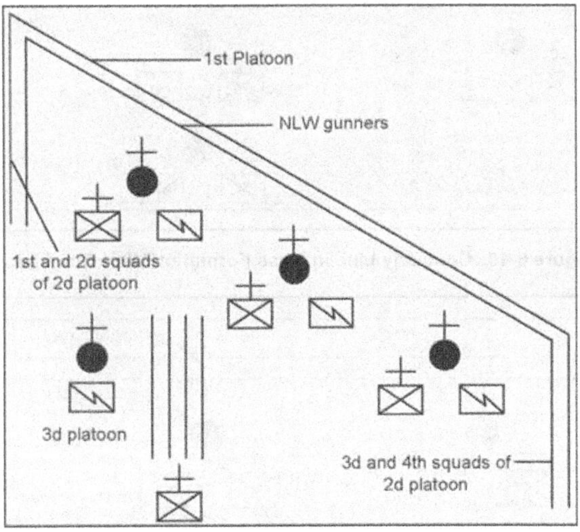

Figure 6-42. Company Echelon Right Formation With One Platoon in Lateral Support and One Platoon in General Support

Company and Platoon Formations With Vehicles

6-103. The commands for vehicles and foot troops are the same as those for foot troops alone. Because of the additional noise and distances involved, the company commander gives the hand-and-arm signal for the line. The command and the signal apply to the lead platoon only. The other platoons remain in general support until further directed (see *Figures 6-50,* page 6-48*;* *6-51,* page 6-49*;* *6-52,* page 50*;* *6-53,* page 6-50*;* *6-54,* page 6-51*;* *6-55,* page 6-52*;* and *6-56,* page 6-53).

6-104. The motor section moves out with vehicle number two moving to the right. Vehicle number two passes the lead vehicle and establishes the position for the center of the line. At the same time, the 3d and 4th vehicles swing out to the left and right. They form to the left and right and slightly to the rear of vehicle number two. The 1st and 4th squads of the lead platoon then move forward and form lines to the left and right, on the front end of vehicle number two. As soon as the troops are in position, the 3d and 4th vehicles close in on the squads and tighten the formation. The 2d and 3d squads of the lead platoon then form a line to the left and right of the 3d and 4th vehicles to complete the company line. The command vehicle (vehicle number one) takes

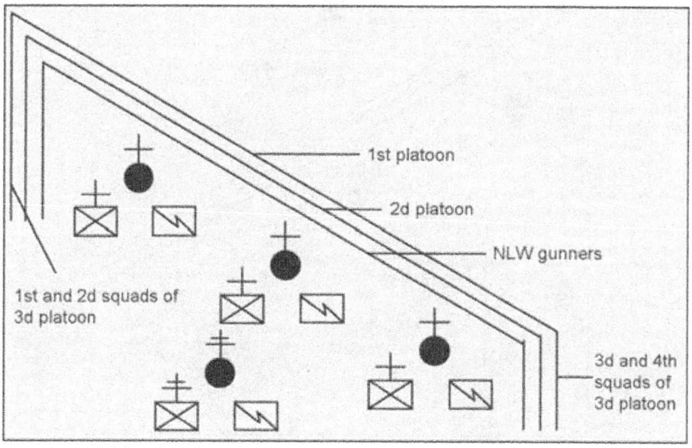

Figure 6-43. Company Echelon Right in Mass Formation With One Platoon in Lateral Support and One Platoon in General Support

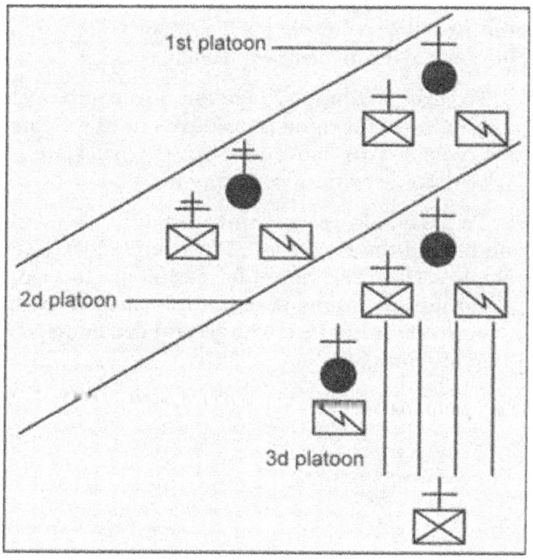

Figure 6-44. Company Echelon Right in Mass Formation With One Platoon in Lateral Support

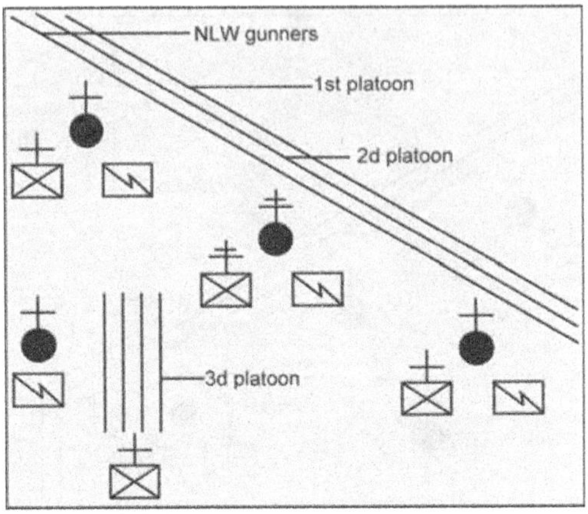

Figure 6-45. Company Echelon Right in Mass Formation With One Platoon in General Support

up a position behind the line where the commander can direct and control the unit. The 2d and 3d platoons move forward and are used in general, lateral, or direct support. In some instances, foot troops move into position first and vehicles join the formation, as needed.

6-105. To form a company echelon or a company wedge with vehicles, the company follows the same procedures used for company formations without vehicles. Vehicles are moved into the appropriate positions using procedures similar to those used for a company line.

6-106. To assemble troops and vehicles, the command vehicle moves to a position behind the formation. The commander faces the formation and gives the vehicle section the signal for assembly. Immediately, the vehicles return to their proper positions in column formation behind the command vehicle while the troops stand fast. The second in command then assembles the troops in the usual manner.

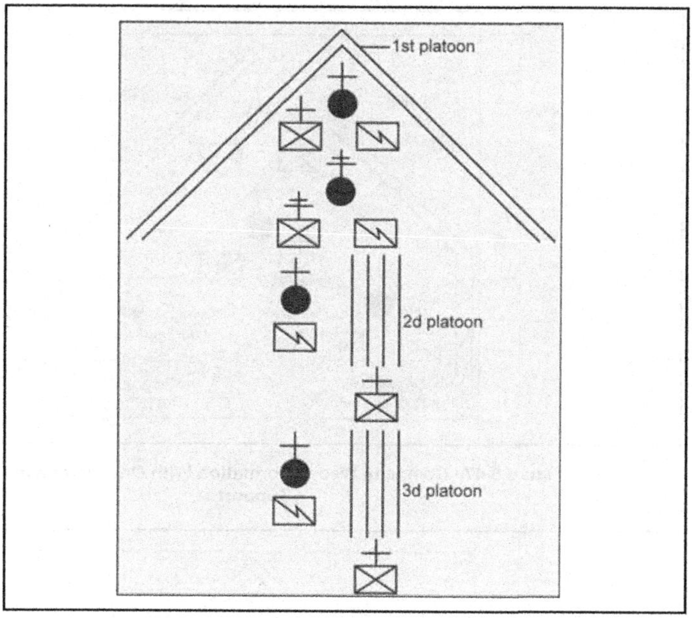

Figure 6-46. Company Wedge Formation With Two Platoons in General Support

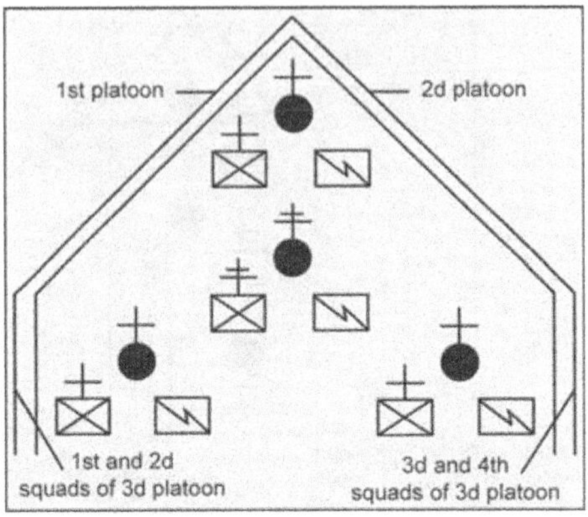

Figure 6-47. Company Wedge Formation With One Platoon in Lateral Support

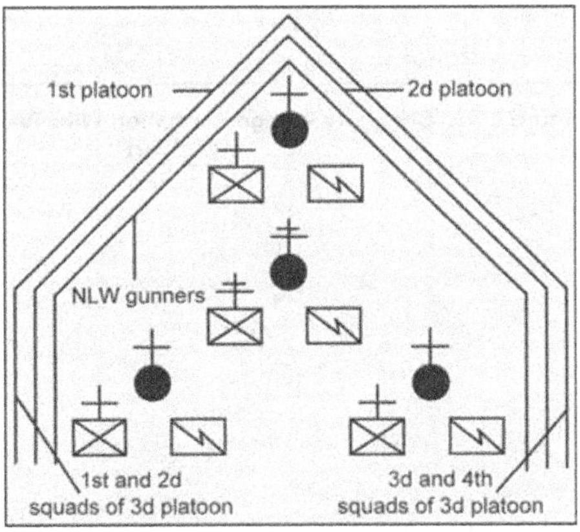

Figure 6-48. Company Wedge Formation With One Platoon in Lateral Support and One Platoon in General Support

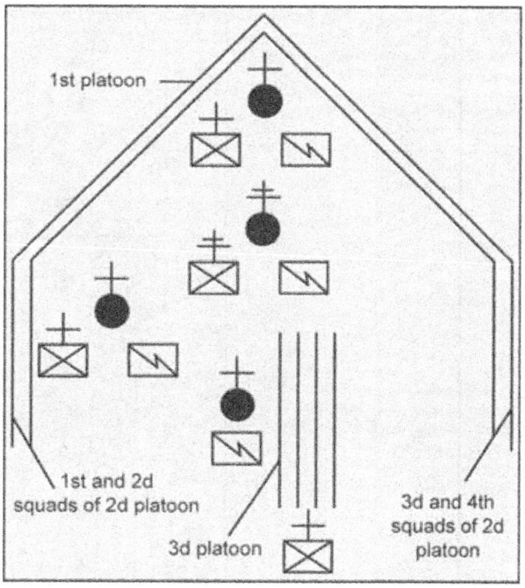

Figure 6-49. Company Wedge Formation With One Platoon in Lateral Support and One Platoon in General Support

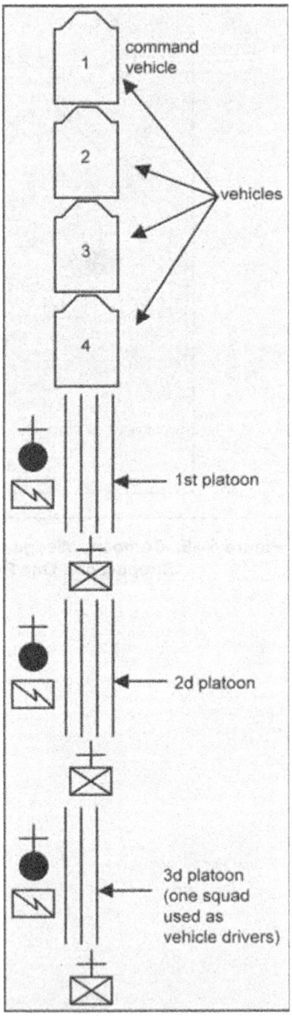

Figure 6-50. Company Column Formation With Vehicles

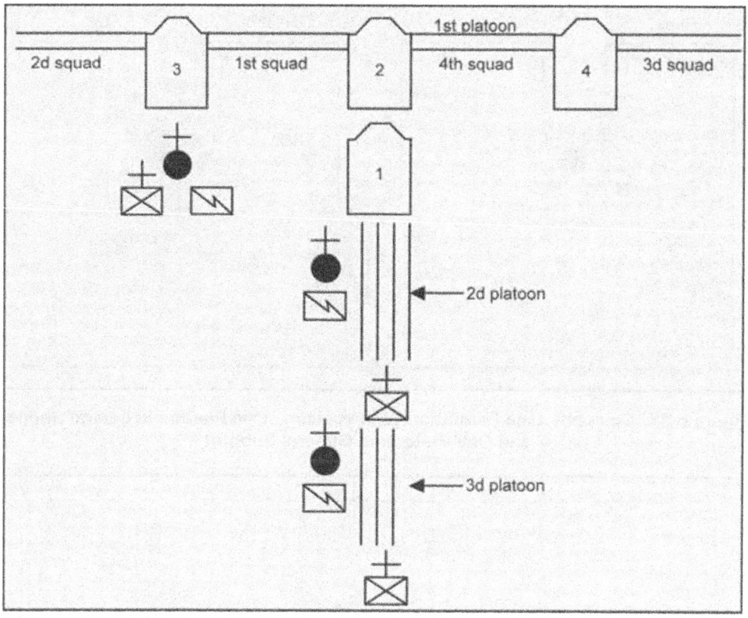

Figure 6-51. Company Line Formation With Vehicles and Two Platoons in General Support

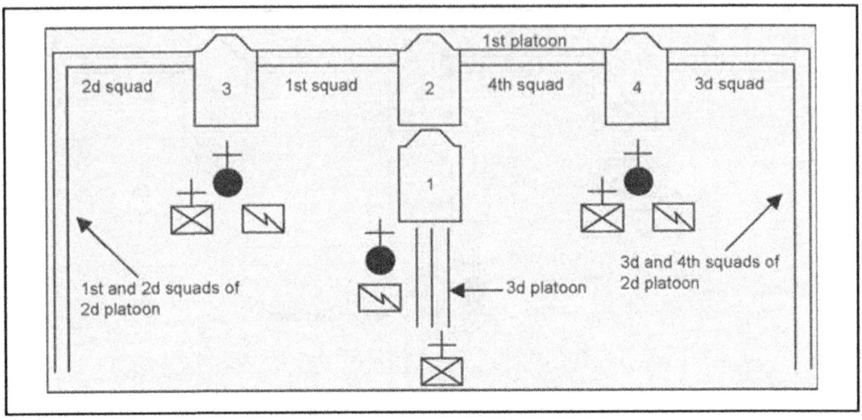

Figure 6-52. Company Line Formation With Vehicles, One Platoon in Lateral Support, and One Platoon in General Support

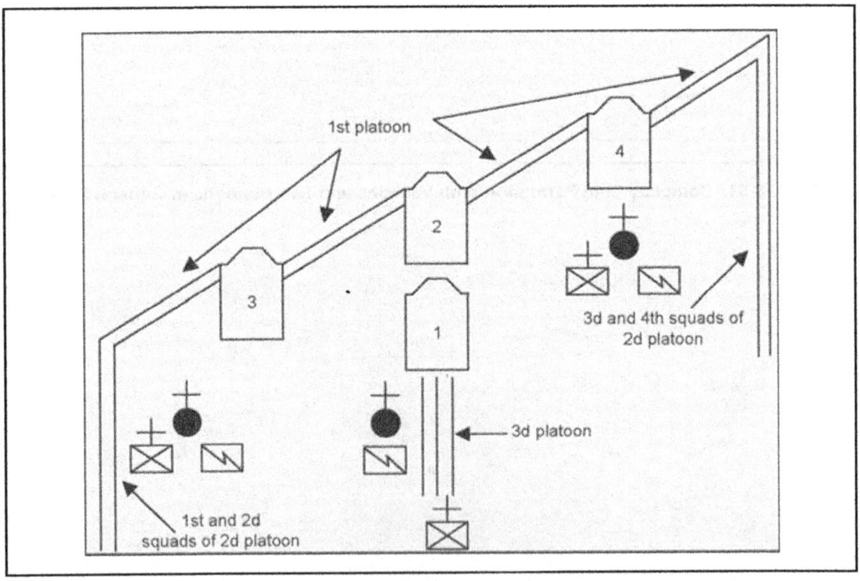

Figure 6-53. Company Echelon Right Formation With Vehicles, One Platoon in Lateral Support, and One Platoon in General Support

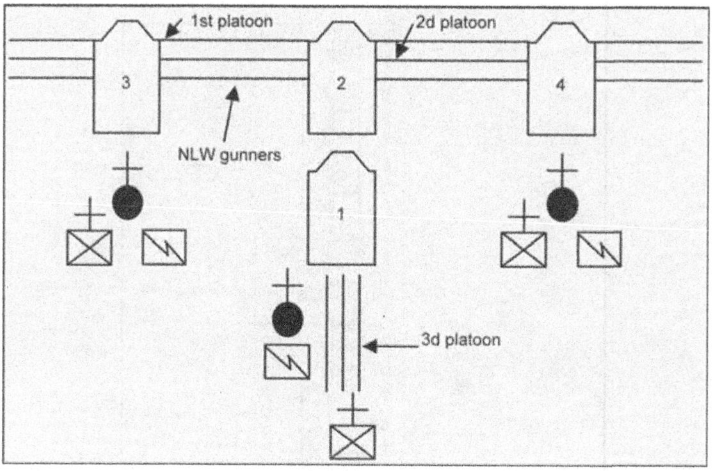

Figure 6-54. Company Line in Mass Formation With Vehicles and One Platoon in General Support

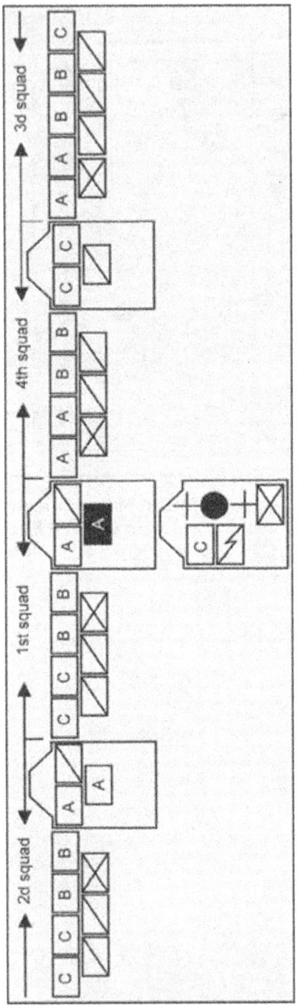

Figure 6-55. Platoon Line Formation With Vehicles

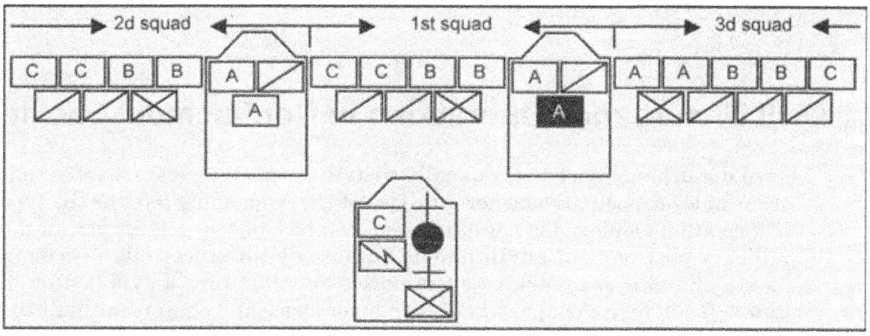

Figure 6-56. Platoon Line Formation With Direct Support and Vehicles

Chapter 7

Civil Disturbance Operations in Confinement Facilities

Civil disturbance doctrine is usually written for quelling disturbances that occur in large open areas where the size of the responding force or the type of formation employed is not hampered by a lack of space. However, in the military confinement environment, inmate disturbances often occur in small, closed areas, such as an inmate's housing unit, a gymnasium, a dining facility, or a chapel where maneuver space and equipment limit the size of the element that may be employed. The layout of each facility is different and presents its own unique problems when employing forces. This chapter covers the use of small military police teams, forced cell move teams (FCMTs), and the formations used to move an unruly inmate from one cell to another. Small generic riot control formations are discussed to provide the commander with a basic employment option that he can tailor to his facility to enable him to regain control of the affected area. The use of NL munitions and RCAs are incorporated into this discussion.

CROWD DYNAMICS WITHIN THE CONFINEMENT FACILITY

7-1. In confinement facility disturbances, inmates employ a number of tactics to resist control or to achieve their goals. Tactics may be unplanned or planned and nonviolent or violent. When a disturbance carries many purposes, it is likely that their tactics are well planned.

NONVIOLENT TACTICS

7-2. Nonviolent tactics range from name-calling to building barricades. Inmates may attempt to distract control force members by shouting at or ridiculing them or by using abusive language, obscene remarks, taunts, and jeers. The inmates' goals are to anger and demoralize the control force. They also want authorities to take actions that may later be exploited as acts of brutality.

VIOLENT TACTICS

7-3. Violent crowd tactics used by inmates are often extremely destructive and can include physical attacks on cadre, other inmates, and property. Use of violent tactics is limited only by the attitudes and ingenuity of the inmates, the training of their leaders, and the materials available to them. Inmates often commit violence with crude homemade weapons. If unplanned violence occurs, inmates will use mops, brooms, chairs, beds, or whatever else is on hand as weapons of violence. During planned violent disturbances, inmates can easily conceal makeshift weapons or tools to use against the control force. Inmates often erect barricades to impede troop movement and/or to prevent the control force from entering certain areas or buildings. Inmates can be expected to vent their emotions on individuals (cadre or inmates), the control force, their equipment, and/or the institution's property. Inmates direct dangerous objects like carts, barrels, liquids, and burning shoe polish at the control force.

GRIEVANCE PROTEST

7-4. A grievance protest may be organized as a riot. However, under normal circumstances, a riot of this type is not of an extremely violent nature. It may turn violent when leaders attempt to exploit any successes or weaknesses found in the control force.

UNORGANIZED RIOTS

7-5. Unorganized riots are characterized as being spontaneous in nature. However, they could be exploited and diverted by leaders at any subsequent stage of the riot, turning it into a different type. They may begin as an isolated assault against authority figures, or acts of civil disobedience. Under determined leadership, an unorganized riot could change to an organized riot.

MULTIPLE RIOTS

7-6. Multiple riots may happen when a disturbance occurs in one area of the confinement facility, causing others to riot in their areas. To control this situation with only one riot control force available, the force commander should first subdue the most violent riot. At the same time, he should employ some of his control forces to contain the other riots until the main control force is able to move to those locations to subdue them.

CADRE BEHAVIOR

7-7. The cadre is susceptible to crowd behavior. They become emotionally stimulated during tense confrontation with unruly and violent inmates. Commanders must be aware of the mood and attitude of the crowd and its effect on the control force cadre. To counteract the effects of crowd behavior on the cadre, the commanders must institute rigorous training and firm and effective leadership. This training must include a complete awareness and understanding of the use of force and the commander's intent. These are all necessary to offset the effect of crowd contagion upon the control force. Even with the best training and leadership, control force members must exercise individual and collective restraint.

PLANNING

7-8. Preplanning is the preparation and training conducted before a crisis occurs. The purpose of preplanning is to have plans and standing operating procedures (SOPs) in place so the cadre may react to an emergency and successfully contain and neutralize the situation. The preplanning process includes plan development, intelligence gathering, and training.

NOTE: Intelligence gathering is essential for the successful containment, neutralization, and/or prevention of a riot.

7-9. Once the control force team has been alerted to a disturbance within the confinement facility, leaders further develop preplans to fit the situation much the same way that DOD develops war plans.

7-10. During the planning process, leaders must be aware of the potential risks involved in quelling disturbances within a confinement facility. Careful planning minimizes collateral damage and risks to control force members and hostages.

7-11. The use of NLW and NL munitions must also be considered when developing plans. There must be strict accountability and control of RCAs, their employment, and other appropriate NL means (such as high-pressure water hoses). When using RCAs, plans must be flexible enough to accommodate changes in the situation and weather because it impacts the effectiveness of some agents. Planners must also consider the—

- Location of the disturbance.
- Estimated number of rioters.
- Accessibility of weapons, tools, or cleaning supplies.
- Inmate's military training.
- Hostage situation.

7-12. Based on an analysis of these factors, the commander makes an estimate of the situation. The estimate must be as thorough as time permits. Using the estimate, the commander considers COAs, selects RCAs, and determines the need for engineer support. In choosing a COA, consider the—

- Number of hostages.
- Number of inmates holding the hostages.
- Leader.
- Attitude and demeanor of the rioters.
- Location of the riot.
  - Inside areas of the confinement facility, such as the dining facility, work site, or housing wing.
  - Outside areas of the confinement facility, such as in exercise or work areas.
- Accessibility of weapons or material to make weapons.
- Accessibility of chemicals, such as cleaning supplies, cooking oils, and shoe polish.

USE OF CHEMICAL IRRITANTS

7-13. The use of chemical irritants can be a valuable NL tool for control force leadership to consider during the planning phase. These chemical irritants can drive a threat from an established, enclosed position or deny the rioters access to a certain area without long-lasting effects to those involved. The proper use of chemical irritants may prevent the control force from having to enter a dangerous area; however, improper use can cause injury, death, or property damage.

USE-OF-FORCE GUIDELINES

7-14. The installation commander authorizes the use of chemical irritants after the control force leader determines the type and dosage. The SOP outlines the procedures for securing authorization and provides guidance to help the control force leader make his decisions. The confinement facility commandant should inform the installation commander of the situation, since the installation commander is ultimately responsible. Team members must be aware of how chemical irritants affect personnel and plan accordingly. Inmates not involved must be removed from the area and the chemical cloud's path. The fire department and emergency medical services (EMS) should be on standby during operations involving chemicals.

PEPPER IRRITANT

7-15. The pepper irritant receives its name from varieties of the capsicum plant that is it made from. The active ingredient, Oleoresin Capsicum (OC), is extracted from the plant and micropulverized to make the irritant. It is neither a chemical nor a gas, but an all-natural organic substance. It is effective against individuals who are emotionally disturbed or under the influence of drugs or alcohol. It is effective against domestic and wild animals. OC is often used to divert the threat from making or continuing an assault. OC does not cause permanent damage. Its effects last 30 to 40 minutes after individuals are placed in fresh air.

7-16. OC can have various effects on people. Some effects are—

- Swelling of mucous membranes and the upper respiratory system; however, it does not shut down the system.
- Burning (intense) and discoloration (bright red) of exposed skin.
- Dilating the capillaries.
- Swelling of the eyelids.
- Burning and involuntary closing of the eyes.
- Coughing uncontrollably.
- Gagging.
- Gasping for air.
- Losing strength and coordination (temporarily).

OC has some disadvantages. They are that—

- The canister depressurizes over time if not checked regularly.
- Inmates may gain access to OC and use it against SRT members.
- The canister must be shaken on a regular basis.

DISSEMINATION OF CHEMICAL IRRITANTS

7-17. There are six methods for disseminating chemical irritants.

Pyrotechnic Method

7-18. In the pyrotechnic method of dissemination, the chemical irritant is placed in a canister with an inert material that is ignited when the device is activated. The agent is then carried into the air on the smoke particles of the inert material.

7-19. There are some disadvantages when using a pyrotechnic device indoors. These disadvantages are that the device—

- May cause a fire.
- Has a very slow saturation time (40 to 45 seconds to burn completely).
- Can be thrown back by the threat, even while burning.

Bursting Method

7-20. In the bursting method of dissemination, the irritant and an inert powder are placed in a saw-toothed canister. After a small detonation, the canister splits and expels the irritant in a cloud.

7-21. The bursting method has its advantages and disadvantages. Some advantages are that—

- There is no risk of fire.
- There is no risk of the canister being thrown back.
- The contents disseminate quickly.
- The canister is effective, easy to carry, and easy to control.

Some disadvantages are that—

- There is a slight possibility of fragmentation when detonated.
- The fuse head may separate from the canister.
- Only 50 to 90 percent of the irritant may be disseminated.
- The direction of the cloud formation depends on where the device is detonated.

Aerosol Method

7-22. In the aerosol method of dissemination, the irritant is suspended in an inert liquid that is located in the rear portion of the device. When the projectile penetrates the structure, the rear portion splits open and the irritant is disseminated in a mist. The aerosol method is most appropriate for tactical operations when chemical irritants are used before entry.

7-23. The aerosol method has its advantages and disadvantages. Some advantages are that it—

- Is easily carried.
- Is an excellent deployment system (uses 40-millimeter, 37-millimeter, or 12-gauge ammunition).
- Disperses the contents quickly.
- Is easily controlled and accurate (fin-stabilized).

NOTE: Practice firing is always necessary.

Some disadvantages are the possibility of—

- Injury to individuals if it is fired from close range.
- Damage to the structure if it strikes a weak portion.

Projectiles Method

7-24. Chemical-containing projectiles can be very effective when used correctly. A soldier must know which projectile to use in each unique situation.

7-25. **Thirty-Seven-Millimeter Projectile.** A 37-millimeter projectile can incapacitate anyone in an enclosed area of 4,500 cubic feet. It can penetrate the following:

- A sheet of ¾-inch-thick plywood at 25 yards.
- An automobile windshield at 33 yards.
- A hollow-core door at 58 yards.

7-26. **Forty-Millimeter Projectile.** A 40-millimeter projectile is more effective at greater ranges than a 37-millimeter projectile because of its rifled bore. It can incapacitate anyone in an enclosed area of 4,500 cubic feet. It can penetrate the following:

- A 1-inch-thick plywood at 10 yards.
- An automobile windshield at 55 yards.
- A hollow-core door at 109 yards.

7-27. **Twelve-Gauge Projectile.** A 12-gauge projectile can be loaded into a shotgun with no modifications to the weapon. It can be fired as a single round, or it can be magazine-fed. The 12-gauge projectile can incapacitate anyone in an enclosed area of 1,000 cubic feet. The angle at which the projectile strikes the barrier can affect its penetration capability. It can penetrate the following:

- A sheet of ¾-inch plywood at 33 yards.
- An automobile windshield at 33 yards.
- A hollow-core door at 100 yards.

Fogging Method

7-28. The irritant is suspended in a fogging solution and dispersed with a fogging machine. It is not recommended for tactical operations.

7-29. The fogging method has its advantages and disadvantages. Some advantages are that it—

- Produces enough chemical irritant to cover an enclosed area of 100,000 cubic feet in 26 seconds.
- Disseminates quickly.
- Provides high-volume capabilities.
- Is effective for riot control or other crowd dispersal situations.

Some disadvantages are that it is—

- Noisy.
- Cumbersome.

Multipurpose Grenade Method

7-30. A multipurpose grenade can be hand-thrown or launched from a shotgun with an adapter. It has an extended shelf life of 6 years and an adjustable fuse delay of 2 to 5 seconds. The irritant is located in the cylinder portion of the device and is forced out the bottom of the grenade upon detonation.

WARNING

A multipurpose grenade deployed incorrectly could result in injury. Safety rules should be observed at all times.

RECORD OF EVENTS

7-31. Reporting procedures for serious incidents should be included in the plans. A record of events must be initiated to provide a basis for the preparation and submission of a formal report to higher HQ. As a minimum, include the—

- Time the incident was reported and by whom.
- Time the incident was reported to the facility commander.
- Time the control force was assembled.
- Time the control force entered the facility.
- Weather conditions as they relate to the use of RCAs.
- Number of cadre and inmates injured or killed, including how they were injured or killed and the medical attention given to them.
- Time the operation was completed.
- Time the riot control force restored order.

TRAINING

7-32. Guard units and associated teams must train on a regular basis in the five basic riot control formations. There must be a continuous training program established, to include—

- Interpersonal communication.
- Self-defense.
- Use of force.
- Use of a riot baton.
- Fatal areas of the body to avoid when using the riot baton.
- First aid.
- Emergency plans.
- Intelligence and counterintelligence techniques.
- The use of RCAs and the various methods of dispersing them.
- The use of NLW.
- The use and effects of high-pressure water.
- Riot control formations.

EQUIPMENT

7-33. The recommended basic riot gear used by the confinement riot control force is as follows:

- Riot baton.
- Riot shield.
- Helmet with face shield.
- Groin protector.
- Flak vest.
- Leather gloves.
- Shin protection.

RIOT BATON

7-34. The riot baton is an invaluable weapon in crowd control situations. A riot baton in the hands of a well-trained control force member is the most appropriate weapon, except with extremely violent crowds. The riot baton is not meant to replace NL munitions, but is instead used in situations in which NL munitions are not needed. Losing a riot baton to the crowd does not present a serious threat.

7-35. Use of the riot baton is based on the commander's appraisal of the situation and his choice of force option. Control forces committed with the riot baton must have teams equipped with RCAs, NLW, and lethal coverage. If the confrontation increases in intensity, the commander has these force options available and may rapidly engage the control force.

7-36. A riot baton is a versatile offensive weapon that varies in length from 26 to 42 inches with the optimum length being 36 inches. It is made of hickory and does not shatter or break easily. Both ends of the riot baton are rounded to prevent unnecessary injury. It is approximately 1¼ inches in diameter. Each riot baton has a hole drilled at the grip end with a leather thong threaded through the hole. The thong helps secure the riot baton to the user's hand.

RIOT SHIELD

7-37. The riot shield (see *Figure 7-1*) is an excellent piece of defensive equipment. It is used by control force personnel in the front and on the flanks of the formation.

7-38. The riot shield has its advantages and disadvantages. Some advantages are that—

- OC pepper spray or high-pressure water can be used to back the control force.
- It provides protection from thrown objects.
- It absorbs some of the impact of rioters hurling themselves at the control force.

The disadvantages of the riot shield are that it—

- Severely limits the use of the riot baton.
- May crack if hit with a heavy object.

Figure 7-1. Riot Shields

RIOT CONTROL FORMATIONS

7-39. The five basic formations for civil disturbance operations are the line, wedge, echelon, diamond, and circle. In the correctional environment the basic formations for disturbances are the line, wedge, or echelon with apprehension teams in support. Generally, the diamond and circle formations are not used in the correctional environment. The minimum size force that the commander should consider is an augmented platoon, with a two-to-one advantage over the rioters.

7-40. Correctional facility cadre must be proficient in all riot control formations. These are usually trained at the squad level and above, but performed by a platoon or company-sized unit. Squad line, echelon left, echelon right, and wedge formations form the basis for platoon and company formations. Each squad must be adept in the basic formations before practicing in platoon-sized or larger formations. For more detailed information on formations, see *STPs 19-95B1-SM* and *19-95B24-SM-TG*. In the correctional environment, the support platoon plays a key role by supporting the control formation with apprehension and equipment teams. See *Chapter 6* for an in-depth discussion on civil disturbance formations.

APPREHENSION TEAMS

7-41. Apprehension teams give the control force the ability to extract inmates. Inmates identified as leaders or agitators and inmates that are injured or no longer want to resist can be extracted from the crowd. Apprehension teams can also be used to enter confined areas where the formation cannot operate.

7-42. The apprehension team will consist of eight members and one leader. Each team will have four shield holders, two nonlethal gunners and two apprehension/search members. The squad leader controls the movement of the team and assists the team in exiting the control formation. He maintains communications with the formation and coordinates the team's operations with the formation. The squad leader also ensures that the team uses the minimum amount of force necessary to extract inmates.

7-43. The apprehension team may deploy through the middle of the formation or from around either of the flanks.

7-44. The apprehension team should be deployed within a distance of approximately 10 meters so that the formation can still provide immediate support. If the apprehension team is deployed farther than 10 meters, the formation must advance to close the gap.

7-45. The team will exit the formation in two columns; shield holders in the first and second ranks, nonlethal gunners in the third rank, apprehension/search members in the fourth rank and the leader last. Team members grip the shoulder or equipment of the member in front of them. The shield holders will envelop the inmate that is to be extracted and form a protective barrier. Once the shield holders have enveloped the inmate, the nonlethal gunners will cover their sides with their weapons while the apprehension/search members control and restrain the inmate. When the team has the inmate under control, the squad leader will give the order for the team to withdraw. The team withdraws in reverse order (apprehension/search members with the inmate, followed by nonlethal gunners and shield holders). Team members then move backwards, face the crowd and grip the shoulder or equipment of the member in front of them.

TEAM EQUIPMENT

7-46. The apprehension team uses the following equipment:

- Flex-Cufs.
- Restraint cutters.
- Pepper spray.
- Helmets with face shields.
- Groin protectors.
- Flak vests.
- Elbow pads.
- Leather gloves.
- Shin protectors.
- Knee pads.
- A video camera.

TEAM COMPOSITION

7-47. The apprehension team is composed of five team members. This team consists of four members and a squad leader. Team member one is the team leader.

Squad Leader

7-48. The squad leader assists the team in exiting the control formation and ensures that the team uses the minimum force necessary to subdue inmates. He is also responsible for signaling the PSG if more help is needed.

Team Member One (Team Leader)

7-49. Team member one's responsibility is to make initial contact with the inmate and use pepper spray and apply Flex-Cufs, as needed. He also carries a set of restraint cutters. Once the apprehension team has returned to the relative safety of the control formation, team member one searches the inmate.

7-50. Team member one receives the command to move forward and secure the inmate from the squad leader. This command is usually given by a signal and verbal command. Team member one then notifies the main force commander where his team exits the control force. Once the team is in position to exit the main formation, team member one signals the member of the formation in front of him. This is usually a prearranged signal, such as a tap on the right shoulder. The control force opens and the team exits the formation and makes contact with the assigned inmate. Team members then subdue and control the inmate. If pepper spray is needed, team member five shouts, "Pepper." This gives the other team members time to react to the command and spray the inmate. Team member one then applies Flex-Cufs to the inmate's wrists and ankles. Depending on the mission, the inmate may be left for follow-on security forces or carried behind the control force to be secured by security forces.

7-51. Should the mission require that the inmate be moved, team member one shouts, "Prepare to Move," this allows each member to grab an arm and position the inmate toward the control force formation. When the team is ready to move, team member two shouts, "Ready," alerting team member one that the team is ready for the last command. Team member one then commands, "Move." The team enters the control formation with the inmate. The inmate is searched by team member one and turned over to the security team.

Team Member Two

7-52. Team member two stands directly behind team member five with his hands on team member five's waist. He protects the left side of team member five by moving to team member five's left and helps to secure the inmate after team member one makes initial contact. He repeats team member five's commands and controls the inmate's head after the inmate is on the ground.

Team Member Three

7-53. Team member three stands behind team member two with his hands on team member two's waist. He protects the right side of team member five by moving to team member five's right and helps to secure the inmate after initial contact.

Team Member Four

7-54. Team member four stands behind team member three with his hands on team member three's waist. His task is to help subdue the inmate in any way possible. He usually concentrates on the inmate's body below the knees. He acts as the eyes and ears for the team during control of the inmate.

Team Member Five

7-55. Team member five gives commands to control the inmate.

FORCED CELL MOVE TEAMS

7-56. The intent of a forced cell move is to move an unruly and/or uncooperative inmate from one cell to another. This is a difficult task and must not be taken lightly. A forced cell move must be thoroughly planned, rehearsed, and resourced to ensure the safety of the team members and the inmate.

7-57. A number of reasons may compel the commander to consider a forced cell move. The inmate may refuse to eat, take medication, accept medical attention, get a haircut, keep himself clean, work, or move or rotate cells when ordered to. These are some of the reasons that the commander may order a forced cell move.

TEAM COMPOSITION AND EQUIPMENT

7-58. The FCMT is composed of six corrections cadre and two support cadre from military police and medical sections specially trained in forced cell movement. Primarily, the team consists of the following personnel:

- One officer in charge (OIC) (E7 or above).
- Five corrections cadre.
- One military police investigator.
- One medic.

7-59. The FCMT uses the following standard equipment:

- Helmets with face shields.
- Groin protectors.
- Flak vests.
- Elbow pads.
- Leather gloves.
- Shin protectors.
- Knee pads.
- A pinning shield.
- Restraints (hand or leg irons or Flex-Cufs).
- A video camera.

ACTIONS PRIOR TO TEAM EMPLOYMENT

7-60. The OIC gives an operations order (OPORD) to team member five (the team leader) of the FCMT. As a minimum, the OPORD should answer the following questions about the incident leading up to the move:

- Did the inmate fail to comply with orders?
- Did the inmate assault cadre or other inmates?
- Is the objective of the move to force a shower or haircut?
- Is OC pepper spray authorized for use?
- Has the inmate been moved before?

- What is the condition of the area where the inmate is currently located?
- Has the inmate tested positive for human immunodeficiency virus (HIV)?
- Does the inmate have a weapon or the materials needed to make one?
- Is the area barricaded?
- Has he smeared himself with body fluids or other slippery substances?
- Can you approach the inmate without risk to him and/or others?

7-61. The team is assembled for inspection by the OIC, and the team members are required to brief the OIC on their individual duties and responsibilities. A line formation is the customary formation for this inspection. It should take place out of site of the inmates.

7-62. Each soldier takes one step forward and briefs the OIC as shown in the following example: *"I am (rank and name), team member one. My mission is to pin the inmate using the minimum amount of force necessary. I carry the pinning shield."* He then faces so the camera can record the number printed on the back of his flak jacket and holds up the pinning shield [see *Figure 7-2*]).

Figure 7-2. Recording Team Members' Duty Position

INDIVIDUAL TEAM MEMBER RESPONSIBILITIES

7-63. Each individual team member has his own responsibilities during a forced cell move. Each team member must do his designated job while working as a team.

Team Member One

7-64. Team member one is the eyes and ears of the team. His responsibility is to carry the pinning shield up to the cell door and describe the scene with as much detail as possible (especially including anything that may cause harm to the team). The pinning shield is used to protect the team from body fluids thrown at the them by the inmate.

7-65. Team member one describes the scene in the cell to his teammates by turning his head to the right and speaking in a loud voice so that all team members can hear him. A slap on the right shoulder by team member two alerts him that all team members have heard him. If he receives a slap on the left shoulder from team member two, he repeats the situation report (SITREP) in a loud, clear voice. The decision to enter the inmate's area with or without the pinning shield is team member one's decision, unless the inmate is known to have a weapon. Should he decide to enter without the pinning shield, team member one then enters with both hands held approximately chest high. When team member one is unable to determine if the inmate has a weapon, the pinning shield is used. Charging through the open cell door, team member one drives the inmate to the bunk or ground with the shield. Once pinned, team member one secures the inmate's head and ensures that it is not twisted or turned in relation to its natural form. He may have to use the various pressure points on the inmate's head to help subdue him.

Team Member Two

7-66. Team member two steps forward and briefs the OIC as follows: *"I am (rank and name), team member two. My mission is to secure the inmate's right arm using the minimum amount of force necessary. I carry one set of hand irons."* He then holds the hand irons up for video camera taping (see *Figure 7-3*) and says, *"I also carry one canister of OC pepper spray to be used at the direction of the team OIC."* He then holds up the canister for video camera taping and faces so the camera can record the number on the back of his flak jacket (see *Figure 7-2,* page 7-13.)

Figure 7-3. Recording Team Members' Equipment

7-67. Team member two lines up directly behind team member one. He keeps contact with team member one until he can secure the inmate's strong arm. Team member two taps team member one on his right or left shoulder after being tapped by team member three. Once inside the cell, team member two secures the prisoners strong hand with wrist restraints and announces, "*Hands secure,*" to team member five.

Team Member Three

7-68. Team member three steps forward and briefs the OIC as follows: *"I am (rank and name), team member three. My mission is to secure the inmate's left arm using the minimum amount of force necessary."* He then faces so the camera can record the number on the back of his flak jacket (see *Figure 7-4*).

Team member three lines up directly behind team member two. Team member three keeps contact with team member two until he can grasp a portion of the inmate's body. He then concentrates on securing the inmate's weak hand and assisting team member two in applying hand restraints. Team member three taps team member two on his right or left shoulder after being tapped by team member four.

Figure 7-4. Recording Number on Flak Jacket

Team Member Four

7-69. Team member four steps forward and briefs the OIC as follows: "*I am (rank and name), team member four. My mission is to secure the inmate's right leg using the minimum amount of force necessary. I carry one set of leg irons.*" He holds the leg irons up for the camera to record (see *Figure 7-2,* page 7-13). He then faces so the camera can record the number on the back of his flak jacket (see *Figure 7-2*). Team member four lines up directly behind team member three. He taps team member three on his right or left shoulder after being tapped by team member five. Team member four keeps contact with team member three until he can secure a portion of the inmate's body. Team member four concentrates on securing the right leg of the inmate. It is his responsibility to apply the leg irons. Once the legs are secure, he announces, "*Legs secure,*" to team member five.

Team Member Five (Team Leader)

7-70. Team member five steps forward and briefs the OIC as follows: *"I am (rank and name), team member five. My mission is to secure the inmate's left leg using the minimum amount of force necessary. I carry a restraint key (hold up for the camera to view). I am also the team leader for this mission and it is my responsibility to ensure that the team uses the minimum amount of force necessary to subdue the inmate."* He then faces so the camera can record the number on the back of his flak jacket (see *Figure 7-4*).

7-71. Team member five taps team member four on the right shoulder when he has heard the SITREP from team member one. If he does not hear the report or needs to hear it again, he taps team member four on the left shoulder. Team member five keeps in contact with team member four until he can secure a portion of the inmate's body. He secures the left leg and assists team member four in applying the leg irons. Once he has heard that the hands and legs are secure, team member five signals for the team to conduct an equipment check. The team members check the equipment in place.

NOTE: At no time is the inmate to be released by anyone.

Medic

7-72. The medic briefs the OIC in the following manner: *"I am (rank and name), the medic. My mission is to treat all injuries that may occur during the move."* He follows the instructions given to him by the OIC/NCOIC. His mission is to observe the inmate for signs of injury before the forced cell move and assess the inmate's condition during and after the move.

Military Police Investigator

7-73. The military police investigator (MPI) is responsible for videotaping the entire forced cell move. This includes the OIC/NCOIC briefing the OPORD, each team member briefing his responsibilities to the OIC, and any medical treatment rendered to the inmate.

ACTIONS DURING THE FORCED CELL MOVE

7-74. There are many actions performed during a forced cell move. Each member of the FCMT must complete his task successfully to accomplish the forced cell move.

Officer in Charge or Noncommissioned Officer in Charge

7-75. The OIC, an E7 or above, is responsible for ensuring that the team uses the minimum amount of force necessary to subdue the inmate. He is also responsible for ensuring that all necessary precautions have been taken to ensure the safety of the team and the inmate. The OIC gives the inmate the initial orders to comply and informs him of the consequences of noncompliance. He does this by placing himself at the front of the inmate's cell (see *Figure 7-5*). The orders are not given until it can be enforced by the FCMT. The following prepared statement is read to the inmate by the OIC: *"I am (rank and name), OIC of this FCMT. At this time I am ordering you to lie face down on your bunk (or other designated location) with your head to the rear of the cell. Cross your hands behind your back and cross your feet as the team enters the cell and uses the minimum amount of force necessary to place you in restraints and move you to (give location). Failure on your part to comply with these instructions or to resist the FCMT will result in the use of force (OC pepper spray if applicable). Do you understand these instructions? If so, lie down on the bunk."*

7-76. After reading the instructions to the inmate, the OIC faces the camera and states whether the inmate complied. He then waits 60 seconds and motions for the team to position itself in front of the inmate's cell (see *Figure 7-6*). The OIC directs the opening of the cell when the team is positioned at the entrance to the cell and prepared to enter. He positions himself so he can observe team entry to and departure from the cell. If OC spray is authorized, the OIC tells the team when to spray the inmate with OC spray. This task is done in silence. Communication between team members should be done through hand-and-arm signals, unless noted otherwise.

Figure 7-5. OIC Reading Compliance Statement to Inmate

Figure 7-6. OIC and FCMT

Team Member One

7-77. Once the team is positioned in front of the inmate's cell, team member one starts his report. At a minimum, he must answer the following questions and report them to the team:

- Does the inmate have a weapon, if so, what kind?
- Does the inmate appear dry?
- Is the inmate clothed or wearing extra clothing?
- Has the inmate smeared anything on his body?
- Can you see the inmate's hands?
- Is the cell barricaded or visibly booby-trapped?

- Is the cell dry?
- Are cell furnishings intact?

7-78. Based on the answers to the above questions, team member one must decide whether to enter the cell with or without the shield (see *Figure 7-7*). Team member one's ultimate goal is to pin the inmate on the bunk or floor.

Figure 7-7. FCMT Preparing to Enter a Cell

Team Member Two

7-79. Team member two keeps contact with team member one until he can secure the inmate's strong arm. If he cannot immediately secure the inmate's arm, he secures a portion of the inmate's body and progresses to the inmate's strong arm. Once the inmate's hands are secure, he announces, "*Hands secure,*" to team member five (see *Figure 7-8).*

Team Member Three

7-80. Team member three keeps contact with team member two until he can grasp a portion of the inmate's body. Team member three concentrates on securing the inmate's weak hand and assists team member two in applying Flex-Cufs. Team member three maintains control of the inmate's hands and looks at team member five for further instructions (see *Figure 7-8*).

Team Member Four

7-81. Team member four keeps contact with team member three until he can secure a portion of the inmate's body. Team member four concentrates on securing the inmate's right leg. He applies leg irons and when the inmate's legs are secure, he announces, "*Legs secure,*" (see *Figure 7-8*).

Team Member Five (Team Leader)

7-82. Team member five keeps contact with team member four until he secures a portion of the inmate's body. Team member five concentrates on securing the inmate's left leg and assists with the application of leg irons. When he is informed that the hands and legs are secure, the team searches the inmate for weapons and contraband. Once the search is complete, team member five commands, "Prepare to Lift."

Figure 7-8. FCMT Securing an Inmate

Movement of the Inmate

7-83. Team member one positions his hands to control the inmate's head during the move, turning the inmate's head toward the wall. Team members two and three position themselves on the inmate's right and left sides. They grasp the wrist or forearm of the team member opposite them to form a platform for the inmate to lie on. Team member four positions himself on the inmate's right side even with the inmate's knees. This places him in position to control the inmate's legs. Team member five's position is on the inmate's left side and enables him to control the inmate's lower legs and feet. When in position, team member five ensures that all team members are in place before commanding, "Ready, Lift."

7-84. On the command "Lift," all team members simultaneously stand with the inmate. They should sandwich the inmate between them to facilitate control. Team member five then commands, "Prepare to Turn, Ready—Turn." The inmate is always turned toward the bunk. If a bunk is not available, team member five should specify which direction to turn the inmate. The inmate is then removed from the cell (see *Figure 7-9*, page 7-20).

7-85. Once the inmate is on the tier, team member five commands, "Prepare to Lower, Team—Lower." The team simultaneously lowers the inmate to the floor with the team keeping positive control of the inmate. The medic asks the inmate, "*Do you have any injuries at this time?*" (see *Figure 7-10,* page 7-20). If the inmate has no injuries, the medic looks at the camera and states, "*Inmate (inmate name) appears to have sustained no injuries.*" If the inmate has sustained injuries, the medic treats those injuries and completes a sworn statement. If OC was employed, the medic decontaminates the inmate using water and paper towels. The medic instructs the inmate to tilt his head to the side and flush his eyes with water for approximately 5 minutes. The medic then dries the inmate's face.

Figure 7-9. Moving the Inmate

Figure 7-10. Treating Injuries

Placement of the Inmate in a Cell

7-86. Team member five orders an equipment check before returning the inmate to a cell. He then commands, "Prepare to Lift, Team—Lift." Team members position themselves exactly where they were when they removed the inmate from the cell. Next, team members simultaneously lift the inmate and return him to the predetermined cell. They move the inmate into the cell head first with his face toward the bunk. Once inside the cell, team member five commands the team to lower the inmate onto the bunk (see *Figure 7-11,* page 7-20). Team members then remove the inmate's restraints, keeping positive control of the inmate at all times. Team members use a nonverbal signal to notify team member five when the restraints are removed. He signals another equipment check. Team members then account for all equipment in place and prepare to exit the cell. The team exits the cell in the same manner as they entered, but in reverse order.

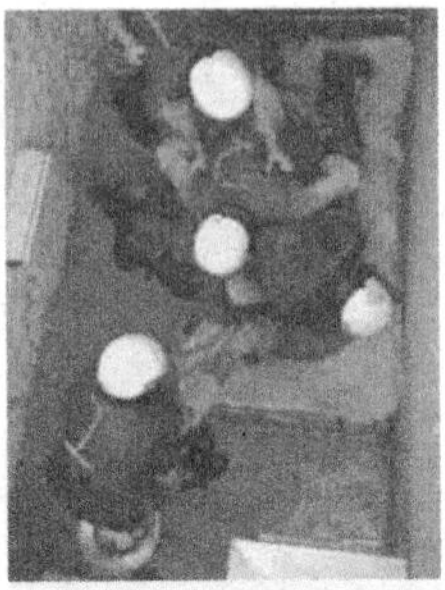

Figure 7-11. Placing Inmate in Cell

7-87. Team member four moves into position where he can control both of the inmate's legs while team member five moves toward the cell door. When team member four is ready to exit the cell, team member three moves to control the inmate's legs, team member two moves to where he can control both of the inmates arms, and team member one maintains control of the inmate's head. Team member four maintains contact with team member three at all times. When team member three is ready to exit the cell, team member two moves into a position where he can control the inmate's lower body and team member one moves into position to control the inmate's upper body. Using a prearranged, nonverbal signal, the team pulls team members one and two from the cell. The FCMT is prepared to reenter the cell if the inmate attempts to attack team members one and two as they exit the cell. The FCMT remains ready to reenter until the cell door is secured.

Chapter 8

Civil Disturbance Training

Training requires planning and preparation. Unit commanders structure their training program according to the mission essential task list (METL) (see *FM 7-1)* and expected contingencies. These training plans include prior training and the experience and proficiency of their unit. Leaders agree that the operational tempo and expected contingencies for the next 10 to 15 years will likely involve low-intensity military operations other than war (MOOTW). Therefore, civil disturbance training must be included in the commander's annual training program and must cover all aspects of that type of operation. To effectively function as a member of a control force, individual soldiers must be trained in civil disturbance operations. Training must be intensive, realistic, and sustained. Any lack of training (from basic soldier skills to leadership skills) can have an adverse effect to the point of endangering lives, especially the lives of soldiers. Leader involvement in the training process is essential for mission success in crowd control operations.

UNIT TRAINING STRATEGY DEVELOPMENT

8-1. Developing a unit training strategy begins with the mission analysis portion of the MDMP and identifying the GRM. Missions that require soldiers and units to enforce treaties or accords, protect the lives of civilians in uncertain or hostile environments (such as NEOs), or provide large-scale humanitarian assistance demand some sort of graduated-response criteria to maintain order and prevent uncertain environments from becoming hostile.

8-2. When developing training strategies, planners should focus training and evaluation on the echelon level above the target element (for example TF) or the unit identified as needing training.

8-3. Planners must agree on the intent of the GRM. The GRM is for most a training and rehearsal tool. It provides leaders with likely vignettes that can be incorporated into COA analysis, predeployment training, and rehearsals. The GRM becomes a handy reference during situations that require graduated responses.

TRAINING PLAN DEVELOPMENT

8-4. In developing a training plan for civil disturbances, many considerations should be addressed in terms of available resources. Considerations include the number of personnel to train, the equipment needed and its availability, the time available to train to standard, the location of where training will be executed, and the incorporation of SOPs (based on the theater of operations

[CONUS or OCONUS]). A personnel assessment is necessary to determine the training and experience level of soldiers. It is critical to identify and use soldiers with previous real-world experience and leaders that have graduated from the Interservice Nonlethal Individual Weapons Instructor Course (INIWIC) as primary instructors.

8-5. Commanders should structure the unit training program based on the METL, expected contingencies, prior training, and the proficiency of the unit. In developing a training program, use the seven-step approach. For in-depth instructions on training development, see *FMs 7-0* and *7-1*.

8-6. The objective is to train and evaluate units in the performance of civil disturbance procedures. Proficiency in performing civil disturbance procedures is reflected by a unit's ability to quickly and deliberately execute the disturbance control procedures required to prevent loss of life and property or to regain control of a civil disturbance in its assigned area of responsibility.

8-7. Training should focus on, but is not limited to—

- Instructors.
- Individual and collective drills.
- Leaders.
- Equipment.
- Weapons.
- Munitions.
- Live-fire range exercises.
- Qualification and familiarization.
- Soldier training exercises (STXs).
- Culminating exercises.

8-8. There are several successful methods to evaluate trained leader skills and trained soldier skills using scenarios, vignettes, lanes, or STXs. Flexibility is an important key in training plan development due to time restraints.

RANGE SETUP

8-9. The key to setting up a successful range for NL munitions is to ensure that they are in compliance with installation safety regulations. Kinetic energy is the force that pushes the projectiles down range. In most cases, kinetic energy will only carry these NL munitions projectiles 75 meters or less. This gives trainers flexibility as to the locations where a live-fire exercise can be conducted, especially in a field-expedient situation. *Figure 8-1* is an example for conducting a NL live fire range setup.

8-10. As illustrated in *Figure 8-1*, mark off 5-, 10-, 20-, 30-, 40-, and 50-meter distances from the targets with markers. Each of the NL munitions has minimum and maximum effective ranges within the distances of 5 to 50 meters. Impact power and accuracy are greater at close distances and depleted at farther distances. Targets can be anything from the standard E-type to more realistic three-dimensional manikins. Evenly divide your shooters into three groups as shown in *Figure 8-1*. Group one will initially be shield holders, group two will be shooters, and group three will be grenadiers.

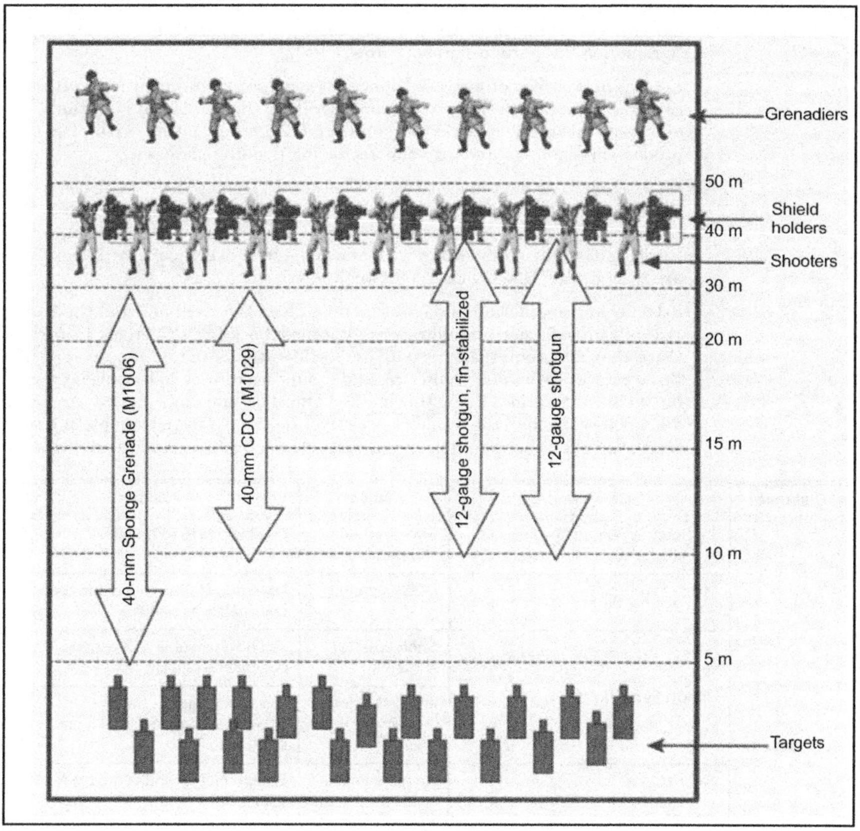

Figure 8-1. NL Munitions Range Setup

8-11. The formation should begin at the farthest effective range firing line (50 meters). After each volley of fire, the formation moves closer to the targets, but stays outside the closest effective range firing line (5 meters). For safety reasons, ensure that the formation stays online throughout each iteration.

8-12. When moving forward to each firing point, it is recommended that the formation always come to a halt before the fire command is given. It is also important to use the tap down drill method (see *Figure 8-1*) rather than having the shooters move forward and fire between the shield holders.

8-13. Range instructors can use verbal commands, such as, "Ready," or "Fire," for each volley. Shooters can be briefed to move forward, tap down the shield holder, and engage targets at the first explosion of distraction grenades thrown by grenadiers. In that instance, the command could simply be "Grenadiers Prepare to Throw, Throw."

8-14. The success of any civil disturbance operation is through effective communication, especially when incorporating the use of NL munitions. For effective training, use verbal commands in conjunction with effective, predetermined hand-and-arm signals during live-fire exercises.

8-15. Each iteration requires two phases of fire. During the first phase, half of the shooters will be firing NL, 40-millimeter rounds using an M203. The others will be firing NL, 12-gauge rounds using a 12-gauge shotgun. In the second phase, shooters switch weapons (safely). It is critical that line safeties are alert during this transition between phases.

8-16. Upon completion of each iteration, clear the weapons and the range. Establish a safe method and return all groups back to the farthest firing line. Once that is accomplished, group one rotates to become group three, group three rotates to become group two, and group two rotates to become group one for the next iteration. Using this method affords a safe and efficient way to get all shooters through the proficiency course. *Figure 8-2* is an example of how to fire each volley based on its characteristics, effectiveness, and capabilities.

Distance	Command	Subject	Actions
10 Meters	12-Gauge Shotgun, Two Point	Shotgun shooters	Load two M1013 rounds.
	M203, Two Point	M203 shooters	Load one M1006 round and have one readily available.
	Ready, Fire	Shotgun shooters	Tap down shields and engage designated targets.
		M203 shooters	Tap down shields, engage designated targets, reload, and reengage.
	Clear Range	Range safeties	Ensure that weapons are clear and safe.
	Safeties, Clear All Weapons	All shooters	Face to right, keep weapons pointed up and downrange.
	Shooters, Right Face, Move	All personnel	Step off shield holder, then shooters, and grenadiers on the command, "Move." Grenadiers reform at 50 meters.

Figure 8-2. Sample Live-Fire Exercise

Distance	Command	Subject	Actions
15 Meters	Shotgun, Two Point	Shotgun shooters	Load two M1013 rounds.
	M203, Two Point	M203 shooters	Load one M1006 round and have another available.
	Grenadiers, Prepare to Throw	Grenadiers	Make distraction grenade ready.
	Throw	Grenadiers	Throw grenade.
		Grenadiers	Throw distraction grenade.
	Ready, Fire	Shot gun shooters	Tap down shields and engage designated targets.
		M203 shooters	Tap down shields, fire at designated targets, reload with M1006 rounds, and engage designated target.
	Clear Range	Range safeties	Ensure that weapons are clear and safe.
	Three Steps Forward	All shooters and grenadiers	Echo preparatory command.
	Move	All shooters and grenadiers	Stay on line and step off three steps, counting each step aloud.
	Clear Range	Range safeties	Ensure that weapons are clear and safe.
	Seven Steps Forward	All shooters and grenadiers	Echo preparatory command.
	Move	All shooters and grenadiers	Stay on line and step off seven steps, counting each step aloud.

Figure 8-2. Sample Live-Fire Exercise (Continued)

Distance	Command	Subject	Actions
20 Meters	Shotgun, Three Point	Shotgun shooters	Load three M1013 rounds.
	M203, One Area, Two Point	M203 shooters	Load one M1026 round and have two M1006 rounds available.
	Grenadiers, Prepare to Throw	Grenadiers	Make distraction grenade ready.
	Throw	Grenadiers	Throw grenade.
		Grenadiers	Throw distraction grenade.
	Ready, Fire	Shotgun shooters	Tap down shields and engage designated targets.
		M203 shooters	Tap down shields, fire area at targets, reload with M1006 rounds, and engage designated target.
	Clear Range	Range safeties	Ensure that weapons are clear and safe.
	Shotgun, Two Point	Shotgun shooters	Load two M1013 rounds.
	M203, Two Point	M203 shooters	Load one M1006 round and have another available.
	Grenadiers, Prepare to Throw	Grenadiers	Make distraction grenade ready.
	Clear Range	Range safeties	Ensure that weapons are clear and safe.
	Three Steps Forward	All shooters and grenadiers	Echo preparatory command.
	Move	All shooters and grenadiers	Stay on line and step off three steps, counting each step aloud.

Figure 8-2. Sample Live-Fire Exercise (Continued)

Distance	Command	Subject	Actions
	Shotgun, Two Area, Two Point	Shotgun shooters	Load two M1012 rounds and two M1013 rounds.
	M203, Two Area, Two Point	M203 shooters	Load one round M1026 and have one M1026 and two rounds M1006 available.
	Grenadiers, Prepare to Throw	Grenadiers	Make distraction grenade ready.
	Throw	Grenadiers	Throw grenade.
		Grenadiers	Throw distraction grenade.
30 Meters	Ready, Fire	Shotgun shooters	Tap down shields, engage targets, and engage designated target with M1013 rounds.
		M203 shooters	Tap down shields, fire area at target, reload with M1006 rounds, and engage designated target.
	Clear Range	Range safeties	Ensure that weapons are clear and safe.
	Seven Steps Forward	All shooters and grenadiers	Echo preparatory command.
	Move	All shooters and grenadiers	Stay on line and step off seven steps, counting each step aloud.

Figure 8-2. Sample Live-Fire Exercise (Continued)

Distance	Command	Subject	Actions
40 Meters	Shotgun, Two Area, One Point	Shotgun shooters	Load two M1012 rounds and one round M1013.
	M203, One Area, One Point	M203 shooters	Load one M1026 round and have one M1006 round available.
	Grenadiers, Prepare to Throw	Grenadiers	Make distraction grenade ready.
	Throw	Grenadiers	Throw grenade.
		Grenadiers	Throw distraction grenade.
	Ready, Fire	Shotgun shooters	Tap down shields and engage targets with M1013 rounds.
		M203 shooters	Tap down shields, fire area at target, reload with M1006 rounds, and engage designated target.
	Clear Range	Range safeties	Ensure that weapons are clear and safe.
	Seven Steps Forward	All shooters and grenadiers	Echo preparatory command.
	Move	All shooters and grenadiers	Stay on line and step off seven steps, counting each step aloud.
50 Meters	Shotgun, One Area	Shotgun shooters	Load one M1012 round.
	M203, One Area, One Point	M203 shooters	Load one M1026 round and have one M1006 round available.
	Grenadiers, Prepare to Throw	Grenadiers	Make distraction grenade ready.
	Throw	Grenadiers	Throw grenade
		Grenadiers	Throw distraction grenade.
	Ready, Fire	Shotgun shooters	Tap down shields, engage targets, and engage designated target with M1013 rounds.
		M203 shooters	Tap down shields, fire area at the target, reload with M1006 rounds, and engage the designated target.
	Clear Range	Range safeties	Ensure that weapons are clear and safe.
	Seven Steps Forward	All shooters and grenadiers	Echo preparatory command.
	Move	All shooters and grenadiers	Stay on line and step off seven steps, counting each step aloud.

Figure 8-2. Sample Live-Fire Exercise (Continued)

NOTE: Shield holders are referred to as group one, shooters are referred to as group two, grenadiers are referred to as group three.

8-17. Currently, there are no qualification standards established in the Army for NL munitions. Area rounds, which contain multiple projectiles designed to affect multiple targets when engaged, only require approximately five rounds. This affords shooters an opportunity to realize the characteristics and capabilities of these rounds.

8-18. Single projectile rounds like the M1013 and the M1006 require a greater degree of proficiency. If situations call for the engagement of targets with these rounds, it is essential that trained shooters be able to hit what they are aiming at. For example, firing 10 rounds from various ranges can provide soldiers with the confidence to be effective shooters. A standard can be established that requires shooters to hit targets with at least 6 out of the 10 rounds used in the example to be considered proficient.

8-19. *Figures 8-3; 8-4,* page 8-11; *8-5,* page 8-21; and *8-6,* page 8-22, give examples of plans to follow if in a situation similar to the following scenario (see *Army Training and Evaluation Program (ARTEP) 19-313-10-MTP*):

A company commander is providing security patrols in support of a joint task force (JTF). Besides providing security patrols for a sector of responsibility, one platoon must always be designated as a quick reaction force (QRF) for civil disturbance operations. For the last 2 weeks the sector has been quiet and the commander has ordered all platoons to conduct civil disturbance training on a rotational basis.

A new order has been sent down to provide a squad-sized element to provide security for the future polling site for an upcoming election. This polling site has yet to open, but intelligence indicates potential problems as the election approaches.

On the day prior to the scheduled election, a crowd of approximately 25 to 30 demonstrators aggressively move toward the polling site, chanting slogans and yelling toward the security forces posted there. While some demonstrators begin throwing whatever they can pick up, the majority are simply yelling, locking arms, and blocking roadways. Hundreds of additional people, not part of the crowd, begin to gather to watch the activities as they unfold.

In an assessment of the demonstrators, it is determined that they intend to stay through the night into the morning of the election with the intention to block access to the polls. It is essential that the crowd is dispersed and that access to the polls is unrestricted in time for the election to begin the next morning.

Evaluation Scheme	
The evaluation will consist of five lanes conducted over a 3-day period. Platoons will execute 24-hour operations during the evaluation. Any retraining will be done immediately following the execution of the lane if time permits. The following tasks will be evaluated:	
Lane 1	Cordon and search operations
Lane 2	Civil disturbance QRF
Lane 3	Security and defense of a critical site (polling site)
Lane 4	Night civil disturbance QRF
Lane 5	Night LP/OP (PIO)

Evaluation Schedule				
Date and Time	**1st Platoon**	**2d Platoon**	**3d Platoon**	**4th Platoon**
	Lane 1	Lane 2	Lane 3	Lane 4
	Lane 2	Lane 3	Lane 4	Lane 5
	Lane 3	Lane 4	Lane 5	Lane 1
	Lane 4	Lane 5	Lane 1	Lane 2
	Lane 5	Lane 1	Lane 2	Lane 3

Figure 8-3. Evaluation Scheme

Unit Training Schedule and Firing Tables

Unit Training Schedule 1

Training Day 1

Time	Subject	Location	Uniform	Instructor	Notes
0800	Administrative time	Classroom	U1		
0830	GRM	Classroom	U1		
1000	ROE and use of force	Classroom	U1		
1200	Chow				
1300	MOOTW	Classroom	U1		
1500	Effective communication	Classroom	U1		

Administration:

Uniforms:

U1: BDU

U2: BDU and LBE with harness and canteens

U3: BDU and LBE with harness, canteens, flak jacket, and Kevlar helmet

U4: BDU and LBE with harness, canteens, flak jacket, Kevlar helmet, and assigned weapon

Transportation:

Weapons:

Ammunition:

Required Equipment for Training:

1. Projection screen
2. Video projector with CPU
3. Television
4. VCR

Miscellaneous:

1. All personnel are required to have a notebook and a pencil or pen.
2. Handouts must be prepared and passed out.

Figure 8-4. Sample Two-Week Plan

Training Day 2

Time	Subject	Location	Uniform	Instructor	Notes
0800	NLCS	Classroom	U1		
0900	PPCT and empty hand	Classroom	U1		
1000	PPCT and empty hand (Practical application)	PT field	U2		
1200	Chow				
1300	PPCT and empty hand (Practical application)	PT field	U2		
1700	COB				

Administration:

Uniforms:

U1: BDU

U2: BDU and LBE with harness and canteens

U3: BDU and LBE with harness, canteens, flak jacket, and Kevlar helmet

U4: BDU and LBE with harness, canteens, flak jacket, Kevlar helmet, and assigned weapon

Transportation:

Weapons:

Ammunition:

Required Equipment for Training:

1. Projection screen
2. Video projector with CPU
3. Training strike bags
4. Flex-Cufs

Miscellaneous:

1. All personnel are required to have a notebook and a pencil or pen.
2. Handouts must be prepared and passed out.
3. Medical personnel must be on location.

Figure 8-4. Sample Two-Week Plan (Continued)

Training Day 3					
Time	**Subject**	**Location**	**Uniform**	**Instructor**	**Notes**
0800	PPCT and empty hand (Practical application)	PT field	U2		
1200	Chow				
1300	PPCT and empty hand (Practical application)	PT field	U2		
1700	COB				

Administration:

Uniforms:

U1: BDU

U2: BDU and LBE with harness and canteens

U3: BDU and LBE with harness, canteens, flak jacket, and Kevlar helmet

U4: BDU and LBE with harness, canteens, flak jacket, Kevlar helmet, and assigned weapon

Transportation:

Weapons:

Ammunition:

Required Equipment for Training:

1. Projection screen
2. Video projector with CPU
3. Training strike bags
4. Flex-Cufs

Miscellaneous:

1. All personnel are required to have a notebook and a pencil or pen.
2. Handouts must be prepared and passed out.
3. Medical personnel must be on location.

Training Day 4					
Time	**Subject**	**Location**	**Uniform**	**Instructor**	**Notes**
0800	Riot baton training	Classroom	U1		
0900	Riot baton training (Practical application)	PT field	U2		
1200	Chow				

Figure 8-4. Sample Two-Week Plan (Continued)

1300	Riot baton training (Practical application)	PT field	U2
1700	COB		

Administration:

Uniforms:

U1: BDU

U2: BDU and LBE with harness and canteens

U3: BDU and LBE with harness, canteens, flak jacket, and Kevlar helmet

U4: BDU and LBE with harness, canteens, flak jacket, Kevlar helmet, and assigned weapon

Transportation:

Weapons:

1. Expandable riot baton with carrier
2. Training riot batons

Ammunition:

Required Equipment for Training:

1. Projection screen
2. Video projector with CPU
3. Training strike bags
4. Flex-Cufs

Miscellaneous:

1. All personnel are required to have a notebook and a pencil or pen.
2. Handouts must be prepared and passed out.
3. Medical personnel must be on location.

Training Day 5

Time	Subject	Location	Uniform	Instructor	Notes
0800	Riot baton training (Practical application)	PT field	U2		
1000	Crowd dynamics	Classroom	U1		
1130	Chow				
1230	Crowd dynamics	Classroom	U1		
1330	CD formations	Classroom	U1		

Figure 8-4. Sample Two-Week Plan (Continued)

1430	RCAs and OC	Classroom	U1
1700	COB		

Administration:

Uniforms:

U1: BDU

U2: BDU and LBE with harness and canteens

U3: BDU and LBE with harness, canteens, flak jacket, and Kevlar helmet

U4: BDU and LBE with harness, canteens, flak jacket, Kevlar helmet, and assigned weapon

Transportation:

Weapons:

1. Expandable riot baton with carrier
2. Training riot baton

Ammunition:

Required Equipment for Training:

1. Projection screen
2. Video projector with CPU
3. Training strike bags
4. Flex-Cufs

Miscellaneous:

1. All personnel are required to have a notebook and a pencil or pen.
2. Handouts must be prepared and passed out.
3. Medical personnel must be on location.

Training Day 6

Time	Subject	Location	Uniform	Instructor	Notes
0800	RCAs and OC (Practical application)	PT field	U2		
1200	Chow				
1300	RCAs and OC (Practical application)	PT field	U2		
1700	COB				

Figure 8-4. Sample Two-Week Plan (Continued)

Administration:

Uniforms:

U1: BDU

U2: BDU and LBE with harness and canteens

U3: BDU and LBE with harness, canteens, flak jacket, and Kevlar helmet

U4: BDU and LBE with harness, canteens, flak jacket, Kevlar helmet, and assigned weapon

Transportation:

Weapons:

Ammunition:

1. MK inert OC
2. MK 4 live OC disperser
3. M36 live CR disperser

Required Equipment for Training:

1. Five large clean trash cans with liners
2. Two water hoses with running water

Miscellaneous:

Medical personnel must be on location.

Training Day 7

Time	Subject	Location	Uniform	Instructor	Notes
0800	NL munitions employment	Classroom	U1		
1030	Chow				
1115	Safety brief	Range	U4		
1130	NL munitions (live fire)	Range	U4		
TBD	Evening chow				
TBD	Safety brief	Range	U4		
TBD	NL munitions night fire exercise	Range	U4		
TBD	Clean up and COB				

Figure 8-4. Sample Two-Week Plan (Continued)

Administration:

Uniforms:

U1: BDU

U2: BDU and LBE with harness and canteens

U3: BDU and LBE with harness, canteens, flak jacket, and Kevlar helmet

U4: BDU and LBE with harness, canteens, flak jacket, Kevlar helmet, and assigned weapon

Transportation:

Yes

Weapons:

1. M16A2 with M203
2. 12-gauge shotgun

Ammunition:

1. Fin-stabilized M1012
2. M1013 multipellet (rubber)
3. M1006 40-millimeter sponge grenade
4. M1026 40-millimeter multiball (rubber)
5. M84 diversionary grenade

Required Equipment for Training:

1. Fifteen to twenty human form targets
2. Eye and hearing protection
3. Fifteen to twenty riot shields
4. Two MREs per student
5. Video recorder
6. Three to five high-intensity lights
7. All students should be issued face shields, shin guards, and ammunition pouches

Miscellaneous:

1. Medical personnel must be on location.
2. Range safety officers must be available.
3. Primary and secondary means of communication with range control must be available.
4. Operational risk assessments must be conducted.

Figure 8-4. Sample Two-Week Plan (Continued)

Training Day 8					
Time	**Subject**	**Location**	**Uniform**	**Instructor**	**Notes**
0800	CD formations (Practical application)	TA 1	U4		
1200	Chow				
1300	CD formations (Practical application)	TA 1	U4		
1700	COB				

Administration:

Uniforms:

U1: BDU

U2: BDU and LBE with harness and canteens

U3: BDU and LBE with harness, canteens, flak jacket, and Kevlar helmet

U4: BDU and LBE with harness, canteens, flak jacket, Kevlar helmet, and assigned weapon

Transportation:

Yes

Weapons:

1. Training weapons
2. Training riot batons

Ammunition:

Required Equipment for Training:

1. Eye protection
2. Fifteen to twenty riot shields
3. Video recorder
4. Three to five high-intensity lights
5. Face shields, shin guards, and ammunition pouches

Miscellaneous:

1. Medical personnel must be on location.
2. Primary and secondary means of communication must be available.
3. Operational risk assessment must be conducted.
4. Twenty to thirty OPFOR personnel must be available.

Figure 8-4. Sample Two-Week Plan (Continued)

Training Day 9

Time	Subject	Location	Uniform	Instructor	Notes
0800	Introduction to M5, PVAB, and M315	Classroom 1	U1		
1200	Chow				
1300	M5, PVAB, and M315 (Practical application/LFE)	Range	U4		
1700	COB				

Administration:

Uniforms:

U1: BDU

U2: BDU and LBE with harness and canteens

U3: BDU and LBE with harness, canteens, flak jacket, and Kevlar helmet

U4: BDU and LBE with harness, canteens, flak jacket, Kevlar helmet, and assigned weapon

Transportation:

Yes

Weapons:

Two complete M315s and installation kits

Ammunition:

1. Ten live MCCMs
2. Ten live M315 rounds of each

Required Equipment for Training:

1. Eye protection
2. Ten inert MCCMs
3. Two complete PVAB systems

Miscellaneous:

1. Medical personnel must be on location.
2. Primary and secondary means of communication must be available.
3. Operational risk assessment must be conducted.

Figure 8-4. Sample Two-Week Plan (Continued)

Training Day 10					
Time	**Subject**	**Location**	**Uniform**	**Instructor**	**Notes**
0800	Tactical consideration	MOUT classroom	U1		
0930	Safety brief	MOUT classroom	U4		
1000	Tactical exercise	MOUT Facility	U4		

Administration:

Uniforms:

U1: BDU

U2: BDU and LBE with harness and canteens

U3: BDU and LBE with harness, canteens, flak jacket, and Kevlar helmet

U4: BDU and LBE with harness, canteens, flak jacket, Kevlar helmet, and assigned weapon

Transportation:

Yes

Weapons:

1. Assigned weapons
2. Inert RCA/OC individual dispersers
3. Training riot batons
4. Fifty practice grenades

Ammunition:

Required Equipment for Training:

1. Eye protection
2. Ten inert MCCMs
3. Face shields, shin guards, and ammunition pouches
4. Fifteen to twenty full-length riot shields

Miscellaneous:

1. Medical personnel must be on location.
2. Primary and secondary means of communication must be available.
3. Operational risk assessment must be conducted.
4. A video recorder must be available.

Figure 8-4. Sample Two-Week Plan (Continued)

Sample Unit Training Schedule				
Training Day 1				
Time	**Event**	**Location**	**Instructor**	**Notes**
0730-0830	Force continuum	Classroom		L
0830-0930	Management of aggressive behavior	Classroom		L
0930-1030	ROE	Classroom		L
1030-1130	Crowd control	Classroom		L
1130-1230	Chow			
1230-1700	PPCT and empty hand	PT field		L, PA
Training Day 2				
Time	**Event**	**Location**	**Instructor**	**Notes**
0730-1130	Riot baton training	PT field		L, PA
1130-1230	Chow			
1230-1530	Riot baton training	PT field		PA
1530-1700	CD formations	Classroom		L
Training Day 3				
Time	**Event**	**Location**	**Instructor**	**Notes**
0730-1130	CD formations	PT field		PA
1130-1230	Chow			
1230-1700	NL munitions, NLW	Range		L, PA
Training Day 4				
Time	**Event**	**Location**	**Instructor**	**Notes**
0730-1130	RCAs and OC	Classroom and PT field		L, PA
1130-1230	Chow			
1230-1700	RCAs and OC	PT field		PA

Figure 8-5. Sample One-Week Plan

Training Day 5				
Time	**Event**	**Location**	**Instructor**	**Notes**
0730-1700	Tactical exercise	MOUT facility		OPS
NOTE: Appointed student leaders ensure that troops are fed and rotated for rest.				

Figure 8-5. Sample One-Week Plan (Continued)

Sample Unit Training Schedule				
Training Day 1				
Time	**Event**	**Location**	**Instructor**	**Notes**
	Shield and riot baton techniques	MOUT site		Individual and collective tasks
	NLCS and munitions	Range Classroom		Leader training
	Formation training	MOUT site		Individual and collective tasks
	Chow			
	Formation training	MOUT site		Individual and collective tasks
Training Day 2				
Time	**Event**	**Location**	**Instructor**	**Notes**
0730-1930, 12-hour operations	Live-fire exercise	Range		M203/12-gauge
	QRF exercise	MOUT site		Individual and collective tasks
NOTE: Once weapons and munitions familiarization is complete, shooters join the QRF exercise to be used. For live-fire range setup, see *Figure 8-1.*				

Figure 8-6. Sample Two-Day Plan

Appendix A

Metric Conversion Chart

This appendix complies with current Army directives which state that the metric system will be incorporated into all new publications. *Table A-1* is a metric conversion chart.

Table A-1. Metric Conversion Chart

US Units	Multiplied By	Metric Units
Cubic feet	0.0283	Cubic meters
Feet	0.3048	Meters
Feet per second	0.3048	Meters per second
Inches	0.0254	Meters
Inches	25.4001	Millimeters
Miles per hour	1.6093	Kilometers per hour
Pounds	0.4536	Kilograms
Yards	0.9144	Meters
Metric Units	**Multiplied By**	**US Units**
Cubic meters	35.3146	Cubic feet
Kilograms	2.2046	Pounds
Kilometers per hour	0.6214	Miles per hour
Meters	3.2808	Feet
Meters	39.3701	Inches
Meters	1.0936	Yards
Meters per second	3.2808	Feet per second
Millimeters	0.0394	Inches

Appendix B

Operations and Legal Considerations in the Continental United States

Providing military assistance to state and local governments to assist them in quelling a civil disturbance or riot requires close coordination through a host of state and federal agencies. It requires a thorough briefing of soldiers at all levels on what they can and cannot do with respect to law enforcement. Civil authorities must be briefed on the restrictions placed on federal forces by the *Constitution of the United States* and federal statutes and laws. The national guard (NG), when not in a federal status, operates under the control of the state governor and the adjutant general (AG). The NG has historically been the first military responder during emergencies.

FEDERAL INTERVENTION AND AID

B-1. Under the *Constitution of the United States* and *United States Code*, the President is empowered to direct federal intervention in civil disturbances to—

- Respond to state requests for aid in restoring order.
- Enforce the laws of the United States.
- Protect the civil rights of citizens.
- Protect federal property and functions.

B-2. Under the *Constitution of the United States*, each state is responsible for protecting life and property within its boundaries. State and local governments use their LEA forces to maintain law and order and to quell civil disturbances. However, if civil disturbance support requirements exceed the resources of a state, federal troops may be requested to help restore and maintain law and order.

B-3. The *Constitution of the United States* and federal statutes authorize the President to direct the use of federal armed troops within the 50 states and territories and their political subdivisions. The President is also empowered to federalize the NG of any state to suppress rebellion and enforce federal laws.

B-4. Federal assistance is provided to a state when the state has used all of its resources, including its NG, to quell a disorder and finds the resources insufficient. Usually, active duty federal forces are used to augment the NG of the requesting state. However, the President may choose to federalize the NG of another state and use them alone or with other forces to restore order.

B-5. The President may employ federal troops to ensure the execution of US law when a state opposes or obstructs US law or impedes the course of justice under those laws. The President may employ armed federal troops to suppress insurrection, domestic violence, unlawful assemblies, and conspiracy if such acts deprive the people of their constitutional rights and privileges and the civil authorities of a state cannot or will not provide adequate protection.

B-6. The president is also authorized to use armed federal troops to protect federal property and functions when the need for protection exists and local civil authorities cannot or will not give adequate protection. The right of the United States to protect all federal property and functions regardless of their locations is an accepted principle of our government.

B-7. As a temporary measure, federal military equipment and facilities may be loaned to law enforcement agencies of state and local governmental bodies for use during civil disturbances. These resources may also be loaned to state NG and non-DOD federal agencies. The requesting agencies are expected to provide enough resources of their own to minimize the need for US military resources. In addition, the loan of the resources must not conflict with US military needs.

ROLES AND RESPONSIBILITIES OF VARIOUS AGENCIES

B-8. Several agencies within the federal and state governments play a key role in supporting the request of local authorities for support during a civil disturbance.

SECRETARY OF DEFENSE

B-9. The Secretary of Defense retains approval for federal support to civil authorities involving the use of DOD forces, personnel, and equipment. Approval may also involve DOD support that will result in a planned event with the potential for confrontation with specifically identified individuals and groups or will result in the use of lethal force.

SECRETARY OF THE ARMY

B-10. The Secretary of the Army, as an executive agent for the Secretary of Defense, is the approval authority for federal emergency support in response to natural and man-made disasters (except weapons of mass destruction [WMD]). Military forces respond to directions and guidance from the director of military support (DOMS) (the action agent for the Secretary of the Army).

DIRECTOR OF MILITARY SUPPORT

B-11. The DOMS plans and executes the DOD domestic support mission to civil authorities. He is a general officer appointed by the Secretary of the Army. The DOMS is the DOD primary contact for all federal departments and agencies during DOD involvement in most domestic support operations.

STATE COORDINATING OFFICER

B-12. The governor appoints the state coordinating officer (SCO) to coordinate disaster operations for the state. The SCO is the primary point of contact for the federal coordinating officer (FCO) in facilitating disaster assistance.

DIRECTOR OF THE FEDERAL EMERGENCY MANAGEMENT AGENCY

B-13. The director of the Federal Emergency Management Agency (FEMA), on behalf of the President, appoints the FCO. The FCO coordinates the timely delivery of federal disaster assistance to the affected state and local governments and disaster victims. He works closely with the SCO to determine state requirements and to coordinate national level issues with the catastrophic disaster response group. There must be close and continuous coordination with FEMA to ensure the proper use of military assets in support of civil authorities.

DEFENSE COORDINATING OFFICER

B-14. The defense coordinating officer (DCO) operates under the authority of the President and is the DOD representative designated to coordinate on-the-scene activities with the FCO. FEMA and other federal agencies that request support from DOD go through the DCO for validation and resourcing.

NATIONAL GUARD

B-15. The NG, in state status (see *Title 32, United States Code {USC})*, is the governor's primary military response organization for emergencies and disasters. The NG responds and operates under state laws unless they are federalized, when they operate under the same constraints and laws as federal units.

UNITED STATES ARMY RESERVE

B-16. The US Army Reserve is capable of providing military support. This assistance and support may include the use of equipment and other resources, including units and individuals.

STATE AND LOCAL GOVERNMENT RESPONSIBILITIES

B-17. State and local government officials, operating under the authority granted by state constitutions and the *Constitution of the United States*, are responsible for daily safety and security issues that impact the quality of life of their citizens. State and local officials have primary responsibility for emergency preparedness.

STATE RESPONSIBILITIES

B-18. Many state government agencies are responsible for coming to the aid of state citizens.

Governor

B-19. A state governor is empowered by each state constitution and the *Constitution of the United States* to execute the laws of the state and to command the state NG while it is serving in state status. Governors are also responsible for issuing executive orders or emergency proclamations, such as declaring states of emergency and ensuring that state agencies plan for potential disasters and civil disorder.

B-20. Once a man-made or natural disaster occurs, the governor assesses its extent and determines if local government requests for assistance should be honored. If appropriate, the governor declares a state of emergency, activates the state emergency response plan, and may order the NG into state active duty. The governor assigns the NG its mission through the office of emergency service (OES)/emergency medical dispatching (EMD).

Plans, Operations, and Military Support Officer

B-21. Each state NG has a plans, operations, and military support officer (POMSO) who prepares and coordinates contingency plans for military support to civil authorities (MSCA) during disaster response and recovery operations. The POMSO coordinates training plans and exercises between the state NG and federal, state, and local emergency management agencies. The POMSO serves as the NG point of contact with DOD officials during a declared federal emergency or disaster.

Office of Emergency Services

B-22. Each state has a specific agency and/or office that coordinates emergency preparedness and serves as the governor's coordinating agency during an emergency. The titles of these offices vary from state to state. Some examples are the Division of Emergency Government, Emergency Management Agency, Department of Public Safety, or Office of Emergency Preparedness. This manual refers to this office using the generic term OES/EMD.

B-23. Generally, the OES/EMD is organized as a stand-alone office under the governor or aligned under the AG or the state police, and in some instances, the AG is the OES/EMD director. It operates the state emergency operations center and coordinates with state agencies and federal officials for assistance.

Adjutant General

B-24. The state NG is the governor's primary military response force in an emergency. The AG, through the state area command (STARC) (specifically the POMSO), develops and coordinates emergency response plans for disasters and emergencies. The AG is in command of state NG forces ordered into state active duty.

State Government Agencies

B-25. State government departments and agencies prepare emergency preparedness plans for their areas of specialization in support of the governor's plan. They also participate in emergency preparedness exercises.

LOCAL RESPONSIBILITIES

B-26. Mayors, city managers, local police, fire protection officials, county executives, sheriffs, prosecuting attorneys, and public health officials are some of the people responsible for law enforcement, safety, health, and fire protection on a daily basis. They are responsible for developing appropriate emergency response plans and responding to emergencies within their jurisdictions. Most local jurisdictions have an OES/EMD to plan and coordinate actions in an emergency. In many cases, local jurisdictions have mutual aid agreements with other jurisdictions that allow for firefighting, medical, and police assistance across jurisdictional boundaries. Once local officials determine that an emergency is beyond the scope of their resources or ability to respond, the senior local official is responsible for requesting additional assistance from the next agency in the assistance request process.

LEGAL CONSIDERATIONS AND CONSTRAINTS

B-27. The *Constitution of the United States*, laws, regulations, policies, and other legal issues limit the use of federal military personnel in domestic support operations. Any Army involvement in civil disturbance operations involves many legal issues requiring comprehensive legal reviews.

B-28. Under the *Constitution of the United States*, Congress has the authority to raise and support an army, provide and maintain a navy, and make rules for governing and regulating the land and naval forces. The *Constitution of the United States* places the military under civilian control and designates the President as commander-in-chief. Statutes provide for civilian leadership in the office of the secretary of defense, service secretaries, and various other civil authorities. Under the *Constitution of the United States*, the civilian government must preserve public order and carry out governmental operations within its territorial limits. The *Constitution of the United States* allows the use of the federal military to execute or enforce the law when necessary to protect federal or civilian property and functions. However, significant restrictions exist on employing federal military forces within the United States.

B-29. The unique capabilities of the military enable it to support federal, state, or local civilian agencies. In most circumstances, the DOD is one of many federal agencies reacting to a domestic emergency or crisis, playing a subordinate, supporting role to a lead civilian agency.

LAW

B-30. In the United States, civilian agencies provide for the needs of citizens. Civilian, federal, state, and local government and law enforcement agencies execute US laws. Laws governing the use of the military in domestic operations are complex, subtle, and constantly changing. Commanders involved in civil disturbance operations should staff plans, policies, programs, exercises, funding, operations, constraints, and limitations with their local SJA to ensure conformity with legal requirements. They should scrutinize each request for aid (whether it is for equipment, individuals, or training) to ensure that it conforms to statutory requirements. The following references are a snapshot of the law on military support to civilian authorities:

- *Section 331, Chapter 15, Part 1, Subtitle A, Title 10 USC.*
- *Section 332, Chapter 15, Part 1, Subtitle A, Title 10 USC.*
- *Section 333, Chapter 15, Part 1, Subtitle A, Title 10 USC.*
- *Section 334, Chapter 15, Part 1, Subtitle A, Title 10 USC.*
- *Section 1385, Chapter 67, Part 1, Title 18 USC.*
- *DOD Directive 3025.12.*

- *DOD Directive 3025.15*
- *DOD Directive 5525.5*
- *DOD Civil Disturbance Plan 55-2*
- *NGR 500-1/ANGI 10-8101*
- *FM 3-19.15*

INFORMATION OPERATIONS

B-31. Information superiority helps forces anticipate problems and requirements. It allows commanders to control situations earlier and with less force, creating the conditions necessary to achieve the end state. Public affairs, PSYOP, and civil military operations are activities that will allow the commander to control situations earlier and with less force.

B-32. Information is available from a multitude of sources, the primary being open sources, law enforcement, and the military. A diversity of sources is the best approach because it prevents biased behavior.

Open sources include—

- Libraries.
- Newspapers and news periodicals.
- Radio and television.

Law enforcement sources include—

- Local law enforcement agencies.
- National law enforcement agencies.

Military sources include—

- DOD intelligence community (most restrictive source).
- Local MI field offices.

B-33. The DOD intelligence community operates under limitations imposed by regulations and executive orders. Attempts to skirt these restrictions may violate regulations or federal statutes.

RESTRICTIONS

B-34. Strict limits are placed on collecting information related to a civil disturbance in order to protect the civil rights of people and organizations not affiliated with the DOD. Civil disturbance plans and materials must not include lists of groups or people not affiliated with the DOD. Information on civilians and civilian organizations can be collected only with specific authorization from the Secretary or the Under Secretary of the Army. Conditions for collecting information include the existence of threats against Army personnel, functions, or property (see *ARs 380-10* and *380-13*). Civil disturbance information (available in public documents) or open-source information may be collected. However, specific rules regarding its storage must be followed. Commanders must coordinate with SJA, MI, and USACIDC before collecting any such information.

B-35. Civilian law enforcement agencies may limit access to law enforcement intelligence. In such circumstances, the military may request that agencies release the unclassified version.

B-36. If the Department of Justice (DOJ) determines that federal intervention into a civil disturbance is likely, then information relating to the disturbance is provided to the Army Assistant Chief of Staff for Intelligence. The information is analyzed and then passed through channels to the TF commander for planning purposes.

B-37. Military intelligence collection efforts, except liaison, may begin only when DA orders. The orders will come through the CSA's personal liaison officer and the TF commander. Covert operations to gather information on non-DOD individuals and groups must be approved on an operation-by-operation basis by the Undersecretary of the Army.

B-38. Once DA approves the collection effort, MI elements establish and maintain liaison with the appropriate local, state, and federal authorities. Then they collect information pertaining to the incident and general situation. Their estimate of the situation is passed to the TF commander and DA.

POSSE COMITATUS

B-39. Generally, federal military forces may not give law enforcement assistance to civil authorities without conflicting with the *Posse Comitatus Act*. However, constitutional and statutory exceptions to this prohibition do exist. The recent emphasis on drug interdiction has led to an increase in those exceptions.

B-40. The *Judiciary Act of 1789* allowed US marshals to call upon the military as a posse comitatus. This continued until after the Civil War, when the federal government used the Army to execute reconstruction era policies. The southern states regarded the use of the military for this purpose as abusive and repressive, and in 1878, President Rutherford B. Hayes signed the original bill ending the practice.

B-41. The *Posse Comitatus Act* prescribes criminal penalties for the use of the US Army or Air Force to execute the laws of or to perform civilian law enforcement functions within the United States. DOD policy extends this prohibition to the US Navy and Marine Corps. Prohibiting the military from executing the laws means that military personnel may not participate directly in—

- An arrest; a search and seizure; a stop and frisk; or an interdiction of vessels, aircraft, or vehicles.
- A surveillance or pursuit.
- A civilian legal case or any other civilian law enforcement activity as informants, undercover agents, or investigators.

B-42. The *Posse Comitatus Act* does not apply to the following:

- Members of the NG when in state service.
- Members of a Reserve Component when not on active duty or active duty for training.
- DOD civilians, unless under the direct command and control of an active duty officer.

- A soldier when off duty and acting only in a private capacity.
- Soldiers taking action for the primary purpose of furthering a military or foreign affairs function of the United States; for example, enforcing military justice, maintaining law and order on military installations, and protecting classified materials.

CONSTITUTIONAL EXCEPTIONS TO THE POSSE COMITATUS ACT

B-43. Under its inherent authority, the US government is responsible for preserving public order and carrying out governmental operations within its territorial limits, by force if necessary. Under the *Constitution of the United States*, exceptions allow the use of military force to execute or enforce the law. Some of the exceptions are to—

- Protect civilian property and functions.
- Ensure preservation of public order and carry out government operations, by force if necessary.
- Protect civil rights where local authorities do not or cannot protect them.
- Protect federal property and functions.
- Quell civil disturbances and labor strife that rises to the level of civil disorder.

B-44. The President may order the armed forces to support state civil authorities suffering from an insurrection or civil disturbance. He must act personally by first issuing a proclamation calling on insurgents to disperse and retire peacefully within a limited time. This occurred when federal forces were called in to counter the Los Angeles riots in 1992.

Protect Civilian Property and Functions

B-45. A sudden and unexpected civil disturbance, disaster, or calamity may seriously endanger life and property and disrupt normal governmental functions to such an extent that local authorities cannot control the situation. At such times, the federal government may use military force to prevent loss of life or wanton destruction of property and to restore government functions and public order. This exception has rarely been used.

Protect Federal Property and Functions

B-46. The federal government may use military force to protect federal property and federal government functions when local authorities cannot or decline to provide adequate protection.

ARMED FORCES

B-47. Basic guidance for the interaction of the military, the Reserve Component, and civilian LEAs is given in *10 USC Chapter 18*. Guidelines on reimbursement and restrictions, direct participation in law enforcement activities, the use of information collected during military operations, and the use of military equipment and facilities are some of the topics covered. Title 10 prohibits the military from directly participating in arrests, searches, seizures, or other similar activity unless authorized by law (such as arrests on military property). The *Fiscal Year 1989 Act* and subsequent *National Defense Authorization Acts* have authorized the DOD to provide more support to LEAs in the counterdrug effort.

NATIONAL GUARD

B-48. For a description of how the secretary of defense may provide funds to state governments (including the District of Columbia, Commonwealth of Puerto Rico, and US territories) for civil disturbance operations performed by the National Guard when not in federal service, see 32 *USC*. The *Posse Comitatus Act* does not apply to the NG when not in federal service. Each state specifies how NG may be employed.

SUPPORT TO CIVIL LAW ENFORCEMENT

B-49. It is DOD policy to cooperate with civilian law enforcement officials to the greatest practical extent. However, cooperation must consistently meet the needs of national security and military preparedness, the historic tradition of limiting direct military involvement in civilian LEAs, and the requirements of applicable law. US military forces are never placed under the command of civilian law enforcement officers or nonfederalized NG. DOD policy concerning the provision of military support to LEAs, including personnel and equipment, are contained in *DOD Directive 5525.5*.

B-50. The Army assists civilian law enforcement by providing personnel, equipment, training, and expert advice within the limits of applicable law. Army NG units in state status (see 32 *USC*) provide the primary source of military assistance. They may be able to provide assistance to civil authorities in instances when federal units are precluded due to the restrictions of the *Posse Comitatus Act*.

ROLES OF THE NATIONAL GUARD AND FEDERAL FORCES

B-51. The preservation of law and order in the civilian community is the responsibility of state and local governments and law enforcement authorities. When a civil disturbance in a civilian community turns to widespread rioting that includes arson, looting, and acts of violence, the civil authorities may decide that they do not have the resources to quell the riot. They may then turn to the NG and federal forces to support the civil authorities in restoring law and order. Civil disturbances in any form are prejudicial to public law and order.

NATIONAL GUARD FORCES

B-52. The NG (as a state organization) responds to the governor according to state law for civil disturbance operations. NG regulations direct planning and training for the civil disturbance mission. During most civil disturbance situations, the NG will be the first military responder and will usually remain in state active duty status (according to 32 *USC*) throughout the operation. In extreme circumstances, the NG can be brought on federal service for civil disturbance operations when ordered to under the appropriate federal statute by the President. During the 1992 Los Angeles civil disturbance, the President federalized the California NG. Once the NG has been deployed to assist the civil authorities with a civil disturbance, it is too far into the deployment cycle to start determining what tasks the NG can do as a state or as a federally controlled force. It is the commander's responsibility not to accept missions that the unit is not adequately trained to do. The following are examples of appropriate missions for the NG:

- Manning traffic control points.
- Providing building security.
- Providing area security and patrols.
- Providing security at custody facilities.
- Providing security and escort for emergency personnel and equipment.
- Protecting sensitive sites.
- Transporting law enforcement personnel.
- Providing show of force.
- Dispersing crowds.
- Employing riot control munitions.
- Providing very important person (VIP) protection and escort.
- Providing quick reaction and reserve force.

B-53. Examples of inappropriate taskings:

- Negotiating hostage situations.
- Barricading suspects.
- Participating in criminal investigations.
- Acting as federal Army forces.

B-54. The commitment of federal troops to deal with domestic civil disturbances must be viewed as a drastic measure of last resort. Their role should never be greater than what is absolutely necessary under the circumstances. Commanders should take every measure to avoid the perception of an invading force. A JTF designated to respond to a civil disturbance should project the image of a restrained and well-disciplined force whose sole purpose is to help restore law and order with minimal harm to the people and property and with due respect for all law-abiding citizens.

B-55. The role of federal Army forces is to assist civil authorities in restoring law and order when the magnitude of the disturbance exceeds the capabilities of local and state LEA, including the NG. Under the provisions of the *Constitution of the United States* and selected federal statutes, the President may order the employment of the federal armed forces to aid local and state civil authorities to protect the Constitutional rights of citizens. Federal military forces may also protect federal facilities and installations in any state, territory, or possession.

B-56. The *DA Civil Disturbance Plan 55-2* gives guidance for Army forces directed to quell civil disturbances. In addition, *AR 500-50* outlines responsibilities, prescribes policies, and provides guidance for the use of Army resources in response to civil disturbances. It outlines statutory restrictions and designates staff and command responsibilities for planning and executing civil disturbance operations.

B-57. The JTF commander exercises control of all federal military forces (including NG in federal status) committed to assist civil authorities. Federal military forces remain under the military chain of command during civil disturbance operations. Federal forces will not be placed under the command of either civil officials or NG commanders in a nonfederal status. Civil authorities retain control of their state and local law enforcement agencies. The JTF commander establishes liaison with the senior representative of the attorney general (SCRAG) and other appropriate federal, state, and local civil authorities.

B-58. When a JTF is being assembled it must deploy with a provost marshal (PM) who will be the POC for situation assessment, operational information, and security matters. He also serves as a liaison with civilian and law enforcement agencies.

B-59. Federal military forces must be tailored to the specific civil disturbance situation. Sufficient combat support and combat service support units will be required to sustain the force throughout the deployment. Coordination with civil officials may allow the force to draw on resources available from state and local agencies. Close and continuous coordination between federal military forces and LEAs will provide the commander with the detailed and timely information required to employ and protect the force effectively.

B-60. In supporting *DA Civil Disturbance Plan 55-2,* intelligence personnel may conduct close and continuous liaison with LEAs and military police to ensure that their units receive the information needed to allow the commander to adequately protect the force. The JTF commander should staff intelligence support missions with his senior intelligence officer and legal counsel before approving the mission.

B-61. Federal military forces must be employed in tasks or missions appropriate to their organization and training; they must not be employed in ways that violate the legal restrictions in effect. Military forces may be used to disperse unlawful assemblies and to patrol disturbed areas to prevent unlawful acts. They may be used to assist in the distribution of essential goods and the maintenance of essential services. Forces may also establish traffic control points, cordon off areas, release smoke and obscurants, and serve as security or quick reaction forces. Certain types of missions (such as gathering intelligence on civilians) are always inappropriate for military forces during civil disturbance operations.

B-62. Requests for specific military missions are typically passed through a single state or federal law enforcement coordinating officer, as approved by the SCRAG. Validated requests are transmitted to the JTF commander and his HQ for staffing and coordination. Approved missions are assigned through the military chain of command to the appropriate element or unit for execution. Units and soldiers will not accept taskings or missions directly from law enforcement or civilian officials, except in a direct support relationship as approved and ordered through the military chain of command.

B-63. It is important that a military liaison be provided to each LEA HQ generating requests for support. This liaison can assist LEA officials in determining the types and quantities of military support to be requested. JTF HQ can facilitate this mission assignment process by providing LEAs with a detailed listing of the types of missions military forces may conduct.

B-64. The AO for a deployed unit should coincide with the jurisdiction or subdivision boundaries of the LEA it supports. This arrangement facilitates liaison and coordination between law enforcement and military chains of command.

LEAD AGENCY CONCEPT AND ROLE OF MILITARY

B-65. The DOJ is the lead federal agency for civil disturbance operations. The Attorney General's on-scene representative is the SCRAG. The DOD has designated DA as their executive agent for military assistance for civil disturbance (MACDIS) (see *FM 27-100.*)

JOINT TASK FORCE

B-66. Military support to civil authorities in disasters and domestic emergencies is a DOD (not a service component) responsibility. The supported commander may decide that the severity and scope of a disaster require a joint response. In civil assistance, as in operational level warfighting, the commander-in-chief (CINC) uses the different and complementary capabilities of each service to accomplish the mission. Such use requires knowledge of both the capability and the availability of all service component assets, to include their agencies and installation.

B-67. The CINC may establish a JTF to provide comprehensive military support. Establishing a JTF may provide the best DOD response to civil emergencies. A JTF is established to execute a specific mission limited in scope and duration. The JTF objective in civil emergencies is to deploy forces to the area rapidly to assist immediately in saving lives and safeguarding property and to continue providing assistance required by the federal response plan (FRP) and the particular situation.

B-68. The JTF is configured for each specific mission. In some civil emergencies, the JTF may require a greater proportion of combat service support type units and capabilities than in typical warfighting deployments. The JTF must be able to provide emergency assistance across all lines of support.

JOINT TASK FORCE CONTROL ELEMENT

B-69. Responsibility for controlling the civil disturbance resides with the civil authorities. The JTF is organized to support and lend assistance to them in their effort to return civil order to the community.

B-70. The control element consists of the command group and the crisis management team (CMT). The command group is made up of city, county, state, and military command personnel. They set policy and issue directives, coordinate the activities of civil and military support agencies, supervise the CMT, and coordinate with outside agencies. The CMT is made up of representatives of civilian and military staff sections. The CMT advises the command group and coordinates operations and support for the action element of the TF. The control elements are located with the emergency operations center (EOC) to facilitate information processing, resource management, and operational control. The JTF command element should colocate with the civilian command element at the EOC. This will expedite the exchange of information. If an EOC has not been set up, the commander may establish one while making provisions for including civil authorities to ensure a unity of effort.

B-71. Not all CMT members are located at the EOC. Some key people may use liaisons to represent them at the EOC. Key personnel can then research and discuss ideas freely, away from the confusion associated with crisis management. The liaison can transmit guidance and answers to the EOC. Some agencies may not be needed in the EOC at first, but they must be prepared for inclusion. The decision to include an agency in the EOC is based on the likelihood that they may have to take an action or a support role on the importance of the agency to the mission.

Appendix C

Urban Terrain Analysis

Of all the environments in which the Army may need to conduct operations, the urban environment confronts commanders with a combination of difficulties rarely found in other environments. The distinct characteristics of the environment result from the combination of man-made features and supporting infrastructure superimposed on the existing natural terrain and the density of civilians in close proximity to combat forces. The human dimension it represents is potentially the most important and perplexing for commanders and their staffs to understand and evaluate. Commanders at all levels must make extraordinary efforts to assess and understand their particular urban environment to plan, prepare for, and execute effective civil disturbance operations.

SIGNIFICANT FACTORS

C-1. Although urban areas throughout the world possess similar characteristics, no two are alike. The sprawl of Los Angeles, for example, bears little physical resemblance to New Delhi. The most significant factors affecting the uniqueness and complexity of each area are its societal characteristics. While the terrain can be extremely complex, information about this aspect, its potential effects on operations, and how it changes over time may be determined with some degree of certainty. However, the human dimension is more difficult to understand and assess, particularly its effect on military operations. The urban environment is not neutral—it helps or hinders. The commander who can best understand and exploit the effects of the urban environment has the best chance of success.

C-2. Whether a metropolitan area or a village, each urban environment has an identifiable system of urban components that constantly change and interact. This system of components consists of the terrain, the society, and the infrastructure that provides the critical link between the two (see *Figure C-1*, page C-2).

C-3. These systems are not separate and distinct categories but instead are overlapping and interdependent. A thorough analysis of these systems contributes to the commander's situational understanding and allows him to develop COAs that apply appropriate resources to control a crowd and prevent a major civil disturbance. This understanding allows commanders to engage decisive points critical to maintaining peace or restoring normalcy to the urban environment.

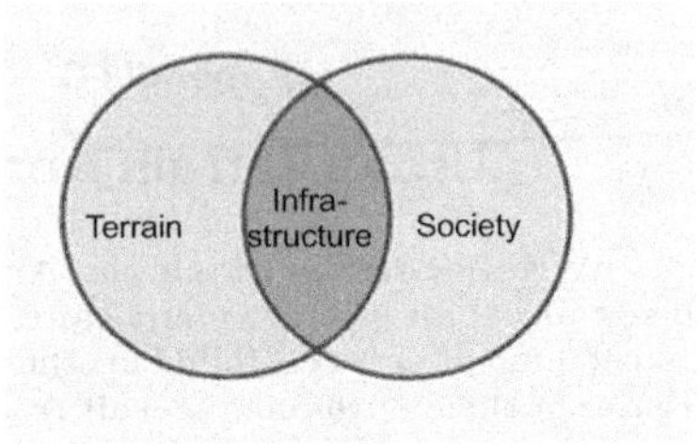

Figure C-1. Keys to Understanding the Urban Environment

C-4. The social aspects and the integrating infrastructure are the most difficult for commanders to understand and assess. These issues require greater dependence on LEAs and local governmental agencies for their information, knowledge, and expertise.

URBAN TERRAIN TYPES

C-5. The physical terrain is the most basic and easily recognizable aspect of the urban area; however, truly understanding it requires comprehending its multidimensional nature, urban and street patterns, urban form and function, and size. The infinite ways in which these factors (see *paragraph C-15)* can intertwine make it impossible to describe a typical urban area. However, these factors provide a framework for understanding the complex terrain within an urban environment. Although the terrain consists of both natural and man-made features, the latter dominates the physical characteristics of the urban environment and has significant effects on military systems and soldiers, tactics, and operations.

URBAN PATTERNS

C-6. The combinations of urban areas, lines of communications (LOCs), and natural terrain result in the identification of four urban patterns (see *Figure C-2*) that will influence civil disturbance operations. An integral part of each urban pattern is the hub or the central urban area around which outlying urban areas radiate. In SOSO, the hub is important, as it is often the economic, political, or cultural center of the surrounding area.

Satellite Pattern

C-7. This pattern consists of a central hub surrounded by smaller, dependent urban areas. LOCs tend to converge on the hub. The natural terrain throughout this pattern is relatively homogenous. Outlying or satellite urban areas often support the principal urban area at the hub.

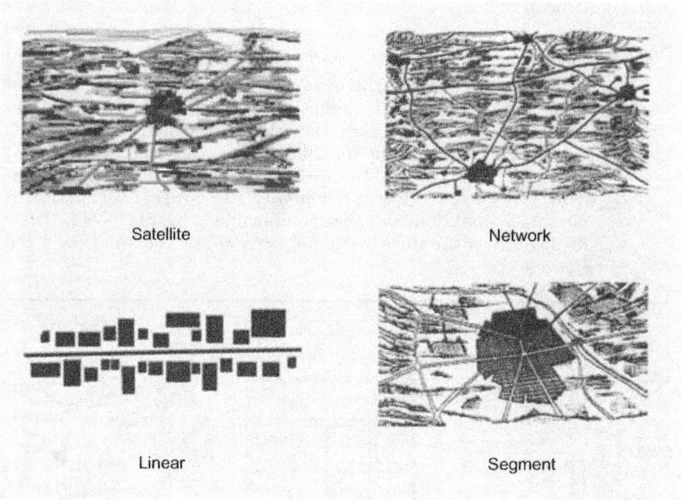

Figure C-2. Urban Patterns

Network Pattern

C-8. This pattern represents the interlocking of the primary hubs of subordinate satellite patterns. Its elements are more self-sufficient and less supportive of each other, although a dominant hub may exist. Major LOCs within a network are more extensive than in a satellite pattern and take a rectangular rather than a convergent form. The natural terrain within a network may vary more than in a single satellite array.

Linear Pattern

C-9. This pattern is a potential subelement of the previous two patterns. It may form one ray of the satellite pattern or be found along connecting links between the hubs of a network. Most frequently, however, this pattern results from the string of minor urban areas aligning a confined natural terrain corridor such as an elongated valley, a body of water, or a man-made communications route.

Segment Pattern

C-10. When dominant natural terrain (such as a river or man-made features [canals, major highways, or railways] divides an urban area, it creates a segmented pattern). This pattern makes it easier to assign areas of responsibility to units. However, if operations are in support of the city LEA, then units should be assigned to a police precinct to avoid crossing several precinct lines, as would happen when assigning units by dominant terrain features.

STREET PATTERNS

C-11. Patterns within the urban area result from the arrangement of the streets, roads, highways, and other thoroughfares. They are a result of the influences of natural terrain, the original designer's personal prejudices, and the changing needs of the inhabitants. The commander must understand the street patterns in the area in which his control force will operate. This will offer force protection, preventing the control formation from being outmaneuvered, flanked, or surrounded by a hostile crowd. Urban areas can display any of the three basic patterns or any combination of the three (see *Figure C-3).*

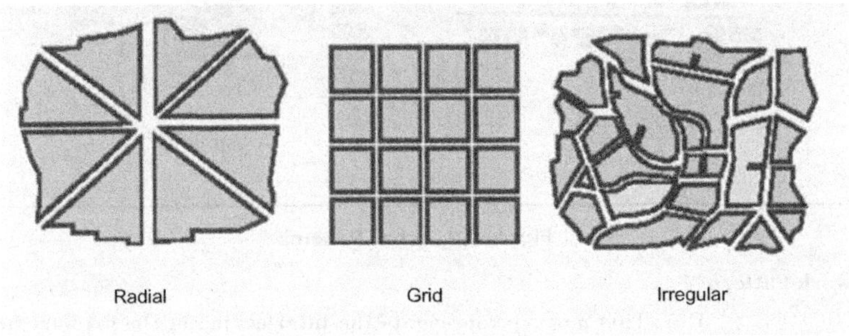

Figure C-3. Basic Internal Street Patterns

Radial

C-12. Societies of highly concentrated religious or secular power often construct urban areas with a radial design with all primary thoroughfares radiating out from the center of power. Terrain permitting, these streets may extend outward in a complete circle or may form a semicircle or arc when a focal point abuts a natural barrier such as a coastline or mountain. To increase mobility and traffic flow, societies often add concentric loops or rings to larger radial patterns.

Grid

C-13. The most adaptable and universal form for urban areas is the grid pattern (lines of streets at right angles to one another forming blocks similar to the pattern of a chessboard). A grid pattern can fill in and eventually take over an original radial pattern. Most urban areas, regardless of the original intent, plan, or vision, emerge from successive plans overlaid on one another.

Areas are well planned to fit with previous plans and others a haphazard response to explosive urban growth. The results may mix different patterns, blend patterns in symmetric combinations, or have no discernable geometric pattern (the irregular pattern).

Irregular

C-14. Irregular patterns can be specifically designed for aesthetic reasons (as in many suburban housing developments). They are sometimes used to conform to marked terrain relief (see *FMs 3-06* and *3-06.11*).

URBAN FORM AND FUNCTION

C-15. Throughout the world, urban areas have similar form and function. In form, urban areas contain like characteristics, readily divisible into distinct sections or areas. Functionally, they tend to be the center of population, finance, politics, transportation, industry, and culture.

- **Core.** The core is the heart of the urban area, the downtown, or central business district. Relatively small and compact, it contains a large percentage of the urban area's shops, offices, and public institutions.
- **Core Periphery.** As the name implies, the core periphery forms the outer edges of the core of the urban area. Generally, it has undergone less change than the core, resulting in buildings of uniform height (generally two to three stories in towns and five to ten stories in larger urban areas).
- **Industrial Area.** Industrial areas generally develop on the outskirts of the urban areas, where commercial transportation is easiest (along airfields and major sea, river, rail, and highway routes). The open/ dispersed pattern of the buildings provides sufficient space for large cargoes, trucks, and materials-handling equipment.
- **High-Rise Area**. High-rise areas consist of multistoried apartments, commercial offices, and businesses separated by large open areas such as parking lots, parks, and individual one-story buildings. The automobile, mass transit systems, and improved road networks encourage these areas to grow and function further from the urban core.
- **Residential Area.** Residential areas can be found dispersed throughout the urban area; however, large suburban areas (or sprawl) normally form on the outskirts. The combined population of surrounding suburban areas often outnumbers that of the urban area. Proper and specific suburbs tend toward homogeneity based on ethnicity, religion, economics, or some other social aspect.
- **Commercial Ribbon Area**. Commercial ribbon areas are rows of stores, shops, and restaurants built along both sides of major streets that run through urban areas. These same types of areas often develop along the roads that connect one urban area to another (strip areas).

URBAN SOCIETY

C-16. Although complex, understanding the urban terrain is relatively straightforward in comparison to comprehending the multifaceted nature of urban society. As urban areas increase in size, they generally lose homogeneity; therefore, commanders must understand and account for the characteristics of a diverse population whose beliefs may vary. The behavior of civilian populations within an urban area is dynamic and poses a special challenge to commanders.

C-17. The center of gravity during an urban operation may be the civilian inhabitants themselves. The side that enjoys the support of the population has many advantages. To gain and retain this support, commanders must first understand the complex nature and character of the urban society. Second, they must understand and accept that every military action (or inaction) may influence the relationship between the urban population and Army forces, and by extension, mission success.

C-18. The identification of groups helps commanders focus on specific segments of the urban society to determine their beliefs and needs and how those beliefs and needs motivate them to future action (or inaction). Groups may be categorized based on race, religion, national origin, tribe, clan, class, party affiliation, education level, union memberships, occupation, age, or any other significant social grouping.

C-19. Commanders should consider political, economic, and historical factors in this analysis. These factors usually affect all groups to some extent and often provide the basis for many of their beliefs, needs (actual or perceived), and subsequent behavior. In most cases, training and discipline, grounded in cultural understanding and sensitivity, will help mitigate many of the potential adverse effects resulting from military and civilian interaction.

URBAN INFRASTRUCTURE

C-20. Urban infrastructures are those systems that support the inhabitants, their economy, and their government. They provide the link between the physical terrain and the urban society. Hundreds of systems may exist. Each system has a critical role in the smooth functioning of the urban area. All systems fit into the following five broad categories: communications and information, transportation and distribution, energy, commerce, and human services.

Appendix D

Practical Application

It is always best to derive lessons learned from practical operations, in the field. These are operations where a measure of success was acquired, along with important lessons learned. The following reflects an actual TF application of the doctrine highlighted in the contents of this manual.

OVERVIEW

D-1. The purpose of this manual is to provide TF units and elements with the latest guidance on providing crowd control and preventing civil disturbances.

D-2. Local civilians often gather in towns or along strategic locations throughout an AO to demonstrate grievances or other causes (for example, protesting the apprehension of a fellow citizen, policies of US forces, and policies of government services). Although civilians have the right to assemble, crowd gatherings are predominantly ethnically based, motivated, and influenced. Local civilians can apply for a permit to legally conduct a rally 72 hours in advance as long as their assembly is peaceful and unobtrusive to government services, freedom of movement, or the rights of others. On occasion, intelligence collection can indicate a possibility for violence. In these instances or during spontaneous crowd formations, the TF must provide forces in support of a safe and secure environment or to assist civilian police in preventing unlawful acts. If unlawful acts are committed, detain violators for adjudication under the legal system.

PROCEDURES

D-3. Take precautions to avoid becoming a catalyst to the civil disturbance. During lawful assemblies, the TF must make every effort to treat HN individuals with respect. This means that the TF must engage in peaceful gatherings using interpersonal communication skills. This ensures that demonstrators understand that TF personnel respect their right to lawfully assemble during peaceful demonstrations. When possible, TF engagement of a peaceful assembly should be unobtrusive.

D-4. Do not display civil disturbance equipment or weapons, overtly prepare defenses against a crowd, or aggressively violate crowd space, unless presented with information, intelligence, or negative indicators of pending violence.

NEGATIVE INDICATORS

D-5. Any response to negative indicators should be sequential, measured, and focused on deescalating rather than escalating the crowd.

NEGATIVE INDICATORS OF A POTENTIAL CIVIL DISTURBANCE

D-6. Predictive intelligence is the best indicator of a potential civil disturbance and may provide insight on internal planning, external influences, environmental conditions, or catalyst events that should influence TF planning and response.

D-7. Other indicators of a potential civil disturbance are people beginning to assemble during a scheduled or spontaneous rally. When people begin to assemble, leaders should pay particular attention to crowd dynamics. As the crowd demeanor or dynamics change from peaceful assembly to unruly behavior or unlawful acts, negative indicators should provide an advance warning and prompt a measured response. Negative indicators of a potential change in crowd dynamics and unlawful activities should prompt immediate countermeasures (see *Figure D-1*).

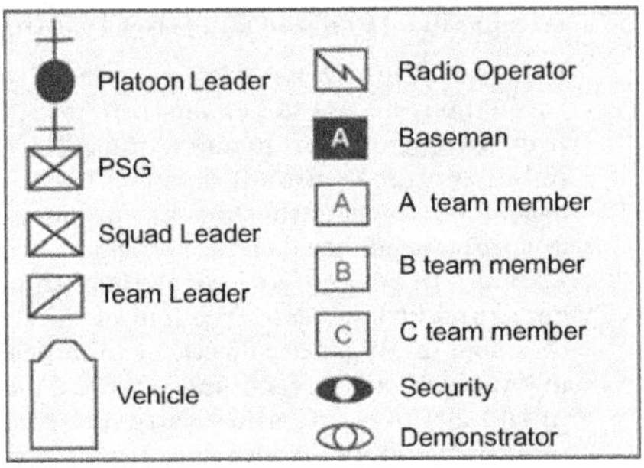

Figure D-1. Symbols for a Formation Element

Orientation

D-8. Crowd dynamics change from people milling around and talking among themselves in isolated groups to a more collective focus concentrated on a single objective (for example, TF personnel, specific agitators or magnets, a government figure, or other activity).

Massing

D-9. The crowd begins to mass or tighten into a large contiguous body from a loose formation to a static congregation located in a more concentrated area. Unchecked, massing can provide anonymity and a collective feeling of invulnerability. This change in group dynamics could provide ideal conditions for a civil disturbance (see *Figure D-2*).

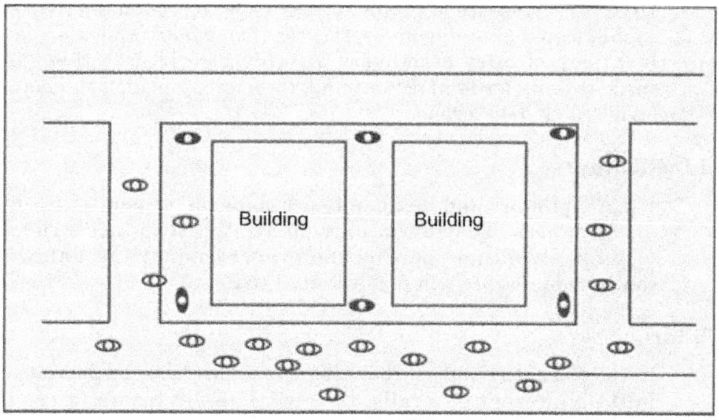

Figure D-2. Crowd Massing

Chanting

D-10. People may begin chanting slogans or yelling ethnically charged rhetoric or obscenities. Because chanting can indicate an increase in crowd intensity, interpreters should be positioned in strategic locations to interpret crowd rhetoric.

Signaling

D-11. Signaling occurs when members of the crowd position themselves to communicate and synchronize crowd actions tied to a plan with specific objectives. Signaling may indicate the intent to escalate the message through the use of violent or unlawful acts.

Blocking

D-12. The crowd may attempt to block public thoroughfares (for example, entrances and exits to public buildings, MSRs, community streets, or other traffic areas that may degrade freedom of movement or similar actions). This escalates the level of attention given to the crowd's message or prevents freedom of movement to opposing persons or groups.

Arming

D-13. Any signs that the crowd is arming itself with objects (for example, bottles, rocks, and sticks) is a clear indication of impending danger to the TF and other civilians. In fact, during a recent civil disturbance, some members of the crowd were seen handing out bottles to others in the crowd. Arming demonstrates intent toward violence is unlawful.

Outside Influence

D-14. The presence of people from outside the community may indicate more sophisticated planning and resources than would otherwise be expected from local community members. Outside people may feel insulated from responsibility for local actions and, therefore, provide a dangerous catalyst for a violent civil disturbance.

Agitators and Instigators

D-15. Agitators and instigators are common elements associated with civil disturbances. They often provide a catalyst to incite the crowd toward violence. Whether spewing rhetoric or committing unlawful acts, their continued presence will fuel potential strife.

Absence of Children

D-16. Recent civil disturbances have shown that even when children were initially present at a rally, they were absent before the civil disturbance erupted with violence. While the presence of children is not a guarantee of a peaceful assembly, the removal of children from a rally may provide early warning for a pending escalation of hostilities or a full-blown riot.

PREVENTION OF CROWD ESCALATION

D-17. If TF personnel believe, based on intelligence or negative indicators, that a crowd is planning to, or may spontaneously, riot they should immediately engage crowd participants using the preventive TTP measures.

D-18. The force executing these preventive TTP can include a number of civil disturbance teams and elements capable of executing the tasks shown in *Figures D-2,* page D-3; *D-3; D-4,* page D-6; and *D-5,* page D-8. These figures demonstrate a full range of crowd control teams and elements, to include camera teams with interpreters, checkpoints, QRF response, riot control formation, NLW teams, snatch and detain teams, blocking teams, and detainee evacuation, as required. *Figures D-2, D-3, D-4,* and *D-5,* illustrate the evolution of a crowd formation and the recommended positioning of teams and elements to engage people as they congregate. Included with each security team is a designated camera or video recorder.

D-19. Depending on the situation and crowd dynamics, it may not be necessary to establish all crowd control teams and elements, but they should be included in planning. Regardless of task organization, teams and elements must be prepared to execute the crowd control and civil disturbance TTP. The order to execute the following TTP should be tailored based on local conditions.

Early Contact

D-20. Engaging people as a crowd forms is essential to prevention. Whether based on intelligence or negative indicators, people should be engaged early, before the crowd masses as shown in *Figure D-2,* rather than waiting until after the crowd masses as shown in *Figure D-3.* Early contact will assist TFs in personalizing their interaction and ensuring that people understand their presence is known, documented, and can be linked to their actions.

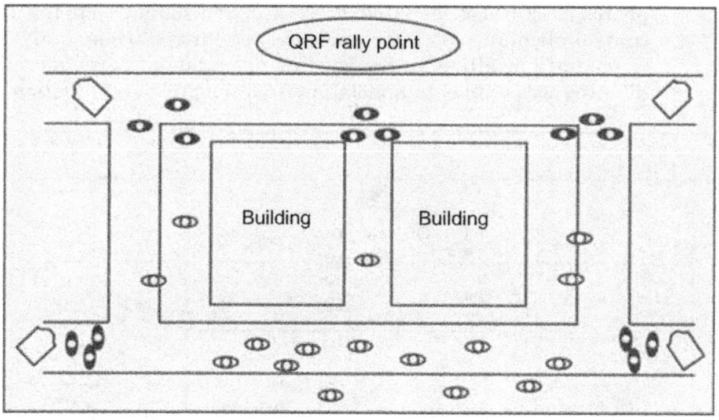

Figure D-3. Early Contact and Unobtrusive Crowd Control Measures

D-21. Cameras are one of the best tools for preventing civil disturbances. The use of cameras to document crowd formations, particularly lens-size groups (small enough to fit in a single photo), will personalize contact and eliminate feelings of anonymity among the crowd. People whose identity has been documented will think twice before committing unlawful acts. Even without documenting names, photo analysis can identify people associated with the civil disturbance, especially those who unlawfully participated as crowd instigators, agitators, or magnets or may commit unlawful acts.

D-22. As people begin to gather in and along streets, camera teams should be deployed as unobtrusively as possible to engage people before they congregate in mass. As depicted in *Figure D-3,* camera teams should consist of two soldiers with an interpreter and be employed in a widely dispersed pattern. Such small, dispersed camera teams are less likely to intrude on lawful rallies or create invasive friction that might become a catalyst to a civil disturbance.

D-23. As the crowd begins to demonstrate negative indicators for a possible civil disturbance (*Figure D-4,* page D-6), camera teams should become more invasive by actively engaging groups to take their pictures. Camera teams should focus cameras on instigators, agitators, or magnets (personnel that garner respect or compliance from the crowd). If a person is purposely avoiding the camera, even if his activities are lawful, the team should take special measures to obtain his picture. Each time a camera team takes a picture of a person or group of people, they must immediately engage them using the warning outlined in *paragraph D-26.*

D-24. Once the crowd has massed and displays negative indicators or begins to riot, camera teams should immediately move behind the riot formation line (see *Figure D-4*). It is essential, however, that camera teams continue to document the event, focusing on people who are committing illegal acts such as arming themselves, throwing objects, and assaulting TF personnel.

D-25. Immediately following a crowd control or civil disturbance operation, photos must be consolidated, processed, and submitted to the TF analysis and control element (ACE) for analysis. A near real-time analysis will provide immediate intelligence for ongoing operations to prevent subsequent civil disturbances or to detain violators from the previous operation.

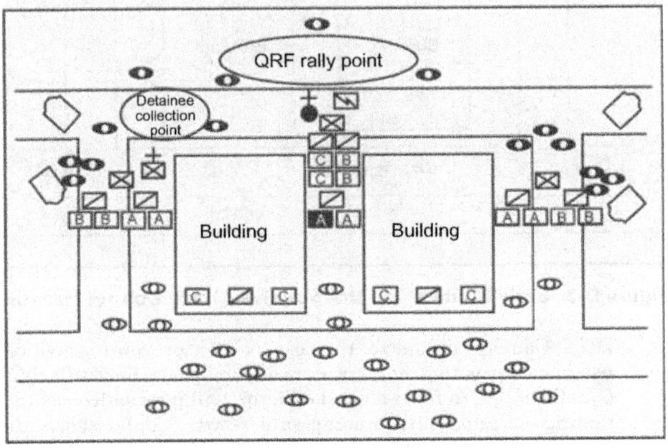

Figure D-4. More Obtrusive Engagement as the Crowd Demonstrates Negative Indicators

Crowd Warning

D-26. Once the team has taken a picture of a person or group of people, they should warn the crowd that as individuals they would be held accountable for their actions. Using an interpreter, the camera team leader will make solid eye contact with the person or group and state the following warning: *"We have your picture. You will be held accountable for your actions. If you commit an unlawful act, you will be detained, if not today, tomorrow. Blocking roads or access to public buildings is unlawful."*

Limit Crowd Size

D-27. Persons seeking to conduct a rally must apply for a permit through civilian authorities at least 72 hours prior to the meeting. If a rally has not been lawfully scheduled through civilian authorities, or if it has, but is demonstrating negative indicators and is turning into a civil disturbance, TF units and elements should attempt to limit the crowd or separate gatherings under 10 people (see *Figure D-3*, page D-5). Walking patrols can assist camera teams in breaking up large groups. Walking patrols should operate in conjunction with camera teams. This will ensure that camera teams can accurately document and provide the aforementioned warning to prevent further escalation of crowd size or behavior.

Clear Routes

D-28. Routes should be considered key terrain and must be controlled by the TF to retain the initiative. Note how a QRF divides in order to clear side streets between protected buildings in *Figure D-4.* Blocking routes or public thoroughfares is not only illegal, but severely restrictive to civil disturbance operations. Routes must remain open to allow movement of TFs and elements, whether it is for camera teams during early contact or for the QRF to respond during an actual civil disturbance. Because blocking routes is illegal, camera teams must document responsible persons for immediate or subsequent detention and legal processing.

Isolate the Area

D-29. As depicted in *Figure D-3,* page D-5, checkpoints should be established to isolate the area. While checkpoints can prevent people from entering the isolated area, they should never prevent people from exiting the area. When trapped, people may escalate the level of violence out of fear, anger, or desperation. When placing checkpoints, leaders should find a location that is both close enough to limit the number of persons being isolated, but far enough away that they will not draw a crowd from people within the isolated area. Checkpoints are essential to preventing others from joining the already gathered crowds. These checkpoints must provide storage room for processing vehicles in both directions; but when limited by space or other considerations, it should simply block traffic trying to enter the isolated area. Improve force protection measures in and around checkpoints as necessary. Strategically placed checkpoints can provide numerous countermeasures to prevent and control a civil disturbance or to identify and detain persons who may have committed unlawful acts during the riot. Use checkpoints to prevent potential participants from joining the crowd or civil disturbance. Checkpoints should also prevent any activity that may commingle ethnic groups. The checkpoint must process all people departing the area by documenting personal and vehicle information with the pictures of all occupants in front of or adjacent to the vehicle.

Establish Lethal Overwatch

D-30. The mentality of "taking the high ground" is just as important in civil disturbance operations as in any other type of operation. Although QRF assets on the ground have both NL and lethal capabilities, it is critical to ensure their safety by deploying marksman/observer teams as shown with the C-team designates on top of the buildings in *Figures D-4* and *D-5,* page D-8. Detain persons committing unlawful acts. The detention and legal processing of persons who commit unlawful acts, against TFs or local laws sends a clear message that all people will be held accountable for their actions. Immediately detain persons who are committing unlawful acts and are isolated or easily accessible, unless the detain operation would create a catalyst event or endanger the apprehension team. If it is not feasible to immediately detain them, a picture with relevant information should be forwarded for analysis and inclusion on the detain list for their detention at a later date and time. Remember, as stated in the warning, they will be detained, if not today, tomorrow. Detainees should be quickly relocated to the forward processing site for detainee evacuation (see *Figure D-5*). Detainee evacuation sites should be behind the control formations to prevent any interference from the local population. Sites should be positioned at two separate locations, such as at the end of a town or located along an MSR. All people who are detained based on

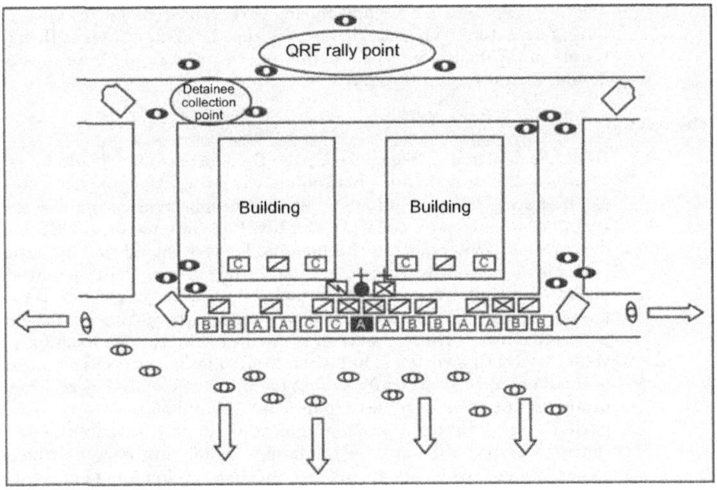

Figure D-5. Full Engagement as Crowd Unlawfully Demonstrates

solid evidence of a crime (for example, pictures of them armed with dangerous objects, blocking a road, assaulting TF members) will be turned over to local authorities for legal processing.

Detain List

D-31. When executing operations based on the detain list, do not show people the list or separate pictures from the list.

Quick Reaction Force Position

D-32. The QRF should be positioned with easy access to decisive points to interdict formations or gathering of crowds, support checkpoints, or provide sustained operations in support of early contact teams. Notice in *Figure D-1,* page D-2, how the QRF is positioned out of direct sight of the crowd until employment becomes necessary. QRF personnel are readily organized, equipped, and easily moved into position to move between buildings and through obstacles (see *Figure D-4,* page D-6).

D-33. At this point, if the crowd ignores verbal orders, NL munitions from appropriate standoff distances could be used to influence and motivate the crowd to comply. Use of NL munitions could prevent the necessity for QRF personnel to come into direct physical contact with the crowd. Early positioning of the QRF could become a catalyst event that may instigate a civil disturbance.

D-34. Assembling in mass (see *Figure D-5*), directly in front of the protected buildings, illustrates to the crowd that there is strength in numbers and the QRF is determined to disperse the crowd from the area. As a reminder, remove the flanking checkpoints that may block the dispersal of the crowd.

Glossary

A2C2	Army airspace command and control
AA	assembly area
AFL-CIO	American Federation of Labor-Congress of Industrial Organizations
AG	Adjutant General
AO	area of operations
ARTEP	Army training and evaluation program
AS	area security
ASV	armored security vehicle
ATTN	attention
BDU	battle dress uniform
CA	civil affairs
CD	combat development
CDC	crowd dispersal cartridge
CFR	Code of Federal Regulations
CINC	commander-in-chief
CMT	crisis management team
COA	course of action
COB	close of business
CONUS	continental United States
COTS	commercial off-the-shelf
CP	candle power
CPU	central processing unit
DA	Department of the Army
DCO	defense coordinating officer
DD	Department of Defense
DM	designated marksmen
DOD	Department of Defense
DOJ	Department of Justice
DOMS	director of military support
EMD	emergency medical dispatching
EMS	emergency medical services
EOC	emergency operations center
FCMT	forced cell move team
FCO	federal coordinating officer
FEMA	Federal Emergency Management Agency
FM	field manual

FRP	federal response plan
FSE	fire support element
FSO	fire support officer
G2	Assistant Chief of Staff, G3 (Intelligence)
G3	Assistant Chief of Staff, G3 (Operations and Plans)
G5	Assistant Chief of Staff, G5 (Civil Affairs)
G6	Assistant Chief of Staff, G6 (Information Management)
GOTS	government off-the-shelf
GRM	graduated response matrix
HIV	human immunodeficiency virus
HMMWV	high-mobility multipurpose wheeled vehicle
HN	host nation
HQ	headquarters
I/R	internment and resettlement
INIWIC	Interservice Nonlethal Individual Weapons Instructor Course
IO	information operations
IPB	intelligence preparation of the battlefield
JMC	joint military commission
JTF	joint task force
L&O	law and order
LBE	load-bearing equipment
LBV	load-bearing vest
LEA	law enforcement agency
LOC	lines of communication
LOS	line of sight
LP	listening post
MACDIS	military assistance for civil disturbance
MANSCEN	Maneuver Support Center
MCCM-GE	modified crowd control munition-ground emplacement
MCM	Manual for Courts-Martial
MDMP	military decision-making process
METL	mission essential task list
METT-TC	mission, enemy, terrain, troops, time available, and civilian considerations
MI	military intelligence
mm	millimeter
MO	Missouri
MOOTW	military operations other than war
MOUT	military operations on urbanized terrain
MPI	military police investigator
MRE	meal, ready to eat

MS	mobility support
MSCA	military support to civil authorities
MTP	mission training plan
MWD	military working dog
NCO	noncommissioned officer
NEO	noncombatant evacuation operations
NG	National Guard
NL	nonlethal
NLCS	nonlethal capabilities set
NLW	nonlethal weapons
OC	Oleoresin Capsicum
OCONUS	outside continental United States
OES	office of emergency services
OIC	officer in charge
OP	observation post
OPFOR	opposing forces
OPORD	operations order
PA	public address
PAO	Public Affairs Office
PIO	police intelligence operations
PM	provost marshal
PLT	platoon
PMO	Provost Marshal Office
POMSO	plans, operations, and military support officer
PPCT	pressure point control tactic
PPE	personal protective equipment
PSG	platoon sergeant
PSYOP	psychological operations
PT	physical training
PVAB	portable vehicle arrest barrier
QRF	quick reaction force
RCA	riot control agent
ROE	rules of engagement
RUF	rule for the use of force
S2	Intelligence Officer (US Army)
SCO	state coordinating officer
SCRAG	senior representative of the attorney general
SITREP	situation report
SJA	Staff Judge Advocate
SOFA	status of forces agreement

SOMA	status of mission agreement
SOP	standing operating procedure
SOSO	stability operations and support operations
SQD	squad
STARC	state area command
STP	soldier training publication
STX	soldier training exercise
TA	Training Area
TF	task force
TLP	troop leading procedure
TM	team
TPT	tactical psychological operations team
TRADOC	United States Army Training and Doctrine Command
TSOP	tactical standing operating procedure
TTP	tactics, techniques, and procedures
UCMJ	Uniform Code of Military Justice
UN	United Nations
UNGA	United Nations General Assembly
UNHCR	United Nations High Commissioner for Refugees
UNSC	United Nations Security Council
US	United States
USAMPS	United States Army Military Police School
VCR	video cassette recorder
VIP	very important person
WMD	weapons of mass destruction
WTO	World Trade Organization

Bibliography

AR 380-10. *Foreign Disclosure and Contacts With Foreign Representatives.* 6 June 2003.

AR 380-13. *Acquisition and Storage of Information Concerning Nonaffiliated Persons and Organizations.* 30 September 1974.

Code of Federal Regulations.

Constitution of the United States.

DA Form 2028. *Recommended Changes to Publications and Blank Forms.*

DA Form 4137. *Evidence/Property Document.*

DD Form 2708. *Receipt For Inmate or Detained Person.*

DOD Civil Disturbance Plan 55-2. *Operation Garden Plot.* 15 February 1991.

DOD Directive 3025.12. *Military Assistance for Civil Disturbances.* 4 February 1994.

DOD Directive 3025.15. *Military Assistance to Civil Authorities.* 18 February 1997.

DOD Directive 5525.5. *DOD Cooperation With Civilian Law Enforcement Officials.* 15 January 1986.

Executive Order 12333. *United States Intelligence Activities.* 4 December 1981.

Fiscal Year 1989 Act.

FM 3-06. *Urban Operations.* 1 June 2003.

FM 3-06.11. *Combined Arms Operations in Urban Terrain.* 28 February 2002.

FM 19.10. *Military Police Law and Order Operations.* 30 September 1987.

FM 3-19.4. *Military Police Leaders' Handbook.* 4 March 2002.

FM 3-19.40. *Military Policy Internment/Resettlement Operations.* 1 August 2001.

FM 3-21.5. *Drills and Ceremonies.* 7 July 2003.

FM 3-22.40. *Multiservice Tactics, Techniques, and Procedures (MTTP) For Tactical Employment of Nonlethal Weapons (NLW).* 15 January 2003.

FM 7-0. *Training the Force.* 22 October 2002.

FM 7-1. *Battle Focused Training.* 15 September 2003.

FM 27-100. *Legal Support to Operations.* 1 March 2000.

Judiciary Act of 1789.

MCM.

National Defense Authorization Acts.

NGR 500-1/ANGI 10-8101. *Military Support to Civil Authorities.* 1 February 1996.

Posse Comitatus Act. 1878.

STP 19-95B1-SM. *Soldier's Manual for MOS 95B Military Police Skill Level 1.* 14 January 2003.

STP 19-95B24-SM-TG. *Soldier's Manual and Trainer's Guide for MOS 95B, Military Police, Skill Levels 2/3/4.* 2 December 2002.

STP 19-95C1-SM. *MOS 95C, Corrections Specialist, Skill Level 1 Soldier's Manual.* 30 October 2003.

STP 19-95C24-SM-TG. *MOS 95C, Corrections Specialist, Skill Levels 2/3/4, Soldier's Manual and Trainer's Guide.* 30 October 2003.

Uniform Code of Military Justice.

USC, Title 32. *National Guard.*

USC, Title 10, Chapter 18. *Military Support for Civilian Law Enforcement Agencies.*

USC, Title 10, Subtitle A, Part I, Chapter 15, Section 331. *Federal Aid for State Governments.*

USC, Title 10, Subtitle A, Part I, Chapter 15, Section 332. *Use of Militia and Armed Forces to Enforce Federal Authority.*

USC, Title 10, Subtitle A, Part I, Chapter 15, Section 333. *Interference With State and Federal Law.*

USC, Title 10, Subtitle A, Part I, Chapter 15, Section 334. *Proclamation to Disperse.*

USC, Title 18, Part I, Chapter 67, Section 1385. *Use of Army and Air Force as a Posse Comitatus.*

Index

Made in the USA
Monee, IL
05 November 2026

67820239R00312